CELL AND MOLECULAR BIOLOGY

CELL AND MOLECULAR BIOLOGY

By
Dr. S. Sundara Rajan
Principal
Reva Institute for Science and Technology Studies
R.T. Nagar
Bangalore – 560 024

ANMOL PUBLICATIONS PVT. LTD.
NEW DELHI - 110 002 (INDIA)

ANMOL PUBLICATIONS PVT. LTD.
H.O.: 4374/4B, Ansari Road, Darya Ganj,
New Delhi-110 002 (India)
Ph.: 23278000, 23261597
B.O.: No. 1015, Ist Main Road, BSK IIIrd Stage
IIIrd Phase, IIIrd Block,
Bangalore - 560 085 (India)
Visit us at: www.anmolpublications.com

Cell and Molecular Biology

PRINTED IN INDIA

Printed at Mehra Offset Press, Delhi.

CONTENTS

Preface

Study of cell known as cell biology is one of the most important branches of biology which has wide ranging impact on many fields of human activity such as medicine, pharmmacy, agriculture etc. A living cell can be studied either at the gross structural level or at the molecualr level. In recent years modern cell biology has gone beyond the light microscopic structure of the cell. By the application of new methods and tools like electron microscopy, ultra centrifugation etc, quite a number of sub cellular structures have been discovered and studied in detail - probing further, biologists have analysed the architectural pattern of the molecules that comprise the living matter.

This has lead to a new branch of biology - **The Molecular Biology** - a science dealing with the ultra structural organisation of the living matter. Molecular biology attempts to explain the phenomena of metabolism, variation, mutation etc., in terms of changes of macromolecules such as proteins, nucleic acids etc.

CELL AND MOLECULAR BIOLOGY deals with both gross and molecular structure of cell in all its structural and functional manifestations. The book consists of twenty seven chapters surveying al aspects of structural and molecular configurations of the cell .There are also chapters on genetic engineering and immunology as the understanding of these are very vital for comprehending the expressinons of cell machinery.

The book is meant to be a text book for students of life science, pharmacy, agriculture etc at the graduate and post graduate level .

I am thankful to Mr J.L Kumar, M.D of Anmol Publications, New Delhi for publishing this book.

Dr. S. SUNDARA RAJAN
Principal

Reva Institute For Science and Technology Studies
R.T. Nagar
Bangalore - 560 024

INTRODUCTION

One of the most important sub divisions of biological sciences is cellbiology dealing with the study of cell and its hereditary potentialities. Cellbiology is in a sense a hybrid science from the historical point of view. Cell and molecular biology comprises study of cell at the structural level and molecular details of the same at the ultra level . Cell and molecular biology is an important branch of biology at the micro level and constitutes an integral part of biology .Studies in cell and molecular biology are reinforced by discoveries arising out of electron microscopy, tissue culture, microbilogy and biochemistry.

Cell biology

Historically the beginnings of cell biology dates back to the earlier part of the 17th century with the invention of microscope and the discovery of cell. For quite some time the microscope was thought to be a curiosity (because of the structure and philosophy of science at that time) rather than a useful instrument for the scientific study of living beings. Nevertheless the observations of Robert Hooke (1635 - 1703) Grew (1641 - 1712) and Malpigi (1628 - 1694) did lead tom an understanding of the fundamental fact that all organisms-plants as well as animals hae a common structural basis and that they are composed of cells. These that all living beings are cellular. Although, it was a Frenchman H.J. Dutrochet who for the first time in 1824 gave the idea of cell theory, the credit for formulating the cell theory generally goes to a German botanist M.I.Schleiden and a German Zoologist T. Schwann. These two scientists clearly outlined the basic concepts of the cell theory in 1839.

The theory of Schleiden and Schwann was simple, direct and elementary and it clearly focussed the study of biology on the cell as the basic unit of all organisms. Thus the study of cell or cytology progressed rapidly and supported the 19th century biologists, it lead to the opening of an immense and varied field of biological research.

In the current sense, cytology or cell biology as it is now called deals with the study of cells from morphological biochemical, physiological, embryological, genetical and even pathological point of view. In recent years modern cell biology has gone beyond the light microscopic structure of the cell. By the application of new methods and tools like electron microscopy, ultra centrifugation etc, quite a number of sub cellular structures have been discovered and studied in detail - probing further, biologists have analysed the architectural pattern of the molecules that comprise the living matter.

This has lead to a new branch of biology - the molecular biology - a science dealing with the ultra structural organisation of the living matter. Molecular biology attempts to explain the phenomena of metabolism, variation, mutation etc., in terms of changes of macromolecules such as proteins, nucleic acids etc.

The following list gives the names of some important scientists and their discoveries with reference to the study of cell.

Name	Year	Contribution
Robert Hooke	1665	Discovered the plant cell

Mirabel	1808	Plants consist of membranous cellular tissues
R.J.H. Dutrochet	1824	All organisms are cellular
F.V. Raspail	1825	Founder of cytochemistry
Turpin	1826	Reported cell division
R. Brown	1831	Described the nucleus in plant cells
Von Mohl	1835	Described the cell division
M.J. Schleiden	1838	Proposed cell theory
T. Schwann	1839	Proposed cell theory
J.E. Purkinje	1840	Named cell contents as protoplasm
K. Nageli	1846	New plant cells arise from pre-existing cells
W. Hofmeister	1849	Studied fertilisation in plants and nuclear divisions in stamens
R. Virchow	1855	All cells arise from pre-existing cells
Shultze	1861	Called protoplasm as the physical basis of life
Gregor Mendel	1865	Founded the science of Genetics
Haeckeel	1866	Named plastids in plant cells
George	1867	Discovered the organnel which was later called golgi complex
F. Mischer	1871	Discovered Nucleic acids
H. Fol	1873	Discovered spindle and astral fibres
O. Hertwig	1876	Discovered that fertilisation involves the fusion of sperm and ovum
E.G. Balbiani	1881	Discovered the giant salivary gland chromosomes in the larva of *Chronomus*
W. FLemming	1882	Introduced the term mitosis
Strassburger	1882	Described mitosis in plant cells
Schimper	1883	Introduced the term chloroplast
Van Benden	1887	Discovered centrioles
Waldeyer	1888	Introduced the term chromosomes
Boveri T.	1892	Described oogenesis and spermatogenesis in *Ascaris* (round worm)
Benda	1898	Named the mitochondrion
G. Golgi	1898	Described the Golgi complex
C.E. Meclung	1902	Identified sex chromosomes in hemiptera
Sutton W.S.	1902	Proposed the chromosome theory of heredity
E. Buchner	1903	Awarded Nobel prize for the discovery of Enzyme
F. Meves	1904	Demonstrated the occurrence of mitochondria in plant cells
J.B. Farmer	1905	Coined the term meiosis along with J.E. Moore
R.G. Harrison	1907	Developed tissue cultural technique
A. Kossel	1910	Obtained Nobel prize for the chemistry of the nucleus
R.M. Willstatler	1915	Got Nobel prize for research work on chlorophyll
R. Feulgen and H. Rossenbeck	1924	Developed feulgen technique for staining DNA
T. Svedberg	1926	Discovered ultra centrifuge and got Nobel prize
O.H. Warburg	1931	Discovered respiratory enzymes and got Nobel prize
M. Knoll and E. Ruska	1932	Invention of first electron microscope
T.H. Morgan	1933	Chromosome function in heredity. Got Novel prize
O.T. Avery C.H. Mcleod and H. Mc.Carthy	1944	Importance of DNA in the inheritance of bacterial cells
F. Zernike	1953	Invention of phase contrast microscope. Got Nobel prize
H.A. Krebs	1953	Discovery of citric acid cycle in aerobic respiration. Got Nobel prize
F.A. Lipmann	1953	Discovery of Co-enzyme A and its role in intermediary metabolism. Got Nobel prize

J.D. Watson and F.H.C. Crick	1953	Proposed the double helix model for DNA molecule and got Nobel prize
G.W. Beadle and E.L. Tatum	1958	One gene one enzyme hypothesis
S. Ochoa	1959	Synthesis of polyribonucleotide *in vitro*. Received Nobel prize
A. Korneberg	1959	Synthesis of polydeoxy ribonucleotides *in vitro*. Received Nobel prize
M. Calvin	1961	Discovered the dark reduction cycle in photosynthesis and Got the Nobel prize.
M.W. Nirenberg and H.G. Khorana	1968	Received Nobel prize for work on genetic code

Cell biology and related sciences : The study of cell has helped biologists to understand various life processes like growth, metabolism, differentiation etc., at the cellular levels. Due to its wide application in various fields, cytology has given birth to various applied branches of biology. Some of these are -

Cytogenetics : This deals with the cytological and molecular basis of heredity, variation, mutation, morphogenesis etc.

Cytotaxonomy : In this branch, cytological details are used in solving taxonomic problems. Individuals of same species or related species have many similarities in chromosome morphology. Cytology also gives proof for the origin of certain taxonomic units or groups.

Cell physiology : This branch deals with functional aspects of cell. Study of nutrition, metabolism, growth, reproduction, division etc., come under the perview of cell physiology.

Cytochemistry : Chemical analysis of the cell is the subject matter of this branch. Analysis of carbohydrates, proteins, fats, nucleic acids, etc., in the cell are dealt within this branch.

Molecular biology : This is the most modern branch of biology. It deals with the molecular configuration and architecture of the living matter and helps to analyse and understand the molecular basis for various life processes. It also attempts to provide a molecular basis for the various inborn errors in metabolism. Molecular biology has provided many significant answers to various problems of biology.

Cytopathology : Study of cells and the analysis of cell structure at the microscopic and the submicroscopic level has helped us to study various diseases. As most of the diseases are ultimately caused by alteration in enzyme functioning, a study of their activity at the cellular level is vital to the understanding of the disease.

Cytoecology : While ecology deals with the study of organisms in relation to their environment, cytoecology deals with cytological characters of individuals in different environmental conditions. Ecological habitat and geographical distribution have their effect on the cytological features of organisms.

BIOMOLECULES

Ever since its origin on this planet, life has manifested itself in myriads of forms and countless functions mainly because of the presence of unique combination of chemical compounds. These chemical compounds, the basic ingredients of which are no different from the ones present in the non living, confer upon the living the unique properties of adaptation and reproduction. The chemical compounds and their various combinations provide the living beings the unique organization and functional versatality.

The 'biomolecules' as these unique chemical compounds are popularly termed, are responsible for the origin, evolution and maintenance of life. These compounds are synthesized by the cell machinery through pathways that are unique to living beings. In this chapter we will study the basic features of three of these biomolecules–carbohydrates, fats and proteins.

CARBOHYDRATES

Carbohydrates are a group of important organic compounds present ubiquitously in plants. By being the basic respiratory substrate and in forming an integral part of the cell walls, carbohydrates are vital to the sustenance of plants, both functionally and structurally. Etymologically meaning hydrates of carbon, carbohydrates are principally composed of carbon, hydrogen and oxygen generally in the ratio of 1:2:1. Some of the carbohydrates have nitrogen and sulphur also in addition to C, H and O. Currently, however, carbohydrates are described as polyhydroxy aldehydes or polyhydroxy ketones and their derivatives.

As has been pointed out already, carbohydrates constitute the principal respiratory substrate that provides energy for all the metabolic processes in plants. They also constitute the major stored reserve in cells. In addition to these, most of the organic compounds that go to make the skeletal frame-work of the cell are composed of carbohydrates.

Classification of Carbohydrates : Carbohydrates may be classified into three principal groups viz. , Monosaccharides, Oligosaccharides and Polysaccharides.

MONOSACCHARIDES : Also known as simple sugars, monosaccharides constitute the simplest among carbohydrates. In other words hydrolysis of monosaccharides does not yield simpler carbohydrates. Monosaccharides are the basic units for the building up of oligosaccharides and polysaccharides.

Monosaccharides have carbon atoms in each molecule ranging from two to eight. This constitutes the basis for the classification of monosaccharides. Basically, all monosaccharides contain either an aldehyde group or a ketone group. Monosaccharides with aldehyde group are known as aldoses while those with a ketone group are known as ketoses.

H
|
C=O C=O
Aldehyde Ketone

Monosaccharides are classified into the following groups based on the number of carbon atoms in each molecule.

(i) **Dioses** ($C_2H_4O_2$) : Two carbon atoms in each molecule, e.g. , Glycollic aldehyde.

(ii) **Trioses** ($C_3H_6O_2$) : These are the principle 3 carbon compounds present in the plant cells. The most important ones being glyceraldehyde and dihydroxyacetone.

H	
|	
C=O	CH_2OH
|	|
CHOH	C=O
|	|
CH_2OH	CH_2OH
Glyceraldehyde	Dihydroxy acetone
(aldose)	(ketose)

Triose sugars exhibit optical activity. This is due to the asymmetrical nature of the carbon atom. The asymmetrical carbon atom imparts isomerism to the trioses. There are two types of isomers (based on optical activity) dextrorotatory and laevorotatory. The former turn the plane polarized light to the right while the latter turn it to the left.

(iii) **Tetroses** ($C_4H_8O_4$) : Four carbon atoms present in each molecule, e.g. Erythrose.

(iv) **Pentoses** ($C_5H_{10}O_5$) : These are 5 carbon sugars which are generally not found in the free state in cytoplasm. They are however present as constituents of complex carbohydrates. Some of the familiar pentoses are Xylulose, Ribose, Ribulose, Arabinose etc. Xylulose and Arabinose are a part of the complex carbohydrates namely Xylans and Arabans.

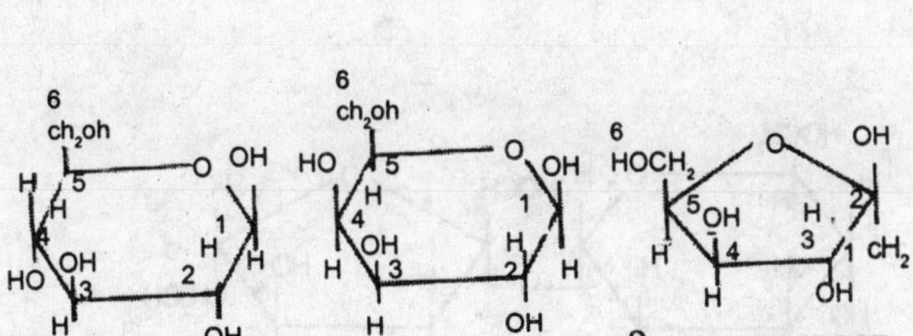

β -D-GLUCOSE β -D-GALACTOSE β -D-FRUCTOSE

Fig.2.1 Biomolecules
Structure of glucose and fructose

```
    H               H
    |               |
   C=O             C=O
 H-C-OH          H-C-OH
   |               |
OH-C-H          HO-C-H
   |               |
 H-C-OH          H-C-OH
   |               |
  CH2OH           CH2OH
 D-xylose       L-arabinose
```

(v) Hexoses ($C_6H_{12}O_6$) : These are the principal sugars that play a very importat role in plant metabolism. Photosynthetic fixation of carbon results in the production of lipids via the glyoxalate cycle. Glucose and fructose are the most important hexoses found dissolved in the free form. Mannose and galactose are the other two found frequently.

```
    H               H
    |               |
   C=O            H-C-OH
   |               |
 H-C-OH            C=O
   |               |
HO-C-H          HO-C-H
   |               |
 H-C-OH          H-C-OH
   |               |
  CH2OH          H-C-OH
                   |
                  CH2OH
 D-glucose      D-fructose
```

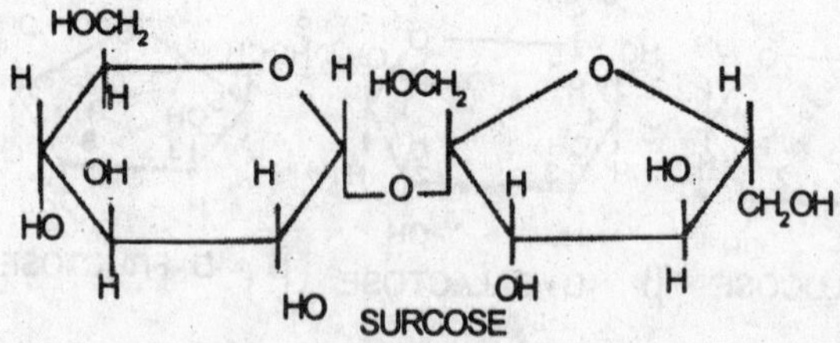

SURCOSE

Fig. 2.2 Biomolecules
Structure of sucrose

Hexose sugars have several asymmetric carbon atoms resulting in the existence of several isomeric forms. Another important feature of hexoses is their ring structure.

(vi) **Heptoses** ($C_7H_{14}O_7$) : Each molecule has 7 carbon atoms. e.g., Sedoheptulose.

OLIGOSACCHARIDES : These are condensation products of two, three or more monosaccharide units and are generally classified based on the number of monosaccharide molecules present. Thus there are disaccharides, trisaccharides, tetrasaccharides etc.

(i) **Disaccharides :** Each disaccharide consists of two units of monosaccharides. Some of the common disaccharides are Sucrose, Lactose, Maltose etc. Sucrose is the common sugar that is the cane sugar, whereas maltose is produced during seed germination by the action of amylase on starch.

(ii) **Trisaccharides :** Three units of monosaccharides condense to form one unit of a trisaccharide. E. g. Raffianose, Gentianose etc.

(iii) **Tetrasaccharides :** Four units of monosaccharides condense to form one unit of a tetrasaccharide. E. g. Stachyose.

METABOLISM OF SUCROSE

Sucrose is the principal sugar found widely distributed in plant tissues. It is made up of one unit of glucose and one unit of fructose with the elimination of one molecule, of water ($C_{12}H_{12}O_{11}$). Sucrose is a non reducing sugar even though the two units that compose it (sucrose) are reducing. Sucrose is of great importance in plant metabolism because it is in this form that carbohydrates are transported in higher plants. Radioisotopic studies have clearly shown that the labelled carbon used for photosynthesis soon finds its way in sucrose molecules.

Sucrose synthesis in plants : The work of Leloir and Cardini (1955), has shown that UDPG (Uridine diphosphoglucse) plays a key role in the addition of glucose to fructose to form a molecule of sucrose. There are at least two principal routes for sucrose synthesis. In the first pathway involving the enzyme sucrose synthetase, glucose is transferred from UDPG to fructose.

UDPG + Fructose ———— Sucrose + UDP

The second pathway is as follows.

(i) Glucose 6 phosphate ———— Glucose-l-phosphate
gluco mutase.

(ii) Glucose I phosphate combines with uridine triphosphate to produce UDPG and pyrophosphate. The enzyme Uridine diphosphoglucose pyrophosphorylase catalyses this reaction.

(iii) UDPG + fructose 6 phosphate ———— Sucrose phosphate + UDP
The enzyme sucrose phosphate synthetase catalyses this reaction.

(iv) Dephosphorylation of sucrose phosphate is brought about by the enzyme phosphatase to yield sucrose.

Sucrose degradation : Sucrose is hydrolysed by the enzyme Invertase to its unit glucose and fructose the enzymatic reaction is believed to proceed in an irreversible manner.

POLYSACCHARIDES : These have a large number of monosaccharide units (polymer). Polysaccharides have a high molecular weight and are generally insoluble in water. Polysaccharides are found in plant cell walls. Starch and cellulose are the principal polysaccharides besides inulin, hemicellulose etc. Starch and cellulose are discussed in some detail in this book.

STARCH : This is the principle storage reserve carbohydrate. Starch is usually stored in the form of granules.

Structure of starch grains : The grains vary in shape, structure and size depending on the source. They have an average size of 1 - 50µ across, have a central point called hilum, around which are wound the chains of glucose units in a helical fashion.

Starch molecules are built up by long chains of glucose molecules joined through 1 - 4 oxygen bridges. On heating, starch yields two components viz amylose and amylopectose; the former is soluble in water while the latter is insoluble.

Each molecule of amylose consists of an unbranched helical chain of large number of glucose residues joined to each other by 1-4 oxygen bridges. Each molecule has a molecular weight from 10,000 - 10,00,000.

Molecules of amylopectin are made up of branched chains of glucose residues with 1-4 as well as 1-6 bridges. Branching is caused by the 1-6 oxygen linkages. Molecular weight ranges from 50,000 to 10,00,000.

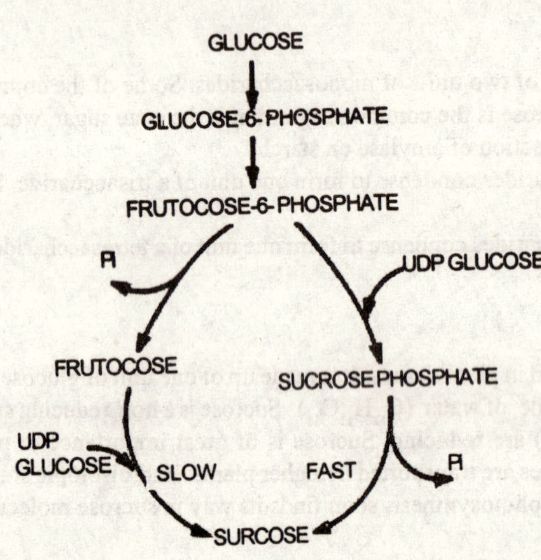

Fig. 2.3 Biomolecules
Synthesis of sucrose

Starch synthesis : Starch is synthesized either in chloroplasts or in amyloplasts. Light is not necessary for starch synthesis provided the glucose residues and the necessary enzymes are available. The following steps are involved in starch synthesis.

(i) Primer : Starch synthesis can only begin in the presence of a primer such as maltose, maltotriose etc. The primer then accepts glucose residues and forms a chain.

(ii) (Glucose 1 phosphate)n + Primer ———— Amylose + (Pi)n

The enzyme starch phosphorylase catalyzes this reaction and helps in the addition of glucose residues one by one to the non reducing end of the primer.

In addition to the above, another enzyme Uridine diphosphate transglycosylase (UDPG glycosylase) also catalyses the formation of 1-4 oxygen bridges between the glucose units. This enzyme was first isolated from bean, corn and potato. The primer molecules could be maltose or maltotriose or even a starch molecule.

UDPG + Primer ———— UDP + 1,4 - glycosyl acceptor

Akazawa, Murata and Minamikawa (1964), found even sucrose can act as a glucose donor in starch synthesis. The reaction takes place as follows.

(i) Sucrose + UDP ———— UDPG + Fructose

(ii) UDPG + Primer Starch ———— UDP

According to Murata et al (1963), ADPG (Adenosine diphosphate glucose) may play an important role in starch synthesis than UDPG.

An enzyme called D - enzyme first discovered by Peat, Whelan and Rees (1953), is known to reversibly transfer glucose units to a primer molecule.

All the enzymes mentioned above help in the formation of 1 - 4 glucosidic bridges and form straight chains. The branches in the chain formed due to 1-6 bridges are however catalyzed by an enzyme called Q - enzyme, which was first isolated from potato extracts (Baun and Gilbert). The Q - enzyme transfers small chains of glucose units on to primers at the 1 - 6 position resulting in an amylopectose molecule.

Starch degradation : Amylases are the principal enzymes which bring about the degradation of starch. They hydrolyse the 1-4 linkages (breaking the oxygen bridge by the addition of a water molecule) between the glucose residues and yield independent glucose moleucles. Starch phosphorylase can also reversibly degrade the starch.

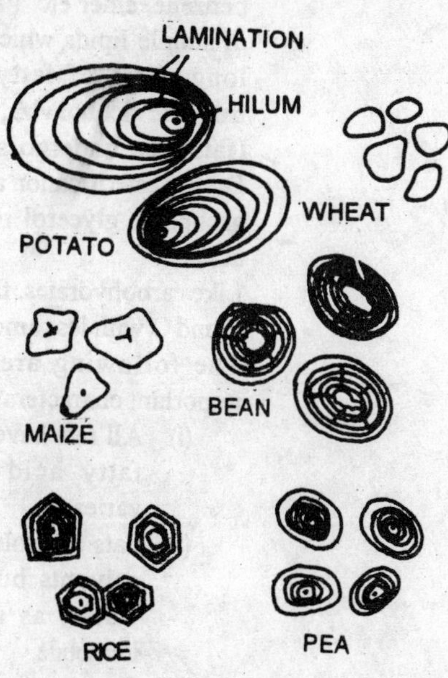

Fig. 2.4 Biomolecules
Types of starch grains

CELLULOSE : This is the principal component of cell wall in most plants. It is a straight chain polymer made up of D- glucose units bound together with 1-4 bonding. Cellulose may be regarded as the most abundant natural product available from plants.

The details of cellulose synthesis and degradation are not available to the same extent as for starch. Studies conducted on Acetobacter (Celulose producing bacteria), indicate that glucose molecules have to be first phosphorylated before they are incorported into cellulose molecule. UDPG may also play a role in cellulose synthesis.

The enzyme complex cellulase atacks the 1-4 linkages in the chain and releases the D - glucose units as given in the figure.

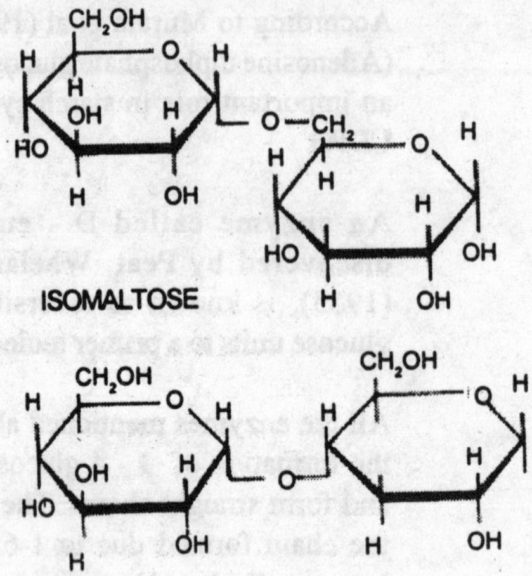

CH_2OH / O / H / H / OH / H / HO / H / OH / O—CH_2 / O / H / H / H / HO / OH / H / OH

ISOMALTOSE

CH_2OH / CH_2OH / H / O / H / H / O / H / OH / H / O / H / HO / OH / H / H / OH

Fig. 2.5 Biomolecules
Structure of isomaltose and maltose

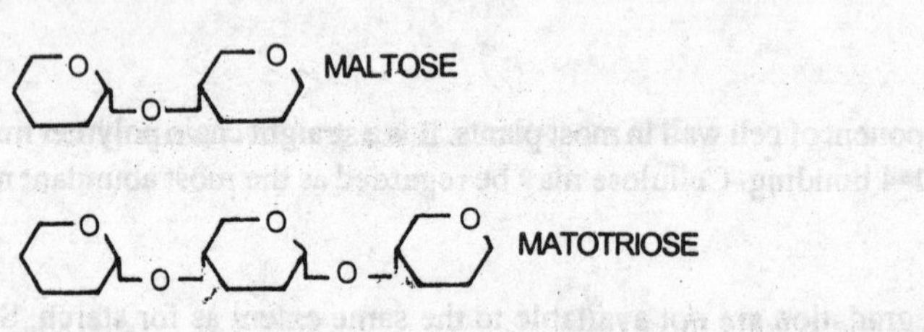

MALTOSE

MATOTRIOSE

MALTOTETROSE

Fig. 2.6 Biomolecules
Primer molecules for starch synthesis

FATS

Lipids constitute an important group of biomolecules playing a vital role in the storage of reserve food. Basically lipids are insoluble in water but soluble in inorganic solvents like acetone, alcohol, chloroform, benzene, ether etc. Fats are a group of simple lipids which are esters of long chain of fatty acids and a trihydric alcohol viz., glycerol. They (fats) can undergo saponification. During esterification all the hydroxyl groups of glycerol react with fatty acids.

Like carbohydrates, fats also have C, H and O with less amount of oxygen. The following are some of the important characters of fats.

(i) All fats have glycerols, bu[t] fatty acid compositio[n] varies,

(ii) Fats are soluble in organi[c] solvents but insoluble i[n] water, as such they ar[e] immobile.

(iii) Oxygen content of fats i[s] less in comparison wit[h] carbon.

(iv) Each cell synthesizes its fa[t] requirement as they (fats) are immobile.

(v) Fats undergo saponificatio[n] i.e., when treated with alkal[i] they release fatty acids ar[d] glycerol.

(vi) Due to their low oxyge[n] content, they (store an[d]) release more energy durin[g] oxidation. On an average, [a] molecule of fat can yie[ld] twice as much energy from a carbohydrate.

(vii) Fats are generally soli[d] under room temperatu[re] those that remain as liqui[d] under room temperature a[re] called oils.

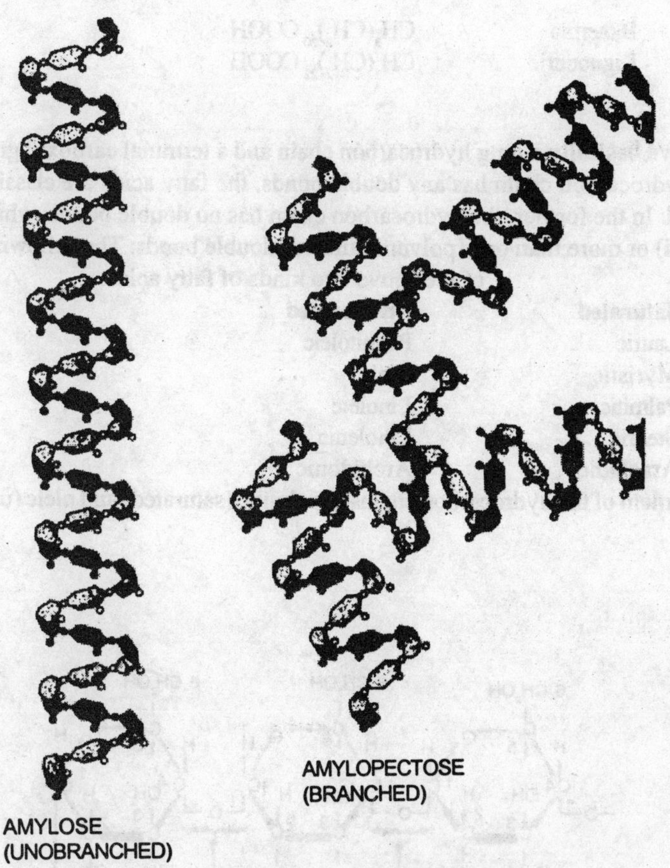

Fig. 2.7 Biomolecules
Structure of amylose and amylopectose

Distribution of fats in plants : Fats are not generally mobilised for respiration under normal conditions. They are usually the reserve food material found in leaves, stems and roots in small quantities. Major quantity of fats are however stored in seeds, either in endosperm or in cotyledons. E.g. Castor, Coconut, Peanut, Sunflower etc.

Fat metabolism : This involves two processes - synthesis and degradation. Fat synthesis involves the following three steps –

(i) Synthesis of fatty acids
(ii) Synthesis of glycerol and
(iii) Condensation of fatty acids and glycerol into fats.

Synthesis of fatty acids : All the naturally occurring fatty acids are long chain molecules having an even number of carbon atoms. The presence of even number of carbon atoms is a salient feature of the fatty acids and the number ranges from C_6 to C_{22}. The following is a list of some of the common fatty acids.

Lauric	$CH_3(CH_2)_{10}$ COOH
Myristic	$CH_3(CH_2)_{12}$ COOH
Palmitic	$CH_3(CH_2)_{14}$ COOH
Stearic	$CH_3(CH_2)_{16}$ COOH
Arachidic	$CH_3(CH_2)_{18}$ COOH

Behemic	$CH_3(CH_2)_{20}\,COOH$
Lignoceric	$CH_3(CH_2)_{22}\,COOH$

All fatty acids have basically a long hydrocarbon chain and a terminal carboxyl group. Based on the fact as to whether the hydrocarbon chain has any double bonds, the fatty acids are classified into – Saturated and Unsaturated. In the former, the hydrocarbon chain has no double bonds, while the latter has one (monounsaturated) or more than one (polyunsaturated) double bonds. The following list gives the examples of the above two kinds of fatty acids.

Saturated	**Unsaturated**
Lauric	Palmitoleic
Myristic	Oleic
Palmitic	Linoleic
Stearic	Linolenic
Arachidic	Archidonic

Structural arrangement of the hydrocarbon chains in palmitic (saturated) and oleic (unsaturated) is given below.

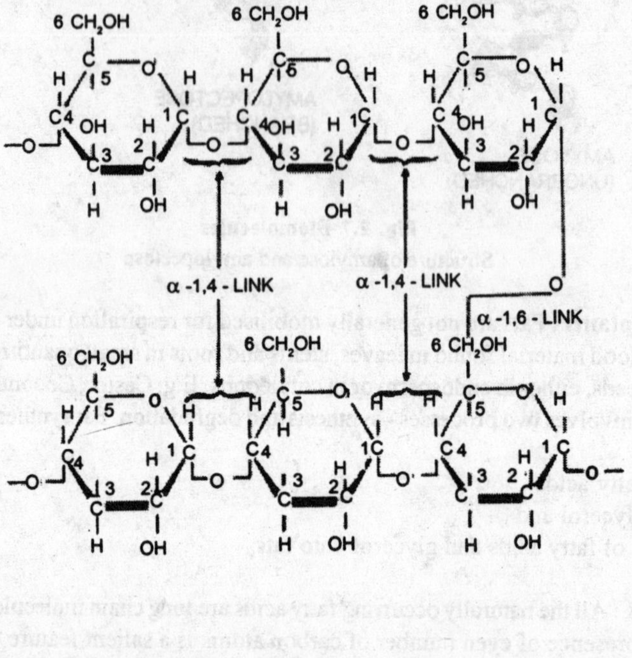

Fig. 2.8 Biomolecules
Structure of amylo pectin (detailed structure)

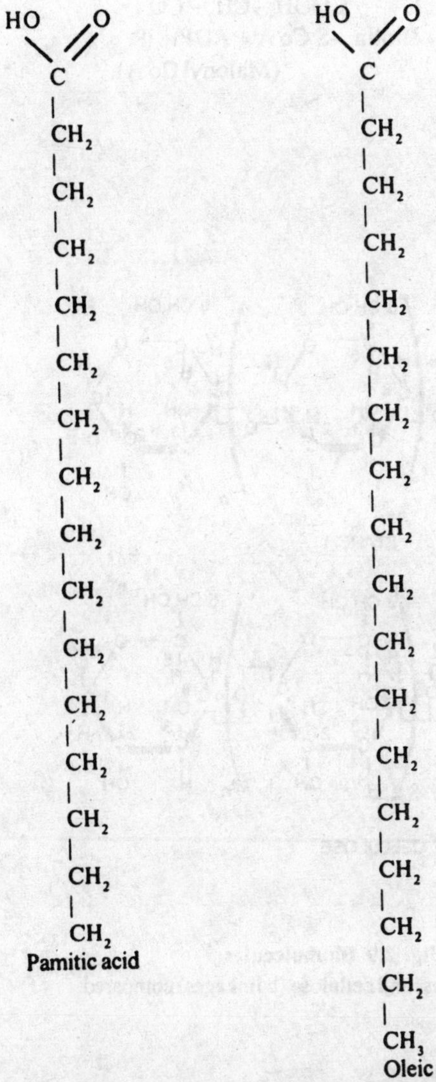

HO O
C
CH$_2$ – CH$_2$ – CH$_2$ – CH$_2$ – CH$_2$ – CH$_2$ – CH$_2$ – CH$_2$ – CH$_2$ – CH$_2$ – CH$_2$ – CH$_2$ – CH$_2$ – CH$_2$ – CH$_2$

Pamitic acid

HO O
C
CH$_2$ – CH$_2$ – CH$_2$ – CH$_2$ – CH$_2$ – CH$_2$ – CH$_2$ – CH$_2$ – CH$_2$ – CH$_2$ – CH$_2$ – CH$_2$ – CH$_2$ – CH$_2$ – CH$_2$ – CH$_2$ – CH$_3$

Oleic acid

Green (1960), has shown that two enzyme complexes are involved in the synthesis of fatty acids. The two enzyme complexes have five co-factors – 2ATP, Mn++, biotin, NADP and CO_2. Basically fatty acid synthesis involves two broad steps –

(i) first enzyme complex together with co-factors like ATP, Mn^{++}, biotin and CO_2 participates in the synthesis of an intermediate compound.
(ii) second enzyme complex together with NADP produces the fatty acid and CO_2.

Long chain fatty acids are synthesized in a stepwise fashion by the addition of 2 carbon compounds starting with the first 2 carbon compound acetyl Co A. Acetyl Co A, which is a key reactant in the respiratory breakdown of carbohydrates first gets converted into malonyl Co A. Malonyl Co A, which is a derivative of malonic acid is the key intermediate compound in the synthesis of fatty acids. Malonyl Co A is a 3C compound (Klein, 1957) and is produced from acetyl Co A as follows –

$$CH_3CO-S-CoA + HCO_3 + ATP \xrightarrow[\text{Biotin}]{Mn^{++}} COOH-CH_2-CO\text{-S CoA} + ADP + iP$$

(Malonyl Co A)

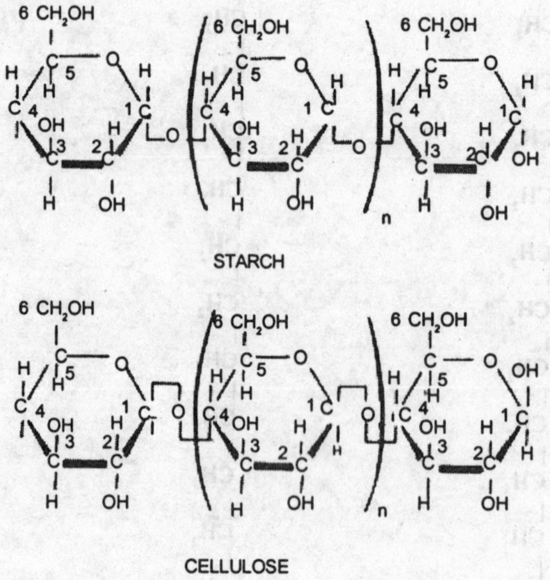

Fig. 2.9 Biomolecules

Starch (a linkages) and cellulose (b linkages) compared

The synthesise of malonyl Co A is catalyzed by the enzyme acetyl Co A carboxylase with biotin as coenzyme. Further reactions involved in the synthesis of a fatty acid chain are as follows.

(i) Acetyl Co A gets attached to a protein called acetyl carrier protein (ACP) to form Acetyl Co A ACP complex.

(ii) The complex reacts with malonyl Co A to form acetomalonyl Co A, a 5 C compound.

$$\begin{array}{l} \quad\;\; H \quad\;\; O \\ \quad\;\; | \quad\;\;\; \| \\ H-C-C-SCoA + CH_3CO \sim S\,CoA\,ACP \\ \quad\;\; | \\ O=C-OH \end{array}$$

$$\begin{array}{l} \quad\; H \quad O \quad H \quad O \\ H-C-C-C-C \sim S-ACP + Co\,a-SH \end{array}$$

| | |
H O C=OH

(Acetomalonyl Co A)

(iii) Acetomalonyl Co A reacts with acetyl SACP to form a 4 C compound. A molecule of CO is released in the process.

Acetomalonyl Co A + Acety – S – ACP ———— Acetoacey – S – ACP + CO_2 + ACP – SH

(iv) The 4 C compound Acetoacety – S – ACP is then converted into a saturated fatty acid – Butyric acid as follows :

(a) Acetoacetyl – S – ACP + NADPH + H^+ ———— Hydroxybutyryl – S — ACP + $NADP^+$

(b) Hydroxybutyryl – S – ACP ———— Crotonyl – S –ACP + OH^-

(c) Crotonyl – S – ACP + NADPH + H^+ ———— Butyric Acid + S – ACP + NADP +

(d) Butyryl – S – ACP + H_2O ———— Butyric acid + S – ACP

The above reaction sequences are repeated to give rise to 6C, 8C, 10C units. When the fatty acid chain reaches the required length, the acyl group is transferred to the SH group of CO A.

Synthesis of glycerol : Available evidences indicate that the synthesis of glycerol in plants involves more than one pathway. One of the well understood mechanisms, involves the 3 carbon compound dihydroxyacetone phosphate (DAP) an intermediate product in the glycolytic breakdown of carbohydrates. In the first step DAP is reduced to glycero phosphate in the presence of NADH. The reaction is catalyzed by the enzyme glycerophosphate dehydrogenase.

(i) DAP + $NADH_2$ ———— glycerophosphate + NAD^+

(ii) glycerophosphate is dephosphorylated as follows to give rise to glycerol.

$CH_2O.H_3PO_3$ | CHOH | CH_2OH + H_2O ———— CH_2OH | CHOH + H_3PO_4 | CH_2OH

Glycerol

Condensation of fatty acids and glycerol : Fatty acids esterify the three alcoholic groups of glycerol to bring about the synthesis of fats. The three fatty acids reacting with glycerol may be same or different. The sequence of reactions is as follows:

(i) Fatty acids combine with Co A to form fatty acyl Co A complex.

(ii) Glycerol undergoes Phosphorylation to produce glycerophosphate. The reaction is catalyzed by the enzyme glycerokinase.

(iii) The acyl residues of the fatty acid acyl Co A complex are transfened to the two free hydroxyl groups of the glycerophosphate to form phosphatidic acid.

CH_2OH | CHOH + 2R – C – ACP → H_2COOCR | RCOOCH + 2ACP
||

|　　　　　　　　　O　　　　　　　　|

$CH_2OPO_3H_2$　　　　　　　　　　　$H_2COPO_3H_2$

(Glycerophosphate) (Fatty acyl CoA) (Phosphatidic acid)

(iv) In the next step, the phosphate residue is removed from the phosptatidic acid molecule to form a diglyceride.

(v) In the last step the free (unesterified) hydroxyl group of the diglyceride reacts with a fatty acyl Co A complex to form a triglyceride as follows :

RCOO CH_2　　　　　　　　　　　　H_2COOCR

|　　　　　　　　　　　　　　　　|

RCOOCH + R – CO – ACP → HCOOCR + ACP

|　　　　　　　　　　　　　　　　|

$H_2PO_3OCH_2$　　　　　　　　　　H_2COOCR

(Diglyceride)　　　　　　　　　　(Triglyceride)

Oxidation of fats : Fats are used generally as storage reserve in plants and animals and are utilized as respiratory substrates during non availability of carbohydrates. The enzyme lipase hydrolyses fats into glycerol and fatty acids by breaking the ester bonds. This process is called saponification and is the reverse of condensation. Glycerol and fatty acids are further oxidized as follows :

Oxidation of fatty acids : There are two pathways for the oxidation of fatty acids. These are α-oxidation and β -oxidation. The type of oxidation depends on whether β -carbon atoms or α -carbon atom undergoes oxidation first.

Oxidation : This process of oxidation is known mostly in plants and is helpful in the oxidation of long chain fatty acids having more than 14 carbon atoms.

The enzyme fatty acid peroxidase brings about the oxidative decarboxyladon of fatty acids. H_2O_2 is the oxidizing agent and the fatty acids are converted into an aldehyde. The aldehyde will have one carbon less than the fatty acid.

The next step is the direct oxidation of aldehyde by an enzyme fatty aldehyde dehydrogenase. NAD is reduced to $NADH_2$ and a molecule of fatty acid with one carbon atom less is produced. The fatty acid molecule can enter the cycle once again for further decarboxylation.

Oxidation : This is the principal pathway of fatty acid oxidation. According to F. Knop (1904) who first studied this pathway, at every stage two carbon atoms are removed following the oxidation of carbon atoms. The 2 C fragments are removed in the form of acetyl Co A. The following are the sequential steps of oxidation.

(i) *Activation of fatty acids* : Fatty acids react with Co A and ATP to form Co A complex.
Fatty acid + Co A + ATP → Acyl Co – A + AMP + 2Pi.

(ii) *Dehydrogenation* : Two H Atoms between and carbon atoms are removed from the fatty acyl Co A to form trans unsaturated fatty acyl Co A. The reaction is catalyzed by the enzyme Acyl CoA dehydrogenase.
Fatty Acyl Co – A + FAD → Trans – fatty acyl Co – A+ $FADH_2$

(iii) *Hydration* : A molecule of water is added to trans – fatty Acyl Co – A to form Hydroxyacyl Co A. The enzyme enoyl hydrase catalyzes this reaction.

(iv) *Dehydrogenation* : Hydroxyacyl CoA undergoes dehydrogenation to form - Keto Acyl Co – A. The enzyme dehydrogenase catalyzes this reaction with NAD getting reduced to $NADH_2$

Hydroxy acyl Co – A + NAD^+ → Keto acyl Co – A + NADH + H^+

(v) *Thioclastic clevage* : Keto acyl Co A undergoes a clevage reaction with the elimination of a 2 carbon compound acetyl Co A. In the process, a molecule of fatty acyl Co A is produced from keto acyl Co A. Fatty acyl Co A now has two carbon atoms less than original. This re enters the cycle once again to undergo cleavage. Thioclastic cleavage is catalysed by the enzyme keto acyl thiolase.

Molecules of fatty acid which undergo oxidation become shorter by 2 carbon atoms for every step and finally get converted into 2 carbon compounds (acetyl Co A).

Acetyl Co A molecules produced as a result of oxidation enter the Krebs cycle to undergo further decarboxylation and release energy.

CONVERSION OF FATS TO CARBOHYDRATES (Glyoxylate cycle)

Acetyl Co A is the key intermediate compound between carbohydrates and fats. Oxidation of both carbohydrates and fatty acids results in the formation of acetyl CoA. While the interconversion between fatty acids and acetyl Co A is reversible, that between carbohydrates and acetyl Co A is irreversible.

(i) carbohydrates → pyruvic acid → Acetyl co A
(ii) Fatty acids → Acetyl Co A

The above reactions show that while fats can be synthesized by carbohydrates via pyruvic acid and acetycl Co A, the reverse is not possible because acetyl Co A cannot form pyruvic acid (the reaction is irreversible). In order to solve this problem, Komberg and Krebs (1957), proposed a new pathway for the synthesis of carbohydrates from fats. This pathway is known as the Glyoxylate cycle. Glyoxylate pathway was first demonstrated in microorganisms and subsequently in germinating seeds of Angiosperms when the reserve food breaks down.

Basically the Glyoxylate cycle is nothing but Krebs cycle with some modifications. The essential features of this cycle are as follows :

(i) Fatty acids get converted into fragments of acetyl Co A by oxidation, while glycerol enters into glycolysis cycle.
(ii) Acetyl Co A combines with oxaloacetic acid of the Krebs cycle to form citric acid.
(iii) Citric acid gets converted to *cis* aconitic acid and then to isocitric acid due to dehydration and rehydration.
(iv) Isocitric acid breaks up into succinic acid and Glyoxylic acid under the influence of the enzyme isocitrase.

Isocitric acid → Succinic acid + Glyoxylic acid

(v) Succinic acid continues further in the Krebs cycle by getting converted into fumaric acid, malic acid and oxaloacetic acid.
(vi) Malic acid either undergoes oxidative decarboxylation (Ochoa reaction), or reductive decarboxylation (Wood/Werkman reaction) to finally form pyruvic acid.
 (a) Malic acid + NADP → Pyruvic acid + $NADPH_2$ + CO_2
 (b) Malic acid + NAD → Oxaloacetic acid + $NADH_2$
 (c) Oxaloacetic acid → Pyruvic acid + CO_2

Pyruvic acid gets converted into PEP (phosphoenol pyruvic''acid) which form glucose and fructose by reverse reactions of glycolysis. Glucose and fructose combine to form sucrose.

Oxidation and glyoxylate pathway occurs in certain membrane bound particles called glyoxysomes, while the conversion of succinic to oxaloacetate (Krebs Cycle) occurs in mitochondria. The site for conversion of fats to carbohydrates is therefore glyoxysomes, mitochondria and cytoplasm (from pyruvic acid to carbohydrates).

PROTEINS

Of the two important macromolecules that mastermind all the activities of the living systems, proteins are one, the nucleic acids being the other. Proteins are present all over the cell and are important both structurally and functionally.

The name protein was first suggested by Berzelius (1838). *Proteios* in Greek means first rank. Proteins are highly complex organic molecules with molecular weight ranging upto many millions. Chemically proteins are made up of C, H, O and N. In addition to these major elements, some proteins may also have sulphur, phosphorous, iron zinc, iodine etc as components. The ratio of major elements in proteins are – C = 50-55%, H = 6-8%, 0 = 20-23%, N = 15-18% and S = 0-4%.

Proteins are made up of building blocks called amino acids (i.e. on hydrolysis, proteins yield these units). The idea that proteins are built up of amino acids was first proposed by Emil Fischer.

Amino acids are carboxylic acids which have at least one COOH group and one NH, group. Other functional groups may also be present. There are atleast 22 different types of amino acids encountered in plants.

The amino acids are bound to one another in a protein chain by means of linkages or bonds called peptide linkages. Peptide linkages are formed between the COOH group of one amino acid and the NH group of the other, with the elimination of a molecule of water. Several amino acids combine together with the peptide bond and form a peptide chain. Several of these peptides constitute a polypeptide chain. The polypeptide chain folds and refolds in a characteristic fashion to form a protein molecule.

Peptides : These are made up of amino acids linked with CONH (peptide) bonds. Peptides are intermediate compounds produced during the synthesis of proteins.

Amino acids $\rightarrow$ peptides $\rightarrow$ proteins

Proteins : These are the final products in the N, metabolism, and are made up of folded peptide chains which have many additional linkages (depends upon the point of folding) in addition to the CONH bond. Proteins have a three dimensional configuration and have a very high molecular weight reaching upto 10^4 - 10^7.

Although proteins contain not more than 20 amino acids, they are of immense variety and complexity due to variations in combinations of amino acids. In general the three following aspects pertaining to the amino acids decide the type of protein. These are – the quality, quantity and sequence of amino acids in the peptide chains.

In other words, the type of amino acid, its quantity and the sequence in which these amino acids are arranged form the basis for the categorization of proteins. Besides, proteins may differ in the number of peptide chains per molecule, type of secondary linkages (SH) etc.

Structure of proteins : Modern biochemists regard four structural levels in the protein molecule which accounts for its complexity. These levels are primary, secondary, tertiary and quaternary structures, in the order of increasing complexity.

Primary structure : This is the simplest and the most basic structure and refers to the linear peptide chain. Besides the CONH bond, disulphide bonds (- S -S -) are found between or among the peptide chain, linking the sulphur containing amino acids.

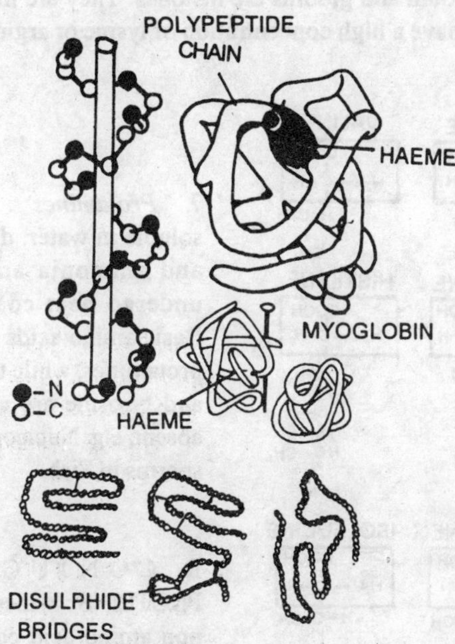

Fig. 2.10 Biomolecules

Protein structure. **A.** Polypeptide chain with helical configuration,
B. Tertiary and quaternary structure of myoglobin,
C. Denaturation of proteins.

Secondary structure : The primary polypeptide chain spirally coils forming the α helical chain. The characteristic bonds of the secondary structures are – hydrogen bonds between the carbonyl and amino groups of peptide links, salt linkages and Vander Walls forces which help maintain the helical structure.

Tertiary structure : Folding and superfolding of the helical chain in a three dimensional orientation constitutes the tertiary structure. Interaction between the amino acid residues situated far apart in the chain brings about the folding.

Quaternary structure: This consists of more than one identical sub units (poly peptide chain) which associate with one another to form a functional unit. Individually each subunit is non functional. Such complicated structures are usually seen in enzymes. For example, the enzyme, Phosphorylase, has two sub units, which are inactive individually, but as dimers, they are active. A protein with identical sub units in the quaternary structure is known to be homogeneous, or if the subunits are dissimilar it is known as heterogeneous. A protein with more than one sub unit (monomer) is referred to as an oligomeric protein.

Classification of Proteins : As the proteins possess, a very complex structure, their classification also poses a problem as many categorizations are possible based on their structure, function etc. The following is a generalized classification. Three major groups are recognized in proteins, based on their solubility and chemical properties.

I. Simple proteins.
II. Conjugated proteins and
III. Derived proteins.

1. SIMPLE PROTEINS : Chemically these have only amino acids. No additional compounds are present in the molecule. Simple proteins are of the .following kinds –

1. *Albumins* : These are soluble in water, dilute salts,acids and alkalies. Heat denatures the protein structure by coagulation. Some of the commonly occuring albumins are –
   Legumelin (from legume seeds),amylase (from barley),
   Leucosin (from cereals) etc.
2. *Globulins* : Less soluble in water than albumins, but soluble in less concentrated salt solutons. Heat coagulation is seen in globulins also. e.g. Legumin (pears), Tuberin(potato) etc.
3. *Glutelins* : Insoluble in water and salt solutions, but soluble in weak acids and alkalies. Glutelins are generally found in plants and are heat immune. Seed proteins found in cereals like wheat (glutenin and rice) and rice (oryzenin) are best examples of glutelins.

4. *Prolamines* : These proteins are water insoluble, but dissolve in 70 - 80% alcohol. Prolamines have a high concentration of the amino acid proline and other amides, e.g. Gliadin (Wheat), Hordein (barley), and Zein (maize)
5. *Scleroproteins* : These are fibrous proteins found as an integral part of the structural framework of plants. These proteins are insoluble in most of the solvents.
6. *Histones* : Most of the nucleoproteins, hemoglobin and globins are histones. They are insoluble in water but soluble in dilute ammonia. Histones have a high concentration of lysine or arginine.

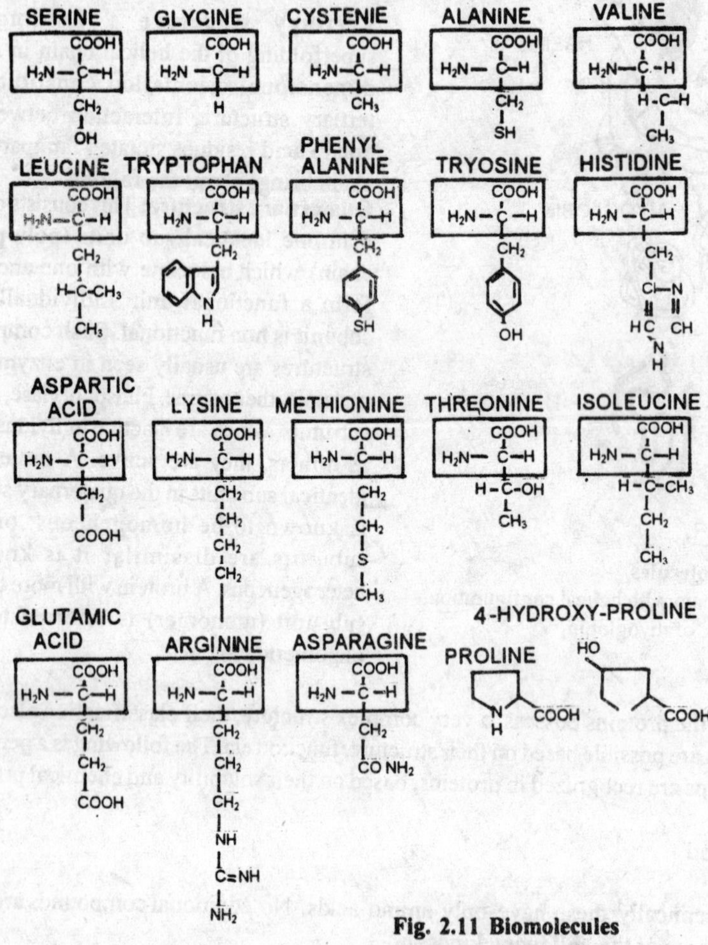

Fig. 2.11 Biomolecules
Structure of aminoacids

7. *Protamines* : These are soluble in water, dilute acids and ammonia and do not undergo heat coagulation. Basic amino acids are rich in protamines, while tryptophan and tyrosine are completely absent, e.g. Nucleoproteins of sperms in Fish.

II. CONJUGATED PROTEINS : In these proteins, non amino acid components are associated with the protein molecule. The non amino components are called prosthetic groups. There are seven major categories of conjugated proteins.

1. *Nucleoproteins* : Proteins are found in combination with nucleic acids. Hydrolysis can separate nucleic acids and simple proteins. Nucleoproteins are water soluble.
2. *Glycoproteins* : Also called mucoids, these are conjugated with carbohydrates like macromolecular polysaccharides, sugars, phosphate etc. Glycoproteins are found in cell membranes.
3. *Chromoproteins* : These constitute the pigments and are coloured, the coloured appearance is due to the presence of metals such as Cu, Fe, Mg etc. Chlorophylls, flavoproteins, carotenoids, cytochromes etc are typical examples of chromoproteins.
4. *Lipoproteins* : Proteins in conjugation with lipids are called lipoproteins. Lecithin, Cephalin etc are the

lipids found in combination with proteins. Lipoproteins are an integral part of all cell membranes.

5. *Metalloproteins* : Certain types of enzymes which require metals as activators belong to this group. Many of the respiratory enzymes belong to this group.
6. *Phosphoproteins* : These are proteins with phosphoric acid as the prosthetic group. Phosphoproteins are insoluble in water, but soluble in alkalies e.g. milk and egg protein (casein and vitellin).
7. *Lecithoprotein* : Proteins with lecthin as the prosthetic group (e.g. white of egg) belong to this group.

III. DERIVED PROTEINS : These are degradation products of proteins. Primary proteins on hydrolysis with acids or alkalies yield derived proteins. There are two types of derived proteins.

(a) Primary derived proteins and

(b) Secondary derived proteins

A. Primary derived proteins : There are two categories of primary derived proteins. These are metaproteins and coagulated proteins.

(i) *Metaproteins* : These are insoluble in water but dissove in dilute salt solutions, acids and alkalies. Hydrolysis of natural proteins on long treatment with alkalies or acids yield metaproteins.

(ii) *Coagulated proteins* : Treatment with heat or alcohol of natural proteins yield these proteins.

B. Secondary derived proteins : Three types of secondary derived proteins have been identified. These are proteoses, peptones and peptides.

(i) *Proteoses* : Prolonged hydrolysis of metaproteins yield proteoses (eg. Albuminoses from albumin). These are water soluble but immune to heat treatment.

(ii) *Peptones* : Peptones are obtained by continued hydrolysis of proteoses. Like proteoses, these are also water soluble and heat insensitive.

(iii) *Peptides* : Prolonged hydrolysis of natural proteins yields peptides. These are also water soluble and not coagulated by heat.

Functions of proteins : Most of the vital activities of the cell are controlled by the two master molecules namely proteins and nucleic acids. Proteins are important, both from the point of view of structure and function of the cell. Structurally proteins constitute an integral part of membranes, pigments etc. Functionally in the form of enzmymes, proteins play such a vital role in cell physiology, that without proteins (enzymes) no metabolic reaction is possible. Even the synthesis of DNA is regulted by proteins. In the form of Chromoproteins (Chlorophyll, Cytochromes etc), proteins mediate all energy transformation reactions. It is because proteins are so important that the nucleic acid directly regulates the protein synthesis.

3

MICROSCOPY

The study of cell biology involves the use of a number instruments. The rapid strides made by cell biology would not have been possible but for the pivotal role played by these instruments in unraveling the mysterious world of microcosm. For instance, if there were to be no microscope the whole of the microbial world would be non available to us and we would not be in a position to explain the various diseases that confront us; for we would not know what microbes are. In this chapter we will try to understand the various instruments and their working mechanism. Microbiological instruments may be basically categorized into two - those that are used for visual observation and those that are used for analytical purposes. Visual observation of microbes is made possible by the use of microscopes while a wide array of instruments such as spectrophotometers, centrifuges, nephelometers, autoclaves inoculation chambers etc., are used for various analytical purposes.

In this chapter we shall study the various optical instruments (microscopes) used in cell biological study.

Microscopy

The most important tool of a biologist is the microscope. The invention of light microscope in the sixteenth century is perhaps the single most important contribution ever made to the advancement of biology.

The present day microscope used in biological laboratories has been developed over the course of several centuries. Ancient Romans, Greeks and Indians had known lenses of glass and quartz. In the 14th century itself spectacles and single lenses were used as magnifiers. Zacharias Jensen, a Dutch astronomer added one more lens and constructed a true telescope. Campini, an Italian scientist found out that lens can be ground to the desired curvature. Since the curvature of the lens increases its magnifying power, microscope thereafter became a very powerful tool in the hands of biologists.

Antony Van Leeuwenhoek (1632 - 1723), some times wrongly credited with the invention of the microscope, used his single lens microscope to observe several minute objects like insects, mold, pond water, saliva etc. It was Robert Hooke (1665) an Englishman who developed an instrument that could truly be referred to as the forerunner of the modern day microscope.

The microscope opened up to biologists (hitherto unknown to them) an exciting microcosm - the universe of microorganisms. Bacteria, cells, detailed structure of tissues, all became visible under the microscope. New fields of biology like cytology, embryology owe their existence to the single most powerful tool of biology - the microscope.

The working of a microscope

The main function of the microscope is to produce a large and clear image of the object. Clarity is very often referred to as the resolving power of the microscope.

In order to understand the functioning of the microscope, it will be necessary to understand as to how our own eye resolves and produces an image of the object.

The lens of the eye focuses the light reflected by an object and projects an image on the retina. The retina is a granular matrix consisting of light sensitive cells called rods and cones. These cells convert the light impulses falling on them into nerve impulses producing a sensation of vision in the brain. The *rods* and *cones* are placed five microns apart, hence the eye cannot resolve details of an object, less than five microns in size. When an object is brought closer to the eye it grows larger, but when it is brought too close, the lens of the eye cannot focus the image and the object appears blurred.

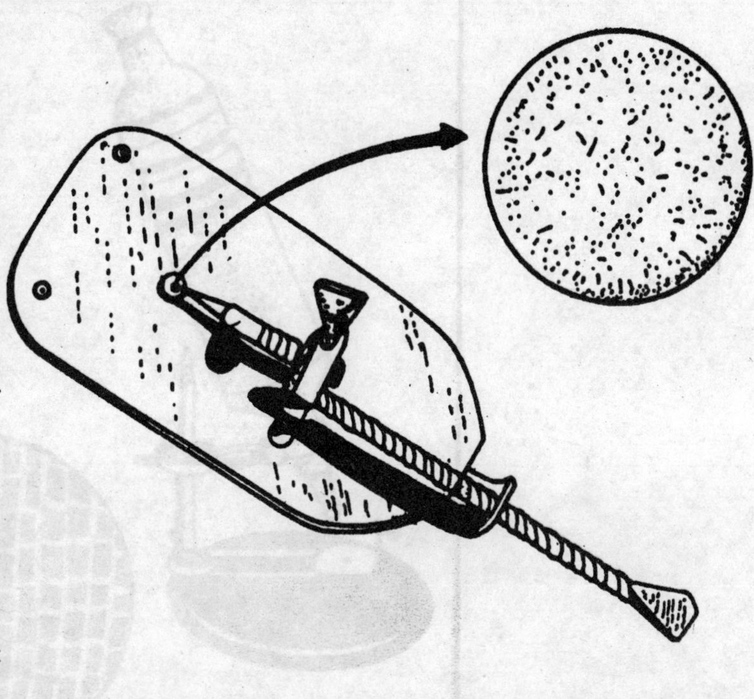

Fig. 3.1 Microscopy

Van Leeuwenhoek's Microscope. A simple lens of high magnification, this was the forerunner of the modern microscope

The shortest viewing distance of human eye is about 25 cms. At this distance, the limit of resolution is about 0.2 mm. Objects lying together closer than 0.2 mm will not be seen separately by the human eye.

Simple Microscope

The simple microscope is nothing but a single convex lens or a combination of more than one lenses functioning like a single lens. This type of lens when placed between an object and the eye allows the object to come closer, thus forming an enlarged image on the retina. Since the image is larger, the eye is able to resolve finer details that could not be distinguished without the lens.

The magnification obtained with a simple lens can be calculated as follows.

$M = 250/f \quad + \quad 1$

Where f is the focal length of the lens in mm, and 250 is the distance of normal vision also in mm.

A single convex lens can be used for magnification upto 3X. Higher magnifications are not possible with a single lens.

A superior quality lens can be constructed by combining two or more lenses into a single unit. Many types of multi-element (unit) lenses have been designed, but even they have limitations. One such multi-element lens is the *triplet* or *aplanat*, which consists of a double convex lens of crown glass sandwiched between a pair of concave lens of flint glass. A magnifying unit of this type can see objects up to 20X. This is commonly used as field magnifying glass by biologists.

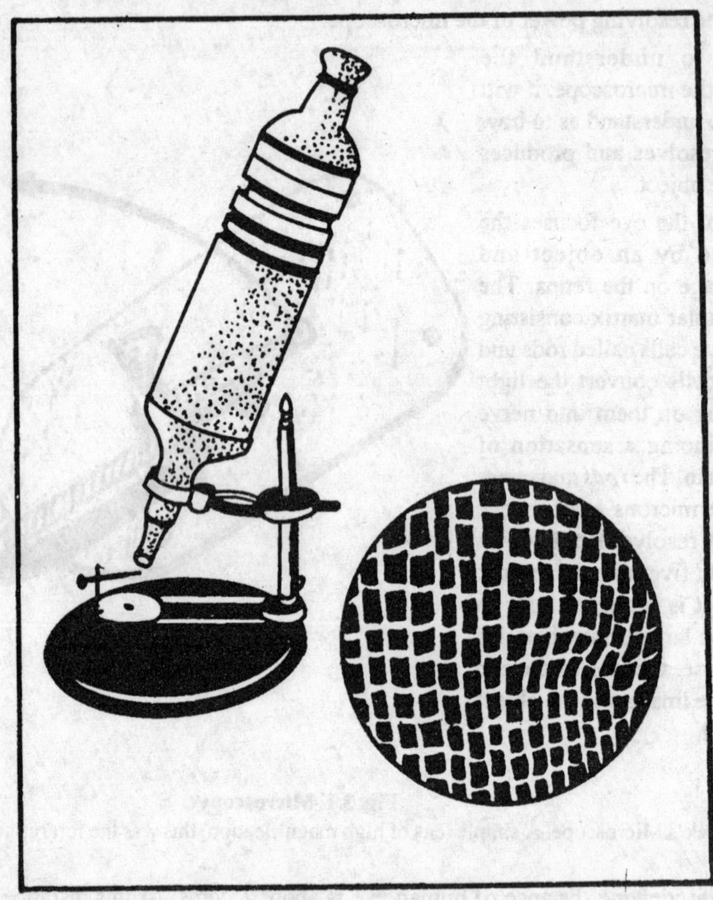

Fig. 3.2 Microscopy
Hooke's Microscope and cork cells

Modern (Compound) microscopes

Modern microscopes, generally called compound microscopes have a complex system of arrangement of lenses that help in higher magnification and better resolution.

Basically the modern microscopes are of two types - light microscopes and electron microscopes. In a light (optical) microscope, the source of illumination of the object is visible light while in an electron microscope, the source of illumination is a beam of electrons.

Light microscopes are of the following types -

1. Bright field (Light) microscope
2. Dark field microscope
3. Phase contrast microscope
4. Fluorescence microscope
5. Ultraviolet microscope and
6. Interference microscope

Electron microscopes are of two types.

1. Transmission electron microscope (TEM) and
2. Scanning electron microscope (SEM)

The following table gives a comparative summary of different types of microscopes.

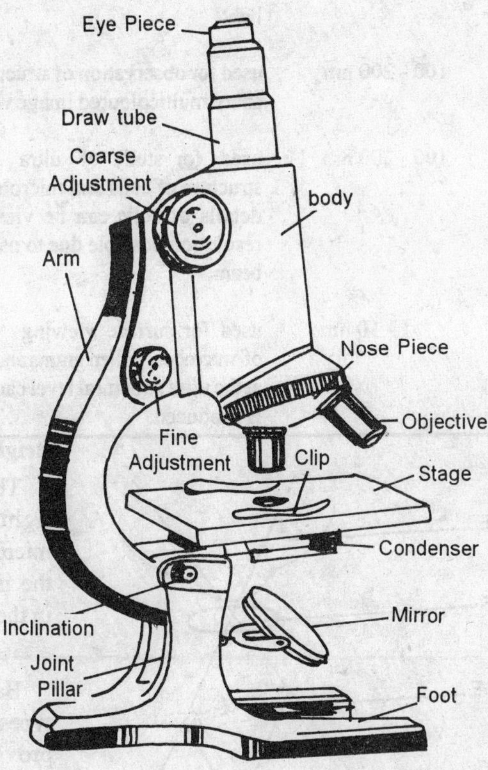

Fig. 3.3 Microscopy
A Compound Microscope

Table Comparison of different types of microscopes

No.	Type of Microscope	Maximum	Resolution Magnification	Remarks (uses)
1.	Bright field	1500 X	100 - 200 nm	used for visual observation of microbes; staining is necessary to view the specimens
2.	Dark field	1500 X	100 - 200 nm	used for seeing live microbes; staining not necessary; specimens appear bright against a dark background.

3.	Phase contrast	1500 X	100 - 200 nm	used for observing the structure of microbes; staining not necessary.
4.	Fluorescence	1500 X	100 - 200 nm	useful in diagnostic identification of microbes; uses fluorescent staining.
5.	Ultraviolet	2500 X	100 nm	better resolution due to the usage of UV (short wave length) light.
6.	Interference	1500 X	100 - 200 nm	used for observation of structural details of microbes; produces sharp multicoloured image with three dimensional view.
7.	TEM	500,000 X- 1000000 X	100 - 200 nm	used for study of ultra structure of cells and microbes including viruses. Anatomical details of cells can be viewed; highest magnification and resolution possible due to usage of short wave length electron beam.
8.	SEM	10,000 - 1000000 X	1 - 10 nm	used for surface viewing of microbes, pollen grains and other objects; surface architecture at the ultra structural level can be seen, three dimensional image is produced.

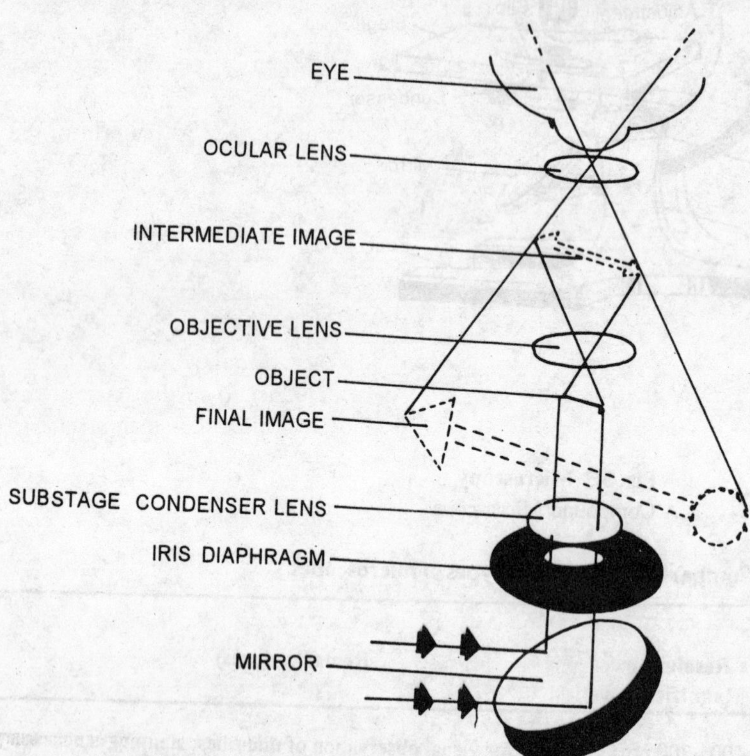

Fig. 3.4 Microscopy
Optical system of the compound microscope

Bright Field (Light) Microscope

The bright field microscope or light microscope or compound microscope is the most common and the most indispensable instrument in the biology laboratory.

Basically the microscope has two types of parts. The mechanical parts provide the structural framework of the microscope and support the optical parts. The optical parts are involved in the magnification of the object. The following is a detailed description of the various parts of a light microscope.

Base, Pillar and Inclination Joint

The base and the attached pillar serve to support the entire microscope. The base is generally *'V'* (horse shoe) shaped or some times 'U' shaped. Usually the pillar and the base are made heavy in order to minimize vibration. The inclination joint joins the arm of the microscope to the pillar. The joint permits the tilting of the microscope to any degree as desired by the observer.

Arm and body tube

The arm is a slightly bent solid piece of iron which is attached at one end to the pillar by the inclination joint. The other end of the arm has the body tube to which the principal optical systems are attached. The body tube is a cylindrical structure and at the point of attachment to the arm there is a graduated rack and pinion which helps in the movement of the body tube up and down.

Draw tube, revolving nose piece

Fitting properly inside the upper end of the body tube is the draw tube, which may be drawn up. "The ocular lens fits into the draw tube. The purpose of the draw tube is to adjust tube length so that the objective lens and the ocular are at a specified distance.

The revolving nose piece is a circular rotating disc attached to the body tube. Objective lenses are attached to this.

Adjustment Knobs

The body tube can move up and down on the rack and pinion and the movement is regulated by two sets of knobs - the course adjustment knob and the fine adjustment knob. The coarse adjustment knob brings in greater degree of movement while the fine adjustment knob is used for slight and fine adjustment of the body tube. The adjustment knobs are meant to bring in the object into proper focus. While the range of the coarse adjustment knob is greater, that of the fine adjustment knob is limited.

Stage : This is a small platform attached to the arm of the microscope. This is meant to keep the objects (on sides) for the purpose of observation. It has a hole in the center exactly in line with the body tube and the condenser below to allow the light to pass through. In most modern microscopes, the stage is fitted with a mechanical device called mechanical stage to help in the vertical and horizontal movement of the slide.

Illumination : In order to collect and reflect the light beams there are two kinds of devices attached to the base of the microscope. These are - mirror or an artificial illumination (built in illumination).

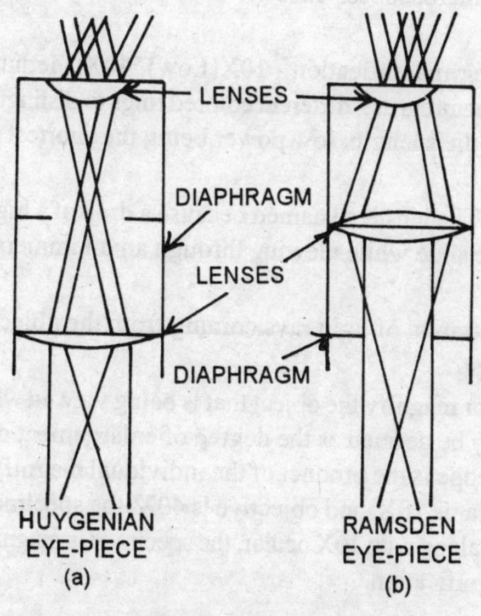

Fig. 3.5 Microscopy
Types of eye-pieces

The mirror collects the light either from a natural source (sun light) or an artificial source (electric light) and reflects the rays into the microscope. The mirror has both sides reflective - one side is plain, the other side is concave. Usually concave side is used when natural light is focussed and plain side is used while focussing artificial light.

In some of the compound microscopes there is provision of built in illumination (instead of a mirror) which has a devise to use 8 or 12 volt bulb and provide direct illumination to the stage.

Sub stage Condenser; Iris diaphragm

The light either reflected by the mirror or by a built in source will be diffuse and does not have sufficient density. In order to condense the light and focus it on to the object, there is a sub stage condenser. This consists of a condensing lens. The condensing lens may be fixed or in some microscopes it can also move up and down for suitable adjustment on a rack and pinion. The condenser helps to focus all the light on the object. Beneath the condenser there is an *iris diaphragm* which regulates the quantity of light entering into the condenser. The iris diaphragm may be closed and opened with the help of a lever to regulate the entry of light.

Optical parts

There are two sets of optics or lenses in a compound microscope. These are eye pieces (ocular) and objectives.

Eye pieces (Ocular) : The ocular or eye piece is a short tube with two lenses which fits into the upper part of the draw tube. The main function of the eye piece is to magnify the image of the object formed by the objective. Eye pieces are marked 5X, 10X, 15X etc.

The eye piece commonly used is the 10X Huygenian eye piece with monocular tube. These eye pieces are also referred to as negative oculars. The positive ocular or Ramsden eye pieces are constructed with convex surfaces of both lenses facing inwards.

Objective lenses : These are attached to the revolving nose piece and are very important as they affect the quality of the image seen by the observer. Based on their type of construction objectives are classified into *achromatic, apochromatic and fluorite.* The first one is simple in construction and less expensive, while the other two are complicated more expensive and used in costly microscopes. These two are also corrected for most of the common aberrations that occur in the lenses.

The objectives are of the following four types based on their magnification - 10X(Low), 40X(Medium), 60X(High) and 100X(Oil immersion). Several microscope manufacturers use different colored rings to distinguish individual objectives. They can also be distinguished based on their length; low power being the shortest and oil immersion, the longest.

The oil immersion lens in used for very high magnification. The lens is so named because a drop of a highly refractive liquid (refractive index same as glass) is added on the slide while viewing through an oil immersion objective.

The main functions of the objective lenses are - 1. Concentration of light rays coming from the object, 2. Forming the image of the specimen and 3. Magnifying the image.

Magnification : A microscope is primarily used to enlarge or magnify the object that is being viewed which can not otherwise be seen by the naked eye. Magnification may be defined as the degree of enlargement of an object provided by the microscope. Magnification by a microscope is the product of the individual magnifying ability of the oculars and the objectives. For example if the ocular is 10X, and objective is 40X, the specimen is magnified 400 times. If an oil immersion objective (100X) is used along with 10X ocular, the specimen is magnified 1000 times. The following factors play an important role in magnification.

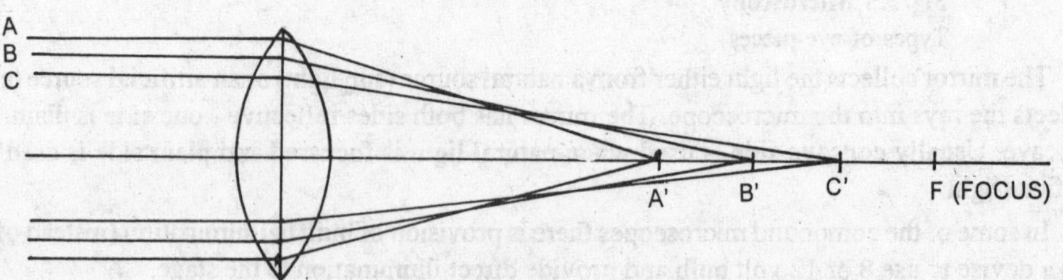

Fig. 3.6 Microscopy
Spherical aberration

(i) Optical tube length
(ii) Focal length of the objective lens
(iii) Magnifying ability of the ocular

Total magnification can be calculated as follows:

Total magnification = Optical tube length / Focal length of the objective X magnification of ocular

Resolving power

Theoretically if the magnifying power of the ocular and objectives are increased it should be possible to get higher and higher magnifications. By using high powered lenses a magnification up to 3000 can be obtained but the image will be blurred and details are not clear. This is due to the fact that in a microscope not only the lenses, but the wave length of the illumination is also important and this decides the resolving power of the microscope.

Resolving power of a microscope is defined as its *ability to distinguish between two particles situated very close.* In a magnified image the object should not only be larger but the details should also be clear. This is possible when a microscope has the ability to see two points situated very close as two distinct entities. In other words, resolving power may be said to be the minimum distance at which two structural entities of an object can be visualized as discrete individual structures even in the magnified image.

(The above explanation may be illustrated clearly with a comparison with the human eye. Human eye functions on the same principal as that of light microscope i.e. we see objects because of the light reflected by them. The human eye has a resolving power about 0.25 mm in the sense, two dots situated 0:25 mm (or more) apart can be seen as two dots; any thing closer will look like a single dot).

The resolving power of a microscope depends on two factors - the wavelength of the light used for illumination and the numerical aperture (NA) of the objective.

In optical light (bright field) microscopes, the wave length of the light used falls in the visible range (400 - 750nm). Within this range if light of shorter wave length is used the resolution will be higher. For example blue light has a shorter wavelength than red light. Greater resolution can be achieved by using a blue light as a source of illumination than a red light.

The second factor that decides the resolving power (RP) is the numerical aperture of the objective. NA is defined as the property of a lens that decides the quantity of light that can enter into it. It depends on two factors.

(i) Refractive index of the medium that fills the space between the specimen and the front of the objective lens and
(ii) The angle of the most oblique rays that can enter the objective lens (The more divergent or oblique rays that an objective can admit, greater is the resolving power).

NA can be mathematically calculated by the following formula

NA = $n \sin f$ Where

n = refractive index of the medium

f = angular aperture. This is defined as the angle between the most divergent rays passing through the lens and optical axis of the lens. Rays cannot enter the objective if their angle of divergence from the normal rays (straight rays) is greater than half the angular aperture.

Using NA, the resolution power of the microscope can be calculated as follows.

RP = Wave length of light / 2 x NA

For instance if a yellow light of wave length 580nm with NA of 1.0 is used RP will be as follows:

RP = 580/2x1 = 290nm

If a blue light is used the RP will be still higher (i.e. it can resolve objects smaller than 290nm).

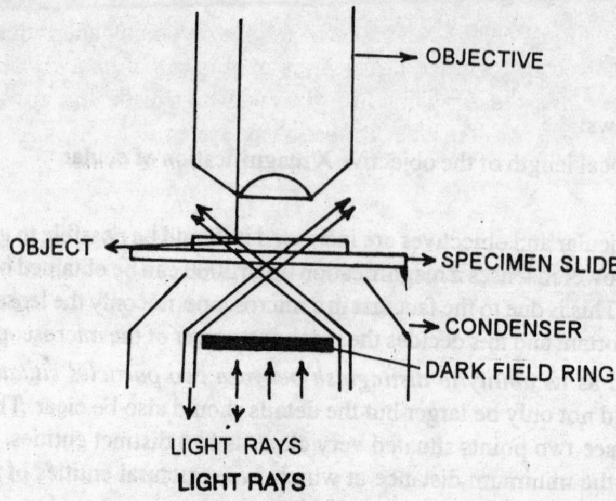

Fig. 3.7 Microscopy
Dark field microscope

Limit of resolution : This may be defined as the shortest distance between two objects when they can still be distinguished as two separate entities. Highest resolution in a light microscope is possible with the shortest wave length light (visible) and with the objective having highest NA. This can be expressed as follows:

$d = \lambda / 2NA$

Where d = resolution and λ = wave length. For instance if we use green light (520nm wave length) and an objective with 1.3 NA the resolution may be calculated as follows:

$d = {}^{520}/_{2 \times 13} = 200nm$

From the above it can be calculated that the smallest particle that can be seen by light microscope is the one not smaller than 0.2 µm in dimension.

Working Principle (higher magnification) **of Oil Immersion Lens**

When an objective lens (less than 100X) other than oil immersion is used it is referred to as a dry objective; in the space between the specimen and the front of objective there is only air. Air has a refractive index of 1.0. Rays whose angular divergence from the center of the optical axis is greater than half the angular aperture can not enter the objective lens and are lost. If a drop of oil (Cedar wood oil) is placed in between the specimen and the objectives lens, many rays whose angle of divergence is more, also can enter the objective lens as the refractive index of oil about 1.5. Thus NA of the objective is increased which results in better resolution and higher magnification.

Types of Condensers : Three types of condensers are in use in microscopes. These are

***Abbe's Condenser* :** This is the simplest one fitted to a microscope. There are two lenses in this, but both are not corrected for any chromatic or spherical aberration. This type of condenser is good only for low power viewing. At higher magnifications, the images of the specimens get blurred due to improper focussing.

***Aplanatic Condenser* :** In this, there is a third lens added to the original two, to correct the spherical aberration only and not chromatic. Aplanatic condenser gives a good image when used with monochromatic light

***Achromatic Condenser* :** This focuses most of the incident light on the object and allows very little light to escape and produce glare. The NA of the condenser generally equals that of the objective. In this condenser both spherical and chromatic aberrations are corrected.

Aberrations in the objectives

The surfaces of objective lenses are spherical and such surfaces can not form perfect images. This is due to variations in refraction and focus by the peripheral parts of the objectives. These cause defects or aberrations in the images. Aberrations in objectives are of two types - *Spherical aberrations and Chromatic aberrations*.

Spherical aberrations result due to passing of lenses-which cannot be brought to the same focus as those which pass through near the optical axis. As a result the image gets distorted.

Chromatic aberration occurs due to the splitting of light into its individual component colours as it would happen when it passes through a prism. These monochromatic beams of various wavelengths are not recombined to the same focus due to varying refraction of lens. The result is a hazy image.

Correction of aberrations : Aberrations may be corrected by using better quality lenses or by the combination of lenses with varying dispersive qualities. A lens system corrected for chromatic aberration of two colours is called *achromatic* and the one corrected for three different colours is called *apochromatic*. Crown glass with low dispersive power is used for convex lenses while flint glass with high dispersive power is used for concave lenses. A combination of the two will result in an achromatic lens system.

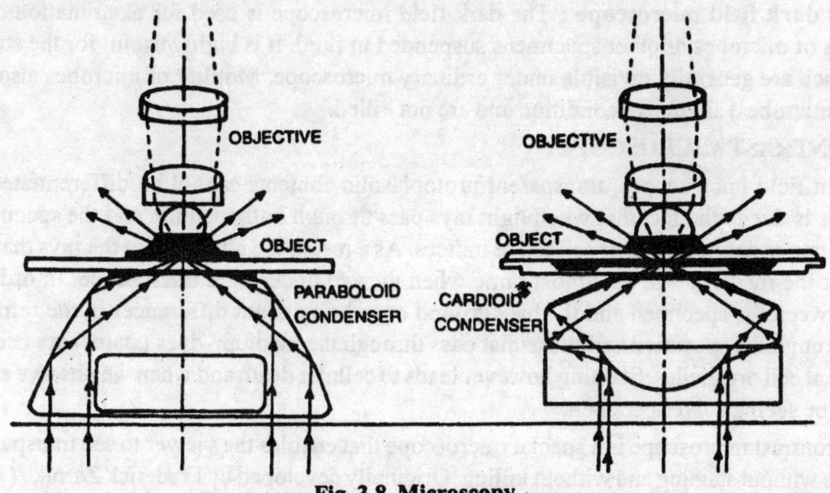

Fig. 3.8 Microscopy
Two types of dark field condensors

Spherical aberrations are corrected by the proper use of iris diaphragm to keep out peripheral rays. Additionally the lens may be ground to a shape which will permit the peripheral and axial rays to have a common focus.

DARK FIELD MICROSCOPE

There are several alternative type of microscopes other than bright field microscope that increase the contrast between the background and the specimen without staining, thus allowing for visualization of living specimens. One such microscope is dark field microscope.

Principle : In a dark room dust particles in the air are practically invisible. But if a beam of light crosses the room the dust immediately becomes visible as bright particles in light. This is due to the fact that the dust particles refract the rays of light and direct some of them towards the observer. By this method of indirect illumination against a dark background, particles are visible.

In a dark field microscope also, the specimen under observation is brightly illuminated against a dark background. In other words, the images of specimens appear as luminary bodies against a dark background. This is in contrast with bright field microscope where the specimens will be darker against a light background.

The dark field condenser : The dark field microscope is essentially a light microscope, but with a different condenser. The normally used Abbe condenser is replaced by dark field condenser. In a dark field condenser, the centre of the top lens is opaque, so that none of the central rays of light can pass through, and the specimen on the stage is illuminated only with the peripheral oblique rays. The light rays are focussed in such a way that they do not reach the specimen directly (as in the ordinary microscope) but instead the light rays traverse across the specimen horizontally almost at right angle to the objective lens. As a result of this the field appears dark but the specimen stands out as a bright refractive structure in the same way as a dust particle in a beam stands out due to its shining.

In normal operations of dark field microscopy a modified version of Abbe condenser with low power objectives would suffice. But for higher magnifications, special dark field condensers with cardoid and paraboloid lenses are employed. In some instances ordinary light may not be enough. Sources of intense light such as *carbon arc* lamps may become necessary. Utmost care should be taken to use perfectly clean and scratch free glass slides and cover slips. Many a time unwanted particles or dust are capable of reflecting light and may illuminate the background which should always remain dark.

Uses of dark field microscope : The dark field microscope is used for examination of live, unstained preparations of microbes or other specimens suspended in fluid. It is highly useful for the study of very small bacteria which are generally invisible under ordinary microscope. Motility of microbes also can be observed since they (microbes) are in live condition and are not killed.

PHASE CONTRAST MICROSCOPE

In bright field microscopes, transparent protoplasmic contents cannot be differentiated unless they are stained. This is due to the fact that when light rays pass through both medium and the specimen, there is very little difference or contrast in their refractive indices. As a result, the alteration in the rays that pass through the specimen or the medium will be almost same when viewed under light microscope. In order to increase the contrast between the specimen and the background or to bring about differences in the refraction of the rays that pass through the specimen and those that pass through the medium, dyes (stains) are used that selectively stain different cell organelles. Staining however leads to cellular death and when we observe a stained specimen we will not be seeing a live specimen.

Phase contrast microscope is a special microscope that enables the viewer to see transparent protoplasmic components without staining and without killing. Originally developed by Frederick Zernike (1933) hence called *Zernike microscope,* Phase contrast microscope is the ideal instrument for observation of living protozoans and other transparent microbes, without staining. Frederick Zerinke was awarded-Nobel Prize in physics in the year 1953 for the discovery of phase contrast principle.

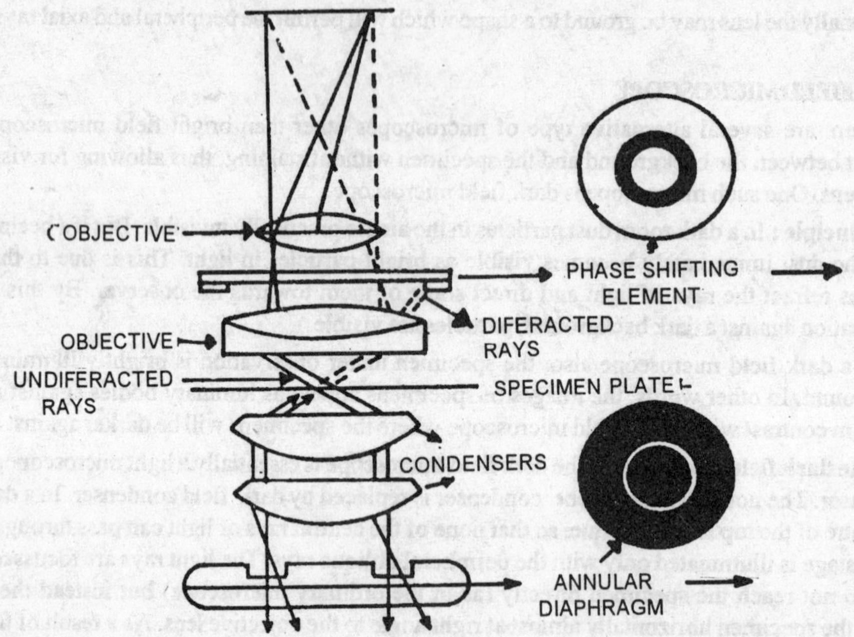

Fig. 3.9 Microscopy
Phase contrast microscope

Principle : In order to understand the principle behind the working of a phase contras microscope we must first understand how image contrasts are produced and the variable behaviour of the light rays as they pass through an object

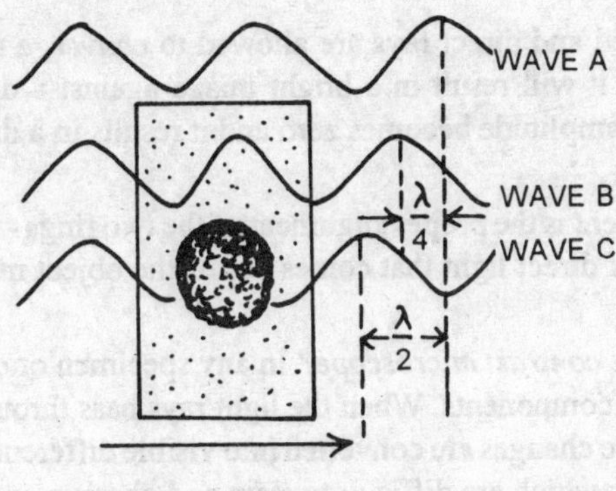

Fig. 3.10 **Microscopy**
The retardation of phases of light waves

Objects in a microscopic field can be categorized into two - *Amplitude objects* and *Phase objects.* Amplitude objects show up as dark objects under the microscopic view because of the reduction in the intensity (amplitude) of the rays that pass through. Phase objects which are transparent on the other hand allow the light rays to pass through without any reduction in their intensity. But when the light rays are passing through a transparent object, some rays will have retardation by about one quarter wave length. This retardation is called *phase shift.* But the phase shift will not cause any change in amplitude; thus the objects appear transparent. The 1/4 wave length phase shift is utilized in the phase contrast microscope to create image contrast.ln order to understand this, First we should understand what happens to the rays when they pass through a transparent object.

Behaviour of rays : Light rays passing through a transparent object emerge out in two different ways - *direct* and *diffracted.* The rays that pass through the object in a straight line are called direct rays and are unaltered in amplitude and phase. The rays that are bent and slowed down as they pass through due to differences in density of the medium, are called diffracted rays. It is this relationship between the rays that is made use of in the phase contrast microscope.

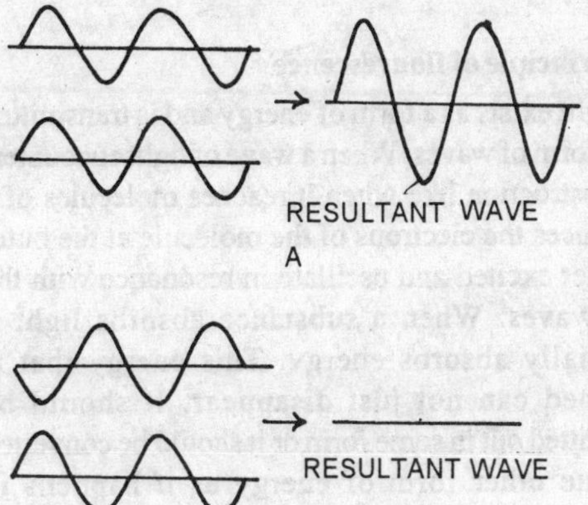

Fig. 3.11 Microscopy
Additive and destructive interference of light waves

If the direct rays and diffracted rays of an object can be brought into the same phase (the crest of both light waves coincide) with each other, the resultant increase in amplitude is due to the sum of both the converaged rays and is called *coincidence.* As a result of the increased amplitude, the object looks very bright in the field. On the other hand if the direct and diffracted rays are out of phase (i.e. the crest of one wave coincides with the *trough* and *not the crest* of another wave) it is said tc be *reverse phase* and their amplitudes cancel each other and the object looks dark. This is called *interface.* Both these phenomena (coincidence and interference) are used in a phase contrast microscope.

The phase contrast set up

The basic construction of a phase contrast microscope is like a bright field microscope only except for two special attachments. These are-(i) a special type condenser and (ii) a phase plate.

The condenser has a special diaphragm consisting of an *annular stop.* This annular stop can be compared to a solid rod kept loosely at the centre of a cylinder. The annular stop allows only a hollow cone of light rays to pass through the condenser and light the object on the slide.

The phase plate is a special optical disc located in the rear focal plane of the objective. It has a special phase ring coated with a material that can either advance or retard the direct rays depending on its construction.

The rays of light (direct rays) that emerge from the object as solid lines converge on the phase ring within the objective. Here, depending on the coated material the desired phase shift (i.e. retardation or advancement) is produced. The diffracted rays (that have already undergone phase shift) that pass through the object on the slide miss the phase ring and are not affected by the phase plate.

The resultant image will depend on whether the diffracted and direct rays are allowed to *converge* and *interfere*. If they converge (it depends on the type of phase ring) it will result in a bright image against a dark back groun (*bright phase microscopy*) or if they interfere the amplitude becomes zero and it results in a dark image against a bright background *(dark phase microscopy)*.

In a phase contrast microscope the most important adjustment is the proper alignment of the two rings - the one in the condenser and the one in the objective. The ring of direct light that comes out of the object must correspond to the phase ring in the objective.

How an unstained preparation can be seen under a phase contrast microscope? In any specimen or cell there will be difference in thickness between the structures of components. When the light rays pass through these there will be variable refraction of the rays and these phase changes are converted into visible differences of light intensity (due to the phase ring). Even some structures which are difficult to stain and observe under light microscope become conspicuous in a phase contrast microscope because small phase changes result in interference of light waves resulting in high contrast images.

FLOURESCENCE MICROSCOPE

Flourescence microscope involves staining of the specimens with special flourescent dyes and is an indispensible instrument in diagnostic microbiological laboratories.

A flourescence microscope differs from an ordinary light microscope in the following respects.

(i) Light source is a mercury vapour lamp

(ii) A dark field condenser is used instead of the normal Abbe condenser

(iii) Three sets of filters are employed to alter the light rays that pass through the instrument and reach the eye.

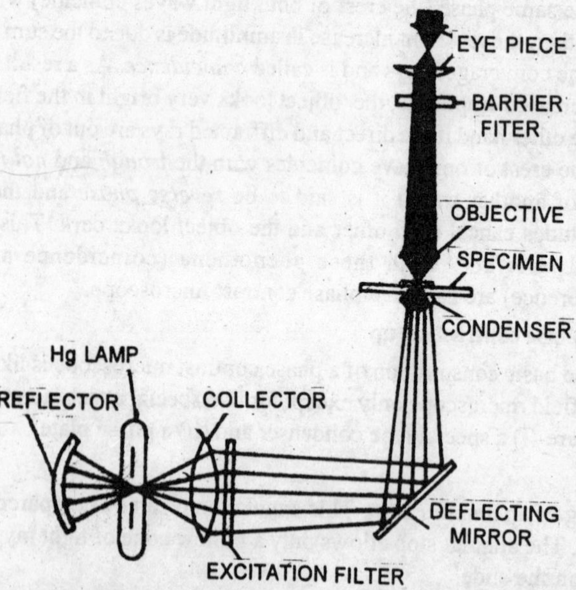

Fig. 3.12 Microscopy
Fluorescence microscope diagrammatic representation of the path of rays and action of excitation and barrier filters

The Principle of flourescence

Light exists as a form of energy and is transmitted in the form of waves. When a wave of light encounters any obstruction like when it reaches molecules of a substances the electrons of the molecule at the outer orbit get excited and oscillate in resonance with the light waves. When a substance absorbs light it essentially absorbs energy. This energy that is absorbed can not just disappear. It should be transmitted out in some form or it should be converted to some other form of energy as it happens in photosynthesis. When light is transmitted by the energized (excited) molecules it is said to be *Photoluminiscence.* In photoluminiscence, the light absorbed is not transmitted immediately. There is always a time lapse between absorption and emission. If the time lapse is greater than 10 (1/10,000) seconds it is said to be *Photoluminiscence.* If on the other hand the time lapse is less than 10 seconds it is known as *fluorescence.*

When a fluorescent dye is applied to a molecule it absorbs a wave length of light (excitation wave length) and emits a light ray of a different wave length (emission wave length). In other words, the wave length of light changes as it emerges out of a fluorescent compound. Usually the emission light will have a longer wavelength. For instance, When the flourescent dye *fluorescein isothiocyanate* is illuminated with blue light, it emits green light. This is due to the fact that light rays of shorter wavelength have more energy than those of longer wavelength and emission light always has less energy.

Microbiological specimens to be observed under a flourscence microscope are coated with fluroscent componds (flourochromes) such as Auramine O, Acridine orange, Fluoroscein, etc.

Parts of a Fluroscence microscope

The principles of magnification and resolution of a fluoroscence microscope are no different from a bright field microscope. The essential components are light source, heat filter, exciter filter.condenser and barrier filter.

Light source : This is provided by a bright mercury vapour arc lamp. As against the normal incandescent bulb, this lamp provides shorter wave length light rays such as UV, violet and blue.

The lamp always comes with a transformer as sometimes a voltage of 18,0000 volts are required.

Mercury vapour lamps produce light rays in range of 200 - 400nm (UV) and visible ravs in the range of above 780nm.

Mercury vapour lamps are not only expensive but also harmful. Hence several precautions are necessary: The bulbs are pressurized hence there is danger of their explosion. Eyes should not be directly exposed to the bulb as the rays are harmful.

Heat filter : Infrared rays produced by the lamp generate considerable heat, besides they are of no use in fluoroscence. In order to eliminate these rays a heat filter is placed in front of the lamp and before the condenser to absorb heat. Heat filter however does not prevent the transmission of UV and the visible rays.

Exciter filter : The light cooled down by the heat filter next passes through the exciter filter which absorbs all but shorter waves that are needed to excite the flurochrome dye coated specimen on the slide. These filters which are dark allow only green, blue, violet or UV rays.

Condenser : For best results always a dark field condenser is used, because in a dark background even mild fluoroscence can be easily detected-Anolher advantage of this condenser is that it deflects majority of the UV rays thus protecting the observer's eyes. In order to achieve this the NA of the objective is always kept at 0.05 less than that of the condenser.

Barrier filter : This is situated in the body tube of the microscope between the objectives and the eye piece to remove all remnants of the exciting light so that only the fluoroscence is seen. When excitation is by UV, the exciter filters are dark and barrier filters will be almost colourless. On the other hand, if blue excitation is used, the barrier filters are either yellow or deep orange in colour.

Hints for using the microscope : The following points are to be kept in mind while using a fluorescent microscope for best results.

1. *Warm up :* Fluorescent micrscopes require a warm up period. The illumination is low when turned on but reaches the maximum in about two minutes. Optimum illumination is obtained in about 30 minutes. It is better to switch on the light 45 minutes before using.
2. *Check of mercury lamp :* The life expectancy of a mercury arc lamp is about 400 hours. Hence it is necessary that a log should be maintained regarding the number of hours that the instrument is used. A check should be made after approximately 200 hours of usage.
3. *Selection of filter:* Proper combination of exciter and barrier filters are necessary to acilieve best results. The most frequently used filters are

   Bluish Schott BG12 - Exciter filter
   Yellowish Schott OG I - Barrier filter

Examination : For preliminary exmination like locating a particular part or cell of the specimen, it is better to use ordinary illumination and use the mercury arc illumination only for precise observation. If the specimen has to be observed under oil immersion, special low fluorsecing immersion oil is to be used. Ordinary immersion oil used for bright field microscopy should not be used.

Safety measures : Fluorescent microscope can be harmful and injurious if not used properly. The following precautions are necessary to avoid any harm to the user.

1. The pressurized mercury arc bulb may explode under certain circumstances. But the equipment is designed to prevent this. However lamp should not be inspected when it is hot. The lamp is safe when relatively cold. Usually 15 - 20 minutes is sufficient to cool the lamps.
2. Do not expose your eyes directly to the mercury lamp. the equipment is designed in such a way as to prevent the scattering of the rays. Please note that the unfiltered rays from the lamp contain UV as well as infrared rays both of which are harmful to the eyes.
3. Make sure that both the filters (of the suitable type) are in their place before looking into the microscope. Removal of any one or both is harmful to the to the eyes.

ULTRAVIOLET MICROSCOPE

In Ultraviolet microscope, the optics are specially meant to transmit or reflect UV light (200 - 350nm). As the source of illumination (UV) is having a shorter wave length, UV microscopes have a better resolving power than light microsopes. It is not possible to look into a UV microscope directly as the rays are injurious. Hence the images of the object are either photographed or projected on using UV sensitive TV cameras.

INTERFERENCE MICROSCOPE (Stereomicroscope)

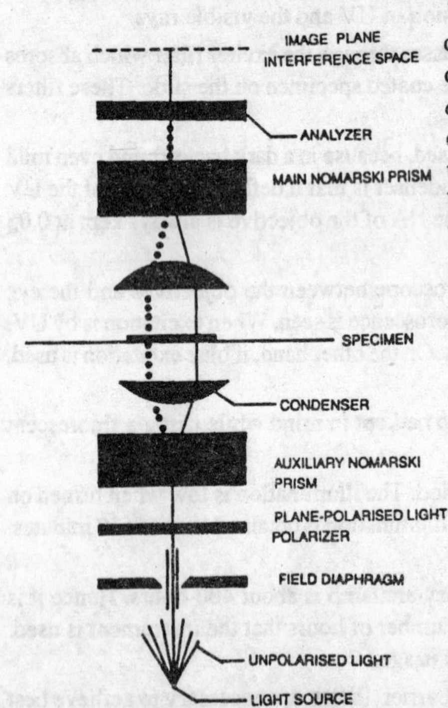

Fig. 3.13 Microscopy
The Nomarksi Interference microscope

The interference is essentially based on the same principle as that of phase contrast microscope. In both the microscopes the principle of increased or decreased amplitude due to the interference between *out of phase* light waves and *in phase* light waves is used to produce contrast in the image. While a single beam of light passing through the object is made use of in phase contrast microscope, there are two beams of light that combine after passing through the specimen. This produces an interference pattern. Interference microscopes are more versatile in comparison with phase contrast microscopes as they have objectives of higher NA and also use colour to increase the contrast.

One of the most frequently used interference microscope is a *Nomarski Differential Interference Contrast(NDIC) Microscope* which can produce a high contrast, almost three dimensional image of transparent objects. The special features of a NDIC microscope are a polarizing filler, and interference contrast condenser and a prism analyzer plate.

In *the-*. NDIC microscope. a beam of polarized light is split into two beams at right angles to each other; are made to travel through the specimen and are combined and forced through an analyzer. The result an tinterference pattern will produce a pseudo three dimensional image. This is due to the fact that the two beams of light passing through the specimen produce a sterescopic effect. The degree or level of three dimensional configuration that is visualized is a function of the varations of refractive indices at the surface of the specimen. In other words, higher the difference of refractive index at the surface, more will be the contrast in the image of the specimen. Edges of specimen such as cell walls with a high refractive index variation are very well defined under a NDIC microscope. Images seen through the NDIC microscope are normally brilliantly coloured (non - stained) because of the phase changes of light waves that pass through the various components of an object such as a microbial cell. Qualitative observations of specimens are best made through an interference microscope, because the images have a high contrast and high topographic relief due to the three dimensional effect.

ELECTRON MICROSCOPY

The microscopes which are based on visible light as a source of illumination have their own limitations. In spite of the numerous modifications there is a limit to the resolving power of these microscopes. The maximum resolution that the light microscopes can achieve is the wave length of the light waves. A particle smaller than the wavelength of visible light can not be resolved under the light microscope.

The development of electron microscope in the late 1940's however has changed the situation dramatically. Using a source of electrons for illumination which have a much shorter wave length than that of light rays, the electron microscope has opened up before biologists a hitherto unknowen exquisite subcellular architecture of the world of microcosm.

The electron microscope is a remarkable research tool of the 20th century. Its power of magnification runs into 100000X or a resolving power of 10^{-9} centimeters. Experiments that eventually lead to the development of electron microscope began as early as 1920. The First prototype model was developed for physical applications.

Principle of working

As the name indicates, in an electron microscope, a beam of electrons is used as a source of illumination, instead of a beam of light as in light microscopes. In order to understand the working mechanism of an electron microscope, we must first understand some elementary facts about electrons and their behaviour.

Electrons are negatively charged subatomic particles having a mass of 9.1 x 10^{-28} grams. An integral part of all atoms, electrons orbit round the atomic nucleus at velocities for exceeding 50,000 kms per second. When the atoms of a metal are excited by heat energy to an extent where the electron velocity exceeds the attraction of the nucleus, they (electrons) fly off from the clutches of the nucleus and are lost. When a melal, like tungsten is heated by applying a high voltage current, electrons come out in a continuous stream and this can be directed to form a high velocity electron beam.

Louis de Bragile (1924) working on electrons opined that electrons being particulate must behave like light waves and should have both a fixed wave length and frequency. He used the following formula to caculate the wave length λ (lambda) of the matter

$$\lambda = h / mv$$

where *h is* planck's constant, *m is mass* and *v* is velocity. Calculations based on this equation have shown that electrons accelerated to a potential of 60,000 volts have a wave length of 0.05 Å (Angstorm) or 105 times smaller than that of visible light. Hence an electron beam should be able to resolve (theroretically) particles 105 times smaller than that of the wave length of light.

The discovery that electrons have wave motion and they can be focussed on to an object in the form of a beam (like light ray) in a magnetic field provided the impetus for the development of electron microscope. By the late 1930's a system of magnetic coils capable of focussing an electron beam had been designed. It was also shown that by regulating the current flowing through the magnetic coils, the magnification can be regulated. Theoretically, an enhancement of resolution power equivalent to the wave length of a beam of electrons should be possible in an electron microscope.

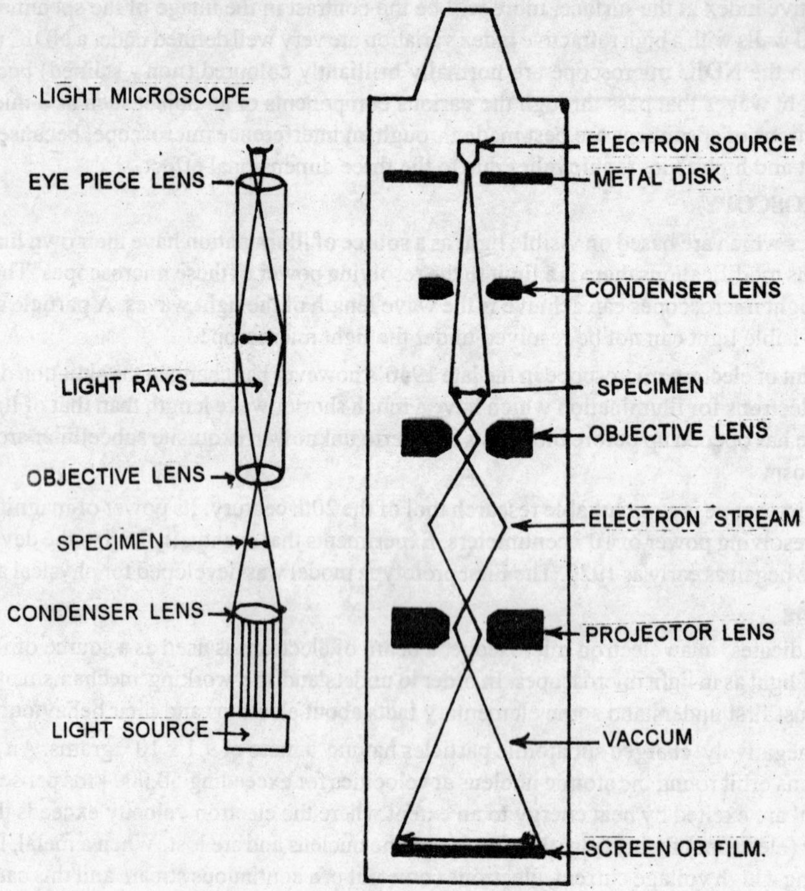

Fig. 3.14 Microscopy
Light and Electron Microscopes (comparison of paths of light)

But there are practical limitations in the design of magnetic coils (equivalent to lenses in a light microscope) used for focussing the beam of electrons. The main problem is the one of the distortion produced when the angle of illumination is more than a few tenths of a degree. As a result, to catch the beam of electrons, the numerical aperture of objectives has to be much lesser than that of light microscopes. The actual resolving power of electron micrtoscopes is only about 0.05nm; far less than the theoretical potential of 0.002nm. But even then, 0.05nm is much much smaller when compared with 100nm that is the limit in a light microscope.

Basic construction of an electron microscope

The modern electron microscopes have a common basic construction pattern with standard components such as

(i) Electron generator (gun)
(ii) Electromagnetic coils (lenses) - 3 sets
(iii) Screen (for viewing) and
(iv) Vacuum pump

The electron generator is the source of illumination. Also called electron gun, it is a tungsten filament, which when heated by electric current emits a stream of electrons. The electrons accelerated by the high

voltage energy are forced through a *collimating* aperture which renders the rays in parallel lines and fashions them into a beam. From there the beam passes through the condenser. The condenser 'lens' is a magnetic coil which corrects the aberrations (bending) in the beam and directs them towards the object. The strength of the magnetic lens depends on the amount of current that is allowed to flow through it. Greater the flow of current, greater will be the strength of the magnetic field and greater will be the bending effect on the rays of electron. In other words, by varying the magnetic field strength,the beam can be properly focussed on to the object.

The specimen to be observed must be exceedingly thin (at least 200 times thinner than the ones used for routine optical observation) and must be supported on an equally exceedingly thin but strong film or metallic grid. As the electron microscope always functions only in a vaccum, the specimen must be absolutely dry. Specimen supporting grid is usually a collodion film or Fonnvar or polymerized plastic plus a screen grid of copper having 200 meshes to an inch. The specimen is kept below the condenser and the electron beam passes through it and is scattered depending on the varying refractive index of the specimen. From there, the electron beam (coming out of the specimen) passes through the second set of magnetic coils (objective lens) which focus the electrons and form an intermediate image. A third set of magnetic lenses (equivalent to the oculars) - the projection lenses, produce the final image.The final image however cannot be directly viewed as the rays of electrons are harmful to the eye. They are either projected on to a fluorescent screen for viewing or allowed to expose a photographic plate kept for that purpose.

The entire electron mcroscope must be in a vacuum or otherwise the electrons get scattered due to collisions with air molecules and fail to get focussed. Hence the equipment unit is housed in a single unit which is continuously evacuated with a high speed vacuum pump.

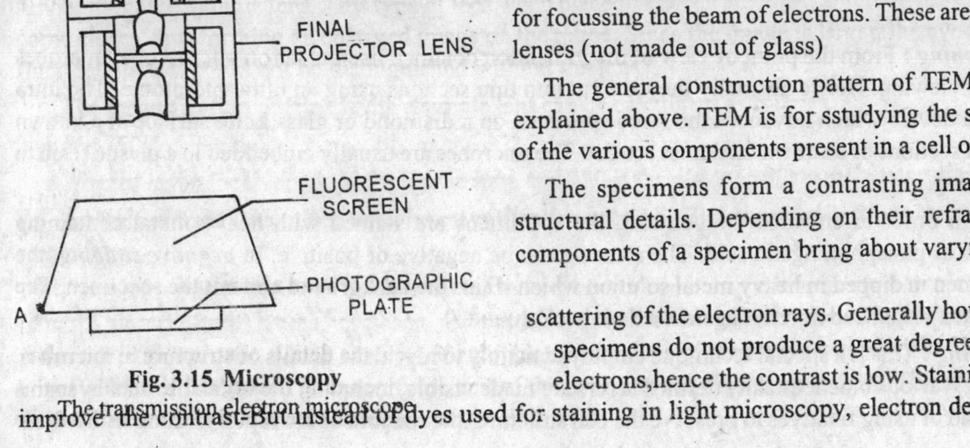

Fig. 3.15 Microscopy
The transmission electron microscope

Types of electron microscopes

There are two basic types of electron microscopes. There are - the *Transmission electron microscope* (TEM) and *Scanning electron microscope* (SEM).

Transmission Electron Microscope (TEM)

In a TEM, the beam of electrons passes through the specimen forming an image of the detailed structure of the specimen. This is similar to the light microscope in construction except for

(i) source of illumination is a beam of electrons (not visible light)

(ii) There are a series (usually three sets) of electromagnets for focussing the beam of electrons. These are called magnetic lenses (not made out of glass)

The general construction pattern of TEM is same as that explained above. TEM is for sstudying the structural details of the various components present in a cell or microbe.

The specimens form a contrasting image with all the structural details. Depending on their refractive index, the components of a specimen bring about varying degrees of scattering of the electron rays. Generally however biological specimens do not produce a great degree of scattering of electrons,hence the contrast is low. Staining is resorted to improve the contrast. But instead of dyes used for staining in light microscopy, electron dense heavy metal salts are used for staining in electron microscopy.

Some problems are encountered while viewing a specimen through TEM. Many a time artifacts appear which can be easily mistaken to be a component of a cell. (An *artifact* is the appearance of something in an image due to causes within the optical system or due to the preparation of a specimen and does not represent any component of the specimen under view). Proper preparation of the specimen, correct adjustment of the electron beams, accurate vaccumization will help in reducing the chances of artifact appearance.

Preparation of specimen for TEM

The biological specimens to be examined have to be specially prepared in order to obtain proper structural details with high magnification, but at the same time avoiding the appearance of artifacts. The following steps are necessary before the specimens are ready for observation under TEM.

Dehydration and Fixation : Under light microscopy specimens are kept in water or oilier liquids for observation. But in an electron microscope, the specimens have to be absolutely dry. Any water present, the specimen would boil (as it is in a vacuum) resulting in disintegrating the structural organization of the specimen. Additionally the specimen also has to be fixed in its proper orientation. Fixation and dehydration have to be carried out in several stages in order to prevent distortion.

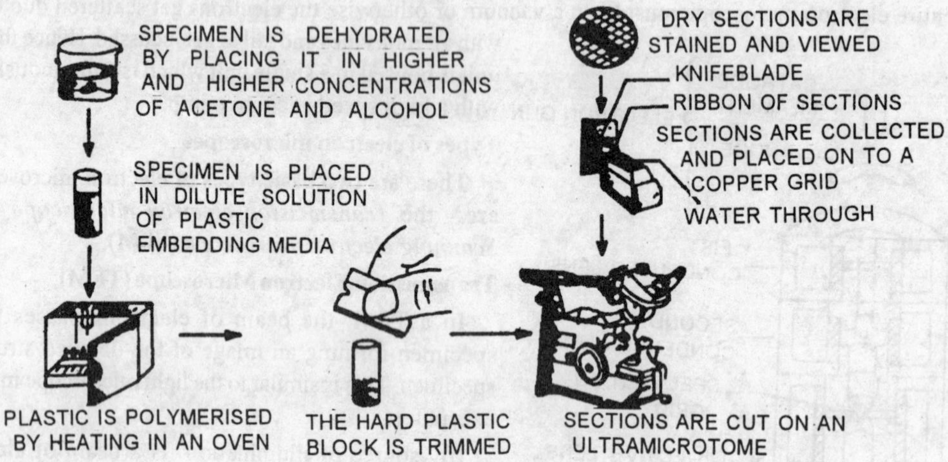

Fig. 3.16 Microscopy
Preparation of specimen for viewing through TEM

Ultra Sectioning : From the point of view of magnification obtained in an electron microscope, microbes are too thick for viewing. Hence normally they are cut into thin sections using an ultra microtome. The ultra microtome has a device which advances the fixed specimen on a diamond or glass knife surface at a known thickness so that sections of uniform thickness are cut. The microbes are usually embedded in a plastic resin to facilitate sectioning.

Staining : In order to enhance the contrast, the specimens are stained with heavy metal containing compounds such as phosphotungstic acid. The staining can be negative or positive. In *negative staining,* the sectioned specimen in dipped in heavy metal solution which sfains the background and not the specimen. The specimens are then visualized in relief against a dark background.

Freeze etching : This is a special technique employed mainly to reveal the details of structure in microbes. In freeze etching, various biochemically defined layers are made visible, including the organelle details. In this technique (instead of using fixatives to preserve the cell structure) the specimens are rapidly struck with a knife blade. At this temperature (freezing) biological specimens are hard to cut, but they crack along the lines of their natural weakness i.e. they are fractured. The fractured specimen is then *etched* i.e. the water (ice) is allowed to evaporate from the surface. This evaporation raises the surface layers of the specimen. A replica, then is made of the freeze etched specimen, by exposing it to vapours of heavy metal (platinum) at 45^0 angle to produce a shadow effect. The specimen then is rotated at 90, and exposed to vapourized carbon. This produces a replica of the surface of the specimen. After removing the remmants of biological material, the carbon replica is viewed under electron microscope. Surface details (internal as well external) of organelles can be clearly visualized by the freeze etching technique.

Scanning Electron Microscope (SEM)

SEM is primarily used for visualizing the surface architecture of the specimen (pollen grains, hairs, membranes, etc.) rather than the internal details. In a SEM, an electron beam is scanned across the surface and a three dimensional image is produced.

The construction plan and working principle of a SEM is different from that of TEM. In SEM, an accelerated beam of electrons is produced from the electron gun and is focussed on the specimen by the condenser lens. The magnetic lenses of a SEM are so constructed as to produce an extremely thin beam of electrons.

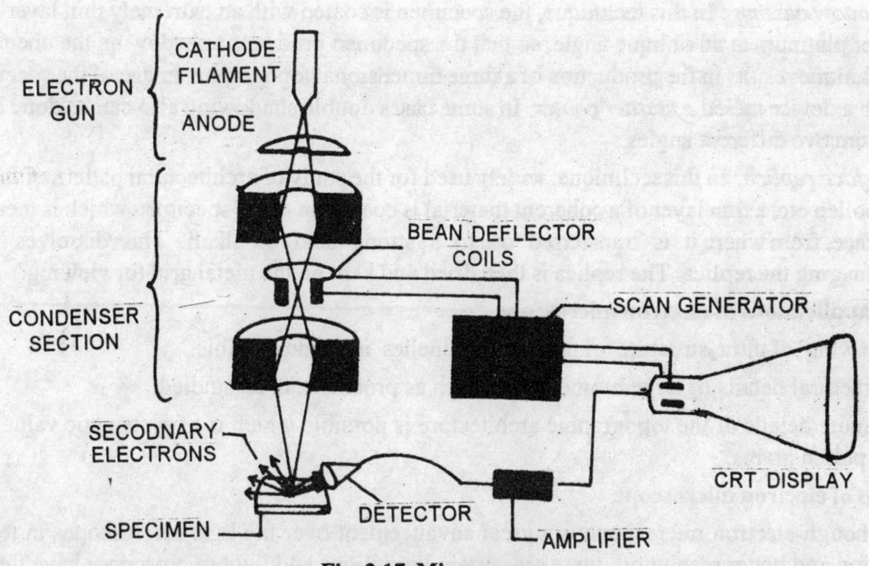

Fig. 3.17 Microscopy
The Scanning electron microscope

The primary electron beam as it strikes the specimen, forces, out electrons from the surface (of the specimen). These are the secondary electrons and are transmitted to a collector. During this process some of the primary electrons also are reflected and transmitted to the collector, but their (primary electrons) number is far less than the secondary electrons. As a result the image signal is developed more by the secondary electrons than by the primary electrons. The electrons are then transmitted from the collector to a detector which has a substance that emits light when struck by electrons. The light so emitted is converted to an electrical current which is used to control the brightness of an image on a CRT (Cathode Ray Tube) screen.

The secondary electrons deflected out of the specimen will be a replica of the refractive index of the surface and thus produce an image on the CRT screen revealing all the topographical details. Image contrast mainly depends on surface topography which determines the number of secondary electrons reaching the detector.

The image on the CRT screen will be three dimensional.

Magnification of the image of the specimen in SEM is not achieved through the lenses as in a light microscope or TEM. It is dependent upon the ratio of the length of the scan across the specimen surface to the length of the scan of CRT. For instance if the electron beam scans 100 nm across of specimen and the image on CRT is 100 nm, the magnification is 100000 times. Thus one can decide the magnification (depending on the requirement) by adjusting the scan distance across the specimen.

SEM also has a resolution equal to that of TEM. A resolution from 1-10 nm is possible with a corresponding magnification from 10000-100000.

Preparation of specimen for SEM

1. *Dehydration* : As SEM also operates on a vacuum like TEM, total dehydration is necessary, but as surface topography is essential, dehydration should be done in such a way as not to disturb the surface configuration. This is achieved by critical point drying which minimizes artifact formation. In critical point drying at a particular temperature and pressure, the liquid changes to gas without any surface tension damage to the specimen. The specimen is first immersed in ethanol or acetone to remove water and then in pressurized liquid of CO_2 simultaneously raising the temperature above 32°C, the critical point of CO_2 . At this temperature range, the liquid vapourizes without surface tension leaving the specimen perfectly dry.

2. *Shadow casting* : In this technique, the specimen is coated with an extremely thin layer of gold, gold-palladium or platinum at an oblique angle, so that the specimen produces a shadow on the uncoated side. The shadow technique results in the production of a three dimensional topographic image of the specimen. Coating is done with a device called *a sputter coater.* In some cases double shadowing also can be done by coating the materials from two different angles.

3. *Surface replica:* In this technique, widely used for the study of architectural pattern of the wall surface of spores, pollen etc, a thin layer of a coherent material is coated on to the specimen which is then floated on to a water surface, from where it is transferred on to a strong acid or alkali. This dissolves the specimen without damaging the replica. The replica is then dried and kept on the metal grid for viewing.

Biological applications of electron microscope

1. Viewing of ultra structure of the cell organelles is made possible.
2. Structural details of large biomolecules such as proteins can be studied.
3. Minute details of the topographic architecture is possible which is of systematic value as in the case of pollen grains.

Limitations of electron microscope

Even though electron microscope is a great advancement over the light microscopes in terms of higher magnification and better resolution, there are certain limitations and light microscopes have their own useful role to play in the microbiological laboratory. The limitations imposed by electron microscope are.-

1. Hydrated specimens are not to be used, hence live specimens cannot be observed
2. Dehydration even if carefully done will cause some distortion
3. Artifacts may appear
4. As electrons have a low penetration power extremely thin sections are necessary.

MICROMETRY

Micrometry or microscopic measurements refer to measurements of microbes (length, breadth) seen under the light microscope. Many a time the size of the microbes is an essential criterion besides their morphological characters. The following are necessary to carry out microscopic measurements.

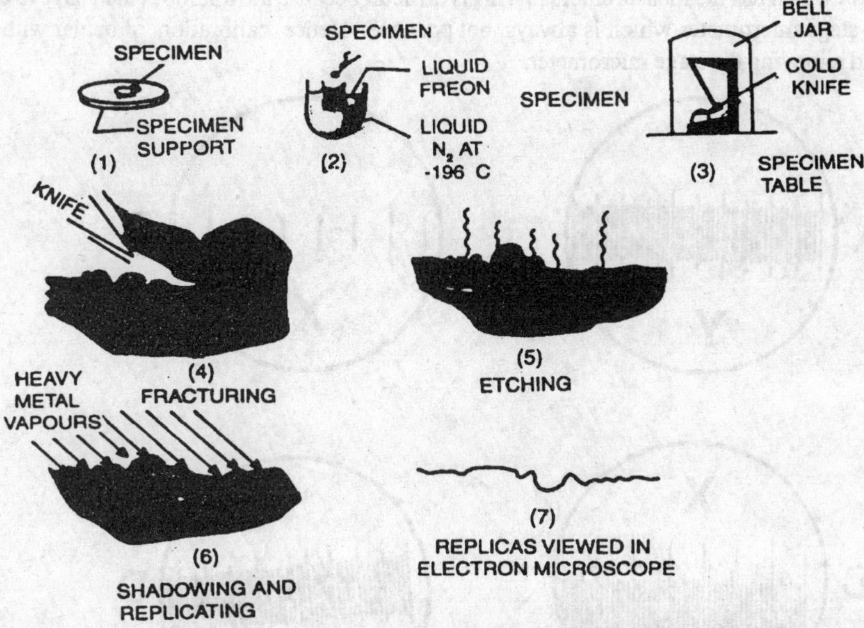

Fig. 3.18 Microscopy
Procedure for freeze etching replicas

A good microscope with 10X eye piece and 10X, 40X and 100X objectives

2. Ocular micrometer
3. Stage micrometer

1. A good microscope (with straight tube) : is necessary to carry out the measurements.

The eye piece may be 10X or 15X with objectives of 10, 40 and 100. It should be noted that when measurements are to be carried out at each magnification, the original calibration (see later) conducted should be at the same magnification.

2. Ocular micrometer : This is a small circular disc of glass which has graduations engraved on one surface. The distance between the lines is fixed, but the lines are arbitarily engraved in the sense it is not a standard measurement of mm, microns etc. In other words the lines on the ocular micrometer do not indicate any measurement of the length.

3. Stage micrometer : Also known as objective micrometer the stage micrometer is a glass slide that has inscribed lines, which are exactly 0.01 mm (10 micrometer) apart. The lines can be easily seen when the slide is placed on the stage and viewed through the microscope.

Procedure for measurement

The first step in the measurement is the calibration of lines of ocular with that of stage micrometer.

What is calibration ? This refers to comparison or super imposition of scales of ocular as well as stage micrometer. This is necessary to determine how many graduations of ocular coincide with one graduation of the stage micrometer. It has already ben said that the lines of the ocular are arbitrary while the lens of stage micrometer are exactly 10 micrometers apart So while using only the ocular micrometer for measurement it becomes necessary to know as to what is the distance between graduations.

At this stage it can be said, that instead of the cumbersome procedure of calibration why not use only the stage micrometer which has the measurements ? This is difficult because the microbes then have to be mounted directly on the stage micrometer which is always not possible. Hence calibration of ocular with the stage micrometer and removing the stage micrometer.

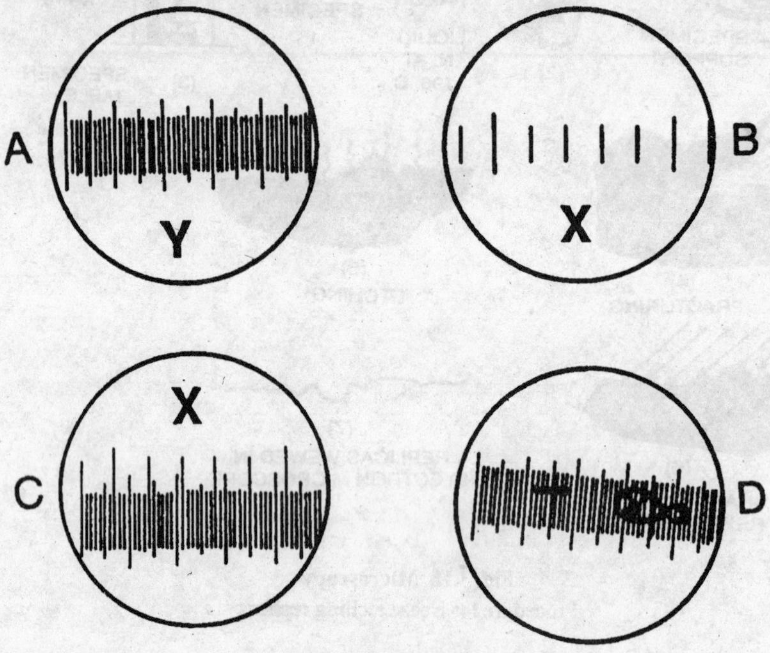

Fig. 3.19 Microscopy
Calibration of ocular micrometer

Procedure for calibration

The first step is the installation of ocular micrometer into the ocular lens. This can be done by unscrewing the top lens in the ocular, placing the micrometer inside thetube and screwing back the lens. In some of the latest microscopes, the ocular micrometer comes with a ring and it can easily inserted into the bottom of the ocular tube.

Next step is placing the stage micrometer on the stage of the microscope and focussing it properly. After focussing, the stage micrometer has to be adjusted so that its graduations align with the graduations of the ocular properly. After the graduations are aligned, count how many ocular divisions equal one division (0.01 mm) of the stage micrometer. If (as shown in the figure) seven ocular divisions equal one division of stage micrometer, the value for each division of ocular is 0.01/7 or 0.0014μm. In other words each ocular division is about 1.43 μm apart. Having known this, the stage micrometer may be replaced with the slide containing microbes for measurements. To determine the size of an organism all that one has to do is to count the number of graduations, the microbe measures and to multiply this number with 1.43 μm to give the actual size.

Note : Calibration has to be made for each pair of ocular and objective lenses as the magnifications vary. Ideally, it is better to use the same microscope used for calibration, also for measurements.

4

THE CELL

One of the marvels of nature is the creation of tiny yet potent biological compartments called cells. There is virtually no living being in the biological world save for Viruses which is not cellular. Either one, two or many millions of cells go to make the body of the living beings. A cell may be defined as the 'structural and functional unit of a living being. It is the minimal biological unit capable of maintaining and propagating itself. The study of cell comes under the perview of cytology, physiology and anatomy; while anatomy deals mainly with the structural aspects, the other two lay emphasis both on structure and function.

History

The word cell has been derived from the Latin word *cellula* meaning a small compartment. The term in the present sense was first used by Robert Hooke (1665) in his book *Micrographic.*

There has been some mention about cell in ancient Indian literature also. Parasara (1st century BC) in his book *Vrkshayurveda* refers to invisible compartments in the leaf and names them *Rasakosha.* He further opines that each cell has a sap *(Rasa),* not visible to the naked eye *(Anavah)* and has two boundaries *(Kalavestitena),* some of the cells have a colouring matter *(Ranjakayukta).*

The credit however of discovering the cell goes to Robert Hooke (1965), since it s difficult to establish the discovery to the ancient Indians in the absence of complete text of Parasara and the general apathy among Indians to their scientific heritage. Robert Hooke who constructed the first compound microscope observed the sections of cork and opined that they contain honey comb like compartments. German Biologists M.J. Schleiden and T.S. Schwann(1838) established the '*cell theory*' that all organisms are made up of cells.

Dutrochet in 1824 separated the cells from *Mimosa* plant. He boiled the leaves in nitric acid to isolate the cells. Turpin and Meyers (1830) observed that every tissue is an assemblage of cells.

One of the significant discoveries of the cell came from Robert Brown (1830). He discovered the presence of a spherical body in the centre of every cell, which he named *'nucleus'*

In the year 1835-37, Purkinje and Mohi independently discovered that protoplasm is an important constituent of every cell and it plays an important role in every cell activity including division. Golgi (1838) discovered the golgi apparatus. Balbiani (1881) discovered chromosomes in the salivary glands of *Chironomus.* At about the same time, Flemming (1882) studied cell division in detail and gave the name *Mitosis.*

Endoplasmic reticulum was discovered by Porter in 1945, while Benda gave the name mitochondria to organelles originally discovered by Hemming. Lysosomes were discovered in 1955 by de Duve.

Shape of Cell

The shape of the cell may be variable as in the case of *Amoeba* or fixed. When the shape is fixed or permanent, all conceivable shapes are found-they may be spherical, rectangular, flattened, polygonal, oval, triangular, cone like, columnar etc. As in *Acetabularia,* the body of the plant made of a single cell is differentiated into a rhizoid like base, a long and slender stalk and an umbrella like cap. Certain cells, like the Ascospores of *Claviceps* are acicular (needle like).

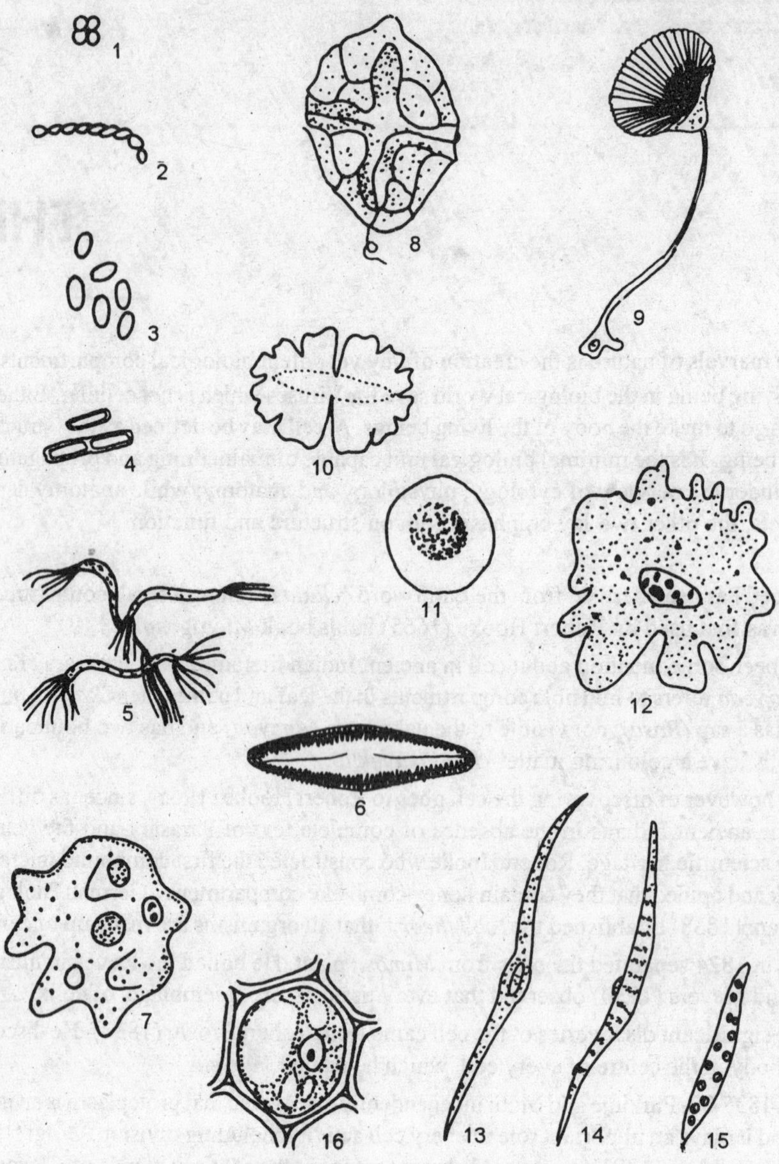

Fig. 4.1 The Cell
1. Tetracoccus, **2.** Streptococcus, **3.** Micrococcus, **4.** Bacillus, **5.** Spirillum, **6.** Navicula (Diatom), **7.** Amoeba, **8.** Dinoflagellate, **9.** Acetabularia, **10.** Desmid, **11.** Human Erythrocyte, **12.** Melonocyte, **13.** Smooth muscle cell, **14.** Striated muscle cell, **15.** Trachied from a vascular plant, **16.** A parenchyma cell

Size

There is a great range of variation among cells in size also. Some of them are not visible to the naked eye, some are barely visible, while some are macroscopic This smallest cell size can be encountered in *Coccus* bacteria (0.2 to 0.5m), while the largest size of cell is seen in Ostrich egg (nearly 15 cms). The table given below summarises the variations in cell size.

Table 1 (Adopted from U. Sinha and S. Sinha, 1976)

Cell type	Size
Fibre cell of *Linum*	550mm (length)
Egg of ostrich	170x35mm
Egg of Hen	60x45mm
Human egg	0.1mm=100μm
Amoeba	100μ
Sea urchin egg	70μ
RBC	7μ
Typhoid bacillus	2.4x0.5μ
E. Coli	1.5x0.7μ
Diplococcus pneumonieae	200 x 100nm
*Influenze virus	100nm
*TMV	300 x 15nm
*T_3 bacteriophage	45nm,

(Even though Viruses cannot be regarded as cells, they have been mentioned as they are independent entities).

The factors governing the cell's size are -a) ratio between the volume of nucleus and volume of cytoplasm, b) ratio of cell surface to cell volume and c) the rate of metabolism.

Gross Morphology of the cell

A generalised plant cell has an outermost envelope called the cell wall. This is absent in animal cells. Internal to this is the plasma membrane. This encloses the nucleus and other cytoplasmic inclusions suspended in cytoplasm. The chief cytoplasmic inclusions are -Ribosomes, Lysosomes, Mitochondria, Plastids, Golgi complex, Endoplasmic reticulum, Vacuole and non living inclusions like crystals, raphides etc.

Prokaryotic and Eukaryotic cells

In primitive organisms like certain bacteria, blue green algae etc., the nucleus is not properly organised hence such cells are called *Prokaryotic,* while in evolved organisms, the nucleus is organised. Such cells are called *Eukaryotic.* The following are some of the fundamental differences between eukaryotic and prokaryotic cells.

1. Nucleus : Nuclear membrane is absent in prokaryotic cells, while eukaryotic cells have a definite bounding membrane. The nuclear material does not form definite chromosomes in prokaryotic cells. The DNA is circular and lies in a tangled mass. In Eukaryotes, the chromenema forms definite number of chromosomes and is associated with proteins.

2. Cytoplasmic organelles : Mitochondria, endoplasmic reticulum, Golgi complex, lysosomes and centrioles are absent in prokaryotes. The enzymatic functions associated with mitochondria are carried out by the infolding of the cell membrane at several points. Eukaryotes have all the organells mentioned above.

3. Cell wall : Chemically the cell wall of prokaryotes is composed of amino sugars and muramic acid These are absent in the cell wall of eukaryotes (plants).

4. Flagella : The flagella of eukaryotes have eleven fibres with (2 + 9 arrangement) 2 central and 9 peripheral fibres, while prokaryotic flagella do not exhibit this arrangement.

5. Plastids : These are absent in prokaryotes. In photosynthetic prokaryotes the chlorphyll is associated with the lamellae, but the lamellae are not enclosed by any membrane. In contrast to this, in all eukaryotes, the chlorophyll pigment is found in definite organelles called chloroplasts.

In addition to the above differences, the streaming movement of cytoplasm may also be added, which is present in eukaryotes, but absent in prokaryotes.

Table 2. Comparison between Prokaryotic and Eukaryotic Cells
(Adopted from De Robertis Et al 1980)

Character	Prokaryotic	Eukaryotic Cells
1. Nuclear membrane	Absent	Present
2. DNA	Naked and circular	Combined with proteins
3. Chromosomes	Single	Multiple
4. Nucleolous	Absent	Present
5. Cell division	amitosis (fisson)	Mitosis or meiosis
6. Ribosomes	70s (50s + 30s)	80s (60s + 40s)
7. Endomembrane system	Absent	Present
8. Mitochondria	Definite organelle absent enzymes of respiration and photosynthesis stored in the folds of plasma membrane	
9. Choloroplast	Absent	Present
10. Cell wall	Non cellulosic (present in plants only)	Cellulosic
11. Exocytosis and endocytosis	Absent	Present
12. Flagella	No definite arrangement of fibrills	9 + 2 fibrillar arrangement

PROKARYOTIC CELL

The prokaryotic cells are generally smaller and vary in size in different members. In mycoplasma it is about 0.12)µm while in *Oscillatoira,* a filamentous blue green the size is 40 x 5\|µ. A great majority of them, however are about µ in size. Among the bacteria the smallest are to be found among cocci (0.1 µ) while the largest are the spirilla (60 x 6~m). The various components of the prokaryotic cell are described below.

Cell wall : Cell wall is absent in some prokaryotic (mycoplasma, L forms of bacteria) members while it forms the outermost boundary in a large majority of them. While the cell wall is non living and does not have the property of selective permeability it can act as a seive as it is porous.

In some prokaryotes the space between the cell membrane and cell wall is filled with some enzymes and metabolites consituting the periplasm.

Chemically the cell wall is made up of strong fibres of hetropolymers called *mucopeptides ofpeptidogycans.* They are also refered to as *glycopeptide, muropeptide, murein etc.*

The peptidoglycans consist of alternating units made up of N acetyl glucosamine and N acetyl muramic acid with 1-4 linkages. This type of a linkage is common to all Prokaryotes. Attached to the muramic acid is a short peptide chain consisting of alanine, glutamic acid and either lysine or diaminopimelic acid (DAP) and muramic acid. DAP and some of D-amino acids are found only in prokaryotic cell walls.

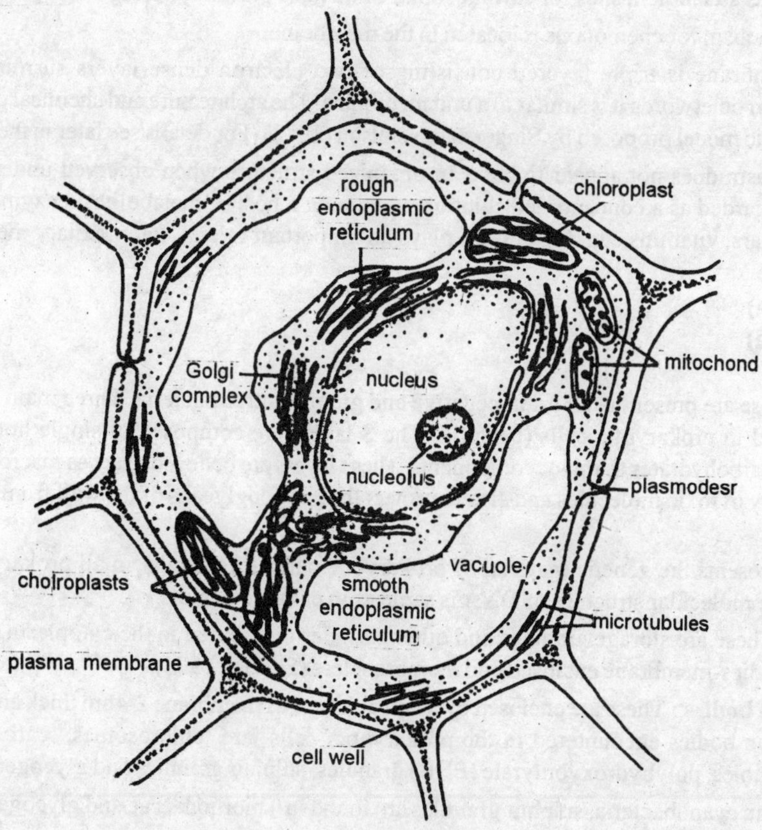

Fig. 4.2 The Cell
Structure of a typical Plant Cell (Electron Microscopic view)

Cross linkage of these polymers assures structural rigidity with the type and extent of linkage being species specific. For example in some bacteria a pentaglycine links the tetrapeptide side chains extending from the muramic acid units while in others the terminal D- alanine of one tetrapeptide may be covalently linked to the DAP of an adjacent tetrapeptide. The peptidoglycans of gram negative bacteria have more cross linkage than those of gram positive bacteria.

In addition to peptiodglycans the wall has many other chemicals. The wall of gram negative bacteria shows as many as five layers under the electron microscope. These are-the periplasm space, Peptidoglycan layer, an intermediate zone, the outer membrane and lipopolysaccharides.

Cell membrane : This is the bounding layer of cytoplasm and is about 6-8nm thick. Unlike in the case of eukaryotic cells it carries out multiple functions. Some of these are -

(i) Presence of enzymes of biosynthetic pathways leading to the synthesis of wall components

(ii) Occurrence of *permease* enzymes responsible for the transport of materials across the membrane

(iii) Membranes of aerobic bacteria have the electron transport chain and oxidative phosohorylation

(iv) The purple bacteria have their photosynthetic apparatus in the membrane

(v) The membrane has attachment sites for chromosomal DNA (also plasmid) during DNA partitioning

(vi) Regulatory mechanism for chemotaxis is located in the membrane.

Structurally the membrane is triple layered consisting of two electron dense layers surrounding an electrontranslucent layer. In otherwords it is similar to a unit membrane. The architecture and chemical coposition is similar to the fluid mosaic model proposed by Singer and Nicolson (1972) (For details see later in the chapter).

Cytoplasm : Cytoplasm does not appear to have an organized structure when observed under electron microscope. It may be regarded as a concentrated solution containing a host of metabolites, enzymes, amino acids, inorganic ions, sugars, vitamins etc. Cytoplasm plays an important role in intermediary metabolism. There is no cytoskelcton.

Flagella (see bacteria)

Capsule (see bacteria)

Pili (see bacteria)

Surface layers : These are present in all gram negative and gram positive bacteria. Three main types of S layers have been reported in prokaryotic cells (bacteria) The S layers are composed of single homogenous peptides with occasional carbohydrates as minor components. These layers are believed to act as macromolecular sieves preventing the entry of toxic molecules and also to protect the peptidoglycans of the wall from the action of lytic enzymes.

Nucleoid : This represents the genetic material of prokaryotic cells. It is similar in all prokaryotes. (for details see - bacteria) The molecular structure of DNA is same as in eukaryotic cells.

Inclusion bodies : These are storage granules and other particles distributed in the cytoplasm. There are two kinds of inclusion bodies-membrane enclosed and membraneless (Shiviey 1974).

Membrane enclosed bódies : These are enclosed by a non unit type of membrane 2-4nm thick and made up of proteins. The inclusion bodies encountered in the prokaryotic cells are chlorosomes, carboxysomes, magnetosomes, gas vacuoles, polyhydroxybutyrate (PHB) granules, sulphur granules and glycogen granules.

PHB granules occur in cyanobacteria, sulphur granules are found in Thiorhodaccae and glycogen granules in *Clostridium* spp.

Chlorosomes have been demonstrated in photosynthetic bacteria belonging to chlorlobiaceae. (Cruden and Stanier, 1970). These are different from chromatophores in being non-membranous.

Carboxysomes are seen in cyanobacteria (Murphy et al 1974).

Magnetosomes are particles which help the bacterium in orienting itself in relation to the magnetic behaviour of the surroundings. Reported to occur in *Aquaspirillum magnetotacticum* (Balkwill et al 1980), the magnetosomes allow the bacteria to punise the most efficient aerotactic behaviour.

Non membranous Inclusion bodies : Phycobilisomes, cyanophycin granules (found in cyanobacteria), polyglucoside granules and several crystalline substances belong to this group.

In addition to the above there are R - bodies (retractile stuctures) demonstrated in bacteria by Lalucat and Mayer (1978). These are supposed to be similar to kappa particles of *Paramecium.*

Ribosomes (see bacteria)

Mesosomes (see bacteria)

Chromatophores : These are present in the photosynthetic bacteria belonging to the families Rhodospirillaceae, Chromatiaceae and Cyanophyceae. Chromatophores vary in form and have vesicles, tubes, bundles, stacks or thylakoids etc.

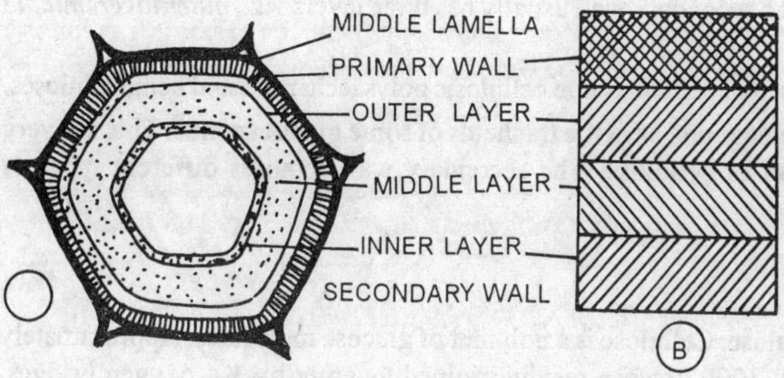

EUKARYOTIC CELL

Cell Wall

The cell wall is the characteristic feature of the plant cells. It is made up of non living matter and is believed to be secreted by the plasma membrane. The structure and composition of the cell wall not only varies in different plants, but in different types of cells of the same plant. Often, there might be variety in the cell wall composition at different stages of growth of the cell. The cell wall is very rigid and maintains theshape of the cell acting as a boundary. A unique feature of the cell wall is its permeability which is due to the presence of small pores.

Fig. 4.3 The Cell
Arrangement of wall layers. **A.** Cross section, **B.** Arrangement of fibrils

Gross structure

Three layers are distinguishable in the cell wall. these are the *primary cell wall, middle lamella* and the *secondary cell wall.* The primary cell walls of adjacent cells are separated by middle lamella. Secondary wall is deposited after a certain period of development.

The primary wall is secreted by the plasma membrane. In some instances like the leaves, fleshy stems, roots etc., the cells have only the primary wall and the middle lamella. The primary wall is formed during the early stages of growth and is about 1 to 3nm in thickness. Chemically the primary wall is composed of cellulose, hemicellulose and pectic compounds. The young primary wall is plastic and is prone to changes. Often it may be lignified.

The Middle lamella is the intercellular substance binding together the primary and the secondary walls. The middle lamella is formed during cell division and is amorphous and colloidal in nature. Chemically the middle lamella is composed of pectin, cellulose, calcium and magnesium pectate.

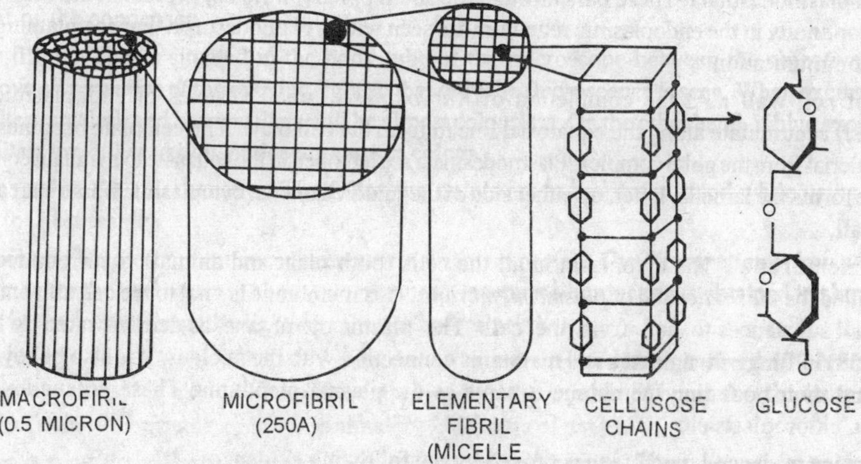

Fig. 4.4 The Cell
Structural elements of Cellulose (An important component of plant cell walls)

The secondary wall is deposited on the primary wall after cell maturation. It is very rigid and does not alter its shpae. The thickness varies from 5 - 10nm. The secondary wall usually has three layers viz., *outer layer, middle layer* and *inner layer.*

Chemically the Secondary wall is composed of cellulose, non cellulosic polysaccharides and hemicelluloses.

Ocassionally a *tertiary cell wall* may also be found as in the tracheids of some gymnosperms. This is a very thin layer and is composed of *xylan,* instead of cellulose. The secondary wall acquires different types of thickenings.

Ultra structure of the cell wall

The cell wall is mainly composed of cellulose. Cellulose is a polymer of glucose molecules. Approximately each molecule of cellulose consists of about 3000 glucose residues joined together by 1-4 oxygen bridges. These cellulose chains are oriented along the axis of the cell. About 100 of these cellulose chains bundle together to form *micellels* or *elementary fibrils.* Each micelle is about 100Å in diameter. Many hundreds of these micelles of these stack together to form *microfibrils.* Each microfibril is 200Å to 300Å in diameter. The micorfibril is visible only under eletron microscope. About 2000 of these microfibrils join together to constitute the *macrofibril* which is about 0.5μ in width. The macrofibrils are visible under the light microscope.

In secondary walls, the microfibrils are more orderly and compact than in the primary walls. The space between fibrils is filled by an interfibrillar substance usually the matrix. The matrix is filled by non cellulosic polysaccharides - such as pectin, hemicellulose, xylans etc.

The primary cell wall of dicotyledons contains many complex polysaccharides such as *xyloglucans, arabinogalactans* and *rhamnogalacturonans.* These are linked to the cellulose fibril and to one another by covalent linkages.

In certain walls, a glycoprotein called extensin is present. It is similar to collagen in being rich in 4 - hydroxyproline.

Extensin has a long protein chain, to the hydroxy groups of which are linked, residues of *arabinose* and *galactose.*

Plasmodesmata : Protoplasts of adjacent cells are connected to one another by means of thin cytoplasmic threads called plasmodesmata. These pass through the small pores (pit fields) present in the cell wall. Tubules which are in continuity in the endoplasmic reticulum are seen within plamodesmata thus maintaining an effective intercellular communication.

Origin of cell wall : At the completion of *Karyokinesis,* small vesicles of endoplasmic reliculum *(phragmoplast)* accumulate along the equatorial line to form the cell plate. The cell plate becomes thick by the addition of material from the golgi complex. Plasmodesmata are left open in the cell plate to maintain communication. This cell plate forms the lamella. Later, on either side of the middle lamella, cytoplasmic fibrils start accumulating to form the wall.

Plasma membrane : The cytoplasm in all the cells (both plant and animal) is surrounded by a living membrane called the *plasmalemma* or *plasma membrane.* This membrane is vital to the cell as it controls the exit and entry of all substances to and from the cell. The plasma membrane, as demonstrated by McAlear and Edwards (1958) in fungi, invaginates and maintains connection with the nucleus. It is also believed that many of the cell and their bounding membrane is same as the plasma membrane These organelles are nucleus, mitochondria, chloroplasts etc.

The presence of the cell membrane is proved by the following evidences.

1. On injury to a cell, when protoplasm flows out, it remains as a discreet unit.
2. Observations under electron microscope have revealed the presence of a boundary membrane over the surface of cytoplasm.

Size of the membrane : It is about 75Å thick and is composed of three concentric layers. The middle layer is about 35Å in thickness, while the outer layers on either side or 20Å thick.

Chemical composition : Membranes from different sources show variations in chemical composition. Essentially however all membranes are lipoproteinanceous. The components are held together in a thin sheet by non convalent bonds. The ratio of protein to lipids varies depending on the type (Golgi membrane, plasma membrane etc) of membrane, type of organism (eukaryote, prokaryote, etc.) and the type of cell (epidermis, phoelm etc.). In adition to these two, certain membranes also have carbohydrates.

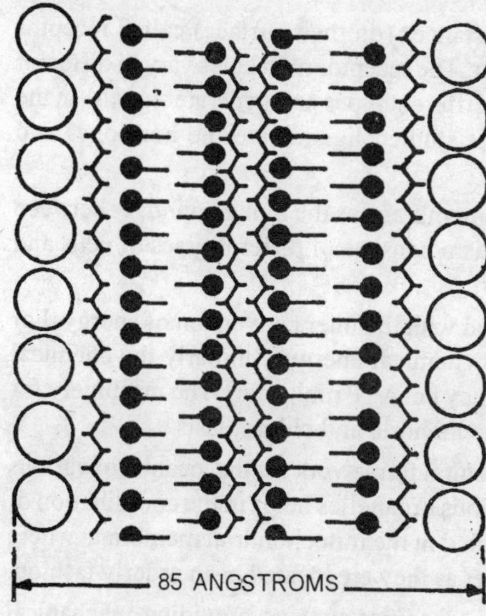

Fig. 4.5 The Cell
Diagrammatic representation of a cell membrane having two unit membranes

Membranes have three dfferent categories of proteins. These are - structural *proteins, functional proteins* (enzymes) and *carrier proteins.* Structural proteins maintain the structural frame work of the membranes. The average molecular weight of structural proteins found in membranes in $3x10^4$. Enzymes catalyzing various physiological activities are found in the membrane of mitochondria, endoplasmic reticulum etc. *Carrier proteins* are known to form a complex with the substances and help them to pass through the membrane.

The protein composition of different membranes vary (as has been pointed out already). For instance plasma membranes have 50% while mitochondrial membranes have 75% proteins.

The proteins present in plasma membrane are categorized into two viz. *integral proteins* and *peripheral proteins.* While integral proteins form the back bone of the membrane, peripheral proteins have a weak association maintained by weak electrostatic interaction. The integral proteins are hydrophobic in nature as they have to penetrate the hydrophobic lipids to form the basic structure of the membrane. These proteins do not easily disassociate from the membrane and it requires harsh chemical procedures to accomplish it. The integral proteins have a hydrophilic portion also which protrudes out of lhe lipoid bilayer. Peripheral proteins are associated either with the hydrophilic head of lipids or hydrophilic part of the integral proteins.

The Lipids of membranes are of several types, but all of them share a common character in being *amphipatic* i.e. having both *hydrophilic* and *hydrophobic* portions within a molecule. There are three categories of membrane lipids-they are phosopholipids, glycolipids and sterols. While most membrane lipids are phospholipids, cholesterol does not have phosphates.

Phospholipids are of two categories viz. *phosphoglycerides* and *sphingophospholipids.* Phosplioglycerides have a glycerol back bone, while sphingophospholpids have the back bone derived from sphingosine an aminoalchobol having a long hydrocarbon chain.

Glycolipids contain one or many monosaccharide units attached to the lipids based on *sphingenine* or *ceramide.* There are three kinds of glycolipids viz., *Cerebrosides, sulphatides* and *gangliosides.*

Cerebrosides contain a single monosaccharide unit (galactose or glucose). Sulphatides are sulphuric acid esters of cerebrosides. Gangliosides are glycosphingolipids having *sialic acid.*

Sterols are steroid alcohols and are of three types - cholesterols (present in animal tissues), phytosterols (present in plants) and ergosterols (present in eukaryotic microorganisms).

Among the glycerophospholipids *lecithin* and *cephalin* are most abundant in higher plants and animals.

Membrane functions

The following are the major functions of membranes.

1. Compartmentalization: Membranes constitute the living boundary for cytoplasm and cell contents. As unbroken sheets they divide the living mater into self sustaining units to effectively coordinate and regulate the activities.

2. Regulation of movement: In a biologically active cell many susbstances enter and many others go out of the cell. However, if the entry and exit is unregulated, it is deleterious to the cell. Hence plasma membrane regulates and decides which substances are to be allowed and which should not. This unique property of the cell membrane called selective permeability or differential permeability is responsible for most of the physiological activities of the cell.

3. Communication: All the membranes have certain special substances (on their surface) called receptor molecules which mediate and control the message from one to another. The receptor molecules vary in different membranes. These molecules combine with substances (ligands) of different types and generate signals (in the form of cyclic AMP) directed to the interior of the cell. The best studied ligands are the hormones and neurotransmitters.

4. Interaction between cells: It is the responsibility of the plasma membrane as the living boundary between cells to effectively coordinate actions between cells since an organism consists of differnt types of cells and many a time the actions have to be coordinated to achieve harmony.

5. Energy transformations : Membranes are intimately associated with the inter conversion of energy. For instance, in green plants the light energy has to be converted into chemical energy. Similarly the chemical energy stored in carbohydrates has to be converted into energy currency i.e. ATP molecules. The machinery for energy capture and conversion is hidden in the membranes of mictochondria and chloroplasts.

6. Locus for biochemical functions : Membranes provide the structural framework for the location of various enzymes. The scaffolding of the membrane interconnecting the various organelles helps in the coordination of activities. Such an effective coordination of activities is best illustrated in the mitochondrial membrane where the various components of the electron transport chain work together as they are located in an orderly fashion.

7. Other activities : Membranes also function in various other capacities such as providing mechanical strength, involving in cell movement secretion etc., and also acting as electrical insulators.

Structure of the membrane : The molecular structure of the membrane has been interpreted in various ways by different cell biologists in order account for their known function. Various models have been proposed for the molecular structure of the membrane. These fall into two basic categories, viz., bilayer models and micellar or subunit models. The bilayer models are believed to have protein and lipid to be arranged in the form of layers, while the subunit models assume the presence of a number of independent subunits. We shall discuss a few of these models. The chart below summarises the list of various models.

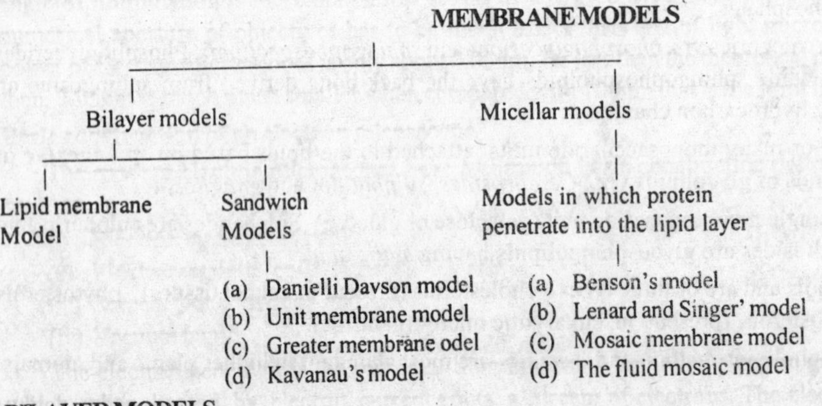

BILAYER MODELS

Lipid membrane model : Overtoil (1895) discovered that fat soluble subsatances easily pass through the membrane indicating their affinity to the lipids. Hober (1910) and Fricke (1925) found out that the low electrical conductivity of the Iving cell is due to the presence of lipids. The experiments of Mudd and Mudd (1931) in placing an oil droplet which adheres to the cell surface provided the presence of lipids in the cell membrane. It was Gorter and Grendel (1925) who suggested a structural model based on lipids. They studied the erythrocyte (RBC) membrane and concluded that it is composed of two layers of liqid molecules whose polar groups are situated on

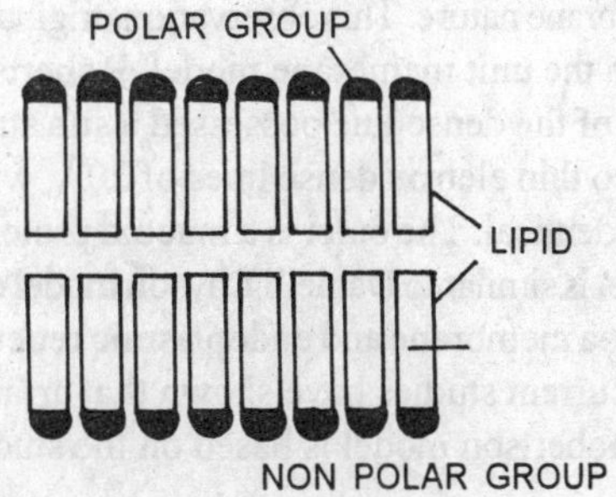

Fig. 4.6 The Cell
The lipid membrane model of Forter

Sandwich models : In these models the lipid layers are supposed to be sandwiched between the two protein layers on either surface. The polar groups of the lipids adhere to the proteins on either side. The following are some of the models based on sandwich principle.

(a) Danielli Davson model : The study of surface tension of cells by Daniell and Harvey (1935) and Harvey and Cole (1931) indicated that there must be more to the structure of the membrane than merely being made up of lipids. In 1935, Hugh Davson and James Denielli proposed that the plasma membrane is a lipid layer sandwiched between layers of globular proteins. In a way the Danielli model was the first structural interpretation to account for the physiological properties of the membrane.

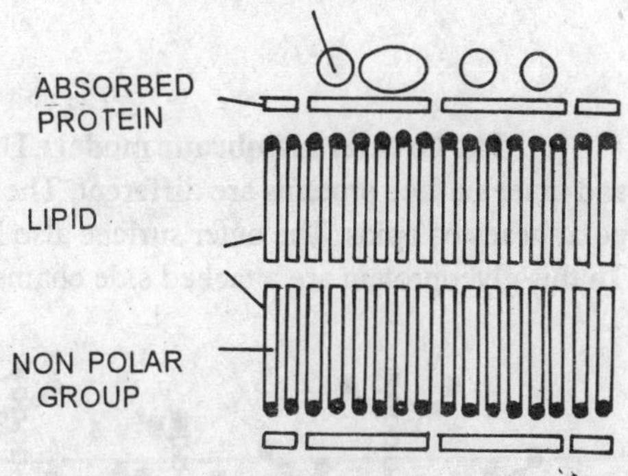

Fig.4.7 The Cell
The Danielli-Davson model

Over the years various modifications have been proposed for the model. Danielli (1938) opined that there could be two kinds of globular proteins on the surface and diagonally arranged proteins in contact with the lipidlayer. In order to account for the permeability, certain small pores were visualised in the lipid layers with an inner lining of protein molecules providing an inner hydrophilic surface.

The following are some other variations in the protein lipid arrangement.

1. Proteins are in the form of folded chain.
2. Proteins are in coiled from on both sides of the lipid layer.
3. Proteins are asymmetrically arranged i.e., one side has globular proteins and the other side a folded β chain.
4. Globular proteins are found on either side.
5. Helical proteins extend into the pores between the lipid molecules.

Nature of lipids in the membrane

The lipids are mainly phospholipids with their polar groups pointing outwards - *lecithin* alternates with a steroid molecule (*cholesterol*). Each lecithin molecule consists of two glycerol chains with the part containing *phosphate* and *choline*, the polar head is bent like a walking stick.

(b) The unit membrane model : The specimens of membrane fixed in osmium tetroxide for the purpose of electron microscopic observation reveal the occurrence of *two lines,* presumably indicating their double membrane nature. This observation originally made in 1950's by Humberto Fernandez Moran and David Robertson led to the unit membrane model. Robertson used a new fixative potassium permanganate. He discovered that each of the dense line possessed a sub structure having a trilaminar appearence. Each membrane is composed of two thin eletron dense lines of 20Å, with the intervening space of about 35Å. The inner and outer layers are non identical. The outer is a mucoid protein, while the inner surface is a non mucoid protein. The unit membrane model is similar to Danielli Davson model except that protein layers are asymmetrical. Membrane of cell organelles, plasma membrane and endoplasmic reticulum are thought to be similar in havingo the unit membrane structure. But current studies have shown that arrangement of lipids and proteins vary in different membranes. Further, the Robertson model is based on the study of myelin which is a non typical membrane.

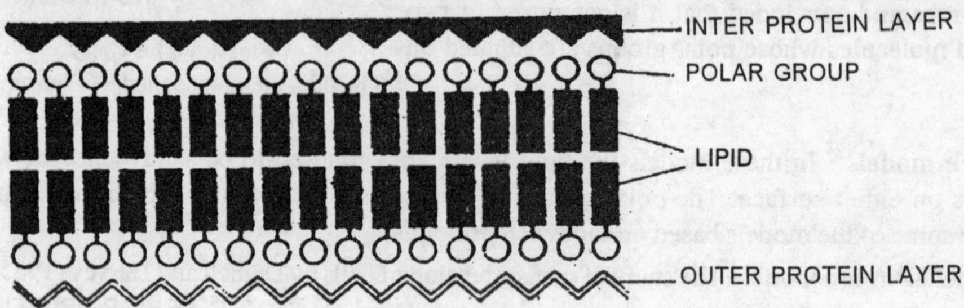

Fig. 4.8 The Cell

The unit membrane model of Robertson

(c) The Greater membrane model : This is similar to the trilaminar sandwich model except that the outer and inner surface proteins are different. The inner surface has an unconjugated structural protein adhering to polar heads of lipids. The outer surface also has structural proteins, on which is superimposed a *glycoprotein.* To this glycoprotein are attached side chains of oligosaccharides.

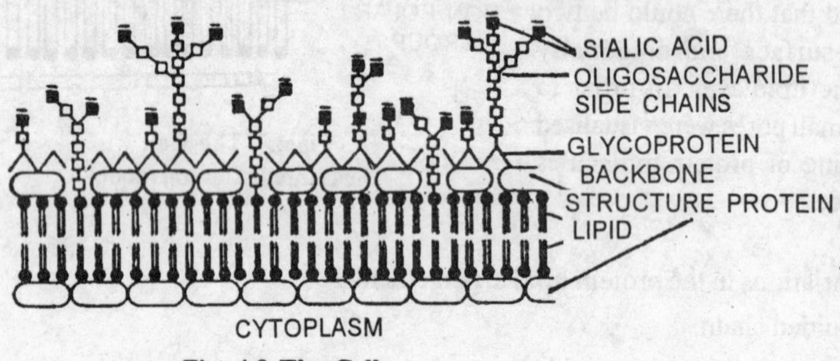

Fig. 4.9 The Cell

The greater membrane model

(d) Kavanau's lipid pillar model : According to Kavanau J.L. (1965), the lipid layer exists in two forms. Kavanau's model is essentially similar to Danielli Davson model. The two types of lipid structures are *pillars* and *flattened discs*. Pores are present between pillars to faciliate the passage of ions. Membranes with pillar lipids are thicker than those with discs. Internally the pillars have non polar tails, while the surface is made up of polar phospholipids. Proteins are present on either side of lipids.

Models based on protein penetration of lipids

(a) Lenard and Singer's model : This is based on the protein arrangement in the membrane. About 25 to 30% of proteins are helical while the rest are irregular coils.

(b) Benson's model : This model was proposed by Benson (1966) based on the study of membranes of chloroplasts. In this, it was postulated that lipid tails are bound to the corresponding hydrophobic points within the coils of proteins. The polar heads of the lipids lie on the surface.

(c) Mosaic membrane concept : According to this model proposed by Baum (1967), the membrane consists of units, each about 80Å in diameter. The wedge shaped protein is enclosed on either side and at edges by the lipids. The polar heads protrude out combining with the hydrophobic areas of proteins.

(d) The Fluid mosaic model : The molecular model of the plasma membrane proposed by Singer and Nicolson (1972) emphasises on the dynamism of membrane organisation and visualises it to be a semifluid like structure with a mosaic arrangement of lipids and proteins. The fluid mosaic model, though essentially based on bilayer models, rejects the idea of continuous bilayer.

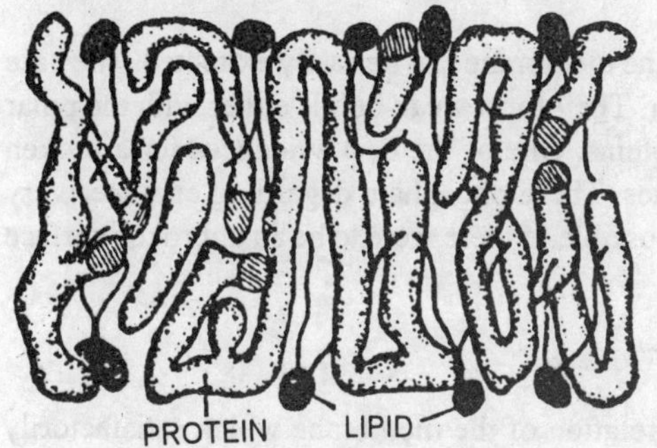

Fig.4 .10 The Cell
Membrane model of Benson

Instead of forming continuous external layer adhering to the hydrophilic heads of the phospholipids, as in the Danielli model; the proteins constitute discontinuous particles often penetrating deeply into the lipid layer. In a way the protein units could be compared to floating incebergs in a sea of phospholipids. The countinuity of the bilayer is not maintained because of complete or incomplete interruption by proteins. While the Danielli model assumes a hydrophilic bonding between lipids and proteins, the Singer Nicolson model regards it to be hydrophobic.

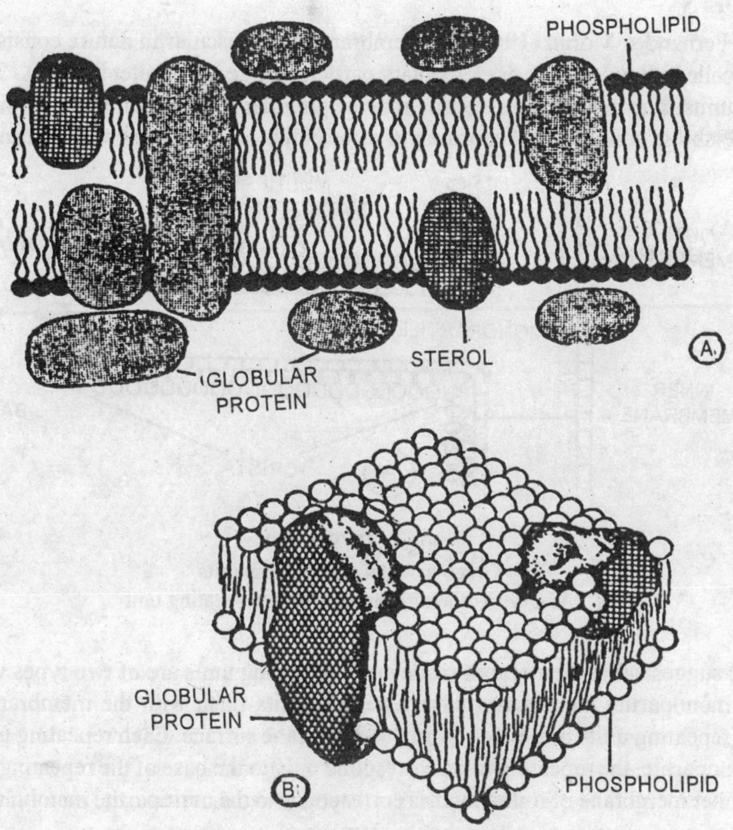

Fig. 4.11 The Cell
Fluid mosaic model of plasma membrane
A. Sectional view of Eukaryotic membrane model
B. Fluid mosaic model of a bacterial membrane

The membrane proteins in the fluid mosaic model are of two types distinguishable by their intensity of lipid association. These are the *integral* or *intrinsic proteins* and the *extrinsic* or *pheripheral proteins.* ITie integral proteins penetrate deep into the lipid layer and are the hydrophobic as they have to coexist in the hydrophobic environment of fatty acids. The integral proteins are *amphipatic.* While their tips are hydrophilic; the central portion is hydrophobic. These proteins are difficult to extract from the membranes.

The peripheral proteins which are weakly associated whith the membrane can be easily extracted. They are hydrophilic and never penetrate the interiors of the lipid bilayer. They form weak bonds either with the polar heads of lipids or with the hydrophilic surfaces of integral proteins. One of the best evidences for a broken surface layer of proteins comes with the using of phospholipases. These phosphate degarding enzymes very easily percolate into the membrane, which would have been impossible, if there were to be an unbroken surface envelope of proteins.

The fluid mosaic model is by far the best structural interpretation of the membrane which satisfactorily explains all its (membrane's) known functions.

MICELLARMODELS

According to Fernandez Moran (1963) the membrane is corpuscular in nature consisting many repeating units. These units called *globular units* or elementary particles range in diamiter form 40 - 70Å and form closely packed repeating units. The membrane structure may be globular or unit membrane in different organelles or under different metabolic conditions. According to some cell biologists, the corpusular nature is an artifact.

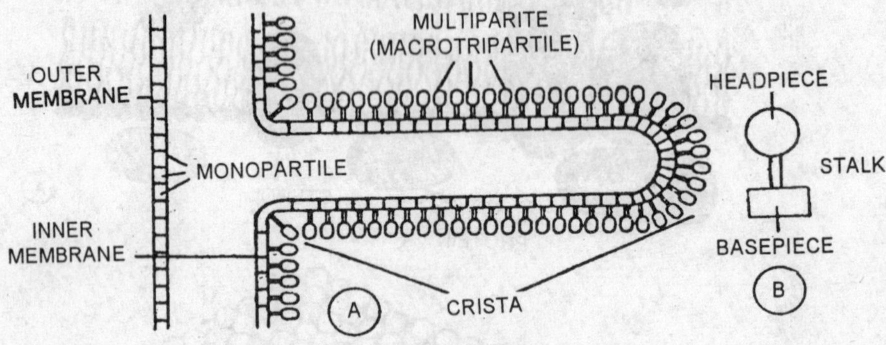

Fig. 4.12 The Cell
Green's model of repeating units
A. Crysta of a mitochondrion, **B.** Repeating unit

Green (1970) suggests that membranes made up of repeating units are of two types viz., *monopartite* and *multipartite* The monopartite membranes have repeating units flush with the membrane surface, while the multipartite have repeating units projecting from the membrane sufrace. Each repeating unit has a base, a stalk and a head. In monopartite, the repeating units correspond only to the base of the repeating unit of a multipartite membrane. The inner membrane of mitochondria corresponds to the multipartite membrane.

Of the two types of membrane stuctures, it is difficult to say which model is correct There seems to be evidence for both the bilayer and micellar models. While electron micrographs have shown the trilaminar structure for the unit membrane, the inner membranes of chloroplast and mitochondria have revealed the repeating globular sub units. It is quite likely that different membranes show different structure or the combination of both actually constitutes the membrane structure.

Cytoplasm

The colloidal substance in the cell, internal to the plasma membrane (without the inclusions) is called cytoplasm. The cytoplasm is usually granular and has many substances either dissolved or suspended in it The suspended particles may be in the solid or semi solid condition. There are several theories regarding the structure of the cytoplasm. Some of these are-

Reticular theory : Cytoplasm is made up of network of fibrils or particles.

Alveolar theory : According to Butschill (1892), cytoplasm consists of a number of droplets or alveoli. It gives an appearence of oil droplets floating on water.

Granular theory : According to Altmann (1893), Cytoplasm is composed of a number of minute granules suspended in a matrix.

Fibrillar theory : This theory was proposed by Fisher (1894) and Flemming (1897). Electron microscopic studies have proved the hypothesis of Fisher. According to this view, cytoplasm is made up of a number of fibrils.

Properties of cytoplasm : Cytoplasm has a number physical and biological proerties.

Physical properties

The colloidal nature of cytoplasm is responsible for the various physical properties A colloidal system may be defined as a system of liqid medium containing suspends particles varying in diameter from 1/1,000,000 to 1/10,000 nm. Some of the physical properties are follows.

Phase reversal : Cytoplasm may be differentiated into a *sol state* and *gel state.* Sol state is the liquid state and the gel state is the semi solid state. These two phases are interchangeable. This capacity is known as phase reversal.

Elasticity : Depending on the circumstances, the cytoplasm can extend or contract subject to a certain limit. This is known as elasticity.

Cohesiveness : The suspended particles in cytoplasm have mutual attraction, thus exhibiting cohesiveness.

Contractility : It is the capacity of peripheral cytoplasm to absorb or remove water from cell to the exterior. This is manifested very well by guard cells.

Viscosity : The suspended particles of cytoplasm are responsible for its viscous nature.

Brownian movement : The suspended particles of cytoplasm are in a state of to and fro movement called Brownian movement

In addition to this, cytoplasm also exhibits movements (seen in the plasmodium of slime molds) and 'cyclosis or streaming movements (seen in the leaf cells of *Elodea).*

Biological properties

Growth, respiration, excretion, nutrition, reproduction, metabolism, irritability etc., are some of the important biological activities of cytoplasm.

Chemical composition of cytoplasm : 90% of cytoplasm is water. The remaining 10% is constituted by organic and inorganic compounds.

About 40 or more inorganic constituents are found in cytoplasm. These are found in the form of ions (either cations or anions). Some of the common inorganic constituents are - Oxygen (62%), Hydrogen (10%), Carbon (18%) and Nitrogen (3%). The less common inorganic contituents are Calcium (0.25%), Phosphorus (1.2%), Sodium (0.1%), Magnesium (0.7%), Iodine (015%) and Chlorine (0.15%).

Among the various organic substances proteins, carbohydrates, lipids, hormones, vitamins and nucleic acids are most important.

Cytoplasmic Inclusions : Cytoplasm has a number of living and non living inclusions. The living inclusions are - Nucleus, mitochondria,. plastids, endoplasmic reticulum, golgi complex, lysosome, centrioles, microbodies, etc.

Nucleus

This is the most important organelle of the cell which regulates all its activities. It was discovered by Robert Brown in 1831.

Position : Usually it is present at the centre of the cell, but in some instances as in mature parenchyma cells, the nucleus is peripheral. In *Acetabularia,* the nucleus occupies various positions, though mainly present at the basal region.

Shape, size and number : The shape is variable depending on the cell type. Though the nucleus is generally spherical, ellipsoidal or flattened ones also may be seen in certain cells. The surface is generally smooth but as in leucocytes it may have small infoldings.

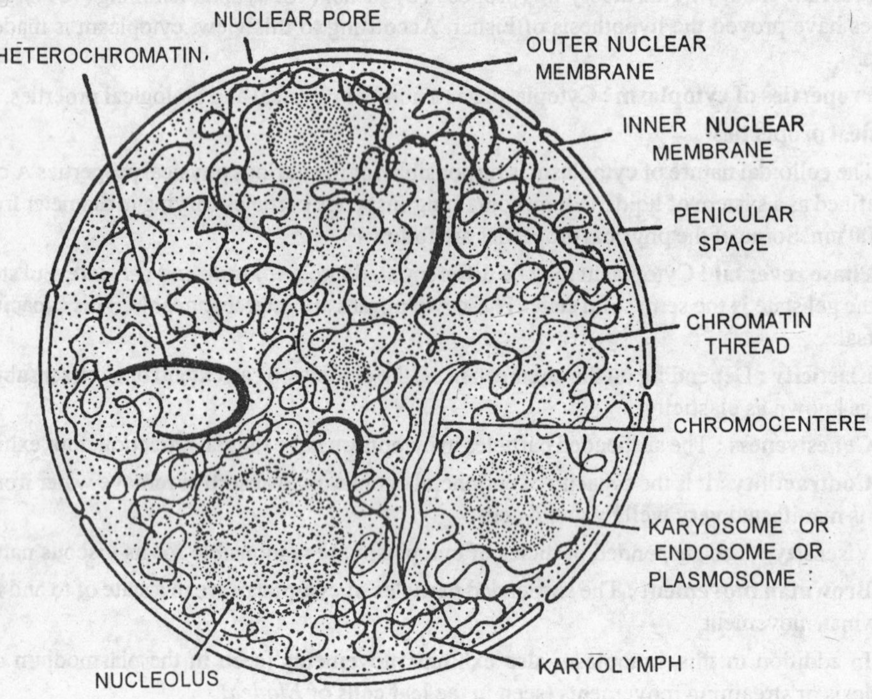

Fig. 4.13 The Cell
Structure of the nucleus

The size of the nucleus is also variable depending on the cell; the size of the nucleus being directly proprolional to that of cytoplasm. The following formula or Heywig may be used to determine the size of the nucleus in a particular cell.

$NP = \quad Vn/Vc \; - \quad Vn$

When NP = nucleoprotein index, Vn is the volume of the nucleus and Vc is the volume of the cell.

Most of the cells are uninucleate, while some *(Paramecium)* may be binucleate or multinucleate *(Cladophora).* In certain fungi like *Mucor,* where there are no septa dividing the plant body into cells, numerous nuclei occur indicating a coenocytic condition.

Ultrastructure of the nucleus : Ultrastructurally the nucleus has the following parts - 1. Nuclear membrane, 2. Karyolymph or Nuclear sap, 3. Chromonemata, 4. Nucleolus and 5. Endosomes.

Nuclear membrane : In all eukaryotic organisms, a limiting membrane envelops the nucleus. This membrane helps in effective communication between nucleus and cytoplasm. The elements of endoplasmic reticulum contribute to the nuclear envelop during cell division.

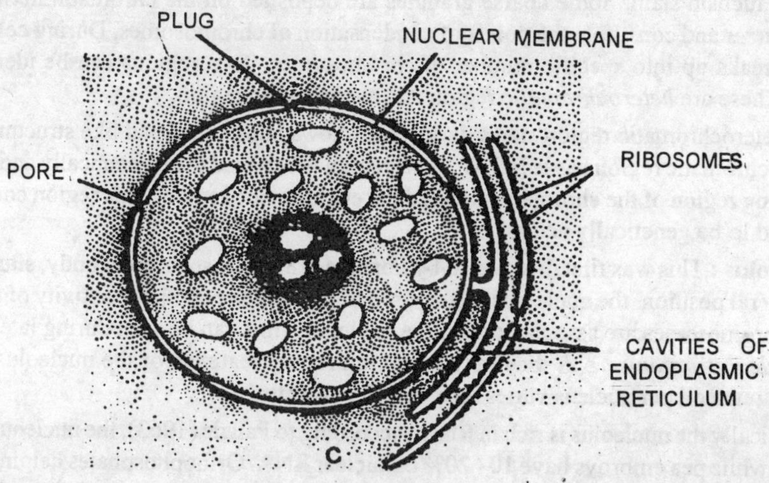

Fig. 4.14 The Cell
Diagrammatic representation of nuclear membrane and pore

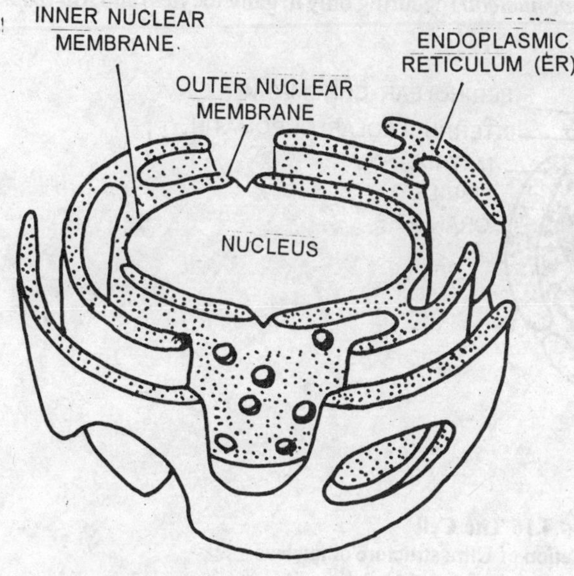

Fig. 4.15 The Cell
Three dimensional view of nuclear membrane complex

Electron microscopic studies have revealed the nuclear membrane to be a double membrane with a number of pores. According to Watson (1959) these pores called *nuclear pores* are octogonal in shape. The outer nuclear membrane is called endokaryotheca. Each membrane is about 75 - 90Å thick and they enclose a space between them about 100 - 150Å (De Robertis et al 1970). This space is called the *perinuclear space* or *cisterna.* The outer membrane is rough due to the presence of RNA particles. Internally the nuclear membrane is followed by a supporting membrane called *fibrous lamina* (300Å thick).

The nuclear pores are 600Å wide and are enclosed by rings called *annuli.* These together constitute the pore complex. In some instances, the pore complex is covered by a membrane to regulate the premeability.

Nuclear sap (Karyolymph) : The interphase nucleus consists of a homogenous matrix of fluid substances enclosing a network. Primarily the Karyolymnh is proteinaceous, but also has nucleic acids, enzymes and minerals. It is quite probable that in plants the nuclear sap contributes to the spindles.

Chromonemata : Enclosed in the Karyolymph and visible in the interphase nucleus are found a number of fibrillar structures constituting a network called *chromonemata* or *chromatin fibrils.* These stain positively with basic fuchsin stain. Some coarse granules are deposited on the chromatin network. These are a called *chromocentres* and constitute the points of condensation of chromosomes. During cell division the chromatin network breaks up into specific number of chromosomes. Two regions can be identified in the chromatin material. These are *heterochromatic* region and *euchromatic* region.

The heterochromatic region stains darkly and shows numerous bead like structures called *chromomeres.* The heterochromatic region has less DNA. This region is believed to be genetically and metabolically inert. The light staining region of the chromatin is called the euchromatic region. This region contains more of DNA and is supposed lo be genetically active.

Nucleolus : This was first discovered by Fontana (1874). A spheroidal body, situated either in the central or pheripheral position, the nucleolus is supposed to regulate the synthetic activity of the nucleus. Usually two or more chromosomes are associated with the nucleolus (this can be seen during late prophase) and these are called *nucleolar organisers* as they play a role in the reappearence of the nucleolus after cell division. The number of nucleoli per nucleus varies form one to two or three.

Chemically the nucleolus is rich in RNA. According to Palade (1963), the nucleous of liver cells contain 3 - 5% RNA, while pea embroys have 10 - 20% of nuclear RNA. Orthophosphates helping in RNA synthesis have been reported in nucleolus. Besides this many enzymes (acid phosphatase, nucleoside phosphorylase, RNA methylase etc) have been reported in the nucleolus.

Types of nucleoli : According to Wilson, there are two kinds of nucleous viz., Plasmosomes and Karyosomes (Ocata). The plasmosomes are positive to acidic stains and have a transparent outer region with a dense interior. The karyosomes stain positively with basic dyes. Three kinds of karyosomes have been identified-*Netknot* type (like nodules attached to threads), *Chromosome nucleoli* (occuring only in gametocytes) and *Karyospheres* (spheriodal bodies with basic chromatin).

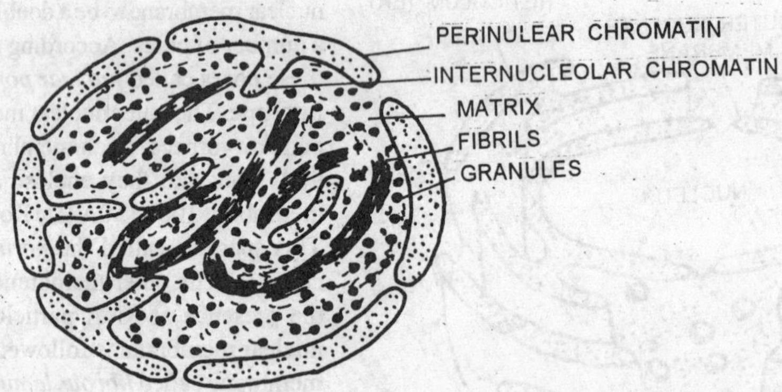

Fig. 4.16 The Cell
Schematic representation of Ultra structure of nucleus

In some instances, plasmosomes and karyosomes combine to form amphinucleoli; these are common ir molluscan eggs.

Fine structure of the nucleolus : Detailed analysis of the nucleolus has revealed the presence of three regions. These are granular region, fibrillar region and amorphous region. The granular region occuring at the peripheral region of the nucleolus is composed of granules of 150-250Å thick. This region has ribonucleic acid and proteins. The fibrillar region also known as nucleolonema is composed of fibrils 50-80Å in length. The amorphous region has low electron density and is the one to disappear first at the time of cell division.

There iş no limiting membrane to the nucleolus and the organisation is maintained by the calcium ions.

Functions of nucleolus

1. It is the active site of RNA synthesis
2. It is the source of ribosoinal RNA
3. It produces precursors of ribosomes
4. According to Maggio (1990), protein synthesis takes place in the nucleolus. But this may not be totally true.

Endosomes : These are granular structures present in the karyolymph and are smaller in size than the nucleolus.

Chemical composition of Nucleus : The main constituents of nucleus are nucleoprolcins, besides enzymes and inorganic salts. The following gives the total distribution of nucleoproteins in the mammalian liver cells- DNA 9%, RNA 1%, Histones 11%, Residual proteins 14%, and other acidic proteins 65%.

Nucleic Acids : These are the most important constituents of the nucleus playing a vilal role in the inheritance of characters from one generation to-the other. There are two types of nucleic acids-Deoxyribose nucleic acid (DNA) and Ribose nucleic acid (RNA). While both constitute the genetic material in different organisms, it is predominantly the DNA which is the genetic material in most of the organisms. The following are some of the important milestones in the discovery of Nucleic acids.

1. Meischer (1869) isolated nucleic acids and called them nuclein.
2. Hertwig (1884) suggested a heriditary role to the nuclein.
3. Kossel (1884) recognised purines and pyrimidines in the nucleic acids.
4. Feulgen (1912) devised basic fuchsin dye positive only to DNA.
5. Astbury and Bell (1938) showed that the bases in DNA molecule are arranged one above the other.
6. Avery, Macleod and Macarty (1944) presented evidence for DNA to be the genetic material.
7. Chargaff (1950) showed that DNA contains equal amount of puriness and pyrimidines.
8. Wilkins et al (1950) identified the molecular distances between the bases.
9. Furberg (1952) opined that DNA is a coil of single chain.
10. Pauling and Corey (1952) suggested that DNA is a helix of three chains.
11. Waston and Crick (1953) demonstrated the double helix model for DNA molecule.
12. Meselson and Stahl (1958) demonstrated the semi conservative replication of DNA molecule.
13. Jacob and Monad (1961) postulated the presence of RNA in protein

Deoxy Ribose nucleic Acid (DNA) DNA is the main genetic material in almost all organisms except for certain Viruses.

Mitochondria : (Gr. *mito* = thread, *chondiron* = granule)

Popularly called 'power house' of cells, these granular or filamentous organelles are present in all eukaryotes. They are characterised by their special structure, and unique physiological function concerning intracellular energy transformation.

Historically, mitochondria were first observed by Altmann (1894) who called them *bioblasts.* The present name was coined by Benda (1897) who developed the crystal violet staining. Bensely and Hoerr (1934) first isolated the liver cell mitochondria. The discovery that mitochondria are seats of cellular respiration was made by Hogeboom *et al* in 1948.

It is difficult to observe mitochondria in living cells due to their low refractive index, but they can be studied under phase contrast microscope after staining with a dilute solution of janus green. Mitochondria may be isolated from cells by centrifugation between 20,000 g -40.000 g.

Shape : Generally mitochondria are granular or filamentous, even through the shape is variable. During certain conditions a long mitochondrion may swell at one end lo become club shaped or look like a tennis racket .(De Robertis et al 1970). Mitochondria may also become vesicular.

Size : Size also is variable. The width is about 0.5µm and the length may reach a maximum of 7µm. The si/ e depends on the osmotic pressure.

Distribution : In general, mitochondria are distributed evenly in the cytoplasm, but in some instances they may accumulate around the nucleus or at the periphery of the cytoplasm. During cell division, mitochondria accumulate near the spindles. The distribution of mitochondria within the cell is in relation to the energy demand. For example in skeletal muscles they are grouped like rings or braces around the I-band of the myofibril.

Orientation : Mitochondria have more or less a specific orientation. In cylindrical cells, their long axis is parallel to the long axis of lhe cell. In leucocytes, mitochondria are arranged radially in respect to the centrioles.

Number : Mitochondria are absent in prokaryotic cells. The quantity of mitochondria per cell varies, but in liver cells it is estimated that mitochondria contains 30-35% of the total protein content of the cell. A normal liver cell contains about 1000-1500 mitochondria. Some oocytes coniain an many as 300000 mitochondria.

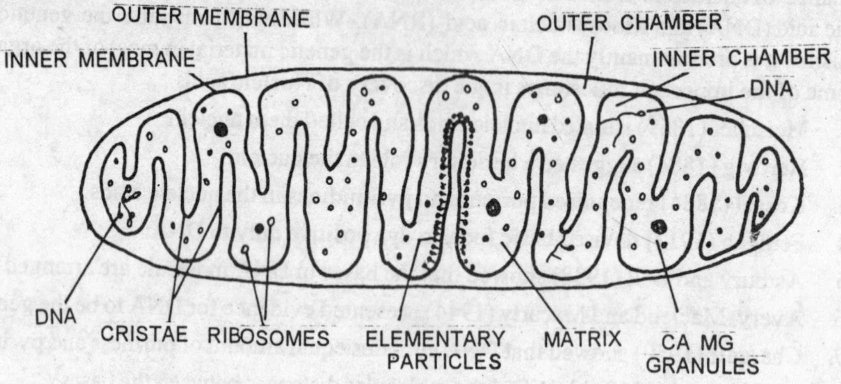

Fig. 4.17 The Cell
Diagrammatic representation of a section through a mitochondrion (ETPs are not shown on all christae)

Structure of mitochondria : The mitochondrion is enveloped by two membranes viz, *inner membrane* and *outer membrane.* The outer membrane is 6nm thick; internal to this is the inner membrane. Separating these two is a space (outer chamber) filled wilh a watery fluid. This space is 40-70Å thick. The inner membrane is folded into finger like projections called *cristae.* This membrane encloses the inner chamber or matrix, usually filled by dense protinaceous material. The matrix may be homogenous or may have a granular substance. The cristae of the inner membrane facing matrix is called the M side, while the other side is called the C side. The cristae have an inner space called the intercristae space which is in continuity with the intermembrane space.

Within the mitochondrial matrix are small ribosomes and a circular DNA. The ribosomes are of 70s type and similar to those of bacteria. These can synthesise proteins (Tyier 1973). The DNA called mDNA may be circular as in most of the animal cells or linear as in plant cells. Rabinowitch (1968) has demonstrated that the mitochondrial

DNA has more G-C content than the nuclear DNA.

The mitochondrial DNA can replicate itself as it has DNA ploymerase. It is due to this, that mitochondria can replicate themselves.

The inner membrane : The inner membrane has numerous nail like or tennis racquet like projections protruding into the matrix. Each particle is about 70-100A⁰ in diameter. The particles are placed apart a distance of 100A⁰ and each is attached to the inner membrane by a short stalk,of 35-50A⁰ long. The number of these particles per mitochondrion ranges from 10^4-10^5. These are called elementary particles, F_1 *particles* or *subunits*. It was formerly believed that these particles contain all the enzymes necessary for electron transport and oxidative phosphorylation. Hence they were named *electron transport panicles* (ETP). Each particle was thought to have four complexes, with complex I and II in base, III in stalk and IV in the head piece. According to current thinking, the head piece has ATPase, the stalk consists of a special protein (oligomycin sensitivity confering protein) and the piece contains the proton channel.

The inner membrane has the respiratory chain consisting of five complexes. Of these, four are involved in the electron transport, while the fifth one is involved in the synthesis of ATP.

Mitochondria play a vital role in respiration. Glycolysis takes place in the *cytosol* (cytoplasm) while the enzymes of the Kerbs cycle are found in the matrix, those of oxidative phosphorylation are found in the inner membrane.

Biogensis of mitochondria : Each mitochondrion lasts for about a week or little more. There are three views regarding the origin of mitochondria-a) origin from cell membranes, b) Division of pre existing mitochondria and c) *de novo* origin.

According to Robertson (1964), mitochondria in muscle cells arise by the infloding of the plasma membrane. This infolding joins another piece and forms a vesicular structure. In the nerve fibres of cray fish, endoplasmic reticulum may form mitochondria.

According io Luck and Rich (1964), mitochondria are capable of self feproduction. DNA, protein synthesis, DNA polymerase and all the other essential, features for self repliciation are present in mitochondria.

Harvey (1951) supports a *de novo origin* for mitochondria. Centrifugal separation of sea-urchin eggs into two halves, a nucleated half (without mitochondria) and a non nucleated half (with mitochondira) showed that the nucleated half produced the mitochondria.

Are mitochondria symbolic organisms? The many similarities between a bacterial cell and mitochondria made many people (Lloyd 1974; Borst 1977, O'Brien 1977 believe that the mitochondrion was once an autonomous entity which later entered into a symbiolic association with an anaerobically respiring host cell. The host cell derived its energy only from glycolysis. Subsequently with the association, the host cell started respiring aerobically. De Robertis (1970) opines that in evolutionary terms it is possible that a symbiotic relationship could have evolved into the present semiautonomus condition of the mitochondria.

The same theory has been Proposed for chloroplasts also, as they also have all (the independent attributes of mitochondria.

PLASTIDS : These are living cytoplasmic inclusions found in most of the plants except for some bacteria and some fungi.

The plaslids are of three categories viz., chromoplasts, leucoplasts and chloroplasts.

Chromoplasts : These are pigmented plastids. The pigments are non chlorophyllous like, carolenes, xanthophylls, fucoxathin, phycocrythrin etc. In photosynthetic bacteria, the chromoplasts have bacteropchlorophyll.

Leucoplasts : They are colourless plastids. They lack pigments and are usually present in cells which do not receive direct light. Leucoplasts may be seen in the storage leaves of onion. Leucoplasts that store starch are called *amyloplasts,* those that store oil are called elaioplasts and the ones storing proteins are called *aleuroneplasts.*

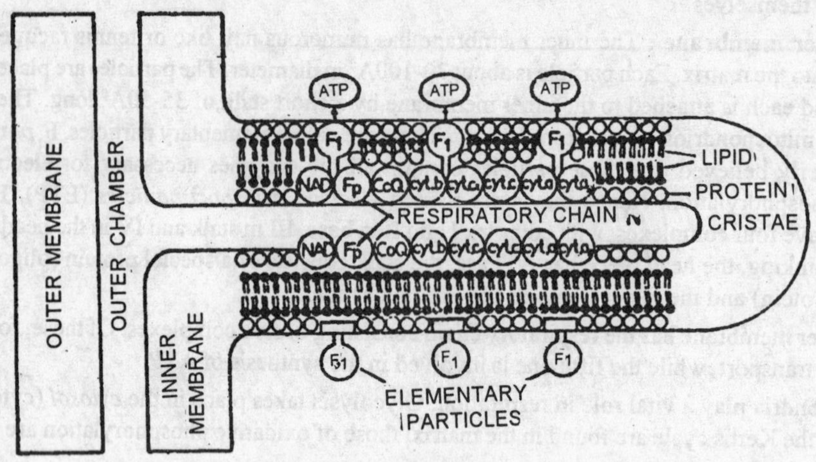

Fig. 4.18 The Cell
Arrangement of complexes (I-IV) in the ETP (old view)

Chloroplasts : These are by far, the commonest and the most plastids. As the primary sites for trapping and converting solar energy they are very vital for the existence of not only the green plants, but for the whole living world.

Shape : Chloroplasts have varied shapes. The algal chloroplasts exhibit an array of shapes. They may be cup shaped *(Chlamydomonas)*, girdle shaped *(Utothrix)*, reticulate *(Hydrodictyon)*, stellate *(Zygnema)*, ribbon shaped *(Spirogyra)* or discoid. They are generally spherical or ovoid in higher plants.

Size : There is a great variety in siz.e. Normally they are about lμm thick and 4-6μm in length. Chloroplasts ofpolyploid cells are generally larger than in the diploid cells. Cells in the shaded area have longer chloroplasts but with less intensity in colouration.

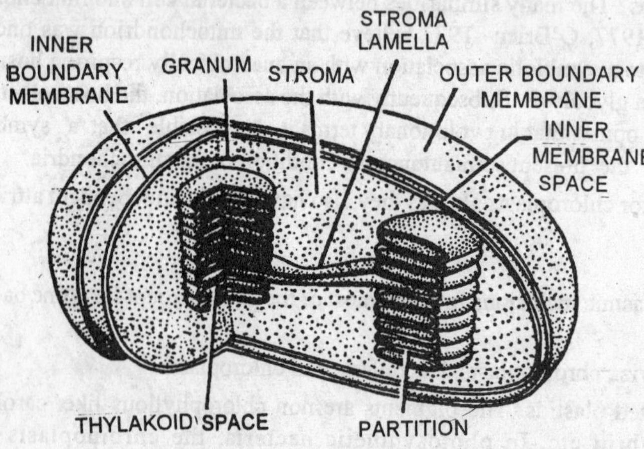

Fig. 4.19 The Cell
Model of Unltrastructure of chloroplasts

Distribition : They arc uniformly distributed all over cytoplasm, but in some instances they cluster towards the nucleus. The concentration of chloroplasts will also depend on light intensity.

Number : The number of choloroplasts per cell varies. In some algae like *Chlamydomonas,* there is only one chloroplast per cell or two as in *Zygnema.* In higher plants, cells have a number of chloroplasts. The leaf of *Ricinus communis* has nearly lour lakh chloroplasts per square millimeter area.

Chemical composition : The chloroplasts contain carbohydrates, lipids, proteins chlorophyll, carotenoid and xanthophylls.

The protein lipid ratio is 40:30. Besides the above, nucleic acids have also been demonstrated in the chloroplasts.

Ultra structure of the chloroplasts : The chloroplast has a covering of two membranes with an inner membrane space. These membranes are smooth and there are no perforations or particles. The membranes are differentially permeable (Mudrak).

A section of the chloroplast reveals an intricate system of membranes enclosed in a granular matrix. These membranes are called lamellae and the surrounding matrix-the *stroma.* In a sectional view, the lamellae can be seen packed and these stacks are called *thylakoids.* In the chloroplasts of higher plants, the thylakoids themselves form highly compact bundles called *grana.* Some thylakoids of granum extend into the stroma and maintain contact with other grana. These are *called-Stroma thylakoid* or *stroma lamellae* or *intergrana.*

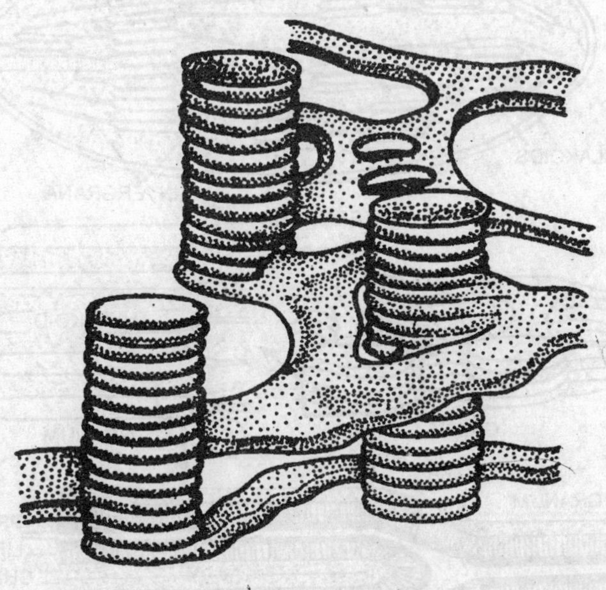

Fig. 4.20 The Cell
Three dimensional view of Grana (enlarged view)

The chloroplasts of algae lack the granum arrangement and in cyanobacteria (blue green algae), the thylakoids lie naked in cytoplasm without any envelope. In such instances, pigments are uniformly distributed on or in the lamellae.

The stroma has in addition to the lamellae - granules (globuli), lipid droplets, starch grains and vesicles.

Structure of the lamellae : All the photosynthetic pigments are concentrated in the chloroplast lamellae. The lamellae as well as the outer membranes are composed of lipoprotein sub units, each sub unit according to Weir and Benson has a protein core surrounded by a lipid sheath. The chloroplast membranes have a single layer of sub units, whereas in thylakoids it is double due to the juxtaposition of the membranes. The appressed areas are called partitions. These are hydrophobic in nature The space between the two membranes of thylakoid (fret membrane) is called *fret channel.* The space between two thylakoids is called *loculus* while the ends of disc shaped thylakoids are called *margins.* These areas (loculus, fret channel) are hydrophilic. The chlorophyll molecules are concentrated in the fret channel ie., in the space between the two membranes of thylakoids. The chlorophyll molecules may be entirely in the space or the *heads* may be in the fret channel while the *tails* are buried in the sub units, occasionally the entire chlorophyll molecule may be found in the fret membrane. Other pigments like cytochromes, carotenoids, etc., are also found in the fret membranes.

Four sub units of the partition constitute a photosynthetic unit or a *quantosome.*

Ribosomes and RNA have also been isolated from the chloroplasts indicating a machinery for protein synthesis.

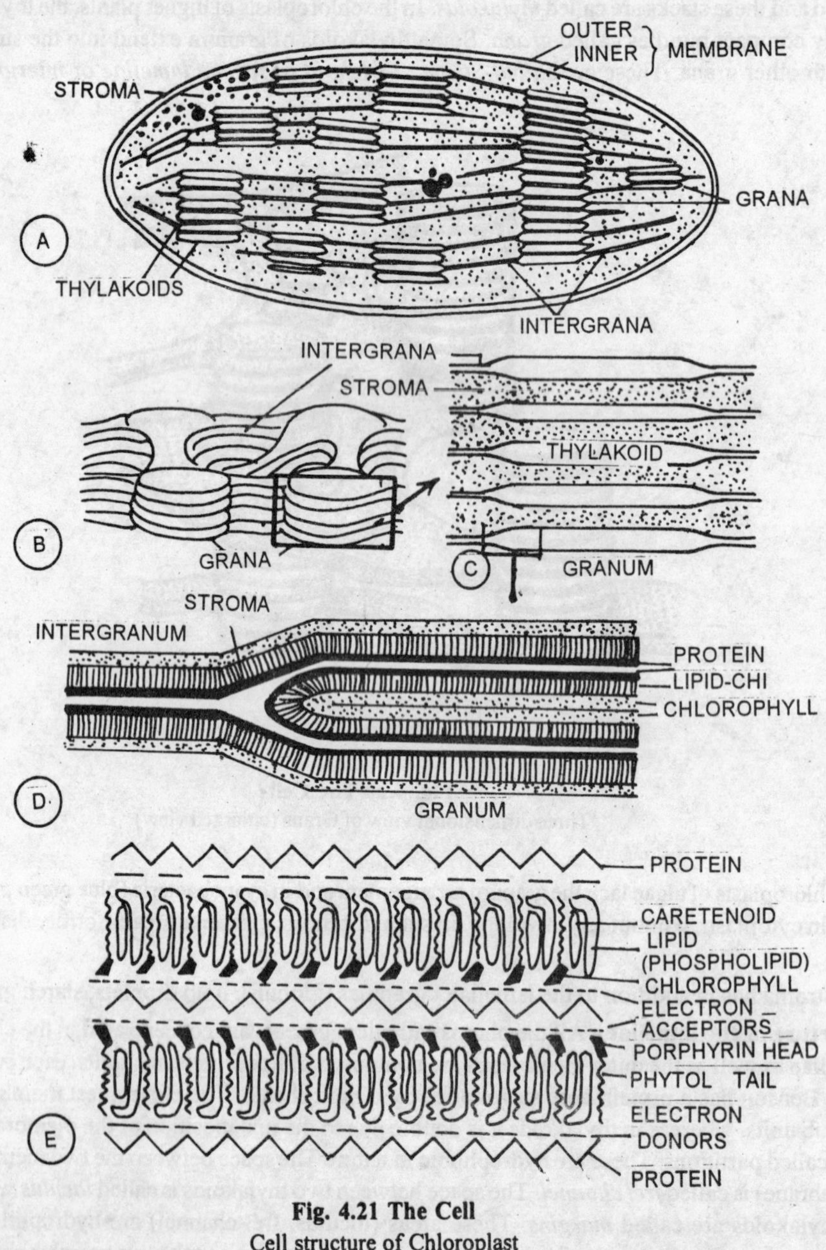

Fig. 4.21 The Cell
Cell structure of Chloroplast

Since the chloroplasts, have their own RNA, DNA ribosomes and protein synthesis, according to Deviln (1975) "One might be inclined to view the chloroplast as a cell within a cell". De Robertis (1980) suggests "It has been suggested that chloroplasts may have resulted from a symbiotic relationship between an autotrophic microorganism, able to tmasform energy from light, and a heterotrophic host cell" But Bogorad (1977) opines that the chloroplast components depend to a great extent on the nuclear genes for their assembly into an integrated organelle.

Pigments in chloroplast : Some of the important pigments present in chloroplasts are chlorophylls, carotenoids, cytochromes etc. In algal cells the thylakoids also have phycobilins.

Chlorophylls : These are the most important pigments involved in the basic photochemical reaction that sustains life. Atleast nine types of chlorophylls have been identified so far - chlorophylls *a. b. c. d* and *e.* Bacterio Chlorophylls a and b and the chlorobium chlorophylls 650 and 660μm.

Among the chlorophylls, chlorophyll *a* and *b* are best known and most widely distributed. These are absent in pigmented bacteria. Chlorophylls *c, d* and e are found only in algae in combination with chlorophyll a. Bacteriochlorophylls and chlorobium chlorophylls are found only in photosynthetic bacteria.

The chlorophyll molecule is a complex structure basically made up of a 'head' and 'tail' resembling a tennis racquet. The 'head' is a *porphyrin* structure made of four *Pyrrole rings* attached to each other at the centre by an isocyclic ring containing Mg atom at the centre. Extending from one of the pyrrole rings is the 'tail' - the alcoholic chain *(phytol).* The emperical formula for the chlorophyll molecule is CH, O,N Mg. The phytol chain is esterified on the C atom of one of the pyrrole rings and has only one double bond.

Chlorophyll *a* and *b* differ structurally in having different atomic groupings at the C atom in the pyrrole ring. In chlorophyll a C atom has a methyl group, while chlorophyll *b* has an aldehyde group; besides the two pigments have different absorption spectra. The peaks are as follows

Chlcrophyll *a* 429, 410 and 660nm
Chlorophyll *b* 430,453 and 442nm
(The above spectra are for chlorophyll in *vitro)*

Carotenoid pigments : These are lipid compounds ranging in colour from yellow to purple. They are widely distributed in both plants and animals. They are also present in microorganisms including red algae, cyanobacteria, photosynthetic bacteria, fungi, etc. Wackenroder (1931) isolated the first carotenoid - carotene from tissue.

Carotenoids are derivatives of *lycopene* a red pigment found in many plants. They (carotenoids) are known to have eight isoprene ($CH_2 = C[CH_3] - CH = CH_2$) like residues.

The main carotenoid found in plants is the orange yellow coloured β carotene with some quantity of α carotene.

The carotenoids are also located in the chloroplast. According to Goodwin (1960), carotenoids and chlorophylls may be combined with the same protein to form a complex known as *photosynthein.*

Carotenoids protect the chorophyll form photooxidation and transfer the light energy they absorb to chlorophyll *a.*

Phycobilins: These are found only in algae. There are two types of pycobilins - the *Phycoreythrin* (red) and *Phycocyanin* (blue) These pigments are strongly associated with a protein and consequently it is difficult to isolate the pigments in the pure state.

Like carotenoids phycobilins are also involved in the transfer of light energy (they absorb) to the chlorophyll.

The absorption spectra of phycobilins presents an intersting study considering the fact that they act as accessory pigments in photosynthesis. R - phycoerythrin has peaks at 495, 540 and 545nm, while R - phycocyanin has peaks at 550 and 615nm.

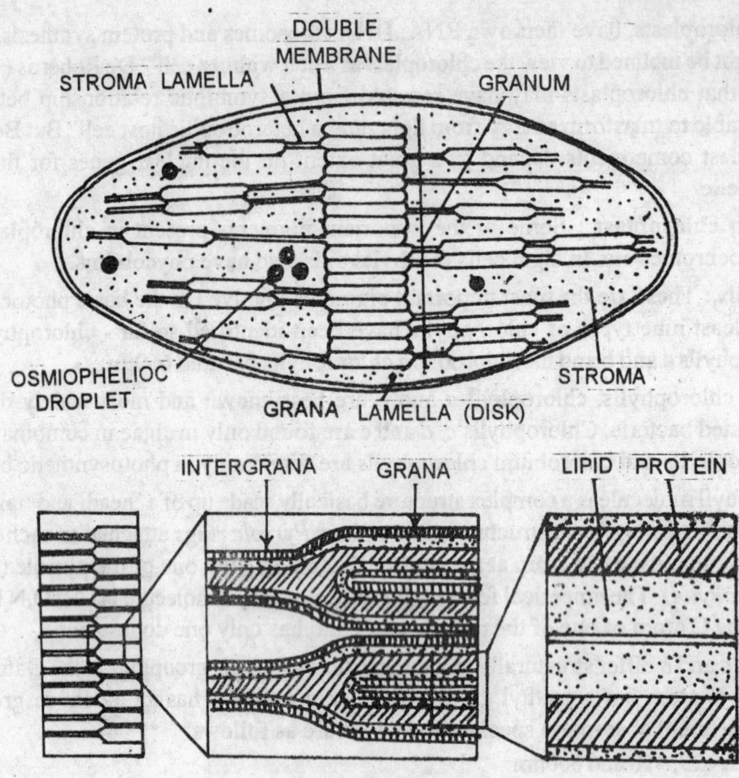

Fig. 4.22 The Cell

Diagrammatic representation of Ultrastructure of chloroplast showing the lamellae.
A. Entire chloroplast, **B-D.** Show successive enlargement of a part of the granum

ENDOPLASMIC RETICULUM

The cytoplasm in the cells of most of the plants and animals is traversed by a network of anastamosing fibrillar structures called endoplasmic reticulum (ER). Originally known as ergastoplasm, the current name ER was given by Porter and Kaallman (1952). The ER is connected to the outer nuclear membrane with the ER space opening into the perinuclear space between two membranes. The ER is known to be connected with the plasma membrane.

Morphologically the ER consists of three types of structures viz., *Cistemae, Vesicles* and *Tubllues.*

The **Cisternae** *are* elongated cyllindrical unbranched structures arranged parallel to one another respresenting a lamellated structure. They are 40 - 50µm in diameter. Cisternae are generally found in cells active in synthesis.

The **Vesicles** or sacs or oval bodies, attached to the membrane and have a diameter of 25-500µm. These are abundant in pancreatic cells.

The **Tubulues** are branched irregularly and are of diverse shapes having a size of 50 - 100µm. The tubules are abundant in cells which are active in the synthesis of steroid compounds. They are also found in the epithelical cells of retina.

Ultra structure : All the three components of ER are bounded by a unit membrane 50 - 60Å thick. The membrane is trilamenar like the other unit membranes. Palade (1956) has observed some secretory granules in the cavity of endoplasmic reticulum. The membrane has two outer layers of protein enclosing two layers ofphospholipids. However the ER membrane is symmetrical unlike in other organelles (mitochondria).

Types of ER : Two types of ER have been observed in the cells (same or different). These are - smooth ER (agranular) and rough ER (granullar).

The membrane of smooth ER are not granular due to the absence of ribosomes. Smooth ER in mainly seen in tubules. The tubules measure about 500 - 1000Å in length. Agranular ER is seen in cells involved in lipid or steroid synthesis. Cells active in protein synthesis do not have agranular ER. Agranular reticulum of muscle cells is called *sarcoplasmic reticulum* and it forms lace like sleeve around myofibrils. In the pigmented cells of retina, the agranular ER occurs in the form of tightly packed tubules and vesicles called *myeloid bodies*.

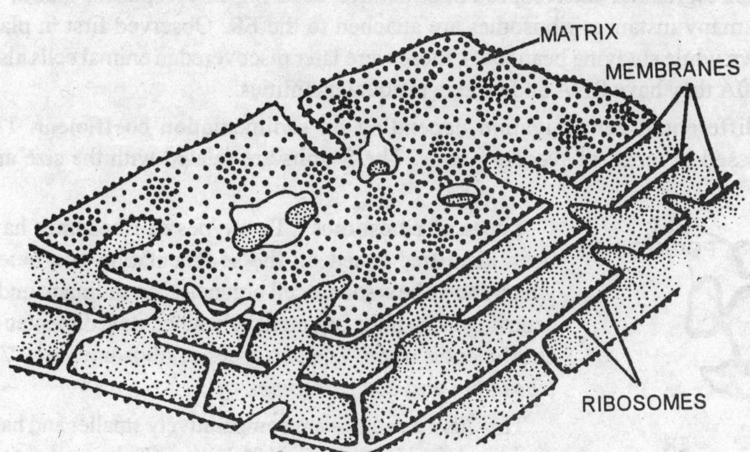

Fig.4.23 The Cell
Three dimentional view of endoplasmic reticulum

The rough ER has a granular surface due to the presence of a large number of ribosomes attached to it. Since ribosomes are actively involved in protein synthesis rough ER is conspicuous in cells that are actively involved in protein synthesis e.g. meristematic cells, plasma cells, goblet cells etc. The rough ER stains positively by basic dyes.

The cells (Spermatocytes) of certain vertebrates show an unnsual type of ER. Usually ER membrane does not have pores or openings. But in the instance mentioned above pores or annuli have been reported resembling the ones found in the nuclear membrane. There is a diaphragm across the octogonal symmetry (Manul 1968).

The ER performs many vital cellular functions. These are

(a) Enzymatic activities
(b) Transport of synthetic products
(c) Storage of metabolities
(d) Formation of nuclear envelope
(e) Intracellular impulse conduction etc.

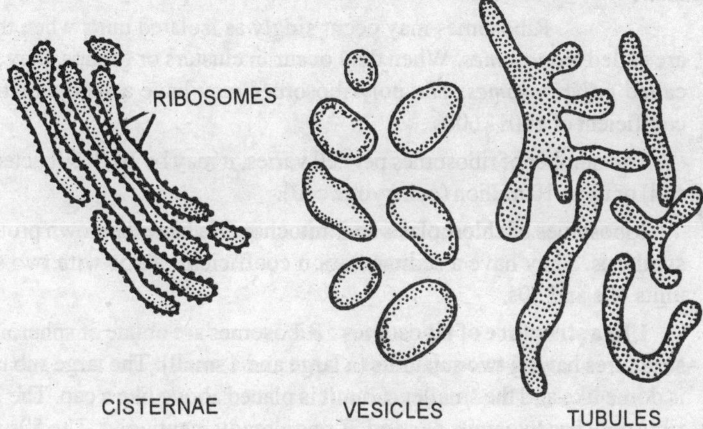

Fig. 4.24 The Cell
Various components of endoplasmic reticulum

Microsomes : These are structures that result due to the homogenization of the membranes. The cell membranes break up into fragments, round off and form microsomes. Microsomes are actually fragments of ER. They can be isolated by high speed centrifugation (100,000g) when they settle after nuclear and ribosomal fractions. Observed under electron microcope they appear as membrane bound vesicles of 500- 1500 Å in diameter. It should be understood however that microsomes are not natural structures found in the intact cell.

They are the result of homogenization of cell. The microsome fraction contains fragmented membranes.

Origin of ER : The origin is uncertain, invagination of the plasma membrane might give rise to a canalicular system. But evidences for this hypothesis are yet to come (Dallmes et al,1966).

RIBOSOMES : These are dense granular nuclecoprotein structures occuring in cytoplasm, matrix of mitochondria and chloroplasts. In many instances ribosomes are attached to the ER. Observed first in plant cells in 1953 by Robinson and Brown while studying bean roots, these were later discovered in animal cells also. Ranging in diameter from 150 - 200Å they have RNA and protein in equal quantities.

Ribosomes are isolated by differential centrifugation depending on sedimentation coefficient. The sedimentation coefficient is expressed in terms of *svedberg units.* The S units are related with the size and weight of the ribosome molecules.

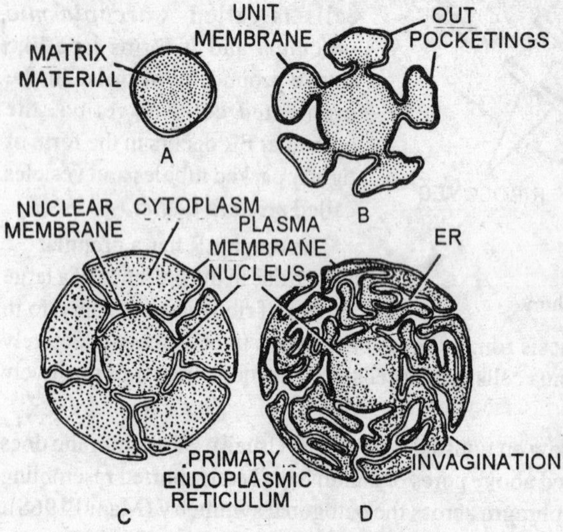

Fig. 4.25 The Cell
Stages in the origin of Endoplasmic reticulum

Types of ribosomes : Two types of ribosomes have been identified based on the sedimentation coefficient which shows how soon a cell organelle can sediment under centrifugation. If the organelle is heavier, its sedimentation coefficient is more. The two types of ribosomes are - *70s ribosomes* and *80s ribosomes.*

The 70s ribosomes are comparatively smaller and have a molecular weight of 2.7 x 10^6 daltons (Dalton is a unit of molecular weight; one dalton = weight of one H atom = 1) According to Huxley and Zubay (1960) the size of a 70s particle is 170 x 170 x 200Å. These ribosomes occur in prokaryotic cells as well as in chloroplasts of eukaryotic cells. The 70s ribosomes have two sub units - a large 50s sub unit and a small 30s sub unit.

The 80s ribosomes have a moleculer weight of 4 x 10^6 daltons, and are predominantly found in eukaryotic cells. There are two sub units in an 80s ribosome. These are - 60 and 40s.

Ribosomes may occur singly as isolated units when they are called *monosomes.* When they occur in clusters or groups, they are called *polyribosomes. Tbe* polyribosomes may have a sedimentation coefficient of 100s - 600s.

The number of ribosomes per cell varies, it may be 10,000 (bacterial cell) or up to 10 million (eukaryotic cell).

Ribosomes of chloroplasts and mitochondria have their own protein synthesis. They have a sedimentation coefficient of 55s with two sub units 40s and 30s.

Ultra structure of ribosomes : Ribosomes are oblate or spheroidal structures having two sub units (a large and a small). The large sub unit is dome like and the smaller subunit is placed above like a cap. The 70s ribosome has two units 50s and 30s as already mentioned. The 50s sub unit has a diameter of 140 - 160A^{0}and is pentagonal in shape. In the center of this sub unit occurs a round area of 40 - 60A^0. According to Fernando (1968), there is a pore like transparent area in the centre which prohibits the entry of proteolytic enzymes.

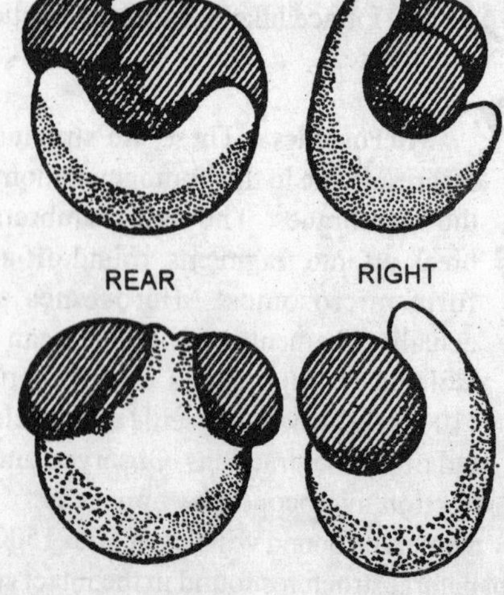

Fig. 4.26 The Cell
Ribosome - different views

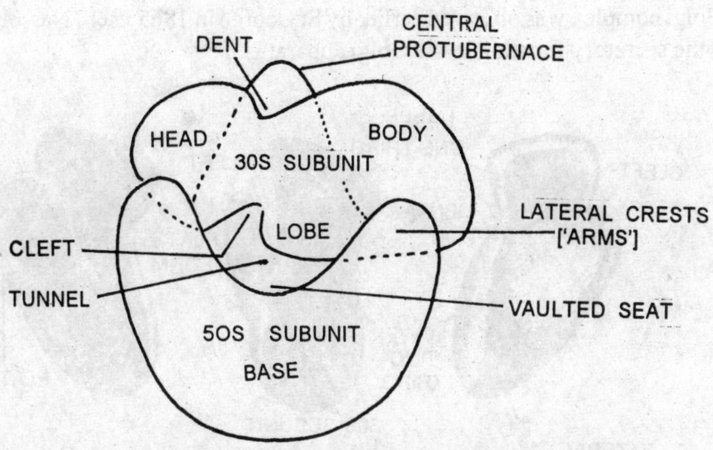

Fig. 4.27 The Cell
Witlmann's model of prokaryotic ribosome

The pores have been found in 50s sub units of 80s ribosomes also. The smaller sub units of both the types of ribosomes are often divided into two portions which remain connected to each other by a strand of 30 - 69A^0 Thickness.

Models for ribosomes structure : There are two models, viz., Wittmann's quasi symmetrical model and Lake's asymmertrical model.

According to Wittmann's model, the 30s sub unit is an elongate slightly bent stucture, having a head and a larger body with a cleft in between; with two parts protruding uniequally. It appears like a telephone receiver. The 50s sub unit is variously shaped - round shaped, maple leaf .shaped etc. The 50s sub unit has three protruberances.

The two sub units of 70s ribosomes have four areas of contact. The mRNA binding. region is located on the right side of the *30s* sub unit.

Lake's model regards the 30s sub unit to be totally asymmetrical. The sub unit is divided into two parts (1/3 and 2/3) unequally. Extending from the lower 2/3 region is a portion called the *platform.* This cleft is important in the sense of location of codon - anticodon interaction. The major difference between the Wittmanns model and Lake's model for the 30s sub unit is that the latter regards the 30s sub unit as totally asymmertrical. According to Lake, even the 50s subunit is asymmetrical.

Biogensis of ribosomes : The synthesis of eukaryotic ribosome is a complicated process involving several areas of the cell. The genes coding for 18s, 28s and 58s RNA are in the nucleolar organizing part of the chromosomes; the 5s genes are localized eleswhere in the chromosome. The ribosomal proteins are synthesised in the cytoplasm. The proteins and 5s gene products move towards the nucleolus, where they are assembled into ribosomes. Evidence for nucleolus being the site of ribosome biogeneis has been obtained by observations on the oocytes of *Xenopus laevis.*

In bacteria, the ribosomes (70s) are synthesised in the cytoplasm as the cistron portion produces a ribosomal RNA which is immediately relased.

Functions of ribosomes : As is very well known, ribosomes (polyribosomes) are the sites of portein systheses. The polyribosomes serve as a platform for the asembly *of* amino acids (on a mRNA template) brought together by specific tRNA from cytoplasm.

GOLGI COMPLEX : Described first by Camillo Golgi in 1890 in the nerve cells of barn owl, the Golgi complex has attracted the attention of cell biologists not only by its complex structure but also by its varied functions. In fact Golgi complex was observed earlier by St George in 1865 itself. Nassonov (1923) and Bowen (1929) established the secretory function in the Golgi apparatus.

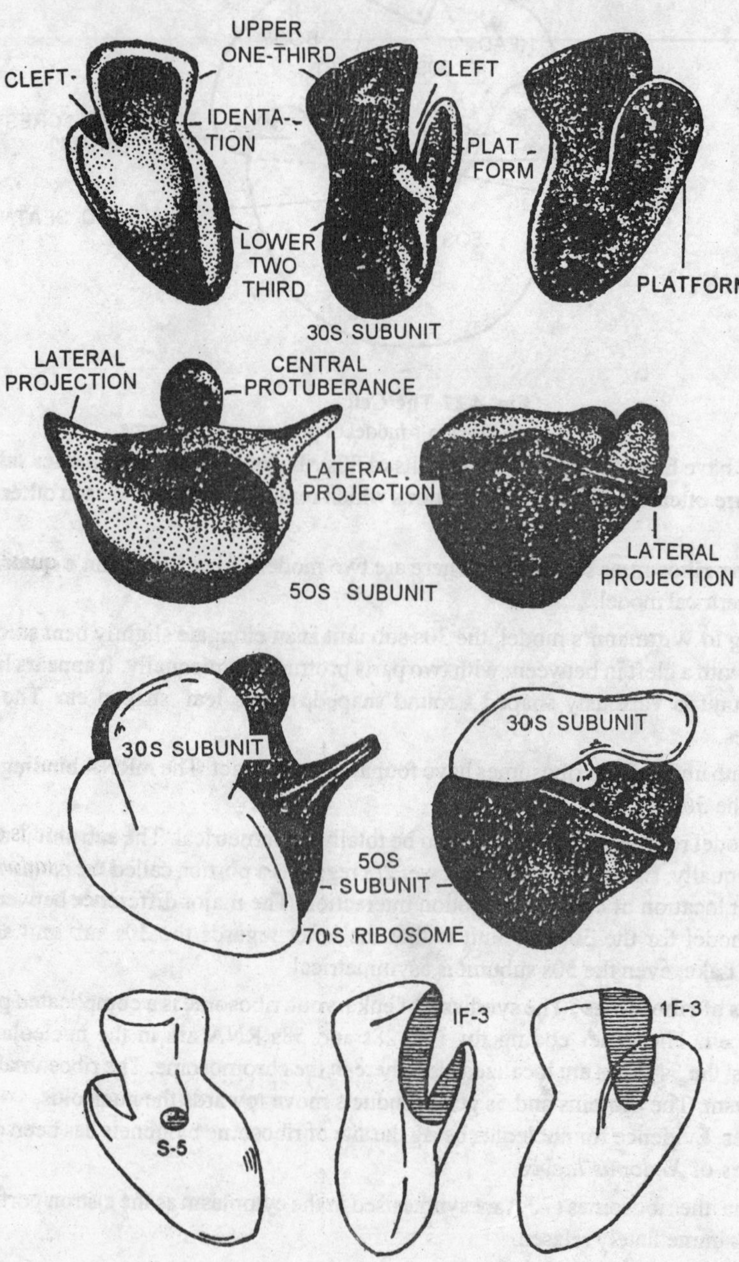

Fig. 4.28 The Cell
Lake's model of 70s ribozome from a prokaryotic cell

Terminoloy : Various terms have been employed to refer to the Golgi complex. Baker (1953) described them as Lipochondria as they have a high lipid content. Golgi complex found in plant cell are often refered to as *Dictyosomes*.

Occurrence and distribution : All the eukaryotic cells have a golgi complex which is absent in prokaryotic cells. Even in eukaryotic cells of some fungi, spenrmatozoids of bryophytes etc. golgi complex is absent.

Distribution : In higher plants, golgi complex is scattered all over the cytoplasm apparently without any localization (Hall 1974). As a contrast in animal cells, the golgi complex is localized into an organelle. In animal cytoplasm, usually there is only one golgi apparatus, but in some cases as in *Storeomyxa* there may be many complexes. In shape, the Golgi complex is variable depending on the cell type; it varies from a compact mass to dispersed filamentous reticulum.

In size, the golgi complex varies from large (nerve cells) to small (muscle cells).

Morphology : The morphology is variable depending on the source. In plants the golgi bodies are about I - 3nm in length and 0.5nm thick. Each golgi body consists of the following parts - *cisternae, tubules, vesicles* and golgian vacuoles.

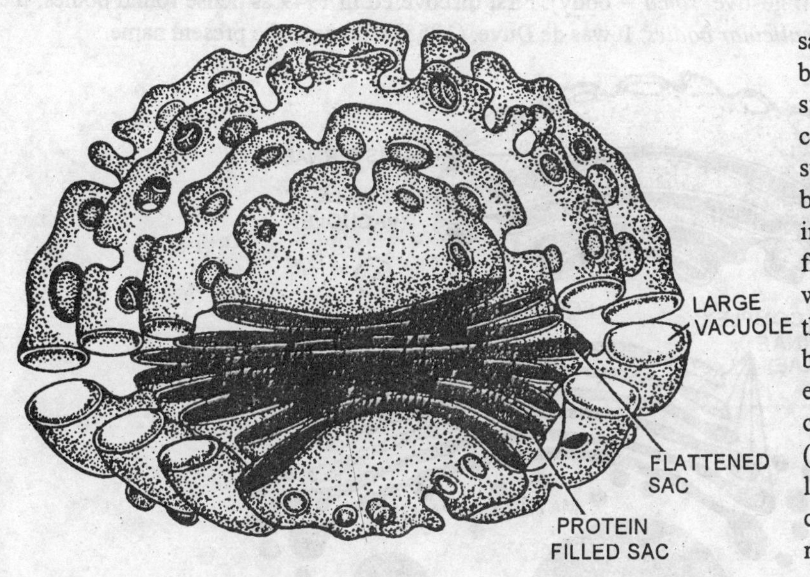

Fig. 4.29 The Cell
Golgi Complex

The *cisternae* are flattened sacs filled with a fluid. Each golgi body consists of 4 - 8 flat structures which are slightly curved and are arranged in stacks. Each cisterna is enclosed by the trilaminar membrane 6nm in thickness and is separated from the other by a 100 - 150Å wide space. In some instances the intercisternal space is filled by fibres called intercistemal elements. The membranes of the cisternae are *fenestrated* (porous). The pores may be localized or present all over the cistemae. Occasionally cisternae may form reticulate structures.

Tubules: Arising from the peripheral region of the cisterane, thetubules form ananastomosing network of 300 - 500Å in diameter.

Vesicles : These are goblet like structures attached to the tubules. There are two types of vesicles *smooth vesicles* and *coated vesicles.* The smooth vesicles are 20 - 80nm in diameter and are often called secretory vesicles. These are pinched off from the ends of cistetnal tubes. The coated vesicles are spherical outgrowths about 50nm in diameter. The function of the coated vesicles is not known.

Vacuoles : These are large spaces occuring at the distal end of the cisternae. They represent the cisternae whose membranes are widely separated.

In the case of Dictyosomes (isolated golgi bodies) of plants, cisternae are extensively fesnestrated and have rough and smooth vesicles attached to them. Parallel fibres fill the intercisternal place.

Chemical composition of golgi complex : Proteins and phospholipids are in equal concentration. Golgi material in the nerve cells consists of cephalin and lecithin. Giroud has reported Vitamin C in the golgi complex.

Functions of the golgi complex : The following are some of the functions attributed to the golgi complex -

1. Absorption of compounds
2. Sites of enzyme production
3. Sites of hormonal production
4. Sites of protein storage
5. Forms acrosome during sperm maturation
6. Forms intracellular crystals
7. Formation of plant cell wall - the golgi bodies synthesise pectin, hemiceullulose and cellulose microfibrils. They also help in the formation of cell plate during mitosis.

Origin of golgi complex : Evidences indicate that the golgi bodies are produced from the membranes of the smooth ER. Earlier it was thought that they arise *de novo* or from the phragmoplasts of the cell wall. Both these theories are found to be incorrect.

Lysosomes : (*Gr.Lyso* = digestive, *soma* = body). First discovered in 1949 as dense round bodies, the lysosomes were called *pericanalicular bodies.* It was de Duve, (1955) who gave the present name.

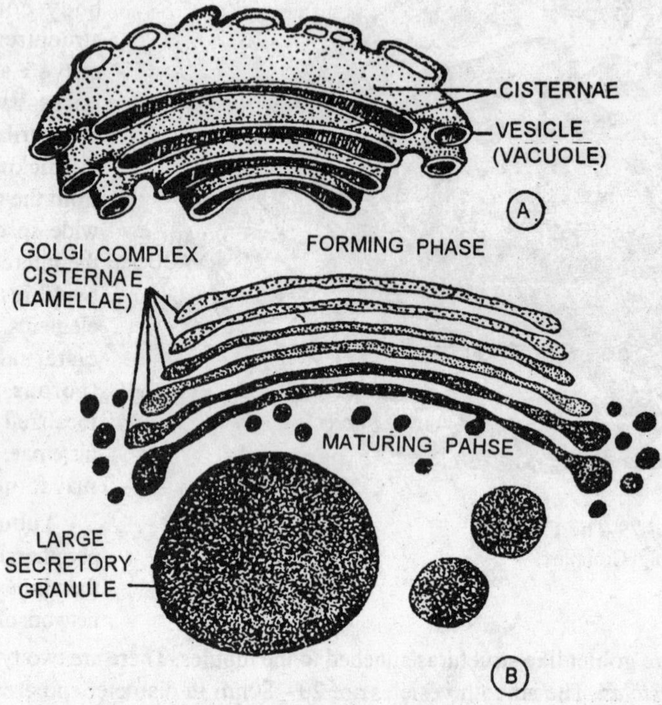

Fig. 4.30 The Cell
Golgi Complex (another view) **A.** Steroscopic view, **B.** Section view

Occurence : Lysosomes are found in most of the animal cells and few plant cells. The hepatic cells, pancreatic cells, kidney cells etc. posses large number of lysosomes. According to Matile (1969) the lysosomes in plant cell are membrane bound. Generally lysosomes are absent in the prokaryotic cells. Slime molds also have lysosomes.

Lysosomes are pleomorphic as seen in the meristematic cells of the root. The diameter may be 0.4 to 0.8μ; but in some mammalian kidney cells lysosomes as long as 5μ have been found.

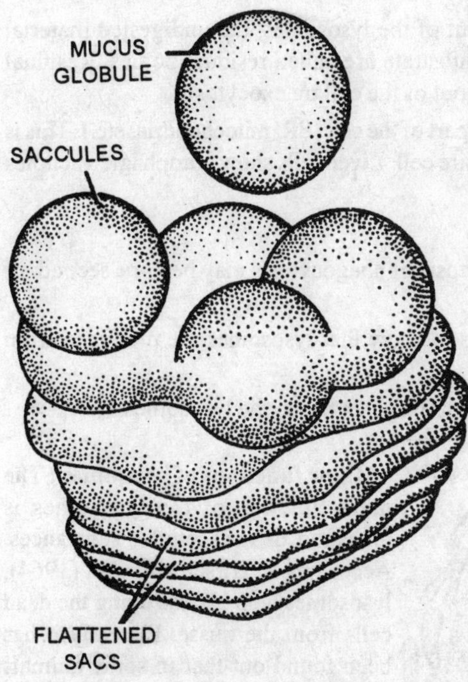

Fig. 4.31 The Cell
Golgi Complex with mucucus globules

Ultra structure : Lysosomes are saccate structures filled with digestive enzymes. Each lysosome has two parts, viz., *limiting membrane* and *inner dense mass.*

The **limiting membrane** is not double membraned like in other organelles. It is single and composed of lipoproteins. The membrane however is bilayered like any unit membrane.

The **inner dense mass** is the core of lysosomes; it is mostly enzymatic. The membrane is impermeable to substrates the enzymes of which are present in the lysosomes. Occassionally certain substances called *labilizers* destabilize the membrane causing the enzymes to come out. Alternatively, there are substance called *stabilizers* which stabilize the membrane. The limited permeability of the membrane prevents excessive digestion of cellular components by the lytic enzymes of lysosomes. The enzymes present in the lysosomes are - Nucleases, Phosphatases, Lipases, Glycosidases, and Sulphatases.

Polymorphism : Lysosomes are ploymorphic assuming various shapes depending on the cell type and situation. Basically there are two types of lysosomes viz. *Primary lysosomes and secondary lysosomes.* The secondary lysosomes are of three categories - *Phagosomes, residual bodies* and *autophagic vaculoes.*

Primary lysosomes : There are sac like structures produced directly from ER from the Glogi bodies.

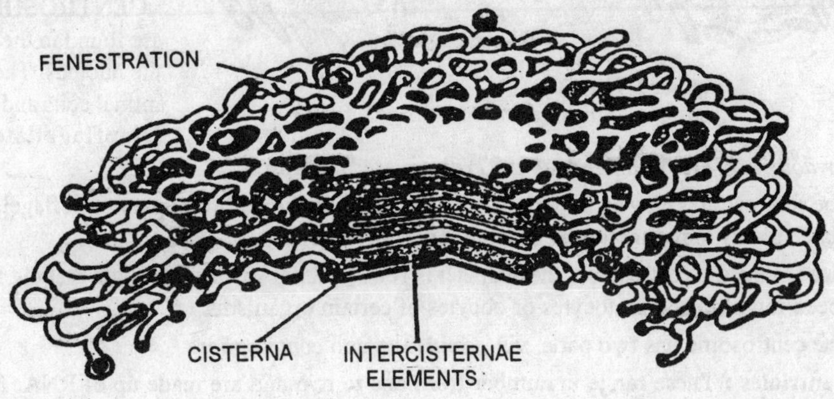

Fig. 4.32 The Cell
Golgi Complex – Diagrammatic representation of a plant dictyosome

Phagosomes : These are also known as heterophagosomes or digestive vaculoes. When a foreign substance enters the cytoplasm, the plasma membrane forms a sac around these substances. These are the phagosomes or pinosomes. The lytic enzymes present in the phagosome digests the foreign substance. The phenomenon is called *Phagocytosis* or *pinocytosis.*

Residual bodies : After the digested material has passed out of the lysosomes, the undigested material remains within the lysososmes. Such lysosomes with residual substrate are called residual bodies. Residual bodies may remain within the cell or very often they are pushed out of the cell by exocytosis.

Autophagic vaculoe : Occasionally, a lysosome may digest a part of the cell (ER, mitochondria, etc.). This is a mechanism to digest a part of the cell without destroying the entire cell. Liver cells show autophagic vacuoles or autophagosomes under starvation.

Origin of lysosomes

(a) *Extra cellular origin* : Vacuoles absorbed due to'pinocytosis or phagocytosis may become secondary lysosomes after fusing with primary lysosomes.

(b) *Origin from golgi bodies* : There are enough evidences to show that lysosomes are produced from golgi bodies at the height of their secretory activity.

(c) *Origin from ER* : According to Novikoff (1965), the rough ER gives rise to the lysosomes.

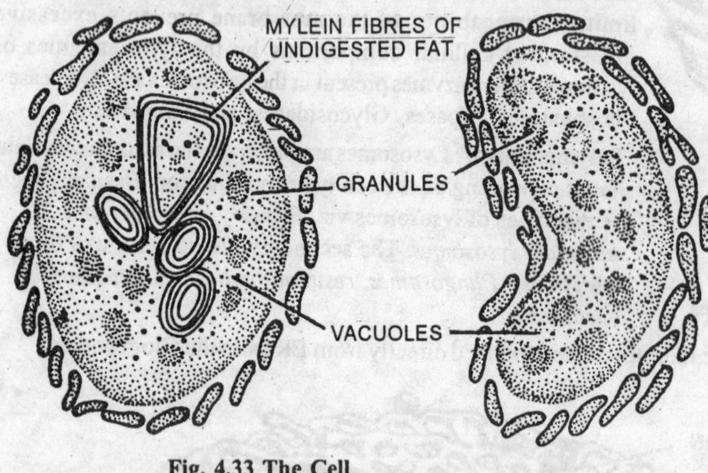

Fig. 4.33 The Cell
Lysosome – Structure

Major functions of lysosomes : The main function of the lysosomes is digestion of intracellular substances. According to Hirsch and Coh (1964), lysosomes help in destroying the dead cells from the tiusse. Recently it has been found out that in some animals (frog), lysosomes play an important role in metamorphosis. Lysosomes also perhaps help in the formation and destruction of bone cells.

CENTROSOMES : The centromes are found in the centre of the cell near the nucleus. These are present in all animal cells and lower plants like Dinoflagellates, *Euglena* and *Chlamydomonas,* etc. Van Benden (1887) disocvered centrosomes.

The centrosomes play an important role in organising the mitotic apparatus in flagellated animals, while in flagellated plants they function as basal granules of flagella.

Size : In mature centrosomes the diameter is 1600Å - 2500Å while the length is about 160A^0. Giant centrioles have been found in spermatocytes or oocytes of certain organisms.

The centrosome has two parts, viz., *centrioles* and *centrosphere*

Centrioles : These range in number from one to two and are made up of RNA. Electron microsocopic observation has revealed the centrioles to be paired cylindrical structures 3000 - 5000Å long and 1200 - 1500Å in diameter. These are open at one or both ends and placed at right angles to each other. Each cylinder is composed *of* nine fibres spaced equally and running parallel to the cylinder axis. Each fibre is in turn made up of three *microtubules (triplet).* In a cross section, these appear to be arranged like the vanes of a pin wheel. The vanes are tiled at an angle of 40^0. The wall of the cylinder is constituted by these nine triplet fibres.

The microtubule triplet in a fibre can be designated as A,B and C of which A is innermost. Each microtubule wall is 50A^0 in diameter and the lumen in 120 - 130A^0 (the total diametre varies from 180 - 200A^0). The length of each microtubule is about 560A°. At the innerside of each triplet fibre a dense line runs and it is called *triplet base* connecting A and C tubules. In addition to this, a dense RNA region called *foot* is present.

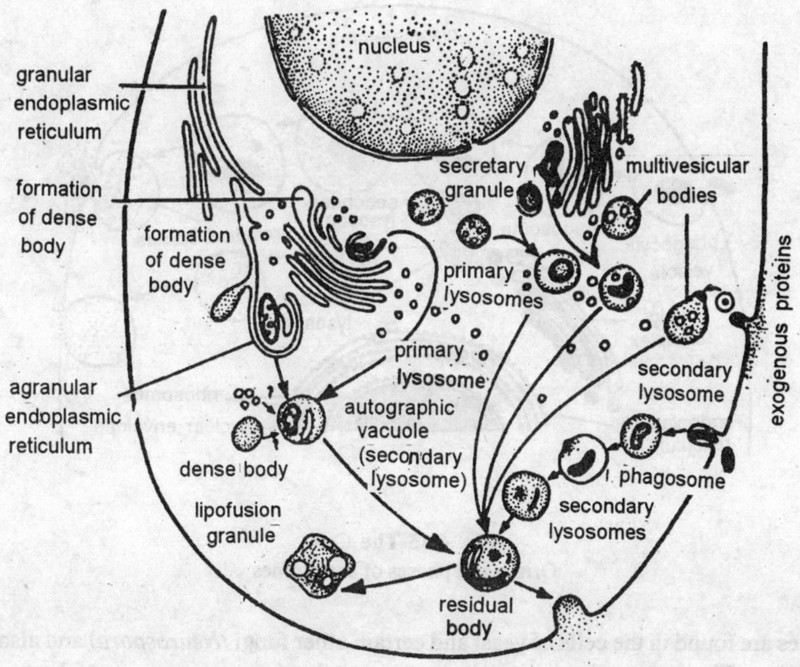

Fig. 4.34 The Cell
Polymorphism in Lysosomes

Centrosphere : The dense granlar region of cytoplasm surrounding the centrioles is called the centrosphere.

Structure : Structurally the microtubules are lipoproteinaceous (Fulton 1971).

Functions : Centrosomes appear to play an important part in cell division. people believe that centrioles play an important part in the formation of spindle. They however constitute the mitotic pole in higher animals. In primitive animals which are ciliated, centrioles seem to help in the generation of cilia.

Microbodies : The cells of plants, fungi, certain protozoa and liver cells of vertebrates contain certain small granular structures associated with ER, mitochondria and chloroplasts. These spherical bodies ranging in size from 0.2μ to 1.5μ have a central crystalline core and are called *microbodies.*

Recent studies have revealed the presence of two types of microbodies viz., **Peroxisomes** and **Glyoxysomes.**

Peroxisomes are found in animal cells and also in the leaves of higher plants. These particles participate in the oxidation of substrates resulting in the formation of hydrogen peroxide. In plants, peroxisomes are involved in photorespiration.

Structurally peroxisomes are variable in shape and size but are generally circular. They have an envelope of a single membrane made up of proteins and lipids enclosing a granular matrix. In certain instances, the matrix is crystalline or amorphous. The peroxisomes catalyze the H_2O_2 production. In plants, the peroxisomes contain enzymes of glycolate pathway.

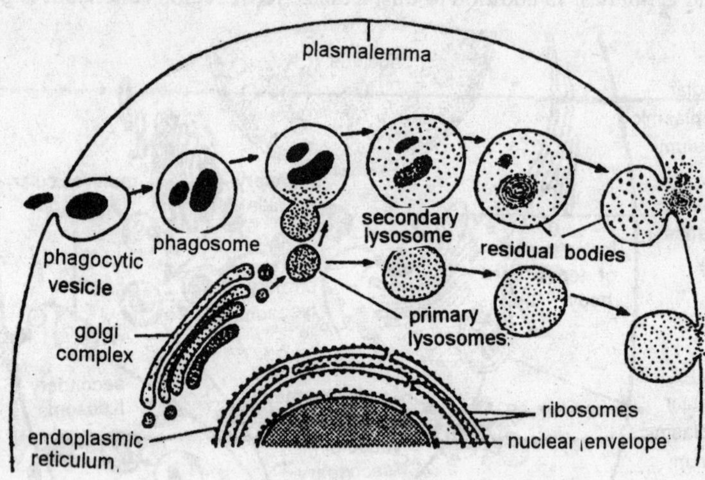

Fig. 4.35 The Cell
Origin and phases of Lysosomes

Glyoxosomes are found in the cells of yeast and certain other fungi *(Neurospora)* and also seeds (oil rich) of many higher plants. They are similar to peroxisomes except that their core consists of dense rods of 6.0μm in diameter.

Functionally the glyoxysomes contain enzymes for fatty acid metabolism, involved in the conversion of lipids to carbohydrates during germination. The *glyoxysomes* play an important role in lhe *glyoxylate cycle*. The enzymes *Urate oxidase* and *allantoinase* are also found in glyoxysomes. Urate oxidase converts uric acid to allantonin, while allantoinase hydrates allantonin to allantonic acid during the degrading of purines.

5

CHROMOSOMES

The coloured thread like structures that appear during cell division have hidden within them all the secrets of genetic characters that are passed on from one generation to another generation. Housing the particles of inheritance,- the genes, these vehicles of heredity are so important for all organisms; because it is these that delimit one individual from another. These are the chromosomes (*Gr.Chromo* = coloured, *soma* = body) which appear like coloured threads on staining. In reality however these are not coloured. In this chapter we will learn about the structure and functions of these unique threads of life.

History : It was Waldeyer (1888) who first coined the term chromosome to the darkly staining bodies of the nucleus. Fleming (1879) actually had described the splitting of the chromosomes, but he had coined the term chromatin to refer to them. Sutton and Boveri (1902) suggested that the chromosomes are the carriers of hereditary particles. Morgan (1933) discovered the role of chromosomes in the transmission of hereditary characters.

Chromosomes occur as a network of Fibrils called chromonema in the interphase nucleus. At the time of cell division the chromonema breaks and condensation takes place forming specific number of chromosomes.

Number of chromosomes : The number of chromosomes is constant for a given species. This number has been evolved during the course of evolution and characterises the species. Chromosome number and morphology is one of the parameters in deciding the phylogeny and interrelationship of the group as well as the individual. The number of chromosome in various species is as varied as the organisms themselves. They range from four to 500.

Ploidy of chromosomes : The total number of chromosomes present in the cell of an organism is called *ploidy.* In all sexually reproducing higher organisms the ploidy is of two types - *haploid* and *diploid*. The somatic cells contain diploid or two sets of chromosomes (2x). In these organisms during reproduction, the chromosome number would be halved in meiosis. As a result, the gametes have haploid or one set (x) of chromosomes. When two gametes fuse, the resultant zygote has two sets of chromosomes (diploid). It is this zygote which develops into the adult individual.

In lower organisms like many algae and fungi, the somatic cells have haploid chromosome numbers. In these organisms only the zygote will have diploid chromosomes. The meiotic division which halves the number of chromosomes takes place in zygote rather than during gamete formation. In some unusual cases (for details see chapter 6.) the cells have many sets (polypoid) of chromosomes, in which case it is a multiplication of the base number called x

Group	Name of the organism	Chromosome number
Protozoa	*Paramecium*	30-40
Coelenterata	*Hydra vulgaris*	32

Nematoda	*Ascaris lumbricoides*	24
Chordata	*Rana esculenta*	26
	Homo sapiens	
Fungi	*Mucor heimalis*	2
Pteridophyte	*Ophioglossum spp*	500
Gymnosperm	*Pinus ponderosa*	24
Angiosperm	*Crepis capillaris*	6
	Allium cepa	

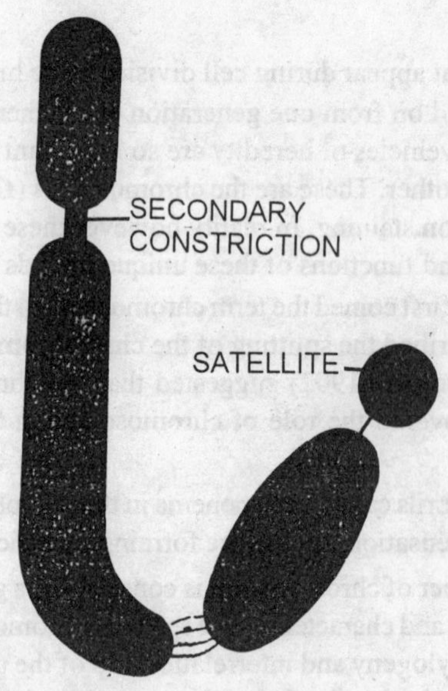

Fig. 5.1 Chromosomes
External view of a metacentric chromosome

Size of the chromosomes : There is a great variety in chromosome size and shape. In size they range from 0.1 μ to about 30 μ in length. In exceptional cases however (polytene chsomosmes) the chromosomes may be as long as 2mm. All the chromosomes in a spicies may be of similar size or they may vary. In *Yucca arakansana*, there are two distinct size groups. In some human beings there is a gradation in size. In some instances (*Mediola*) the cells at different regions (root and shoot tip) show different sizes of chromosomes. In general, the chromosomes of plants are larger than those of animals. Monocotyledons have larger chromosomes than dicotylendons.

Structure of the chromosome : The following are the parts usually associated with the chromosome.

Pellicle and matrix : Light. microscopic studies have revealed that each chromosome is surrounded by a membranous structure called *pellicle*. This membrane encloses a jelly like substance called *matrix*. The presence of matrix has been shown in *Luzula campestris* (Juncaceae). Electron microscopic studies however have revealed, that there is nothing like a pellicle enveloping the chromosome.

Chromonema : When the nucleus is resting, the chromosomal material will be in the form of a network called *chromonema*. At the time of cell division these break up and condense to form chromosomes. The metaphase chromosomes have two thread like structures called chromatids. The two chromatids are held together at the region of primary constriction (for details see later in the same chapter). Treatment of chromosomes with a protein hydrolysing enzyme (trypsin) will reveal each chromatid to have two *subchromatids*, but there is no unanimity on this. Various chromosome models have proposed chromonema coils ranging from 2 to 32. The chromonema coils (subchromatids) were first observed by Baranetzky (1880) in the pollen mother cells of *Tradescantia*. It was Vejdovsky (1912) who First gave the name chromonema to these coils. The number of chromonema fibrils not only varies with different species, but at different growth stages in the same cell.

The chromonemal fibrils are coiled around each other in two different ways. These are *paranemic coils* and *plectonemic coils.*

The paranemic coils are loosely coiled and the threads are easily separable from one another. The plectonemic coils however are intertwined very closely so that they are not easily separable.

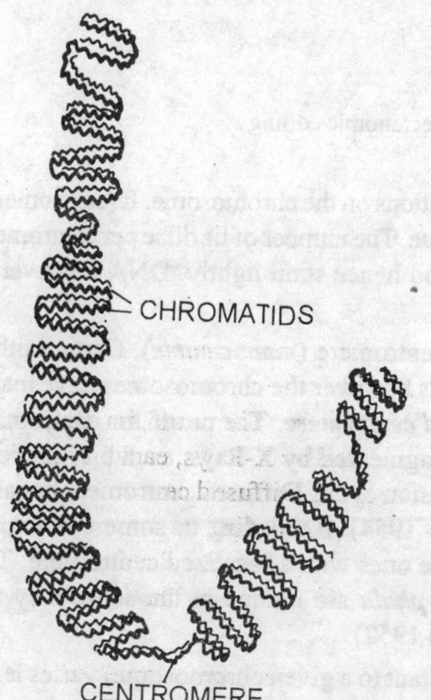

Fig. 5.2 Chromosomes
Internal structure of a metacentric chromosome

The extent of coiling in the chromonema fibrils during cell division depends on the length of the chromosomes. These coiling are of three categories-

Major Coils : Each coil consists 10-30 gyres

Minor Coils : These are at right angles to the major coils and have many more gyres than major coils

Standard or somatic coils : These are coils found in the chromonema of cells during mitosis. They are similar to major coils.

Chromomeres : The chromonema in the chromosomes particularly during mitotic prophase appears to contain alternating thick and thin regions giving a bead like appearance in a chain. The thick bead like regions are called *chromomeres* and the region in between the two chromomeres is called the *interchromomere*. The position of chromomeres in the chromonema appears to be specific to a given chromosome in an individual.

What are chromomeres? Cytologists have given various interpretations; while some regard it as the regions of condensed nucleoplasmic material, others interpret it as regions of superimposed chromonemal coils. This view is supported by electron microscopic observations. Previously the chromomeres were mistaken for genes.

The chromomeres were first described by Balbiani in 1876 and later by Pfitzner in 1881. Belling (1928) had equated chromomeres with the genes in liliaceous plants. But McClintock (1944) has shown that genes are located in the interchromomere region of the chromonema. However in case of polytene chromosomes the chromomeric bands are associated with certain regions.

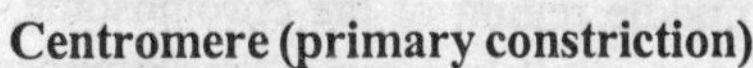

Centromere (primary constriction)

The stained chromosomes reveal a certain region or constriction which stains lightly. This is called *centromere* or *kinetochore or primary constriction*. The centromeres occupy a constant position relative to the ends of chromosomes and are responsible for the various shapes of chromosomes.

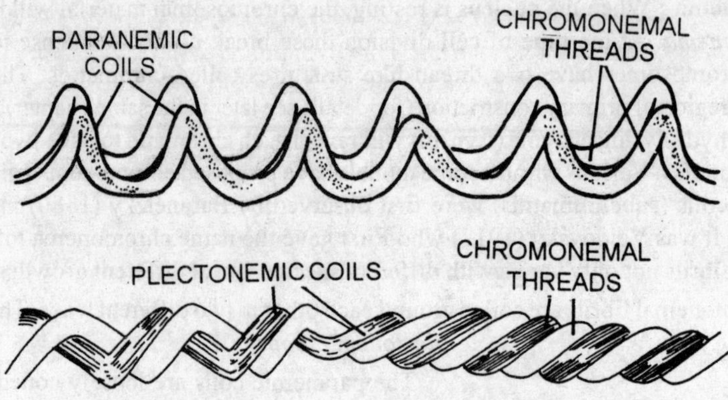

Fig. 5.3 Chromosomes
Diagrammatic representation of a paranemic and piectanomic coiling

The centromere has a diametre of 0.2 μ and may occupy various positions on the chromosome. It (centromere) consists of spherules or granules (*Kinosome*)and interchromomeral fibrillae. The number of fibrillae per centromere varies. The chromonema are less coiled in the centromeric region and hence stain lightly. DNA however is present in the centromeric region (Lime-de-Faria 1955).

Most of the organisms have their chromosomes with a single centromere (*monocentric*). Certain other chromosomes have two centromeres (*dicentric*). In hemipteran insects however the chromosomes have many centromeres (polycentric). Such a centromere is said to be a *diffused centromere*. The proof for centromere being diffuse comes from the fact that when such a chromosome is fragmented by X-Rays, each bit functions like a typical chromosome, replicates itself and participates in the division cycle. Diffused centromere has also been reported in *Luzllla purpurea* of the family Juncaceae (Brown 1954). According to some cytologists chromosomes with a diffused centromere are more primitive than the ones with a localized centromere. The polycentric chromosomes found in the nematode *Ascaris megalocephala* are unique in the sense they are restricted to only those cells which constitute the germ line (Swanson 1957).

The position of the centromere in the chromosome (which is constant to a given chromosome) varies ie., it may occupy different positions. Based on this,four morphological shapes have been identified in chromosomes. These are -

1. Metacentric

The Centromere occupies a middle position with reference to the length of the chromosome. The two arms thus resulted are almost equal in length. During anaphasic movement in cell division metacentric chromosomes appear 'V shaped, eg: *Trillium, Tradescantia* etc.

2. Sub metacentric

When the centromere is located some distance away from the middle region of the chromosome, the position is said to be sub median and the chromosome sub metacentric. As a result, one arm of the chromosome will be shorter than the other. During anaphasic movement, sub metacentric chromosomes appear 'L' shaped eg: Human beings.

3. Acrocentric

In this case, the centromere is situated almost near one end of the chromosome. As a result, one arm of the chromosome will be extremely short and the other one very long. The centromere is said to occupy a subterminal position, Eg: Grass hoppers.

4. Telocentric

When the centromere issituated exactly at one end, the hromosome will be having only one long arm. Telocentric chromosomes are very rare. Truly telocentric chromosomes have been identified by Marks (1957) in certain plants, protozoa and certain mammals. Cleveland (1949) has also reported in certain protozoa inhabiting the digestive track of woodtermites,theoccurrence of telocentric chromosomes. The earlier report regarding chromosome IV of *Drosophila melanogaster*, being telocentric has been disproved as a second arm has been seen.

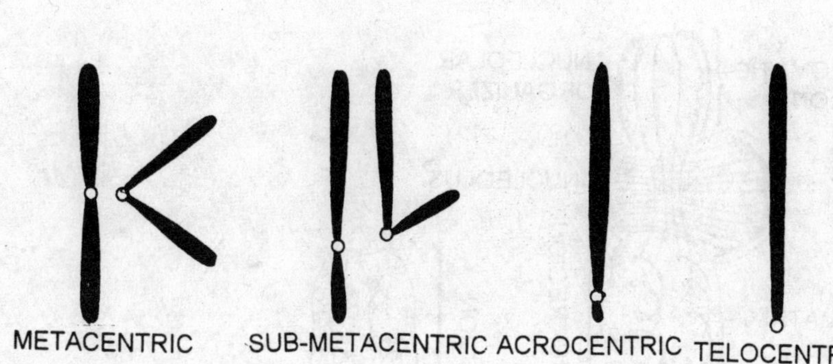

Fig. 5.4 Chromosomes
Types based on centromere position

Acentric chromosome

In certain instances, chromosomes may undergo fragmentation resulting in one fragment without a centromere. Such a chromosome fragment is said to be *Acentric*.

Arm ratio

The ratio of length between the long arm and short arm is said to be the arm ratio, which is constant for a given chromosome. The arm ratio is less in metacentric chromosomes and high in acrocentric chromosomes.

Centromeric chromomeres

The centromere should not be mistaken or equated with a chromomere as the former stains lightly and is a constriction while the latter is a swelling on the chromosome found throughout its length. Metaphase chromosomes which have two chromatids exhibit in their centromere four granular structures. These granules called *centromeric chromomeres* are about 0.5 μ in size and are arranged in a square.

Functions of centromere

1. Centromere provides mobility to the chromosome. Chromosomes lacking in centromere fail to orient themselves properly on the anaphase plate, lag behind and finally get eliminated.
2. Centromeres decide the shape of the chromosome.
3. Centromeres attach themselves to the spindle fibres during anaphasic movement.
4. The centromere divides along the longitudinal axis of the chromosome at the beginning of anaphase. Very rarely the centromere may divide at right angles to the chromosome giving rise to the pieces of centromere, each bit holding the arms parallel to each other and genetically alike (because of duplication) as they represent the replicated parts of a single chromatid. Such chromosomes with genetically identical arms (having same genes on both arms) are called *isochromosomes* (Darlington 1939).

Secondary constrictions

These are lightly staining areas on the chromosome other than centromere. One or more secondary constrictions may be found in the chromosomes. The part of the chromosome (arm) present beyond the secondary constriction is called a satellite. There may be one or two satellites depending on the number of secondary constrictions.

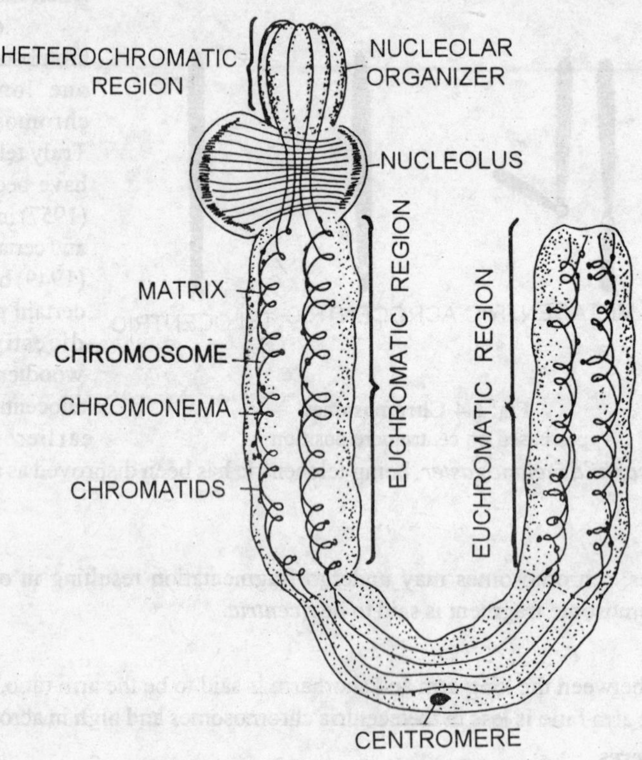

Fig. 5.5 Chromosomes

Structure of a typical chromosome showing a nucleolar organiser

Chromosomes with satellites are called *SAT- Chromosomes* (sine acid thymonucleinico - without DNA). Generally in a genome atleast one chromosome will have a satellite.

Secondary constrictions associated with the nucleolus are called *nucleolar organisers*. Usually in a diploid organism two homologous chromosomes have additional constrictions where nucleolus will be located during the interphase stage of the nucleus. During prophase, the nucleolus disappears to appear again in the identical area. The nucleolar organising region is best seen during the pachythene stage and it is heteropycnotic in maize (McClintock). In polyploid organisms, nucleolar organisers will be more than two in number. Even though it was believed earlier that the nucleolar organisers lack DNA, subsequent studies have shown that it has about 0.3% of nuclear DNA. The gene loci at this region are responsible for the formation of 28*s* ribosomal RNA.

Telomeres : The terminal part of chromosomes are called telomeres which have polarity, preventing the tips of other chromosomes from fusing with it. Telomeres however attach themselves to the nuclear envelope during interphase.

CHEMICAL COMPOSITION OF CHROMOSOMES

The two main chemical components of the chromosome are nucleic acids and proteins.

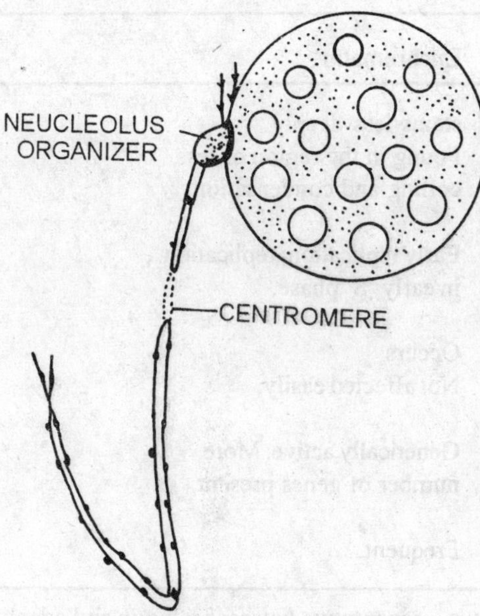

Fig. 5.6 Chromosomes
Nucleolar organiser (enlarged)

Proteins : Besides nucleic acid, there are three types of proteins in the chromosomes. These are *protamines, Histones and non histone proteins.*

Protamines are basic proteins with a molecular weight less than 4,000. They are rich in amino acid, arginine and are found in the spermatozoa of salmon,herring and other fish, snails and squids. The protamines are helically bound to the DNA. Each polypeptide of protamine consists of about 28 amino acids of which 19 residues are arginine.

Histones are basic proteins associated with eukaryotic nuclear DNA in the ratio of 1:1 in quite a number of plants and animals. They are the main structural proteins of chromosomes.

Based on the amino acid sequence they are categorized into five principal classes HI(H5),H2a,H2b,H3 and H4.

Histones are charecterized by the total lack of *tryptophan,* but have an abundance of *arginine* and lysine. Other modifications of histone proteins are acetylation, methylation and phosphorylation.

Functions of Histones : They act as gene repressors; but this may be a general activity rather than being specific as most of the histones have identical composition. Histones perhaps provide a structural framework for encompassing the DNA molecule.

Synthesis of Histones : In the interphase nucleus, histones are synthesized during the 'S' phase. Some histones may also be synthesized at other stages. The eukaryotic DNA has certain genes called repetitive genes coding for histones.

Non histone proteins : These are acidic proteins with a very high molecular weight when compared with that of histones. NH proteins show great variation in structure in different species. NH proteins are known to inititate genetic activity, while histones are known to depress genetic activity.

DNA, protein association in the chromosome : There are different views regarding the association of DNA with the proteins. It is possible, according to one interpretation that the protamines occupy only the small grooves in the DNA molecule, while histones occupy both the grooves (small and large). Another view is that the histones follow the two helices of the DNA molecule coiling in the same way. According to Dupraw, the histones hold together the DNA molecule in the supercoiled position.

Euchromatin and heterochromatin : Based on the staining reaction two regions may be identified in the chromatin material. These are - *Euchromatin* and *Heterochromatin.*

The euchromatin stains positively with the DNA specific stains (basic fuchsin) indicating a concentration of DNA. This region is genetically active and stains lightly. The euchromatic regions are supposed to represent areas of less condensation.

Heterochromatin stains more deeply than euchromatin and represents highly condensed regions on the chromosome. In the interphase nucleus, the heterochromatic regions form condensed structures called *chromocenters* or false nucleoli. The following table summarises the differences between euchromatin and heterochromatin.

	Heterochromatin	Euchromatin
1. Staining	Stains deeply	stains less deeply
2. Coiling	Found in the region of more coiling and condensation.	Foung in the region of less coiling and condensation.
3. Replication	Late replication replication in late 'S' phase.	Early replication; replication in early 'S' phase.
4. Acetylation	Does not occur	Occurs.
5. Effect of temperature	Pronounced	Not affected easily.
6. Genetic effect	Inert genetically but few genes are present	Genetically active. More number of genes present.
7. Cross over frequency	Less	Frequent.

Types of Heterochromatin : There are two main types viz., *constitutive heterochromatin* and adaptive *heterochromatin*. The first type shows dark staining reaction (Heteropycnosis) always and in all cells, while the second type shows heteropycnosis only in certain cells or only al certain stages in the cell cycle. Constitutive heterochromatin was originally named satellite DNA (S-DNA). It is mostly inactive during protein synthesis and is repetetive from the point of view of base sequence. Adaptive heterochromatin is metabolically inert and comprises of about 2.5% of genome.

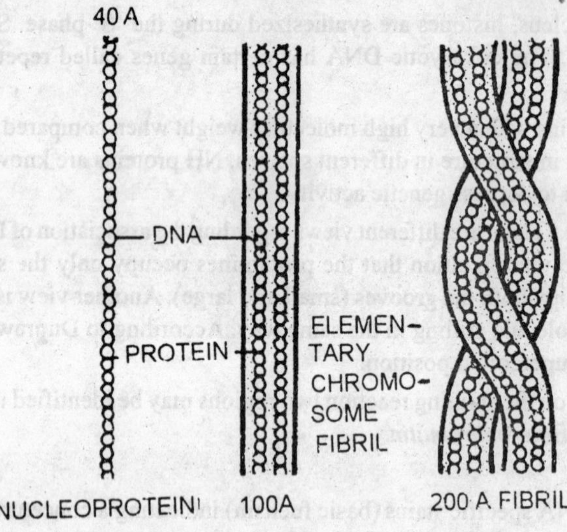

Fig. 5.7 Chromosomes
Ris' multistrand model

Heteropycnosis : This is the staining phenomenon seen in the chromosomes with reference to heterochromatic and euchromatic regions. The heterochromatic regions which stain deeply are said to exhibit heteropycnosis as against euchromatin which stain sightly. However it should be noted that the heterochromatin always does not stain deeply with reference to euchromatin. While the euchromatic region is always constant in its behaviour (non heteropycnotic) towards staining (has constant coils), the heterochromatic region may stain deeply or lightly and this is known as *differential heteropycnosis*. Differential heteropycnosis is of two types-positive and negative.

The areas of heterochromatin showing positive heteropycnosis are more condensed and deeper staining than euchromatic region. In areas of positive heteropycnosis there is more of nucleic acids and proteins. Chromosomes with positive heteropycnosis in prophase appear like metaphase chromosomes.

The areas of heterochromatin showing negative heteropycnosis are possibly less condensed and stain lighter than euchromatic region. Chromosomes with negative heteropycnosis have a hazy outline and metaphase chromosomes appear like prophase chromosomes.

CHROMOSOME MODELS

The chromosome consists of DNA and proteins, but the way in which the two are arranged in the chromosome is a matter of debate. The main aspects of the chromosome structure are a) the number of strand or strands of DNA; b) whether the DNA molecule is continuous throughout the length of the chromosome or whether it is interrupted and c) what is the mode of association of protein and DNA.

Various chromosome models have been proposed and these fall into two basic categories viz. *Multiple strand models* and *single strand models*.

Multiple strand models : According to this, the chromosome consists of several DNA and protein strands. The work of Nebel (1938) on radiation effect on chromosomes have shown that there are at least four sub units. Similarly some models suggest the existence of 4,8,16,32 or even 64 fibrils with each fibril having about 20-40 DNA molecules. The following are some of the multiple strand models-

Simple multistrand model : According to this model proposed by Steffensen (1961), the chromosome has 64 molecules of DNA arranged parallely and twisted around each other. If one agrees with this model, the replication of DNA would be a difficult process.

Ris multistrand model : In this model, the DNA and histone form a complex 40Å wide fibril. Two such Fibrils according to Ris (1961) combine to form an *elementary chromosome fibril*, two of these again combine to form a fibril which is 200 Å wide. These fibrils form the chromatids. Each chromatid has 16 to 32 elementary chromosome fibrils. But there is no experimental proof for this work. In fact Ris himself rejected this model later.

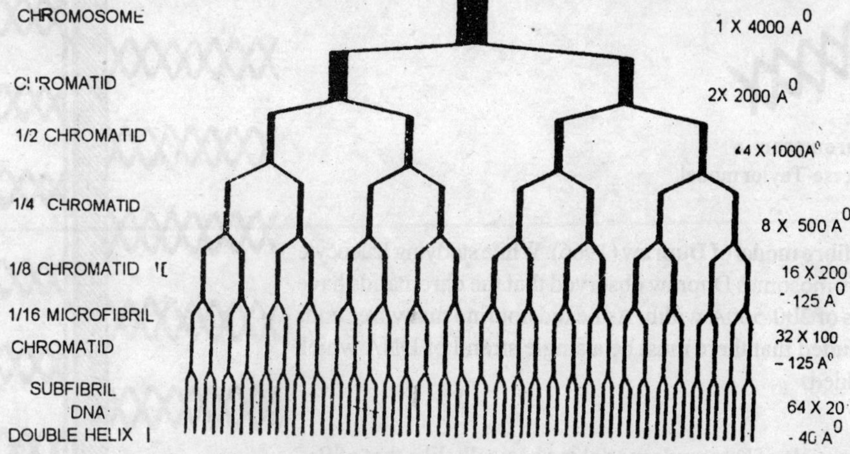

Fig. 5.8 Chromosomes
Arrangement within a chromosome based on multistrand model

Single strand models : Most of the current evidence available indicates that the chromosomes (barring the abnormal chromosome) are made up of a single strand of DNA. Experimental evidence by X-ray radiation have suggested the whole of the chromosome is composed of one single coiled molecule of DNA. The following are some of the single strand models.

The Freese-Taylor model : According to this model originally proposed by Freese (1958) and later modified by Taylor (1959), there are two chains of proteins with the DNA molecule attaching itself to these like the rungs of a ladder. The protein linkers maintain the shape of the DNA molecule. The two protein chains get closely appressed in the chromosome resulting in the coiling of DNA.

The original model suggested by Taylor (1957) is called the *side chain model* or *centipede model*. According to this model, the chromosome consists of a central backbone of protein from which the loops of DNA spring up like the legs of a centipede. According to him, replication is possible when the protein backbone made up of two chains, separates longitudinally into two and in the process separates also the DNA coils, with each strand having a single strand of DNA. But this is not in conformity with the linear arrangement of the genes on the chromosome.

Nebel's coiled coil model : According to this, the chromosome is made of a single fibril which is coiled in a complex manner.

Ris' modified model : After discarding his earlier multistranded model Ris (1967) suggested, that the chromosome fibril (see his earlier model) is folded to form a 100Å wide nucleo - histone fibril. The nucleohistone fibril folds because of calcium bridges to form a 250Å wide *basic fibril.* Further folding of the basic fibril forms the chromosome.

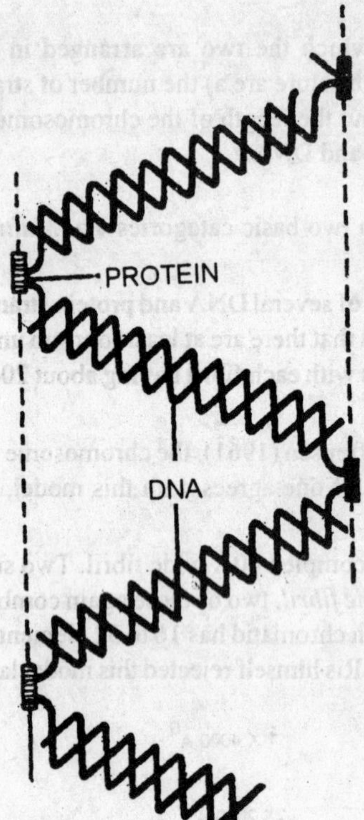

Fig. 5.9 Chromosomes
The Freese-Taylor model

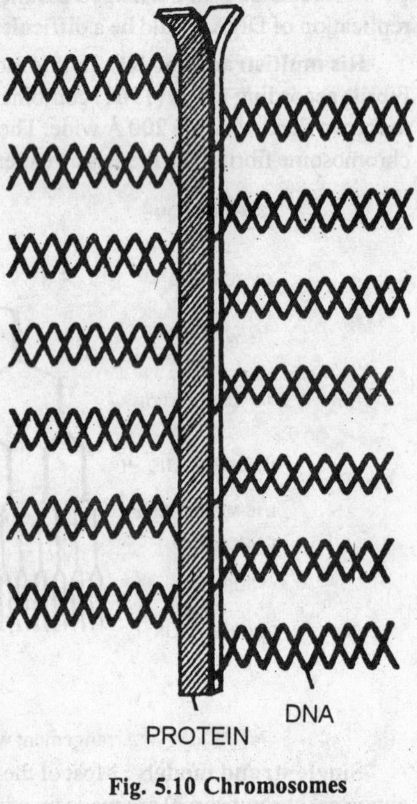

Folded fibre model of Dupraw (1966). While studying leucocyte (human) chromosomes Dupraw observed that the chromatids have folded fibres of 200-500Å width. As he did not find many free ends it was concluded that there must be a single strand of DNA which is highly folded.

The essentials of Dupraw's model are basically like that of Ris' model with some variations. According to this, the 20Å DNA is packed in protein to form a fibril. The fibril gets coiled and forms a 100Å wide *type A fibre.* The DNA is packed in this fibre in the packaging ratio of 6:1 (packaging ratio is the ratio between the condensed length and the extended length). The type A fibre is coiled in the ratio of 10:1 to form type B fibre which is 200-250Å wide. The type B fibre is extensively folded to form the chromatid. This has been supported by many workers (Me Dermot 1968, Fedric 1969, Lamport 1969 etc). Though many scientists have supported Dupraw's view, White, M.J.D (1973) strikes a discordant note when he does not agree with the folded fibre model and he considers it extraordinary that the DNA of all the chromosomes are linked in the form of a ring. Dupraw's model is now regarded as untenable in view of the fact that the DNA itself forms a wrap around the histones to form *nucleosomes*.

Fig. 5.10 Chromosomes
The centipede model of Taylor

Nucleosome model : This model proposed by Woodcock (1973) suggests that the chromatid is a string of

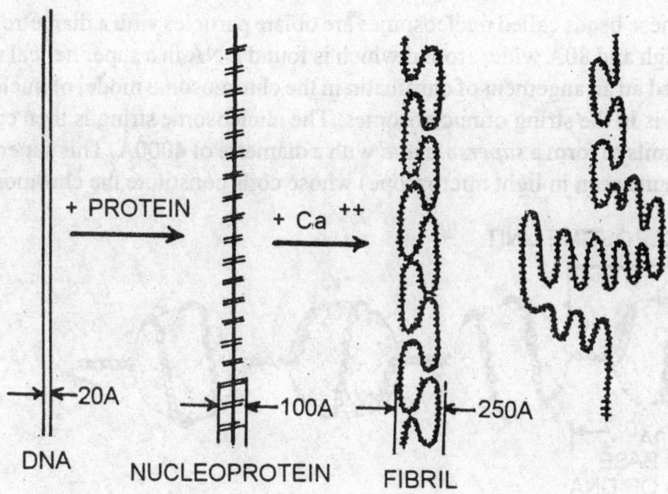

Fig. 5.11 Chromosomes
Ris' modified model

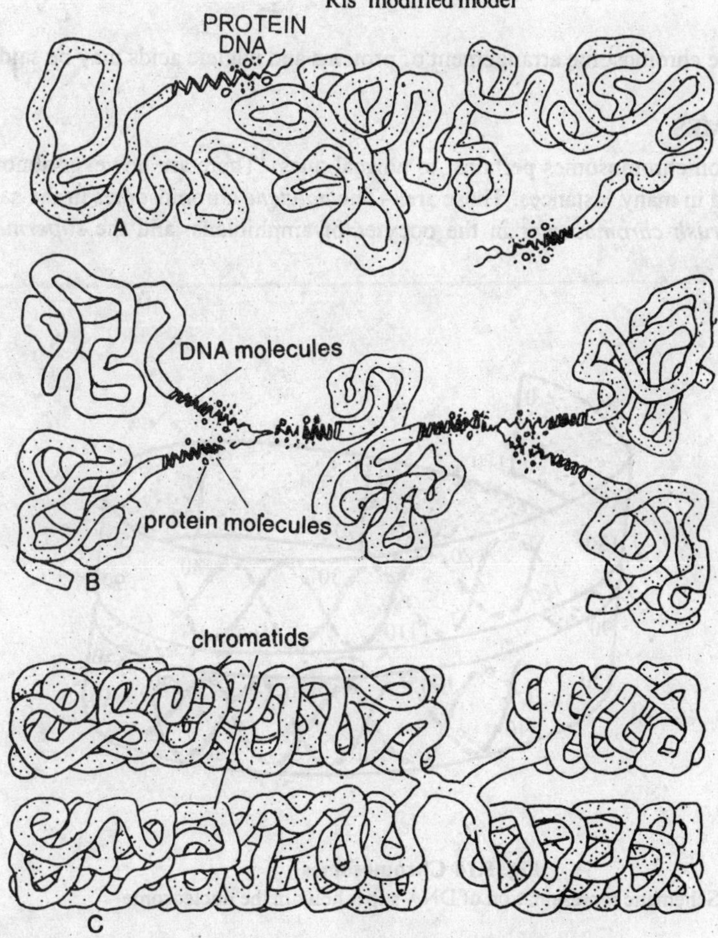

Fig. 5.12 Chromosomes
Folded fibre model of Dupraw

beads made up of repeating units. These beads called nucleosomes are oblate particles with a diametre of 110Å. They have a core of histones 40Å high and 80Å wide, around which is found DNA in a super helical structure. Finch and Clug (1976) have proposed an arrangement of chromatin in the chromosome model of nucleosomes. The coiled DNA according to them is in the string of nucleosomes. The nucleosome string is then coiled into 300Å wide solenoid, which in turn coils to form a *supersolenoid* with a diametre of 4000Å. This super solenoid structure is the unit fibre (chromonema seen in light microscope) whose coils constitute the chromosomes.

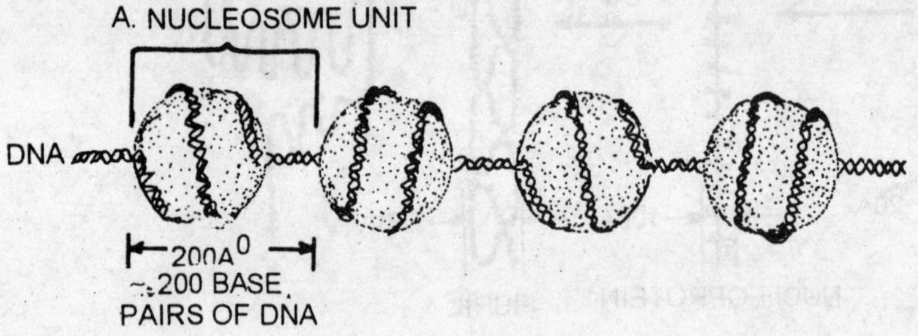

Fig. 5.13 Chromosomes
Nucleosome model

The exact nature of the chromosome arrangement of proteins and nucleic acids may be said to be still an open question.

UNUSUAL CHROMOSOMES

The above discusion on chromosomes pertains to normal ones. There are however abnormal or giant chromosomes encountered in many instances. These are – The *polytene* chromosomes in the salivary glands of fruit flies, the *lamp brush chromosomes* in the oocytes of amphibians, and the *supernumerary* or *B chromosomes*.

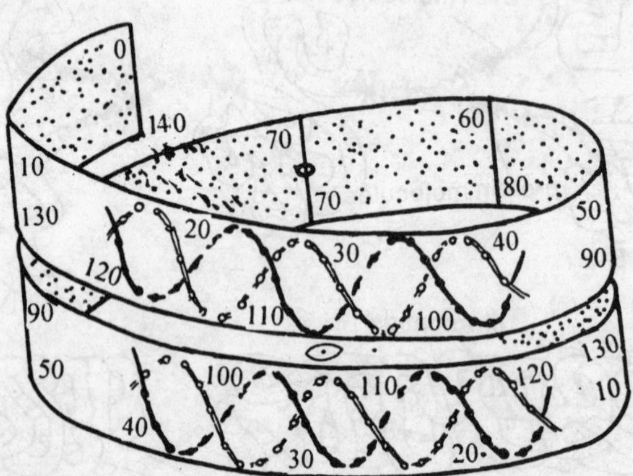

Fig. 5.14 Chromosomes
Schematic representation of DNA super helix in the nucleosome

Polytene chromosomes : These multithreaded chromosomes were first reported by Balbiani in 1881 while observing the squash of salivary cells of *Chironomous*. Subsequently they have been seen in the larvae of Dipterans (*Drosophila* etc). The name polytene was suggested by Kollar due to the appearance of many threads. Besides salivary glands the polytene chromosomes also occur in rectal epithelium and maiphigian tubules.

The nuclei in the salivary gland cells are much larger than the normal ones and are about 25 μ in diametre with the giant chromosomes larger by 50-200 times than the normal chromosomes. The polytene chromosome in the salivary cells of *Drosophila melanogaster* (Painter 1933) are about 200 μ in length. The enormous size is due to the duplication of chromonema which do not separate. According to an estimate, the polytene chromosomes have 1000 times more DNA than the normal somatic chromosomes.

Structure : Each polytene chromosome is transversely striated by alternating *bands* and *inter band regions*. In *Drosophila*, over 5000 bands have been found in the four chromosomes of the salivary gland cells. The bands take an intense stain (Feulgen) while the inter band regions stain less intensely.

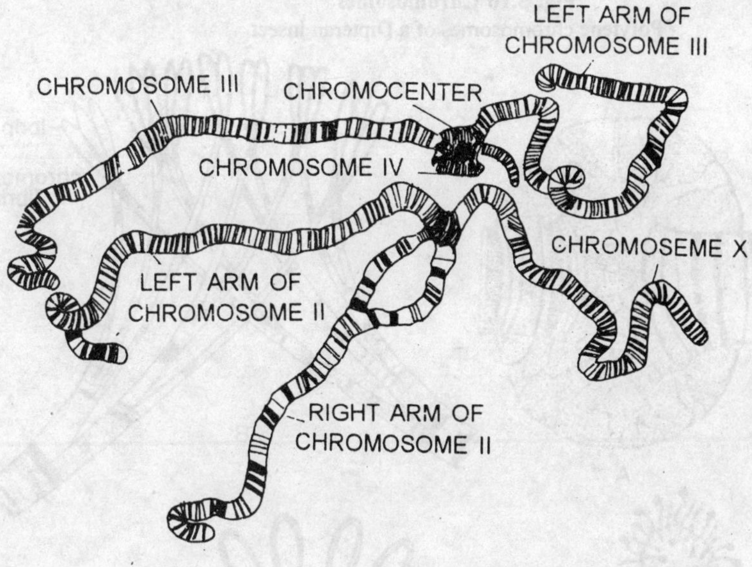

Fig. 5.15 Chromosomes
Polytene chromosomes of drosophila

The bands have chromomeres arranged at right angles to the long axis of the chromosome. The bands actually represent regions of tight coils of chromonema Van Herwerden (1911) interpreted these bands as *gyres* of a continuous spiral. This view was later supported by Kostoff (1930). The bands have a high DNA content and absorb U.V. light at 2600Å.

The interbands stain lightly and they are negative to feulgen stain indicating a low concentration of DNA. The chromonema in the interband region are less coiled and absorb very little U.V. light.

Painter (1933) and Bridges (1936) have shown that in *Drosophila* the bands are clearly associated with the genes. Association of certain defects due to mutated genes have also been associated with the bands. Certain recent studies however have shown that together with the bands, the interband regions also have genes.

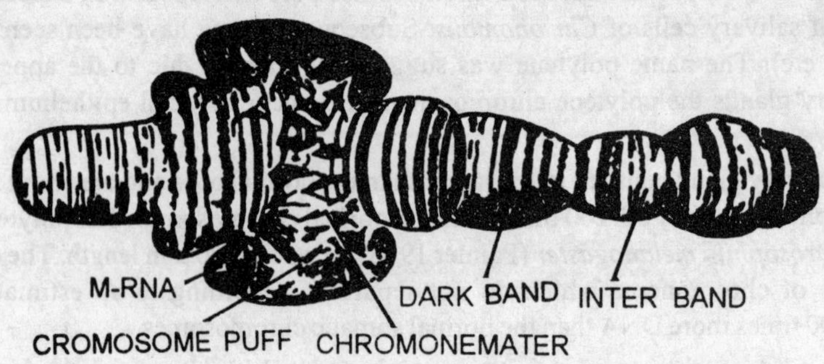

Fig. 5.16 Chromosomes
Polytene chromosomes of a Dipteran insect

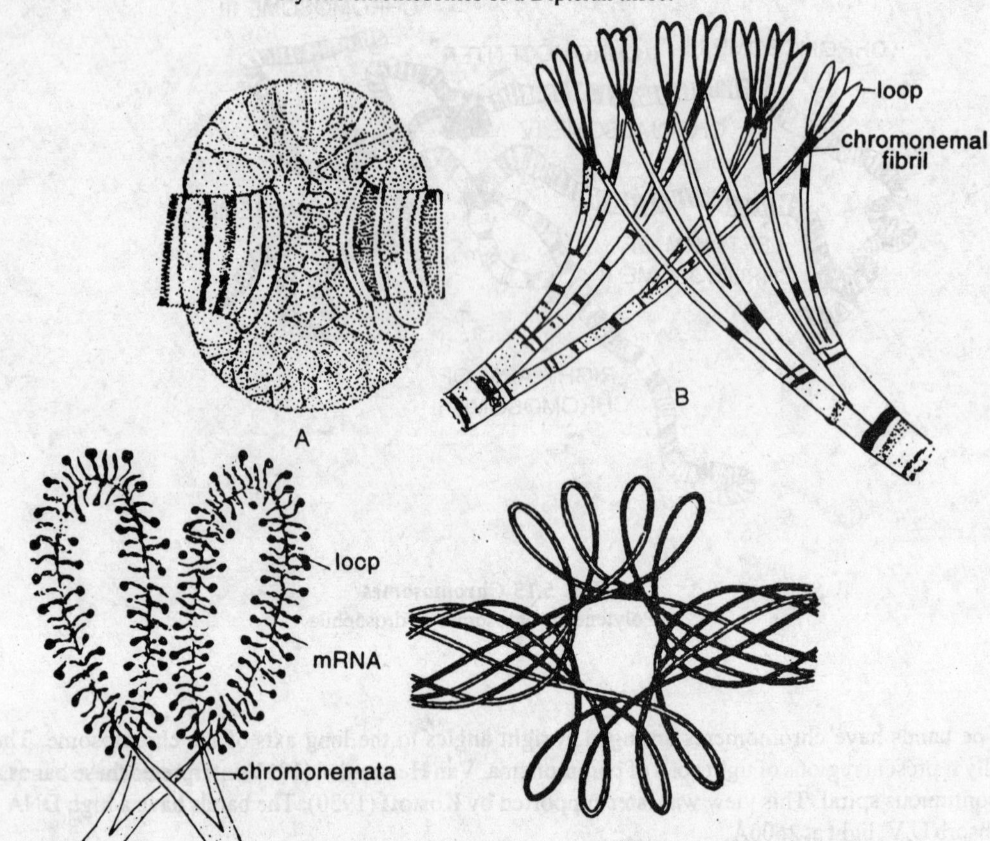

Fig. 5.17 Chromosomes
Balbiani ring in a polytene chromosome

Chromosome puffs : These are certain interesting structures associated with the bands in the giant chromosomes called *Chromosome puffs* or *Balbiani rings*. The swellings in bands at certain stages are given the name chromosome puffs. It is quite possible that these puffs are associated with the metabolic activites and represent areas of active RNA synthesis. According to Beerman and Bahr (1954), the coils in the puff region open out to form many loops. The larger chromosome puffs are called *Balbiain rings* and they show a high DNA and mRNA content.

Lamp brush chromosomes : The giant chromosomes found in the oocytes of fish, amphibians, reptiles and aves look like the brush used to clean the chimney of a kerosene lamp-hence the name *lamp brush chromosomes.* These are very large and are characterised by loops arising out of the main axis laterallv. Discovered first by Flemming is 1882, in amphibian oocytes, they were studied in great detail by J. Rucket (1892) in the oocytes of shark. Rarely lamp brush chromosomes have also been found in some invertebrates (Sepia, Echinaster, insects etc) but they are not as large as those of vertebrates. It seems they are present in certain plants also (Grun 1958).

In some organisms, lamp brush chromosomes as long as 1000 microns and 20 microns wide have been reported during early prophase I. Towards the end of the prophase, they get reduced in size. There are reports of lamp brush chromosomes as long as 5900 microns in some salamander oocytes. These chromosomes are elastic and can be stretched.

Structure : There are two main parts viz., the main axis and lateral loops.

The main axis has two chromatids (the homologues have four) and is rich in DNA. The axis is in continuity with the loop axis. The tips of the axis are without loops and are the regions of telomeres. The centromeric region is also without loops. At certain regions along their length the chromonema of the axis get coiled and form the chromomeres. The chromomeres are found in pairs. Electron microscopic studies of the lamp brush chromosomes of the salamander *Triturns* (Miller and Beatty (1969) have shown the axial Fibre containing dense granules which are probably the molecules of the enzyme RNA *polymerase*.

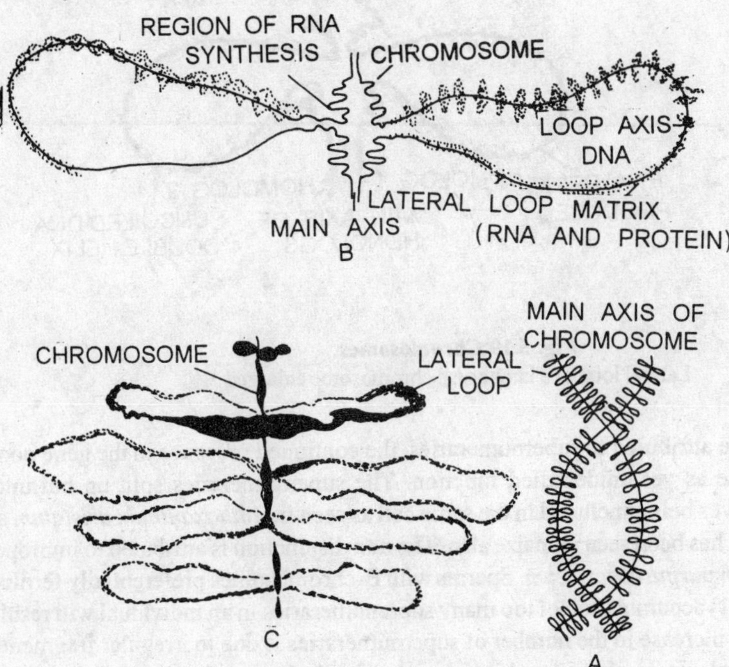

Fig. 5.18 Chromosomes

Lamp brush chromosomes, **A.** Paired homologues with two chaismata, **B.** Variation in loops, **C-A.** Chromomere with an attached loop.

The loops are of two main types viz. *typical* and *special*. Most of the loops which are typical consist of a bent central axis from which are given out RNA fibrils. These RNA fibrils are of progressively increasing length as a result one side of the loop is pronouncedly thick. The special loops are symmetrical and have granules at the end of RNA fibrils. Each loop axis has an axial fibre covered with a matrix. The diametre of the loop axis varies from 30-50Å. The axial Fibre is made up of DNA, while the matrix has RNA and proteins.

Lamp brush chromosomes are involved in the synthesis of RNA and proteins. Each loop is believed to represent one long operon consisting of repetetive cistrons (structural genes). Each gene locus codes for RNA. The loop is supposed to synthesise at high rate because of repetetive gene sequence. There are also reports that the lamp brush chromosomes help in the formation of yolk material in the egg.

Super numerary or B-chroroosomes : In many plants and animals, the cells contain in addition to the normal complement, one or more extra chromosomes called supernumerary or accessory orB- chromosomes.

These were first discovered in Hemipteran insects (Wilson 1905); since then they have been found in a number of insects and plants.

The supernumerary chromosomes are generally smaller than the others and are genetically inert hence produce little or no phenotypic expression. This suggests that these chromosomes may largely be heterochromatic which is supported by their differential staining activity. But this is not universal, for in *Tradescantia* they are entirely euchromatic, while in maize they are part euchromatic and part heterochromatic.

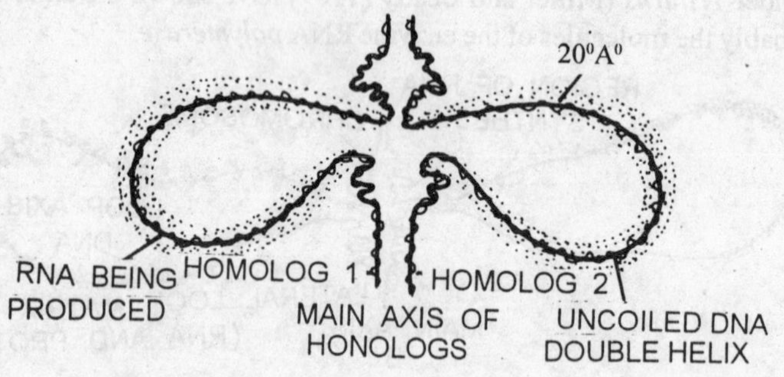

Fig. 5.19 Chromosomes
Lateral loop of a lampbrush chromosome enlarged

While no function can be attributed to supernumeraries, the continued presence in the gene pool suggests that they may perform some as yet unidentified function. The supernumeraries split up but undergo non disjunction thus both the halves being included in the same cell as seen in *Anthoxanthum aristatum* and *Secale cereale*. A similar behaviour has been seen in maize also. The non disjunction is attributed to improper division of the centromere in *Sorghum purpureosericeum*. Sperms with B-chromosomes preferentially fertilize the egg. According to Randolph (1941) accumulation of too many supernumeraries in an individual will result in loss of vigour as seen in maize. The increase in the number of supernumeraries is due to irregular fragmentation. The origin of supernumeraries in plants is unknown, while in some animals they can be traced to the Y chromosome (Wilson 1905). This is the case in *Metapodius*, where the chromosomes have a diffused centromere, hence any fragment can survive.

6

CHROMOSOMAL ABERRATIONS

The chromosomes - often called vehicles of heredity are sufficiently characteristic to provide the cytologist morphological criteria of an individual that can be used for identification much in the same way as a taxonomist uses phytographic (morphological characters of organs) characters to delimit the species. The constancy of karyotype is due to the fact-as Swanson (1957) observes "at a given stage of cell division and in a given tissue each cell of an organism has a constant number of chromosomes of reasonably definite volume, length and shape........" This constancy of chromosome number, volume etc., is determined in turn by the number arrangement, and sequence of genes in a definite order.

It becomes clear from the above discussion that constancy in chromosome number and structure are very essential for the individual not only to maintain its morphological characters, but many a time even for its own survival. Occasionally however, there occur variations in structure as well as number in the chromosomes producing alterations variations etc., in the individual. Depending on the magnitude of these changes, the individual undergoing these changes either will survive with little or no change or may barely survive with great changes. In some instances however, the changes are of such a magnitude that they may be lethal to the individual. These changes are called *chromosomal aberrations.* Very often these aberrations are also called *chromosoma mutations,* as they bring about change in genetic characters of the individual. Chromosomal mutations or aberrations should not be mistaken with gene mutations as the latter are involved with single genes, while the former affect either segments of chromosomes or sometimes entire chromosomes or even the entire set of genome (the total chromosomes present in an individual).

In natural populations, the chromosomal aberrations occur spontaneously but less frequently. In grass hoppers about one in a thousand individuals is known to carry some form of chromosome rearrangement. Naturally occuring chromosomal abberrations are found in human beings also. After the initial discovery of a naturally occurmg aberration, geneticists learnt that they can induce these aberrations artificially by using mustard gas, X ray, colchicine, etc.

Of what use are the aberrations to a geneticist? Chromosomal aberrations in an individual in comparison with the normal chromosomes provide a wealth of data to geneticists to understand the structure of a chromosome and also to compare any morphological expression with a particular segment of the chromosome. If the changes brought about are not good, the geneticist may try to find a way out to overcome these aberrations. On the other hand if these aberrations produce a new desirable trait, he may learn to produce those aberrations artificially.

For a cytologist, chromosomal aberrations, particularly structural are a gold mine as he can study by camparison (with the normal chromosome) the arrangement of genes, their order, frequency etc. It helps him to map the chromosome and locate a gene in a particular segment of the chromosome.

On a broad classification, chromosomal aberrations are of two kinds-*Structaral* and *Numerical.* Structural aberrations bring about a change in the chromosome structure in terms of length, shape or sometimes in the form of altered, or loss of genes and their arrangement. When the chromosomal changes affect a large segment, the changes may even be visible morphologically. Numerical changes on the other hand do not affect the structure of the chromosome, but are only involved in alteration in the number which may be either increase or decrease when compared with the normal number. This change also has many drastic genetic influences on the individual.

STRUCTURAL ABERRATIONS

As has been pointed out already, any change in the structure and architecture of the individual chromosome comes under the perview of structural changes.

The structural aberrations are of the two following categories -

(i) **Intra chromosomal aberrations**

(ii) **Inter chromosomal aberrations.**

The alterations of the genes or gene sequences influencing only one chromosome or taking place in a chromosome are called Intrachromosomal aberrations. This should not be mistaken to mean that more than one chromosome will not undergo alteration, but it means to say that all the events leading to a particular aberration take place within a chromosome.

Inter chromosomal structural aberrations on the other hand involve at least two chromosomes with reference to an aberration. Both intra and inter chromosomal aberrations are further classified as the chart below shows.

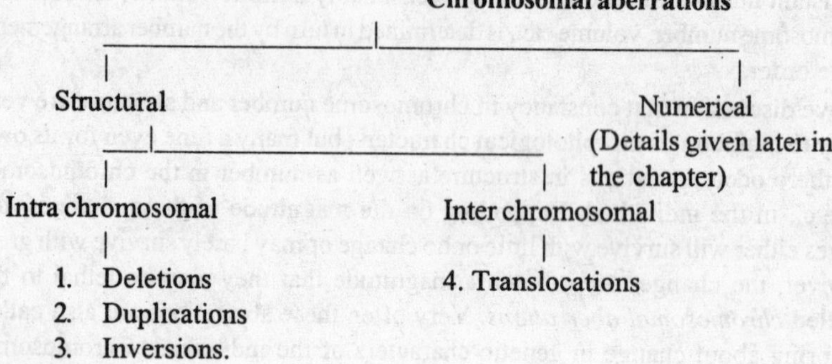

DELETIONS

Also known as deficiencies, these involve loss of a segment or loss of a portion of the chromosome, but the entire chromosome is never lost. The deleted portion which comes out of the remainder of the chromosome will not survive if it does not have a centromere. But in cases of chromosomes with a diffused centromere the fragment or fragments that come out of the chromosome do survive and this leads to increase in the chromosome number, but no loss of genes. The portion of the chromosome with the centromere (in the case of chromosomes with a single centromere) functions as a deficient chromosome.

How does a deletion occur? A break in the chromosome at any one point results in the tearing away of a part of the chromosome. This break may occur naturally or by the influence of external agencies like chemical gas, ionising radiation, drugs etc.

Types of Deletion

Two types of deletion have been noticed in organisms. These are *Terminal deletion* and *Interstitial deletion* (Intercalary deletion).

In a terminal deletion, a break occurs in a chromosome a little away from one end and a segment comes out. For instances if a chromosome has gene loci ABCDEFGH, the deleted portion will be GH and the chromosome will now have only ABCDEF. The terminal deletion produces two raw ends-in the main chromosome. The raw end heals up, while the deleted segment is lost in the next cell division as it lacks a centromere.

In an interstitial deletion, a segment comes out of the chromosome from the middle. This is brought about by two breaks at two points in the chromosome, removal of an intercalary segment and fusion of the two broken ends in the original chromosome. For instance, in a chromosome with gene loci ABCDEFGH, there will be two nicks-one between C and D and the other between E and F, with the result a segment of the chromosome carrying gene loci DE comes out, the two ends-at C and F unite to form the deleted chromosome ABCFGH.

Between the two types of deletions, terminal deletions are generally rare as they involve the loss of a telomere. The healing end of the chromosome must be such that, it should not fuse with any other chromosome.

Deletion Heterzygote

Deletion may affect one or both of the homologous chromosomes in a diploid organism. If both the homologues lose an identical segment, it will be known as deletion homozygote or as it happens quite frequently, one of the homologues may remain normal, while the other may undergo deletion, such an individual is said to be a deletion heterozygote. Generally individuals with homozygous deletion will not survive.

Occurrence

Deletions have been reported from a wide variety of organisms both in the plant and animal kindgom maize (McClintok 1941), *Drosophila,* Grasshopper, man etc, where several phenotypic traits have been associated with deletions.

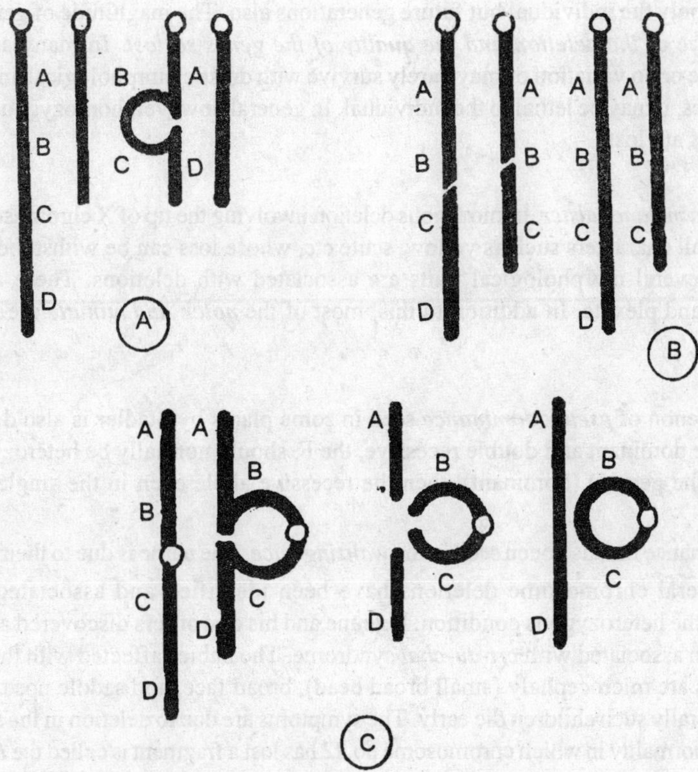

Fig. 6.1 Chromosomal Aberrations

Deletions A. Intercalary deletion and passing in a deletion heterozygote,

B. Terminal deletion and pairing in a deletion heterozygote,

C. Deletion and origin of ring chromosome

Cytological identification

A homozygous deletion (if the individual survives) is difficult to identify morphologically at the chromosome level. However if the normal chromosome has been studied previously, the deleted chromosome can be pointed out because of its reduced length and also change in shape, since a terminal deletion in a metacentric chromosome would reduce the length of one arm and produces a sub metacentric chromosome.

Heterozygous deletion on the other hand is easy to identify cytologically during meiosis, when the normal and the deficient homologues undergo pairing. The pairing homologues produce specific configurations and can be easily located. In the case of a terminal deletion, the deleted chromosome will be shorter in length, hence there is no pairing in that portion and one can see a long and short chromosome undergoing linear pairing with the terminal part of the normal chromosome left out of pairing. In the case of interstitial deficiency, the normal chromosome forms a loop in that portion (of its chromosome) where the genes are lost in the other homologue.

Deletions can be easily identified by visual observation in instances like salivary gland chromosomes of *Drosophila,* where a deletion may lead to the loss of several bands. The same thing is possible in lampbrush chromosomes also where several specific loops will be lost.

Genetic consequences of Deletion

A deletion is involved in the loss of genetic material, hence it is imperative that there will be consequences which affect not only the individual but future generations also. The magnitude of genetic effect will however depend on *the size of the deletion and the quality of the genes so lost.* In many an instance the organism survives with little or no variation or may barely survive with drastic morphological and functional alteration or as in extreme cases, it may be lethal to the individual. In general however, homozygous deletions rarely survive as both the alleles are lost.

In *Drosophila melanogalster,* homozygous deletion involving the tip of X chromosome can survive provided they are very small characters such as yellow, scute etc, whose loss can be withstood without much problem. In *Drosophila,* several morphological traits are associated with deletions. These are blond, pale, beaded, carved, snipped and plexate. In addition to this, most of the *notch and.minutte* are also known to be due to deletions.

The phenomenon of *pseudodominance* seen in some plants by Stadler is also due to deletion. In a cross between a double dominant and double recessive, the F_1 should normally be heterozygous dominant. But if a deletion affects the gene D (dominant), then the recessive allele even in the single dose will express itself morphologically.

Pseudodominance has also been seen in the *waltzing mice.* The name is due to their (rat's) erratic movement.

In man, several chromosome deletions have been identified and associated with many congenital abnormalities in the heterozygous condition. Lejeunc and his coworkers discovered a chromosome deletion in man that has been associated with *cri-du-chat* syndrome. The babies affected with this disorder cry like cats. Other characters are microcephaly (small broad head), broad face, and saddle nose. There is general mental retardation. Generally such children die early. The symptoms are due to deletion in the short arm of chromosome no.5. Another abnormality in which chromosome no.22 has lost a fragment is called the *Philadelphia chromosome* named after the city where it was discovered by L.M. Tough and his collegues. This deletion brings-about chronic granulocytic (myeloid) leukemia and is identified with the loss of a part of long arm of chromosome no 22 as identified in bone marrow cells.

Deletion and chromosome mapping

In *Drosophila,* deletions have been employed to locate genes on the salivary gland chromosome. First of all a correlation is made between the loss of a band and the occurence of a particular morphological trait. For instance the red eye color is dominant over white eye. When the Band 302 is lost due to deletion, the eye color becomes white in the next progeny due to the loss of dominant allele, hence the white eye colour. It appears then that the loss of band 302 is near or on the 302 band. In the same way the *notch* wing appears, when ever the band 30, is lost due to deletion. In this way deletion can be employed as markers to locate the genes on the chromosome.

Evolutionary importance of Deletions

It is highly unlikely that deletions have a profound evolutionary impact, since the loss generally diminishes the potentiality of the organism. But theoritically speaking, if deletion removes some genes that suppress other genes, which can then express themselves, it may lead to new and desirable phenotypic variations being prefered by natural selection. In annual herbs, the loss of chromatin material is known to lead to specialization like adaptation to a habitat, to an insect pollinator etc. In *Crepis* (Comopositae) the loss seems to be restricted to only heterochromatin material.

DUPLICATIONS

This is another kind of structural chromosomal aberration affecting a single chromosome. In the affected chromosome, a segment of the chromosome containing several gene loci duplicates itself so that the chromosome has those genes in double dose. Duplication may occur at the tip of the chromosome or in the middle. In some instances the duplicated portion may attach to the chromosome *i.e* the same sequence or in a reverse sequence. Usually terminal duplications are rare due to the presence of the telomere. The following are the types of duplications in a chromosome with a normal gene complement ofABCDEFGHI. Let us assume that the segment DEF is duplicated.

Tandem duplication

In a tandem or sequential duplication the duplicated segment is attached to the chromosome in the same locus and in the same sequence. In the present example, the chromosome with tandem duplication will have the gene sequence *ABCDEFDEFGHI*.

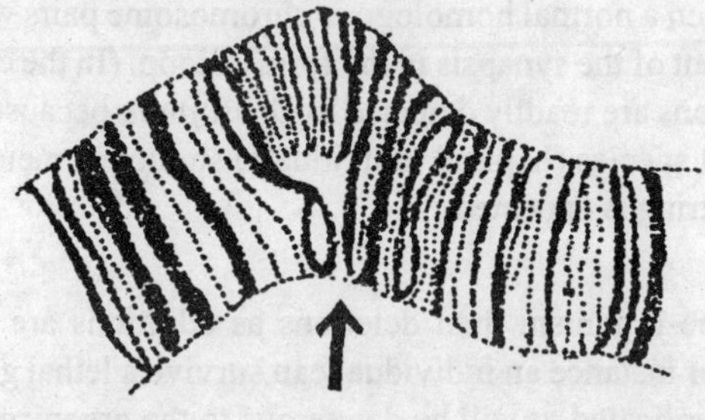

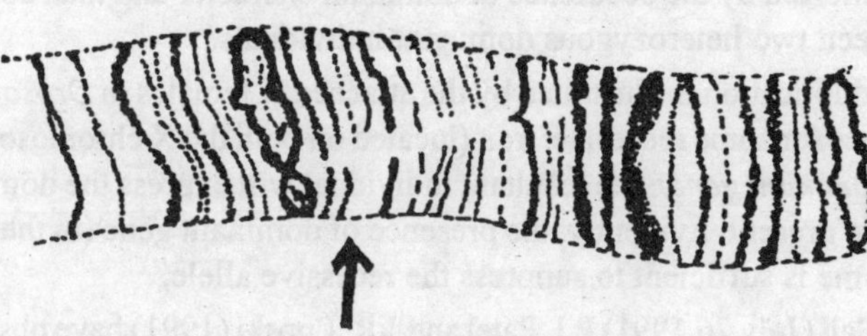

Fig. 6.2 Chromosomal Aberrations
Diagram showing loss of certain bands in the chromosome due to deletion

Reverse tandem

Here the duplicated part will attach itself to the chromosome at the same locus but in a reverse sequence. For e.g. *ABCDEFFEDGHI*

Displaced duplication

Here the segment DEF (duplicated portion) will attach to the same chromosome but in a different locus other than its original position. For e.g. RSTDEFUVW. Since the duplicated segment retains the sequence it is called tandem, and since it joins the chromosome in a different locus it is called *displaced tandem.* In a displaced reverse tandem duplication, the gene sequence would be RSTFEDUVW. Here the duplicated segment joins the chromosome in a different locus in the reverse order.

The displaced duplication (reverse or tandem) may take place on the same arm *(Homobrachial displacement)* or an a different arm *(Heterobrachial displacement)*. Unusually, the duplicated segment may join another non homologous chromosome, in which case it is called *transposition.* (This is different from translocation where there is no duplication of any segment).

How duplication occurs

In a tandem duplication there is a break and the duplicated part joins the point of breakage. In a displaced duplication, a break occurs in a point releasing the duplicated part and then rejoins. In another part another break occurs and the duplicated segment is accommodated. Duplication may also be produced by unequal crossover between homologues. In the same way as in deletions, either one or both the homologues may undergo duplication. The former is called duplication heterozygote, while the latter is called duplication homozygote.

Duplications have been reported more frequently than deletions in a number of organisms. One reason for this is perhaps duplications are less deleterious than deletions.

Fig. 6.3 Chromosomal Aberrations

Types of duplications **1.** Normal Chromosome, **2.** Tandem duplication, **3.** Reverse duplication, **4.** Displaced duplication

Cytological identification

The peculiar chromosome configuration observed during meiotic pairing in a duplication heterozygote provides a demonstrable visual evidence for duplication. When a normal homologous chromosome pairs with a duplicated chromosome, the duplicated fragment remains out of the synapsis in the form of loop. (In the case of deletion the normal chromosome forms a loop). Duplications are readily detected in *Drosophila* because of large size of salivary gland chromosomes, but in some plant species (haploid individuals) studies in meiosis have indicated pairing (in duplicated segments) where no pairing is expected.

Genetic effects of duplication

Generally speaking duplications are less harmful to the organism than deletions as additions are not expected to cause drastic changes. This is not true always. For instance an individual can survive a lethal gene in the heterozygous condition, but when the lethal gene is duplicated, it will be dangerous to the organism.

While deletions can be recognised genetically by the occurrence of recessive phenotypes, when they are not expected, similarly duplications can be inferred by the occurence of dominant character and total absence of recessive phenotype even in a cross between two heterozygous dominant individuals.

The phenotypic expression produced by duplication is illustrated by the attached X females in *Drosophila.* If we consider such flies which are homozygous for some reccesive, trait (located on both the X chromosomes), and these are fertilized by a male (having a *dominant gene),* the resultant individual will express the dominant character, eventhough two recessive genes are present. Evidently, the presence of dominant gene (in the same locus) in double dose on the same chromosome is sufficient to suppress the recessive allele.

In one of the recent reports published **in Cell** (July 26,1991) P.I. Patel and J.R. Lupski (1991) have observed that in many of the patients suffering from mental disorders such as Schizophrenia and Aizhiemer's disease, there are three copies of DNA (instead of two) in certain portions of Chromosomes no. 17. They believe that the duplication of DNA sequence might be the cause for mental disorders.

Position effect

A classical example of duplication and its genetic effect has been the study of *bar eye* locus in *Drosophila melanogaster* (Bridges 1919). The wild eye with the normal shape has an average of 779 ommatidia or single eyes. Genetically it is represented by two bar regions one on each X chromosome. However on a mutation in this locus a bar eye appears. As a result of mutation the bar region in one of the X chromosomes undergoes duplication, the size of the eye is reduced and it is called *bar eye* (BB,B). The individual heterozygous for bar eye has three bar regions and has 358 facets in the compound eye. In some bar individuals one more duplication may occur in the bar region (BBB, B) producing an *ultra bar* or *double bar* having totally four bar regions, each eye reduced to about 45 facets. Sturtevent and Bridges who studied these inheritance patterns of eye shape in *Drosophila* have concluded that during crossing over occuring at different levels in the chromosomes carrying bar regions, bar eyed individuals may give rise to normal individuals or even to *ultra bar.* Ultra bar is produced when crossing over joins together three bar regions. The fly having four bar regions and is homozygous bar has both the X chromosomes with two bars. Even though there are four bars, the eye will not be *ultrabar* (it has 68 facets in comparison to 45 for ultra bar), where as in heterozygous individuals with four bars, (BBB, B) there are three bar regions produced by double duplication on one X chromosome and the other normal X chromosone has a single bar locus. This individual will be *ultra bar,* because the three bar regions are located together on a single chromosome (BBB, B) while the single bar region is on another X chromosome. In the previous instance even though there are four bar regions (BB, BB) they did not produce an ultra bar, because on any one chromosome there are only two bar regions. This is due to a phenomenon called *position effect,* which is brought about by juxtaposition of bar sections (3,1) in the same chromosome having a stronger effect than the same quantity of bar regions (2,2) distrbuted on two chromosomes.

Catcheside, has found similar position effect due to duplications in *Oenothera.*

An instance of duplication and position effect in human beings is represented by the variants in haemoglobin. Haemoglobin A having 2 alpha and 2 beta chains is the most commonly found one. Haemoglobin B differs in the beta chain being replaced by delta chains. An abnormal haemoglobin (Hb- Lepore) has 2 alpha chains and one delta and one beta chain. Balgioni who studied this, reports that unequal crossing over taking place in beta and delta loci have deprived one chromosome of both the loci, while the other got both, resulting in the production of beta and delta chains.

Evolutionary significance of duplication

In contrast to deletions, duplications are certainly of some evolutionary consequence as they can produce viable and heritable morphological traits. Duplication seems to provide a feasible method for the acquisition of new genes and hence new functions. Duplications also provide the organism the survival of the original gene (in case it has any selective value) in case mutation occurs changing the original gene.

INVERSIONS

This is a kind of intrachromosomal aberration where there will be no change in quality or quantity of genes, but there will be alteration in their sequence. In order to bring about an inversion, there will be two breaks in the chromosome, with a segment getting separated from the chromosome. The segment joins back the chromosome but in a reverse sequence. For eg., in a chromosome with a gene sequence ABCDEFG, the segment BCDE is inverted, after the inverted portion rejoins the chromosome the altered chromosome would have the gene sequence AEDCBFG

Inversions have also been reported in a wide variety of plants and animals like deletions and duplications. These perhaps have a wider distribution than the other two aberrations in natural populations. Among plants, inversions have been reported in *Paris, Tradescantia* etc. Among animal species, the favourite fruit fly, *Drosophila* exhibits quite a number of inversion types. *D.willstoni* and *D.psuedo obscura* show a variety of inversions that can be studied in great detail in salivary gland chromosomes. Several types of inversions have also been reported in maize (McClintock 1938).

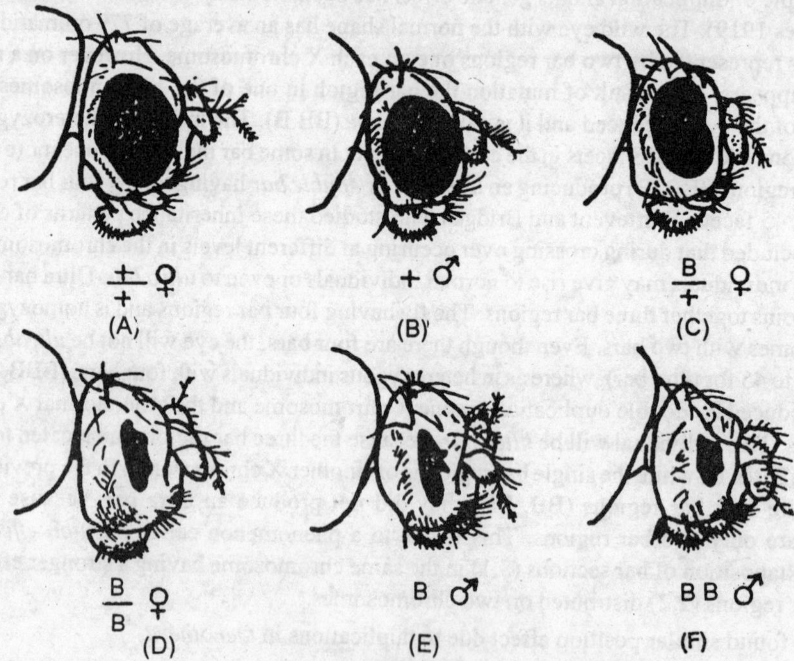

Fig. 6.4 Chromosomal Aberrations
Effect of duplication on the size in Drosophila
A. Normal female, **B.** Normal male, **C.** Normal (Heterozygous) female,
D. Bar eyed female, **E.** Bar eyed male, **F.** Double bar male.
(Note : In D even though two bar loci are present, it is only bar eyed as the bar loci are present on different chromosome. In F, two bar loci are present on the same chromosome and the eye becomes double bar due to position effect).

Anopheles mosquito, urodeles Amphibia and many species of grasshoppers are other organisms where inversions have been reported.

Types of inversion

Inversions are classified into two basic categories viz, *paracentric inversion* and *pericentric inversion*. Paracentric inversion does not involve the centromere. In other words, paracentric inversion does not alter the morphological shape of the chromosome. In pericentric inversion however, the centromere is included and as a result quite frequently leads to changes in the appearance of chromosome in somatic cells. For e.g. as it happens *Drosophila melanogasiner,* a metacentric chromosome (2nd and 3rd) undergoing pericentric inversion, will have two breaks at different distances from the centromere. When the segment rejoins the chromosome in the reverse order, the metacentric chromosome becomes submetacentric because the centromere is shifted a little away from the centre (see diagram for details). Inversions can be cytologically detected; inversion homozygotes with some difficulty, and inversion heterozygotes easily. An inversion homozygote is one in which both the homologues have undergone inversion, while in inversion heterozygote one of the homologue would have undergone inversion while the other is normal.

Identification of inversion is possible in inversion homozygotes, only by means of genetic tests as meiosis would show normal pairing between homologous chromsomes. If the inverted portion contains known genes and is linked to genes outside inversion, the linkage will break and thus results in changed phenotype. In inversion heterozygotes on the other hand, cytological demonstration is possible visually as the normal and inverted homologues pair in a peculiar loop type configuration. While the inverted chromosome forms an overlapping loop, the normal chromosome forms a loop surrounding the overlapping loop.

Paracentric inversions : As has already been pointed out, here the centromere does not figure in the inversion. Let us assume a chromosome with gene loci *abcdefg* with the centromere between *a* and *b*. After inversion in the segment *bcde* the inverted chromosome will be *aedcbfg,* the centromere however will not be altered and it is located even now between *a* and *b*. How does pairing occur between a normal (ABCDEFG) and an inverted chromosome of this type? While the normal chromosome forms .a loop outside, the inverted chromosome will form an overlapping loop (ie one thread overlaps the other) so that all gene loci are matched. In such a configuration a cross over might take place wherein the two non sister chromatids exchange segments. In the given example let us assume a crossover occuring between C and D. What will be the type of separation and the fate of chromatids? Will all of them have same gene loci or will they be different? When the anaphasic movement begins there will be an unusual sight at the equatorial plate. Out of the four chromatids -two will move to the opposite poles, one will remain in the centre without moving to any of the poles while another will strecth between the two poles forming what is classically known as the *anaphasic bridge.* The gene loci of the above chromatids will be as follows:

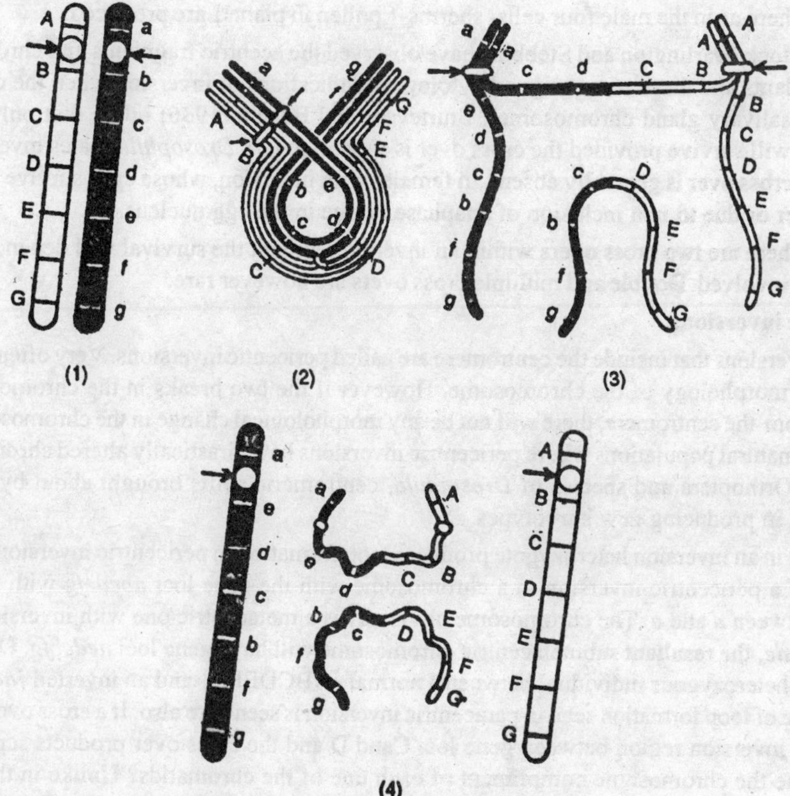

Fig. 6.5 Chromosomal Aberrations

Consequence of crossing over in a heterozygote with a paracenteric inversion

1. Homologus chromosome (Dark chromosome) shows inversion
2. Pairing and overlapping loop forming in an inversion heterozygote
3. Movement of chrosomes during anaphase (Note in the chrosome bridge formed by a dicentric chrosome and an acentric chromosome lying in the centric
4. Meiotic products showing duplications and deficiencies

Chromatids at	1 aedcbfg
Either poles	2 ABCDEFG
Chromatid at the centre	3 gfbcDEFG
Chromatid streching between poles	4 aedCBA

(If one carefully traces the diagram (at Diplotene) following the chromatid with different colours, it will be easy to find out the gene loci that the four different chromatids will get.)

In the chromatids mentioned above, No. 1 will be an inverted chromosome; it will have a centromere and the gamete that encloses this chromatid will be viable as it has all the gene loci. Chromatid two is a perfectly normal one. The gamete that gets this will also survive. Chromatid 3 will not survive because not only it has duplications and defficiencies for genes (deficiency for A and duplication for G and F) but it also lacks a centromere. Hence the gamete which will include this chromatid will not be viable. In fact the chromatid will normally be not included in the gamete as it fails to move during division. The chromnatid No. 4 will have two centromeres and is streched between the two poles. It has duplications and deficiencies (deficiency for A and duplication for A) This chromatid also is not viable. Normally however the oocyte where there is inversion and a cross over occurs within the inversion, it (egg) will not survive due to meiotic irregularities mentioned above; because after meiosis I the products may not participate in the second division. The chances of survival of gametes however is only 50% in the case of pollen mother cells.[ln the female side after meiosis, only one cell *ie* egg will be produced where as in the male four cells- sperms-{ pollen in plants} are produced.

McClintock, Darlington and Stebbins have observed the acentric fragments and chromosome bridges in a variety of plants and these serve as good cytological indications of inversion when the chromosomes are not as large as salivary gland chromosomes. Sturtevant and Beadle (1936) opine that only the non cross over chromatids will survive provided the cross over is suppressed. *In Drosophila* males inversion does not pose a problem as cross over is generally absent. In females with inversion, whose eggs survive it is either due to lack of cross over or due to non inclusion of anaphase bridge in the egg nucleus.

When there are two cross overs within an inverted segment the survival will depend upon the number of chromatids involved. Double and multiple cross overs are however rare.

Pericentric inversions

The inversions that include the centromere are called pericentic inversions. Very often pericentic inversions change the morphology of the chromosome. However if the two breaks in the chromosome occur at equal distance from the centromere, there will not be any morphological change in the chromosome. While there is no estimate of natural populations where pericentric inversions have drastically altered chromosome morphology, in certain Orthoptera and species of *Drosophila,* centromeric shifts brought about by inversion have been responsible in producing new karyotypes.

Pairing in an inversion heterozygote produces loop formation in pericentric inversion also. Let us study the example of a pericentric inversion in a chromosome with the gene loci *abcdefg* with the centromere being situated between *d* and *e* .The chromosome obviously is a metacentric one with inversion taking place in the segment *bcde*, the resultant submetacentric chromosome will have gene loci *aedcbfg.* During meiosis pairing occurs in a heterozygous individual between a normal {ABCDEFG} and an inverted *{aedcbfg}* chromosome. Similar type of loop formation seen in paracentric inversion is seen here also. If a cross over takes place {see Fig. 1.90 in the inversion region between gene loci Cand D and the cross over products separate during anphase what will be the chromosome compliment of each one of the chromatids? Unlike in the case of paracentric inversion no anphase bridge or acentric fragment is produced; all the chromatids will have centromere but they {50%} suffer from duplications and deficienicies. The chromosome compliment of each of the chromatid is as follows.

↓

Chromatid 1 ABCDEFG (Metacentric)

(non cross over) ↓

Chromatid 2 ABCdea (Submetacentric)

(cross over) ↓

Chromatid 3 gfbcDEFG (Submetacentric)

(cross over) ↓

Chromatid 4 aedcbfg (Submetacentric)

(non cross over)

(Arrows among gene sequence indicate the position of the centromere)

In the above instance, if the inversion has occured on the male side 50% of the pollen (in plants) or 50% of the sperms (in animals) would survive. Balance would abort due to duplications and deficiencies. In the case of female however, it depends as to which of the chromatid would be included in the egg. The egg would survive if it gets chromatid 1 or 4, if it gets 2 or 3 it would not survive. In the case of embryo sacs of higher plants generally the intact chromatids (1 and 4) are so oriented that they will be included in the outer megaspores and the irregular chromatids (2 and 3) will be in the central megaspores. Consequently, the basal megaspore receives either a normal or an inverted chromosome and the gametophyte survives.

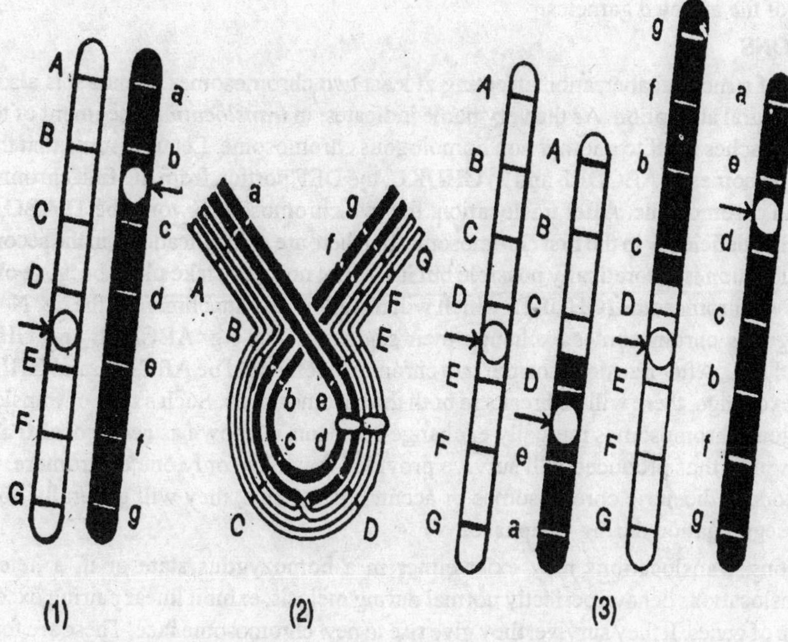

Fig. 6.6 Chromosomal Aberrations

Consequence of crossing over in a heterozygote, with pericentric inversion

1. Homologous chromosomes
2. Crossing over with loop formation
3. Meiotic products

Overlapping inversions

In some instances a second inversion may take place within the first inversion; such a type of inversion is said to be overlapping inversion. For example the normal chromosome is ABCDEFG and the inverted chromosome is AEDCBFG; a second inversion taking place within the inverted segment EDCB between B and E will result in a chromosome of the gene loci - AECDBFG.

Role of inversions in evolution : Interest in inversion having a bearing on evolution centres round paracentric inversions. Overlapping inversions studied in *Drosophila pseudoobscura* and *D.persimilis* have shown that the inversions are interrelated. An analysis of the natural populations at mountain ranges and elsewhere has shown that, populations in cool weather have a high frequency of inversions than those found elsewhere. This indicates that inversions are not only tolerated but may actually have selective value in natural populations. For the inversion to have any selective value, it must not be too long to have cross over effecting the allelic sequence but must be long enough to accumulate a genetic difference. It is possible according to Swanson (1967) that inversions can serve as foci for species divergence given sufficient time to accumulate genetic differences and the erection of barriers to prevent breeding within the population.

Inversion at the molecular level

What happens in the DNA molecule when an inversion takes place? Obviously the two stands of a DNA molecule are involved and two breaks occuring would switch the nucleotide sequence resulting in alteration. But there is one problem here due to the antiparallel nature of the helix; the segment cannot rejoin the main helix in the reverted position; it must also rotate 180 before reinsertion. Because of this rotation, there is some sort of a hybridization in both the helices in that they get a portion of the opposing helix. This will drastically alter both the helices and consequently the RNA and protein synthesis. It is this that is perhaps mainly responsible for the non viability of the affected gametes.

TRANSLOCATIONS

This is a type of structural aberration affecting at least two chromosomes. Hence it is also known as inter chromosomal structural aberration. As the very name indicates, in *translocation* a segment of the chromosome is removed and it attaches itself to another non homologous chromosome. Let us assume that there are two non homologous chromosomes 1) ABCDEF and 2) GHIJKL; the DEF portion from the first chromsome will attach itself to the second chromosome. After traslocation, the two chromosomes would be 1) ABC and 2) GHIJKL DEF. There would be deficiency in the first chromosome, but there are no duplications, in the second chromosome. This type of translocation is theoretically possible but it will not normally take place because of the presence of telomere in the other chromosome (GHIJKL) which will not allow any attachment at the tip. Normally however two non homologuous chromosomes exchange their segments. For e.g. ABCDEF and GHIJKL exchange segments DEF and JKL. After translocation the two chromosomes would be ABCJKL and GHIDEF. In order to have this type of exchange, there will be breaks in both the chromosomes. Such a type of translocation in which two non homologous chromosomes mutually exchange segments is known as **reciprocal translocation**. The two new chromosomes thus produced will survive provided they have only one centromere. If the reunion is such that as to produce dicentric chromosomes or acentric fragments, they will generally not survive due to failure of proper seggregation during anaphase.

Like inversions, translocations may exist either in a homozygous state or in a heterozygous state. Homozygous translocations behave perfectly normal during meiosis, exhibit linear pairing except for the fact of new linkage group of genes. It they survive, they give rise to new chromosome race. These are found sporadically in many organisms, but more commonly in *Oenothera, Paeonia, Datura* etc., in plants and scorpions and roaches among animals.

Translocation heterozygotes have only one of the homologues (of 2 chromosomes) affected while the other is normal. For e.g. considering the above example the pollen mother cell or MMC (or oocyte or spermatocyte) will have the following chromosomes.

Homologue I ABCDEF and ABCJKL
Homologue II GHIJKL and GHIDEF.

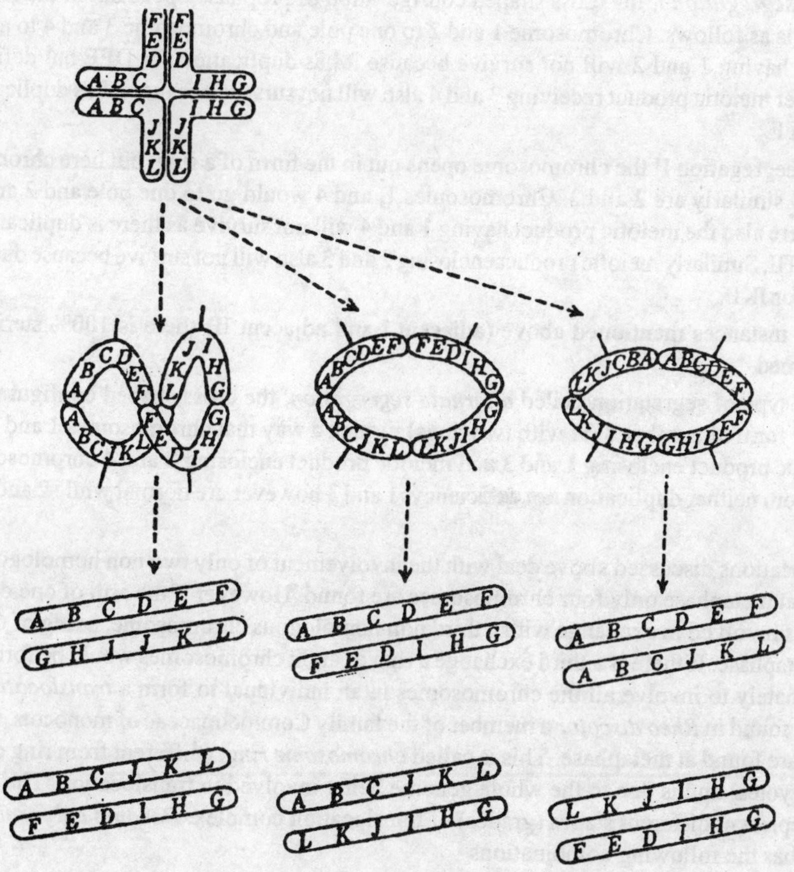

Fig. 6.7 Chromosomal Aberrations

Consequences of meiosis in a translocation heterozygote

In such a situation how can pairing take place between ABCDEF and ABCJKL; GHIJKL and GHIDEF. If there is a linear pairing roughly 50% of the length of the chromosome in both cases will remain out of synopsis. This is not conducive to the individual. Hence in order to have pairing, at all gene loci, all the four chromosomes combine and form a peculiar cross shaped or + shaped configuration at pachytene with the pairing partners being changed at the point of translocation break. In other words one half of the chromosome pairs with its homologue and the other half pairs with the allelic loci attached to a non homologous chromosome.

After pairing and cross over they have to separate out during metaphase. At that time a ring of four chromosomes is formed as the cross shaped configuration opens out. When the time of segregation comes as during metaphase and early anaphase how do they segregate to the two opposite poles?

The arrangement of the chromosomes on the metaphase plate determines the type of segregation and this has far reaching genetic consequences. There are three possible types of seggregation. *Adjacent I, Adjacent II* and *altemale* segregation. If we give the chromosomes, numbers as follow:

ABCDEF 1, GHIDEF 2
GHIJKL3 ABCJKL4.

In *adjacent segregation I,* the cross shaped configuration of prophase opens out in the form of a ring and the segregation is as follows. Chromosome 1 and 2 to one pole and chromosome 3 and 4 to another. Here the meiotic product having 1 and 2 will not survive because it has duplication for DEF but deficiency for JKL; similarly the other meiotic product receiving 3 and 4 also will not survive because it has duplication for JKL but deficiency for DEF.

In adjacent segregation II the chromosome opens out in the form of a ring; but here chromosomes 1 and 4 are adjacent and similarly are 2 and 3. Chromosomes 1, and 4 would go to one pole and 2 and 3 would go to another pole. Here also the meiotic product having 1 and 4 will not survive as there is duplication for ABC and deficiency for GHI. Similarly meiotic product enclosing 2 and 3 also will not survive because duplication for GHI and deficiency for JKL.

In both the instances mentioned above (adjacent I and adjacent II) there is 100% sterility as no viable gametes are formed.

In the third type of segretation called *alternate segregation,* the cross shaped configuration of prophase opens out in the form of number 8 (ie with two rings) in such a way that chromosomes 1 and 3 and 2 and 4 are adjacent. Meiotic product enclosing 1 and 3 and meiotic product enclosing 2 and 4 chromosomes will survive as they suffer from neither duplication nor deficiency. 1 and 3 however are normal while 2 and 4 are traslocated chromosomes.

The translocations discussed above deal with the involvement of only two non homologous chromosomes and in the ring at metaphase only four chromosomes are found. However if the arm of one of the translocated chromosomes is involved in exchange with a third non homologous chromosome, a ring of 6 chromosomes is found at the metaphase. If there is a third exchange a ring of eight chromosomes would be formed. This process can go on ultimately to involve all the chromosomes in an individual to form a *translocation complex.* This phenomenon is found in *Rheo discolor* a member of the family Commelinaceae of monocots, where a ring of 12 chromosomes are found at metaphase. This is called *chromosome ring.* (different from ring chromosome seen in some prokaryotes) and is due to the whole genome being involved in translocation. The plants *Oenothera* and *Paeonia* represent different stages (grades) of translocation complex. *Paeonia californica* with 5 pairs of chromosomes has the following combinations

(a) Ring of 4 Chromosomes and 3 bivalents
(b) Ring of 6 Chromosomes and 2 bivalents
(c) Ring of 8 Chromosomes and 1 bivalent
(d) Ring of 10 Chromosomes and no bivalent.

In *Oenolhera* also, different grades of complexes are found; while *O.hookeri* has all 7 normal bivalents, *O.lamarckiana* has all the 14 chromosomes forming a ring. (more details of *Oenothera* are discussed later)

Genetic detection of translocation

Heterozygous translocation may be detected by crossing two such individuals. The resultant progeny is of three types *viz.* Normal homozygotes, Interchange heterozygotes and Interchange homozygotes.

Translocation both in the homozygous and in the heterozygous condition affects the gene linkage. In homozygotes, the linkage is disrupted, if the translocated segment is not linked to its original group. Study of this linkage can be used to detect translocation and identify the chromosome involved. In the translocation heterozygote all the genes in the involved chromosomes are linked. This is because only the perfectly balanced gametes survive and take part in fertilization. Heterozygotes can also be detected by their semisterility as in the male side only 50% would survive. In the female side however there can be partial or complete sterility.

Evolutionary implication of translocation

The evolutionary implications of translocation are quite perplexing in that they produce new chromosome linkages. Among plants *Datura* and *Oenothera* have provided a wealth of information. In a translocation heterozygote involving two chromosomes, a ring of four is formed at the metaphase plate. The four chromosomes do not have an independent assortment. Sheer pressure of survival, forces on them alternate seggregation, resulting in two normal chromosomes forming a group and so are the two translocated chromosomes. In other words the two chromosomes which were initially independent, now are linked in spite of the fact that chromosomes exist as independent entities. As the translocation complex increases in number, the independent behaviour of the chromosomes decrease and as it happens in *Oenothera lamarckiana,* all the 14 chromosomes form one large translocation complex (a large chromosome ring is formed at metaphase involving all the 14 chromosomes). In this plant, the translocation complexes are called the **Renner complexes** named after the discoverer. Of the 14 chromosomes found at the metaphase plate, once again the chromosomes can not afford the luxury of free or independent seggregation. Survival instinct again forces them into two groups (or complexes) of 7 each. In effect there are two linkage groups with 7 chromosomes each. In a seggregation of this type there is no lethality. These two chromosome groups are called *Renner complexes.* Each ring having 14 chromosomes is actually a dual entity, because each has two complexes which may differ in genetic content. In *O.lamarckiana,* the two Renner complexes called **Gaudens** and **Velans** yield entirely different projenies when crossed with other forms. The individuals of *O.lamarckiana* are generally self pollinated and breed true and retain the structural complexity. A further refinement of the Renner complex is the *Renner effect,* when the two complexes are trasmitted in different ways ie. one *via.* egg and the other *via* sperm.

The evolutionary path in *Oenothera* is via the translocation complexes. The translocation having once arisen in the course of evolution has been stabilized in the gene pool by making it workable in the genetic machinery. The *Oenothera* is a clear example as to how a genetic irregularity deleterious to the organism to a certain degree has been incorporated into the system and made to function well. Similarly, the absence of cross pollination and retention of only self pollination, which to a certain extent is disadvantageous to the organism, is actually beneficial here as it helps in maintaining the complex, while cross pollination can break the complex by bringing in new and unknown genes into picture.

NUMERICAL ABERRATIONS

In the biological heirarchy, species form the basic genetic unit capable of surviving and maintaining its structure. This constancy (subject of course to mutational influences) of character in a species has as its basis, the chromosomal background. The number and kinds of chromosomes in a species are sufficiently constant to delimit it from others, But as the genes mutate or change, so also chromosomes are subject to variations. The process of alteration in chromosome numbers seen very often in plants and to a less extent in animals is called numerical aberrations.

Numerical aberrations in chromosomes produce two types of cells or individuals. These are 1) those whose somatic genome (chromosome complement) is the exact multiple of the basic number characteristic of the species and 2) those in which the somatic number is an irregular multiplication of the basic number. Occasionally there could be reduction in the basic number also. The first category is said to be *Euploid* and the second one is called *Aneuploid.* Euploids may be Haploid (monoploid), diploid, triploid, tetraploid and polyploid. Haploid is the basic number of the species with one set of chromosomes in the body cells, diploids have 2 sets, triploids have 3 sets, tetraploids have 4 sets and further multiplication series are refered to as polyploid.

Individuals whose genome complement is aneuploid have one or two extra chromosomes added or one or two chromosomes deleted from the normal number of diploid in the body cells. For instance a single chromosome may be missing from the diploid complement when it is called 2n -1: The following is the classification of numerical aberrations.

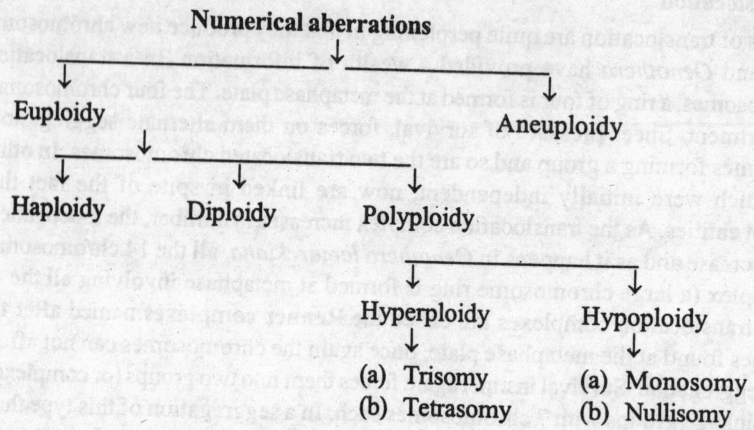

ANEUPLOIDY

In aneuploidy, there is either reduction or partial increase of the chromosome number. Reduction is said to be hypoploidy, while increase is said to be hyperploidy. Hypoploidy is due to loss of one or two chromosomes. If a single chromosome is lost from a diploid complement it is said to be Monosomy (individuals are called *monosomics)* and is symbolically designated as 2n-l. If two chromosomes are lost (ie. both the homologues in a pair) it is said to be *Nullisomy* (2n-2). Similarly the addition of a third chromosome to one of the bivalents is called *Trisomy* (2n + 1) and addition of 2 chromosomes to a bivalent is said to be *tetrasomy* (2n + 2).

Monosomy : In monosomic diploid individuals, (in the genome) one of the pairs of homologues would undergo a loss of one chromosome and it becomes a univalent while all others are bivalents. The diploids cannot generally withstand such a loss, while polyploids can withstand it. This is due to the fact that polyploids (see later in the same chapter) have more than two homologues for every pair. Theoretically the number of monosomics possible in an individual equals its haploid number. In common wheat where there are 21 pairs of chromosomes; there are 21 possibilities for monosomics. E.R. Sears has artificially induced these 21 monosomics in a wheat variety called *Chinese spring*. Monosomics have been induced in cotton (Endrizzi *et al)* and tobacco (Clausen and Cameron).

While diploids with monosomics do not survive, there are instances as in tomato, where monosomics could be produced. Here the chromosome number is 2n = 24. Double and triple monosomics (2n - 1 - 1) have two homologues lost from two different bivalents. During meiotic prophase, together with the normal bivalents there will be two separate univalents indicating that both these have lost their partners. Similarly in triple monosomy one homologue each is lost from three different bivalents and three univalents appear at the meiotic prophase.

Monosomy can be detected by comparing it with the normal individual and associated traits. In a cross between two monosomics, we will get normal, monosomics and nullisomics in a mixture. Nullisomics (if they survive) will not exhibit any of the morphological traits seen on the missing chromosome; there by specific traits and genes can be located on this chromosome.

Aneuploidy has been noticed in human beings also. A classical instance is the *Turner's syndrome* named after the discoverer H.H. Turner (1938). It occurs in 0.23 infants for every 1000 births. Here there is a loss of one of the X chromosomes and the females would be XO instead of XX. Adult females with Turner's syndrome have no ovaries and no secondary sexual characters and are sterile. They have abnormal jaw, hypoplasia and webbed neck. Turner's syndrome arises by the fertilization of an egg with a defective sperm (having no X chromosome)

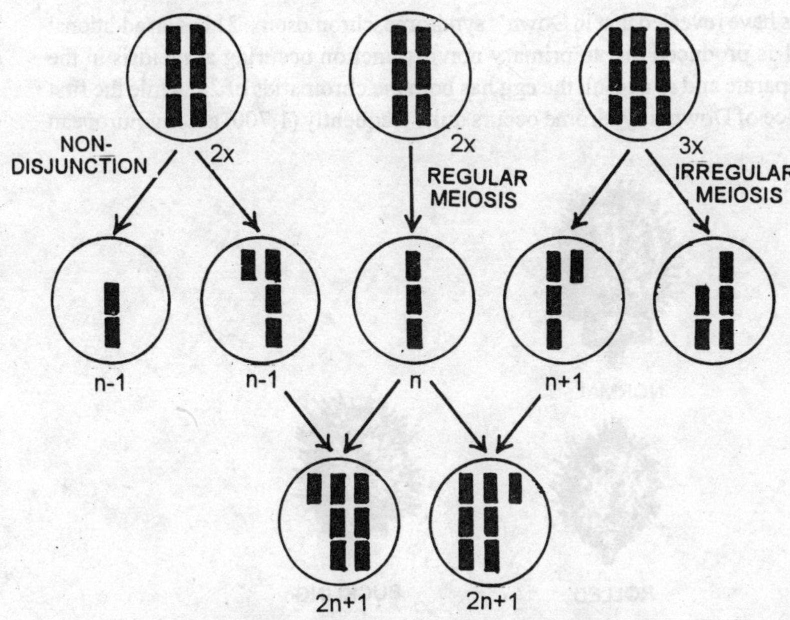

Fig. 6.8 Chromosomal Aberrations
Origin of different types of aneuploidy

Nullisomy : In this, an entire pair of homologous chromosomes will be missing (2n - 2). This should not be mistaken with double monosomy where 2 chromosomes are lost from two different pairs. In nullisomics, the metaphase plate at meiosis I will apparently show no abnormality and only a comparison with the normal can detect the loss of one whole pair of chromosomes. Nullisomics have been identified in wheat and other plants.

Trisomy : In trisomic individuals one of the pairs of homologues will have an additional member. Trisomic individuals exhibit a trivalent together with the normal bivalents during meiotic metaphase. These are designated as 2n + 1. There could be a double trisomy in which case two of the bivalents will have one additional member. At the metaphase of meiosis I, there will be two trivalents. The number of possible trisomy in individuals equals the number of haploid set. For eg. in barley where the haploid number is 7, seven trisomics are possible.

Trisomics are classified into primary, secondary and tertiary based on the type of the additional chromosome. If the extra member is identical to the other two homologues, the trisomy is said to *be primary trisoinm* (there is a triplicate of a chromosome). In secondary trisomy, while two are normal chromosomes, the third member is an isochromosome (with two arms genetically identical). In tertiary trisomy, the extra chromosome is the product of translocation.

In plants, trisomy has been studied extensively *in Datura,* maize, tomato and *Nicotiana.* In animals, trisomy has been reported in *Drosophila* and others including human beings.

Blackslee and his associates have described a series of trisomy in the jimson weed *(Datura stramonium)* belonging to the family Solanaceae. The base chromosome number of this species is 12 (n = 12) and 12 primary trisomics are possible. In this case instead of 12 bivalents at the metaphase of meiosis I, there will be 12 trivalents. Each one of this is associated with a distinct fruit character and can be separated from the others. Blackslee has named these trisomics (as represented by the fruit character) as follows - *Rolled, Glcssy, Buckling, Elongate, Echinus Cocklebur, Microcarpic, Reduced, Poinsettia, Spinach, Globe* and *Ilex*. Secondary and tertiary trisomics have also been reported from *Datura.*

In *Drosophila,* trisomy has been reported for the 4th chromosome. The fly can survive this and breed also.

Trisomy has also been reported in human beings leading to many abnormalities, both in the physical features as well as in mental faculties. *Down's syndrome* (mongoloid idiocy) was first described by Langdon Down in 1866. Affected individuals are short in stature, have slanting eyes, broad skulls, and stubby hands. They have low mental faculties. It was Lejuene (1959) who first associated this disorder with a chromosomal abnormality. Cytological investigations have revealed that in Down's syndrome, chromosome 21 has an additional member. Trisomy in chromosome 21 is produced due to primary non disjunction occuring at meiosis in the mother. The two chromatids do not separate and as a result the egg has both the chromatids of 21, while the first polar body receives none. The incidence of Down's syndrome occurs quite frequently (1:700) among European people.

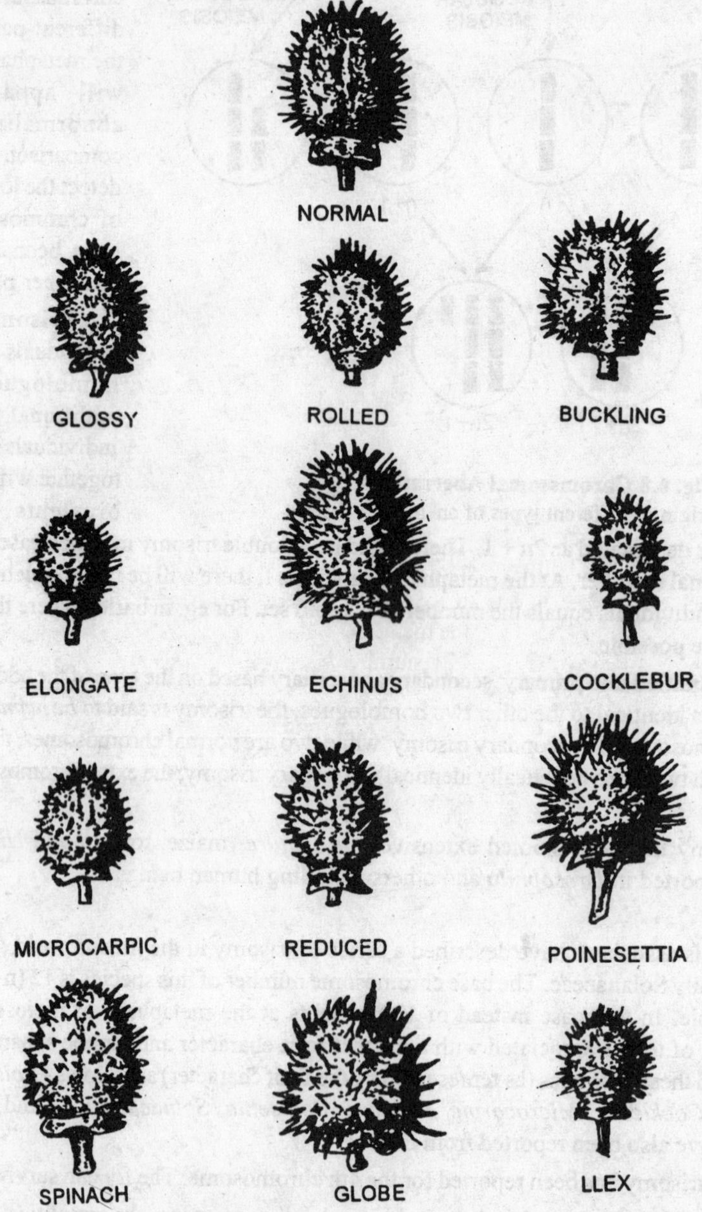

Fig. 6.9 Chromosomal Aberrations
Trisomy in Datura, resulting in different types of fruit shapes

Other trisomic abnormalities in human beings include the *Patau syndrome* (Patau 1960), *Edward's syndrome* (Edward J.H. 1960) and *Kleinfeter syndrome* (H.F. Kleinfelter 1942). The first two are autosomal trisomics while the third one is a trisomy for the sex chromosomes.

Kleinfelter syndrome occuring one in every 500 births, results in sexual malfunctioning. Individuals are morphologically male, but have small testes and are effiminate in their behaviour. Enlarged breasts, lack of body hair are other characters. Cytologically the cells show a Y chromosome in addition to 2X (XXY). This also indicates that Y chromosome determines the maleness in human beings.

Tetrasomy : In tetrasomy, a particular bivalent has two extra members; in other words there will be four homologous chromosomes in one pair, Symbolically tetrasomy is represented as 2n+2. Tetrasomics have been identified in wheat (E.R. Sears).

EUPLOIDY

Euploid individuals could be haploids, diploids or polyploids and represent the multiplication of the entire genome.

Haploids : Haploid or *n* number of chromosomes in the body cells may be normal or abnormal. In lower plants and animals, haploid condition is normal in the sense, here the individuals have only one set of chromosomes in the body cells, while diploid condition is represented only in the zygote. This is of normal occurrence. Haploidy is of interest when it occurs in diploid individuals. Here normally the body cells have 2 sets of chromosomes. Reduction division takes place and haploid gametes are formed. These gametes however have no morphological expression ie. they cannot directly develop into new individuals. They fuse with another gamete resulting in the formation of a diploid individual. Rarely however haploid individuals are produced as during parthenogenesis when the egg directly develops into a new individual without fertilization. Haploids are of two kinds *euhaploids* and *aneuhaploids.* While the former are derived from euploids the latter are derived from aneuploids.

Origin and production of Haploids : In some insects haploid individuals are produced naturally due to parthenogenesis. In plants also parthenogenetic development of egg results in haploid individuals. Such haploids have been seen in cotton, tomato etc.

In rare instances, pollen tube, or microspores may develop into new individuals. Such individuals are called *androgenic haploids* as they are produced from the male tissue. Occasionally synergids, antipodals etc may also develop into haploid embryos.

Haploids can artificially be obtained by tissue culture of the anther, pollen, X ray treatment, colchicine treatment, distance hybridization etc.

Anther culture has yielded excellent haploids in *Nicotiana* and Potato. Similarly Inter specific crosses may also result in haploid individuals. According to Kasha (1970), in a cross between *Hordeum vulgare* X *H. bulbosum'*. Chromosomes *of H. bulbosum* are eliminated in early zygotic division. Embryosacs of these can be cultured to obtain haploid individuals.

Morphology of haploids : There is general reduction in size and stature of the individual. This is also reflected in their physiology and generally they are not vigorous.

Cytology of haploids : In a haploid set all the chromosomes are non homologous and hence at metaphase I, they retain as univalents. Consequently these are distributed randomly at anaphase I for e.g. in maize a haploid (n = 10) individual produces gametes whose chromosome number ranges from 0-10. As a result there will be partial or complete sterility. Restitution nucleus may be formed.

Use of Haploidy : In normal diploids, haploid individuals help us to understand the functioning of recessive genes and also the expression of other genes in the haploid state. Doubling of haploid chromosomes will also bring about a completely homozygous diploid. This is of particular interest in plant breeding as this would bring about the effect of several generations of inbreeding.

POLYPLOIDS

In polyploid individuals, the body cells have more than two sets of chromosomes and consequently multivalents are formed at the metaphase I in meiosis.

There are mainly three categoreis of polyploids - *Autopolyploids, Allopolyploids* and *Autoallopolyploids.* The following chart gives the derivation of these from three diploids with genomes AA and B_1B_1 and B_2B_2.

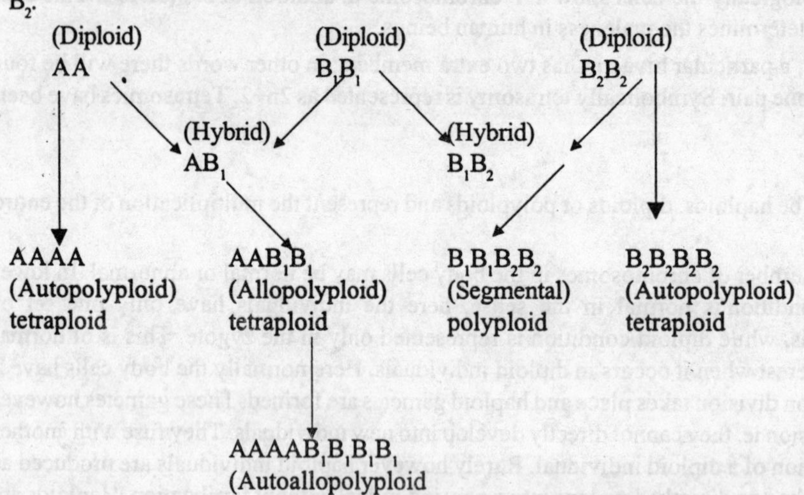

Autopolyploids : Here the same basic genome set is multiplied. For instance if the diploid genome is AA, autotriploid would be AAA, autotetraploid would be AAAA, autooctoploid - AAAAAAAA etc. An auto tetraploid would result either by doubling of somatic chromosomes or by the union of two unreduced gametes. Autotriploid can arise by a cross between a tetraploid and a diploid as shown below.

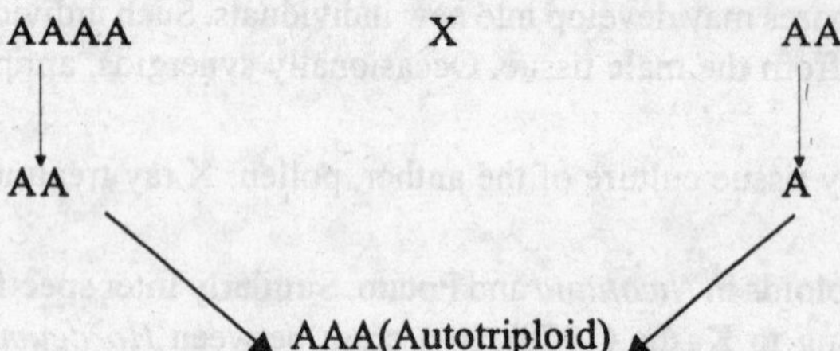

Distribution ofAutopolyploids : According to Muntzing (1936), these are widely distributed among the plants. But recent studies have shown (Clausen, Stebbins etc) that they are not as frequent as they were thought to be. Autopolyploidy has also been reported in the common Indian grass *(Cynodon dactylon).* This is known to be an autotriploid and survives mainly due to vegetative propagation. Autotriploids are known in water melons, sugar beet, tomato etc. Similarly autotetraploids are known in rye, corn, red clover, marigolds, snapdragons, flox, grapes etc.

Autopolyploidy can be artificially induced by various treatments. These include, ionizing radiation, chemicals, mustard gas etc. Since polyploid individuals have large stature, bigger fruits, increased yield etc, various techniques have been employed to induce polyploidy. Of all the methods the use of *Colchicine* is most sought after. Colchicine is an alkaloid, obtained from the bulbs of a liliaceous plant *Colchicum autuinnale.* The drug was first isolated by Houde (1887). Doubling of chromosomes due to the treatment with Colchicine was first reported by Levan (1938). The drug however is not effective against the *Colchicum* plant due to the presence of *anticolchicine.* Colchicine treatment may be given in one of the following ways.

a) *Seed treatment* : Seeds are soaked in 0.1 to 0.5% solution.

b) *Injection* : Solution is injected into the seedlings.

c) *Bud treatment* : Cotton dipped in solution is placed on the buds.

Effect of doubling of chromosomes : Polyploids are generally larger is size, but they bring in many undesirable characters also, like lack of resistance decreased rate of cell division, slow blooming etc.

Cytology of autopolyploids : Cytologically autopolyploids produce multivalents at metaphase I. The number of multivalents is 4 for tetraploids and 3 for triploids. While there is no abnormality in the case of tetraploids as diploid gametes can be formed, the meiosis is irregular in triploids. During disjunction, the trivalents separate as 2:1, or 1:1 with one chromosome left at the centre. This depends on the type of cross over. Since there are three homologues one chromosome may have cross over with two others at different levels. This leads to irregularity in disjucntion and deficiencies in the gamete. In tetraploids, the disjunction is usually 2:1 or rarely 3:1. Hence autotetraploids may show partial sterility, while triploids are generally sterile.

Commercial use of autopolyploidy is possible as in the case of seedless fruits (Produced by Kihara in Japan). Similarly, in crop plants also polyploidy has been known to produce plants with a greater vigour.

Allopolyploids : This is produced by crossing two different species with no chromosome homology.

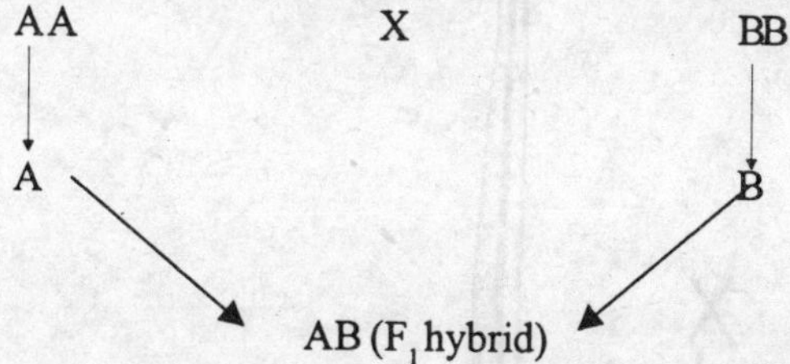

If one speices has AA genome and the other BB genome, the resultant F_1 will be AB. This individual is completely sterile because there is no pairing and all the chromosomes spread randomly at metaphase I and distribute irregularly at anaphase I. No gametes would survive. If the chromosomes of this F_1 hybrid (AB) are doubled, the resultant individual is called an *Allotetraploid (Allopolyploid).* As distinct from autopolyploids, allopolyploids have multiplication of different genomes (in autopolyploids the same genome is multiplied). The Allotetraploid is fertile because the two genomes A and B will find their homologues (AA and BB) and a perfect pairing takes place Since A and B genomes are different, there is no multivalent formation (since there are only two homologues for a kind of chromosome). At metaphase only bivalents are formed. This can be easily mistaken for a diploid if one does not known the origin. It is because of this that very often Auotetraploids are called *Ainphidiploids* (tetraploids functioning like diploids).

A classical example of Allotetraploid is provided by *Raphanobrassica* reported in 1927 by the Russian scientist Karpachenko. He crossed *Raphanus sativas* (2n = 18) *and Brassica oleracea* (2n = 18) and obtained an F_1 hybrii which was completely sterile. The sterility was due to complete non homology between the genomes of *Raphanus and Brassica.* But some fertile plants were obtained which had 2n == 36, obviously due to doubling or the incorporation of all the chromosomes at metaphase into one gamete (after the division of the centromere).

There are several examples of allotetraploids artificially produced by crossing two different species and doubling the chromosome of the F_1 hybrid.

1. *Primula verticellata* (18) X *Primula floribunda* (18)

   = *Primula kewensis* (Autotetraploid) (36)

2. *Galiopsis speciosa* X *Galiopsis pubescens*
(16) (16)
= *Galiopsis tetrahedra* (Autotetraploid)
(32)

3. *Spartina alternifolia* X *Spartina stricta*
(70) (56)
= *Spartina townsendii*
(126)

(Numbers in parenthesis indicate the 2n chromo number)

Examples of Amphidiploidy (Allotetraploid) among cultivated plants are seen in cotton, wheat, barely etc.

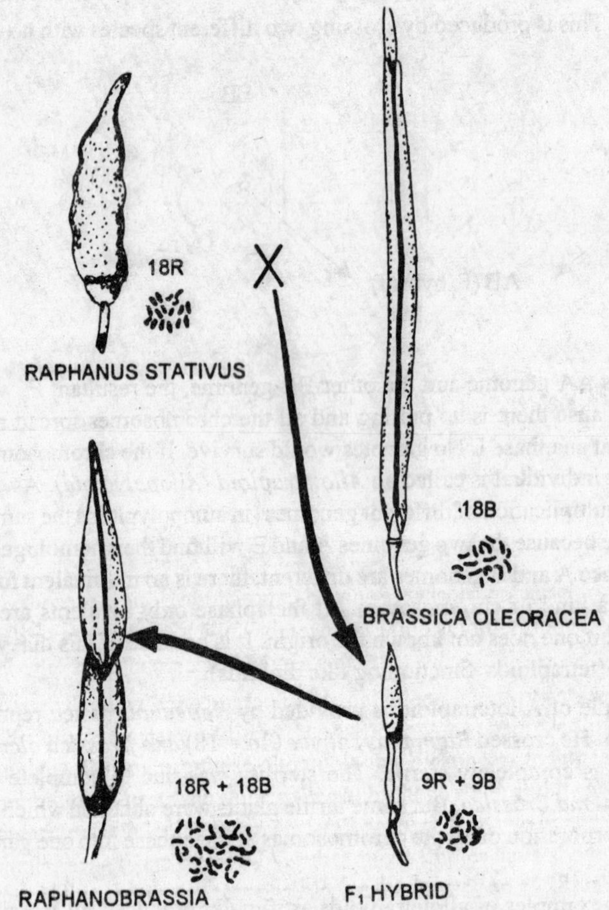

Fig. 6.10 Chromosomal Aberrations
Origin of Raphanobrassica

Segmental allopolyploids : In certain allopolyploids, the genomes coming from different parents are not altogether different. As a result there is some degree of pairing and multivalents are formed. This means that segments of chromosomes and not entire chromosomes are homologous. Such allopolyploids are called segmental allopoyploids and they are intermediate between auto and allopolyploids. The common bread wheat is regarded as a segmental allopolyploid.

Auto allopolyploid : This is a higher series of polyploid (octoploid). It can be produced as follows. A cross between two species (AA, BB) would yield a F_1 sterile hybrid (AB); on doubling, this becomes an amphidiploid (AA BB). If there is doubling of chromosomes in an amphidiploid it results in an autoallopolyploid (AA AA BB BB). The name autoallopolyploid is due to the fact that allopolyploid genomes are doubled to give rise to a polyploid having characters of both auto and allopolyploids.

ROLE OF POLYPLOIDY IN EVOLUTION

Polyploid individuals are different from their diploid relatives in many characters; yet they retain some degree of genetic semblance (allopolyploids) with the diploids. Polyploidy influences not only the morphological attributes of the individual but also has a profound effect on the reproductive potentiality and genetic variability of the progeny. All these obviously play an important role in evolution.

Comparison of natural polyploids with their diploid relatives and artificially produced polyploids provides us an indepth knowledge into the genetic machinery of polyploids. The resynthesis of polyploids from their diploid ancestors has provided additional proof about the origin of polyploids. Natural hybridization or introgression between species and doubling of the chromosomes has contributed a great lot in the evolution of species.

Polyploidy is widely distributed among the various groups of plant kingdom, but they are relatively scarce among animals. Even though this seems to be a mystery a careful analysis of the karyotype of animals and plants will readily provide an answer for the paucity of polyploidy among animals. In majority of the animals, there is a sharp contrast of two sexes resting upon a delicate balance between autosomes and sex chromosomes. This does not permit the addition or deletion of any member of the genome as is easily done in the case of hermaphroditic plants. In animals, wherever polyploidy has been reported it is associated with parthenogenetic development. There is one exception however as in the case of the saw fly, where an authentic case of polyploidy has been reported without the accompanying parthenogenesis.

Effect of polyploidy on structure and function of the organisms

A polyploid individual arising from a diploid population can survive and stabilize itself in the population provided the genetic alterations are such that it can fit into the existing genetic pool, while still showing variations. Between allopolyploids and autopolyploids, the former have a greater chance of survival. The effect of polyploidy on the morphology and physiology of the individual to a great extent will determine, whether the individual in question will survive in the environment. One of the favourite effects of polyploidy is gigantism, but it may not be always so. The other morphological and physiological effects of polyploidy have already been discussed.

Distrubiton ofpolyploids in the plant kingdom : Polyploidy has a wide distribution in the plant kingdom except perhaps for lower plants. Among the lower plants, tetraploidy perhaps exists in bakér's yeast *(Saccharomyces cereviseae)*. Tischler (1950) records a number of Algae as being polyploid. According to Manton (1952) the Horsetails, Clubmosses and the Psilotales represent the remanants of an ancient polyploid system. The Filicales also record a very high chromosome number *(Ophioglossum valgatum* has 500 chromosomes). Among the Gymnosperms polyploids are encountered in Gnetales.

Among the flowering plants however, polyploids are quite extensively distributed. According to Stebbins (1950), Tischler (1950), Darlington andJanakiammal (1945), about 30-35% of flowering plants are polyploids (allopolyploids). In grasses about 75% are polyploids. Rosaceae, Polygonaceae, Malvaceae, Crassulaceae among Dicots and Cyperaceae, juncaceae,Iridaceae and Orchidaceae among Moncots exhibit polyploid individuals.

The propensity of allopolyploids and the apparent paucity of autopolyploids in flowering plants is a clear indication that natural hybridization or Introgression and the subsequent doubling of the chromosomes have been going on in plant populations since a long time and have greatly contributed to evolution.

Polyploidy and speciation : In the evolutionary heirarchy, the lowest stable morphological unit is species. Two species are delimited not only on the basis of morphological traits, but underlying this is the chromosomal divergence. When two individuals have sufficient non homology among their chromosomes as to make a cross impossible, they are then said to have evolved into new species. If this statement is true completely; after the establishment of a species, the only factor for variation would be mutation. But evolution cannot depend on only one mechanism for its functioning. Even though species cannot be crossed generally, it does happen in nature and nature itself (or the selective forces) has found an answer to the resultant sterility in the F_1 hybrid. The chromosomes are doubled and the individual becomes fertile; has basic morphological and genetic characters sufficient enough to raise it to the status of a new species. It has happened in the case of *Primula kewensis.* Ever since then several new polyploid species have been created such as *Galeopsis tetrahedra, Spartina townsendii* etc.

Several of our cultivated crops represent a polyploid series indicating their origin from diploid ancestors. The two classical examples are those of wheat and cotton.

The wheat represents an allopolyploid series of diploid, tetraploid and hexaploid species. These three groups are called *Einkorn, Emmer* and *Vulgare* groups. The einkorn group has two diploid (2n = 14) species viz. *Triticum monococcum* and the wild *I. boeticum.* The einkorn speices are single seeded and not of much use for human consumption, because the grains are tightly held in glumes. In some parts of middle east and Europe, they are used for making dark bread.

The emmer group consists of seven species of tetraploid wheats of which the important ones are *Triticum dicoccum* and *T.durun.* The emmer wheat has arisen as a cross between an einkorn wheat and a wild grass *Aegilops* as shown below.

Triticum boeticum X *Aegilops speltoides*
(2n = 14) (2n = 14)
AA BB

AB (Diploid sterile hybrid)

Doubling

AABB
Triticum dicoccooides
(2n = 28) Fertile alloplyploid
evolved
Triticum dicoccum
(2n = 28) A'A'B'B'

The vulgare group consists of five species of hexaploid wheats (2n = 42) including the familiar bread whea *Triticum aestivum.* This is said to have been originated as a result of a cross between *Triticum dicoccur* (A'A'B'B') and another species of goat grass *viz. Aegilops squamosa* (diploid DD). The bread wheat therefor has genomes from 3 different sources. All the three wheat groups are interrelated and it is very obvious as t how polyploidy and associated chromosome doubling would have resulted in different species of wheat. Bu for polyploidy, the other evolutionary force (mutation) could not have produced such variety among wheat

In cotton, the upland cotton *Gossypicum hirsutum* is a tetraploid and is a result of crossing (an subsequent chromosome doubling) of *G.herbaceum* and *G.raimondii.*

Several such polyploid series are found in *Bromus, Sanicula* etc. Among the Rosaceae the entire trib pomoideae seems to have been arisen through poplyploidy. According to Sax (1931), the group Pomoideae with n = 17 seems to have been the result of an ancient cross between Spiraeoideae (n = 9) and Prunoideae (n = 8).

Among the polyploid series, the allotetraploid perhaps represents a stage very advantageous to the individual. Allotetraploidy combines the best of both worlds; it has the advantage of higher chromosome numbers, but the genetic and pairing behaviour of a diploid. No wonder then among the polyploids which survive, allotetraploids rank first. The higher series of polyploids however have gone too far in polyploidy and have committed themselves to a polyploid existence with its attendant advantages and disadvantages. Thus they lack the plasticity that is noticed in allotetraploids.

According to Swanson (1975), the sequence of evolution where polyploidy has been an important consideration is as follows. Gene mutation, aberrations and recombination will provide the principal raw material at the diploid level for natural selection. This results in a number of ecotypes from which arise divergent lines leading to higher taxonomic categories. In this scheme of things, autopolyploids have little role to play for they are no different genetically from their diploid parents.

Introgression and doubling of chromosomes however will lead to allopolyploidy and at once creates (evolves) a new species.

Triticale - a man made cereal : This is produced by a cross between wheat and rye. *Triticum durum* (2n = 28) *andSecalecereale* (2n = 14) are crossed producing a F_1 sterile (2n = 21). On doubling, this becomes a hexploid triticale (2n = 42). An octoploid triticale can also be produced by a cross between *Triticum aestivum* (2n = 42) and *Secale cereale* (2n = 14). The resultant F_1 hybrid (2n = 28) is sterile; on chromosome doubling this will lead to an octoploid triticale (2n =-56).

There is some acceptance for the hexaploid triticale from the point of view of grain quality and taste. These hexaploid Triticales are crossed again with wheat to bring about improvement. The products of these crosses are called *secondary Triticales* and are known to be better than *primary Triticales* (produced from wheat and rye). In producing a new species Triticale, man has just imitated the forces of nature which have brought in,an extensive array of biological diversity on earth.

7

CELL DIVISION

The unique and fundamental property of protoplasm is its power to grow and multiply by cell division. Thus new cells arise only by the division of preexisting cells in plants and animals. In unicellular organisms like bacteria, some algae, and fungi and protozoans cell division is the fundamental means of asexual reproduction. In multicellular organisms division of the single celled zygote contributes to the growth of the body. The gametes of the sexually reproducing organisms develop by means of cell division. Three types of cell division occur in living organisms. These are

1. Amitosis or direct cell division
2. Mitosis or indirect cell division
3. Meiosis or reduction division

Amitosis : This is found mainly is prokaryotic organisms like cyanobacteria, bacteria, yeasts etc., (among plants) Protozoa among animals are also known to divide by amitosis. Mostly found in unicellular organisms, amitosis is a means of reproduction. During amitosis, the nucleus elongates and a constriction appears in the centre resulting in a dumb bell shaped structure. The constriction deepens and the nucleus gets divided into two bits. There is no spindle formation and no formation of chromosomes. The nuclear division is followed by cytoplasmic division and the cell gets divided into two daughter cells. Thus amitosis is only a quantitative division.

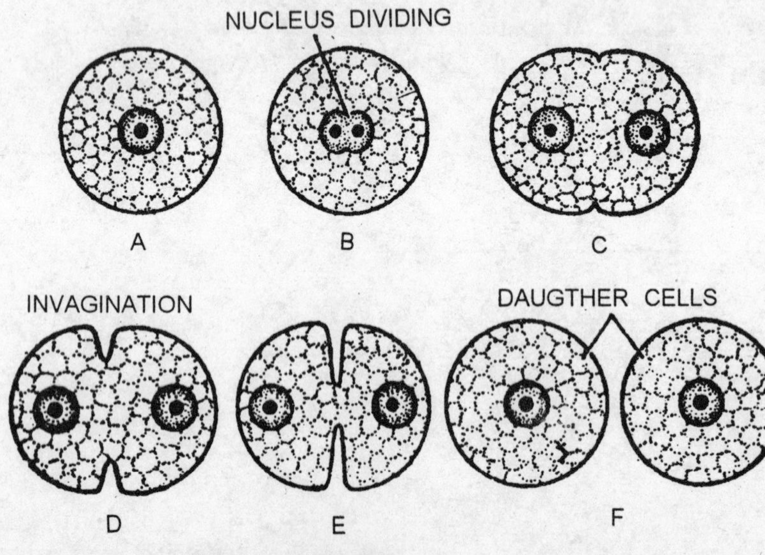

Fig. 7.1 Cell Division
Stages in Amitosis

MITOSIS

Occuring in somatic cells of multicellular organisms mitosis is the chief means of multiplication of cells contributing to the increase in size of the individual. It is the chief means of embryonic development into the adult. Thereafter the capacity for division is restricted to the cells at the tip of the stem, root tips and the cells of the cambium in flowering plants. In animals it is limited to the replacement of worn out cells like blood cells, skin, intestinal epithelium, and the germinal epithelium of the gonads. Mitotic division also occurs during the repairing of wounds in plants and animals.

Definition : The division of the somatic cells is generally called mitosis (Gr *mitos* = thread). It is a complex process involving series of changes leading to the formation of two equal daughter cells from a single parental cell. The term mitosis was first proposed by W.FIemming (1882). It is also called somatic cell division.

Cause of cell division : Why do cells divide? It is very difficult to obtain an answer to this question, although Kinetin a hormone has been isolated which acclerates cell division even in extremely small quantities. The possible reasons are - a) To maintain a relative volume of nucleoplasm and cytoplasm, b) To maintain the nuclear surface to nuclear volume.

Initiation of cell division : Chromosomes may release certain substances which initiates the nuclear changes and cause cytoplasmic division. Cell division may also be resulted by a periodically released hormone called cell division hormone. According to Heilbrunn and Wilson (1948), changing viscosity of the cell may initiate division, the viscosity being lowest at the beginning and gradually increasing and once again becoming low. While it is true that viscosity changes occur during division whether these changes are cause or effect is still not known.

Stag^s of Mitosis : For the convenience of study, mitosis maybe divided into two stages viz. Karyokinesis or nuclear division and Cytokinesis or cytoplasmic division. Karyokinesis or nuclear division is again divided into the following phases - a) Interphase b) Prophase, c) Metaphase d) Anaphase and e) Telophase. Some cytologists call the Interphae as the preparatory phase and the other stages as the distributive phase. During preparatory phase, the cell is supposed to prepare its components ready for division and during the distributive phase the chromosomes duplicate and distribute themselves to respective poles.

INTERPHASE

This was orginally called the resting phase of the cell and the nucleus, between two cell divisions. But recent physiological studies have clearly revealed that resting phase is a misnomer, because even though the nucleus appears to be at rest morphologically, physiological activity will be at its peak. The same is true for cytoplasm also. The interphase nucleus is large and conspicious, so also the nucleolus.

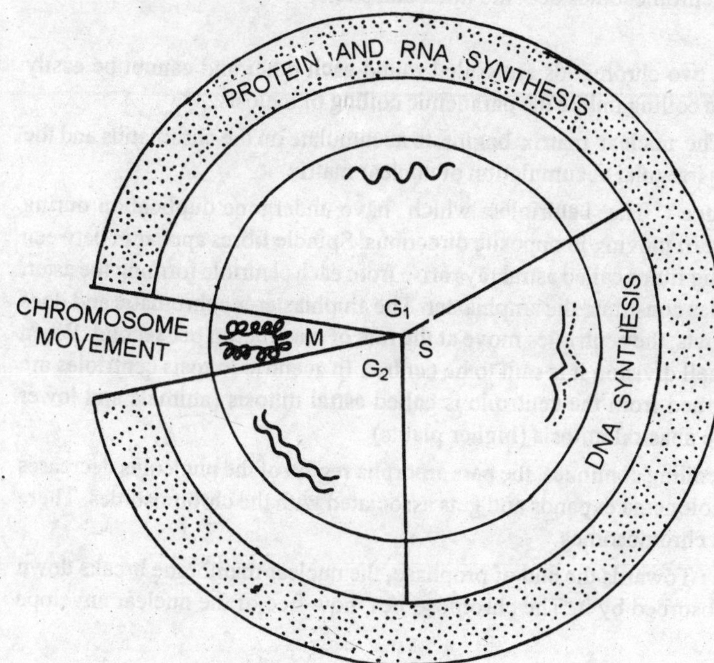

Fig. 7.2 Cell Division
Stages in the inter phase (between mitotic circle)

Three important stages are recognised in the interphase. These are G_1, phase, S phase and the G_2 phase. The G_1 phase also known as the post mitotic gap phase occurs at the end of every cell division (at the beginning of subsequent division). During this period the daughter cells grow in size. Proteins and RNA are synthesised during this period, but there is no synthesis of DNA.

The S phase, also called the synthetic phase is the most active stage from the point of view of DNA synthesis. It is in this phase that new DNA is formed by replication. Cell division will not proceed if S phase is inhibited.

In the G_2 phase also called the premitotic gap phase, RNA and protein synthesis continues. The volume of the nucleus increases with the increased synthesis of various types of RNA. DNA synthesis however stops in this phase.

In the division cycle G_1 phase is the longest with G_2 and S phase lasting equal amount of time.

Some cytologists use the term M phase or Mitotic phase for the entire distributive phase involving prophase, metaphase, anaphase and telophase and the following cytokinesis.

PROPHASE (Gr. *pro* = before, *phase* = appearance).

This is the first stage of nuclear division. This is the longest stage in the mitotic phase lasting from one to several hours. In the case of onion root tip cells, it lasts for 71 minutes, while in grasshopper neuroblast-102 minutes and in the cells of pea endosperm about 40 minutes. The following are the important features and events taking place in prophase.

1. The cell becomes spheroid (in animals) with increase in its viscosity and refractivity.

2. **Differentiation of chromosomes :** Chromonemata start condensing and as aresult a long coiled net work called spireme is formed. Continued coiling of the spireme produces thick threads called chromosomes. By the end of prophase the chromosomes contract up to about 1/25 of their length. Specific number of chromosomes are formed (specific to each species).

3. **Duplication of chromosomes :** Each chromosome splits longitudinally into two identical chromatids connected by a centromere. Two types of coils develop in the chromatids namely somatic coils and minor coils. While the former are larger, the latter are smaller.

4. **Shortening of the chromosomes :** As the prophase progresses, the somatic coils increase in size but decrease in number. As a result of this,the chromosomes become thick and short.

5. **Coiling of the chromatids :** The two chromatids are coiled round each other and cannot be easily separated. The coiling is called plectonemic coiling unlike the paranemic coiling of meiosis.

6. **Condensation of chromatids :** The nuclear matrix begins to accumulate on the chromatids and the chromosomes assume a hazy outline due to irregular accumulation of nuclear matrix

7. **Production of achromatic figure :** The centrioles which have undergone duplication during interphase and situated above the nucleus start moving in opposite directions. Spindle fibres appear in between the two centrioles. At the same time, radiating fibres called astral rays arise from each centriole forming the aster. The spindle fibres between the two centrioles constitute the amphiaster. The amphiaster is achromatic and does not take stain. Taylor has shown that in Newts, the centrioles move at the rate of one micron per second. When the centriole is present and particpates in cell division it is said to be centric. In acentric mitosis centrioles are absent. Cell division in which aster is formed from the centroile is called astral mitosis (animals and lower plants), when aster is absent it is said to be anastral mitosis (higher plants)

8. **Dissolution ofnucleolus :** As the coiling continues, the pars amorpha region of the nucleolus decreases in size and finally disappers. But the nucleolonema expands and gets associated with the chromosomes. There is also increase in the RNA content of the chromosomes

9. **Dissolution of nuclear envelope :** Towards the end of prophase, the nuclear membrane breaks down and gets dispersed into cytoplasm to be absorbed by it. The chromosomes, now free of the nuclear envelope get collected inside the spindle fibres.

PROMETAPHASE

Some cytologists (Threadgold 1968) regard this to be the last stage of prophase, when nuclear envelope disintegrates.According to White (1963), prometaphase is the stage when the spindle is formed and the chromosomes try to reach the equatorial region of the spindle fibres.

METAPHASE

In the cell cycle the metaphase lasts only for a short duration. The following are the main events during metaphase

1. In early metaphase the highly shortened and condensed chromosomes show characteristic radial arrangement on the equatorial plane of the spindle. This pattern of orientation is known as the metaphasic plate or the equatorial plate. It is formed at right angles to the spindle axis.

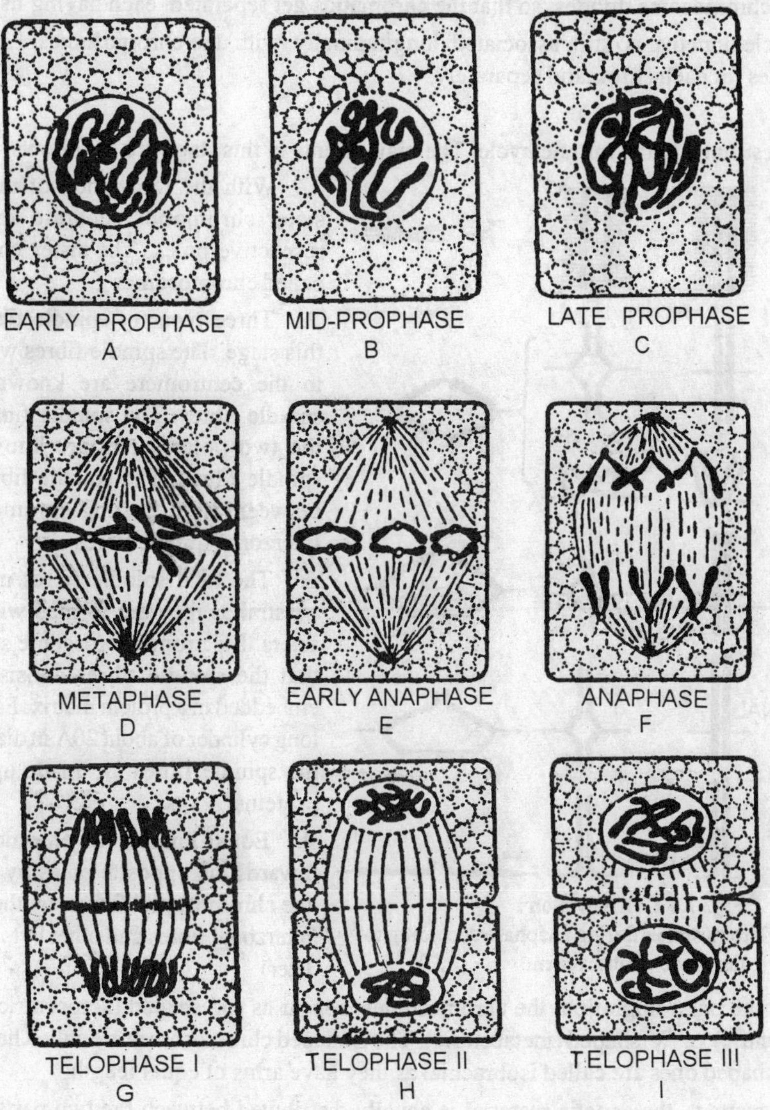

Fig. 7.3 Cell Division
Stages of mitosis in a plant cell

2. The chromosomes lie at the equator with the centroiles pointing towards the poles attached to the spindle fibres, and the arms lying freely in the cytoplasm. In plant cells, chromosomes are irregularly scattered on the equatorial plate. Smaller chromosomes lie at the centre, while larger ones are towards the periphery. According to Taylor, the spindle fibres are definite physical entities as can be proved by the inability of a microdissection needle to move across the spindle fibres easily. RNA synthesis ceases during metaphase.
3. The chromatids become separated from each other as a result of further shortening. Finally the centromere of each chromosome divides, so that the chromatids get separated, each having its own centromere
4. The nucleolonema is still associated longitudinally with the chromosomes. Later it enlarges and duplicates for each chromatid separately.

ANAPHASE

This is the shortest stage in the mitotic cycle. The main events in this stage are as follows-

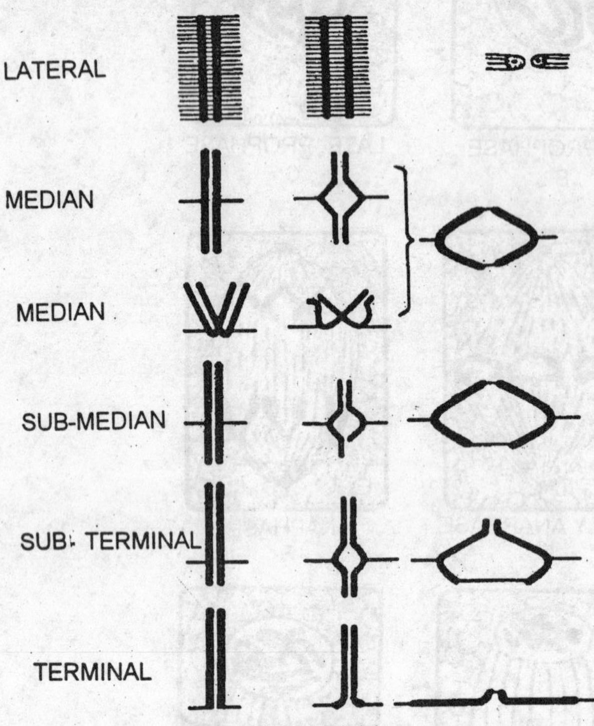

Fig. 7.4 Cell Division

Movement of Chromosomes during anaphase according to the mode of attachment

1. With the replication of the centromere, the sister chromatids separate and migrate to the respective poles. The sister chromatids are now called chromosomes
2. Three types of spindle fibres are distinct at this stage. The spindle fibres which are attached to the centromere are known as chromosomal spindle fibres. The spindle fibres which connect the two polar asters are known as continuous spindle fibres. The spindle fibres which form in between the departing chromosomes are called interzonal spindle fibres.
3. The total spindle fibres in plants are called anastral as they are formed without centrioles or asters. Electrom microscopic studies have shown that the spindle fibres consist of microtubules embedded in a protein matrix. Each microtubule is a long cylinder of about 20Å in diametre. Chemically, the spindle fibres are made up of monomers of proteins.
4. Equal number of chromosomes are pulled towards the opposite poles by the contraction of the chromosomal fibres, by the stretching of the interzonal fibres and other cytoplasmic events (see later)
5. During the polar migration, the chromosomes, appear as rod shaped (acrocentric), L shaped (sub metacentric) or 'V shaped (metacentric). The L shaped chromosomes are called heterobrachial, while the 'V'shaped ones are called isobrachial as they have arms of equal length.
6. In this manner, the genetic material is equally distributed between the two parts of the cell, which eventually become two daughter cells.
7. Since the centromere is essential for movement, those chromosomes which are without a centromere fail to move to the opposite poles. Such chromosomes are called laggards.

8. The velocity of movement of chromosomes during anaphase ranges . from 0.2 μ per minute to about 5 μ per minute. The rate of movement is independent of the size of the chromosomes.
9. Generally all the chromosomes move simultaneously except for the sex chromosomes which lag behind. Some chromosomes move very early before the group has started migrating. Such movement is said to be precocious movement.
10. End of the anaphase is marked by the appearence of clusters of chromosomes at each pole of the cell.

Chromosome movement during anaphase : Several theories have been proposed to account for the anaphasic movement of the chromosomes. Some of these are-

Electrostatic force : According to Lillie (1905), the chromosome movement is due to electrical charge. The cytoplasm is electronegative, while the nucleus (not chromosomes) is electropositive. Due to this difference in charge, the chromosomes move. This theory however does not explain how chromosomes having same charge as cytoplasm can move.

Hydrodynamic theory : According to this theory proposed by the physicist Bjerknes (1919), two spheres pulsating in a liquid medium set up hydrodynamic forces resembling electric or magnetic forces. In the cell, the spindle poles represent the pulsating forces and they attract the neutral spheres ie., chromosomes. this theory has not been acepted as it will not explain the formation of metaphasic plate.

Tactoids : Long fibres suspended in an unoriented medium tend to line up parallely in a spindle. Several workers have tried to explain chromosomal movement on this basis. (Bernal 1940, Schrader 1951) There is however no evidence for this.

Streaming hypothesis : According to this theory, viscous streams ernnating from the poles would push the chromosomes towards the equatorial plate, while reverse forces (ie streaming towards the poles) direct the chromosomes towards the poles. In this theory the spindle fibres were thought to be lamellar structures, but the fact that the spindles are tubular has negated this hypothesis.

Present views : The tubules of the spindle, (each one) consist of 13 filaments arranged in a ring. Chromosome movements are brought about by the action of chromosomal spindle fibres or continuous fibres or both. As evidence has indicated, in some orthopteran insects, chromosomal spindle fibres shrink and simultaneously the continuous fibres elongate resulting in the movement. In plants, movement takes place solely by the shortening of the chromosomal spindles, where due to rigidity of the cell wall, continuous fibres cannot elongate.

Elongation of continuous fibres takes place by the addition of protein unit and the reverse is true for the shortening of the spindle fibres.

TELOPHASE

When the chromosomes reach their respective poles reorganisation of the daughter nuclei and cytokinesis set in. This differs in plant and animal cells. In plants, cytokinesis does not occur simultaneously with Telophase as in animal cells. The following are the nuclear events in telophase.

1. The chromosomes begin to uncoil into their strands and become indistinct.
2. Nucleolous begins to appear as the nucleolonemal strands become thicker and accumulate.
3. The *gel* state of the spindle reverts to the sol state and the spindles disappear.
4. In animal cells, the asters lose their fibres and reconverted into centrioles. The centrioles take up their natural position outside the nucleus.
5. The nuclear membrane reappears forming a regular nucleus.

CYTOKINESIS

In plants, cell plate is formed between the two groups of chromosomes, by the accumulation of phragmoplasts

at the equatorial region. The cell plate is formed at the centre, they grow on either side and join the cell wall. The cell plate formed first will have only middle lamella.

In animal cells, cytokinesis takes place by furrowing, a cleavage furrow appears at the early telophase which becomes deeper as the spindle starts disappearing. The furrows begin from either side and both join at the centre separating the two daughter cells. Various views have been put forward to explain the mechanism of furrowing. Some of these are -

1. **Contractile ring theory :** (Swann and Mitchison 1958) The ectoplasm contracts like the non motile part of amoeba resulting in the formation of a furrow.

2. **Expanding surface theory :** According to this view, some secretions from the nucleus allow the cytoplasm to enlarge at the poles resulting in their contraction at the equational region and a furrow results. But the phenomenon of nuclear secretion has been questioned by Naehtwey (1965).

3. **Spindle elongation theory :** (Dan and Dan 1947, 1958) According to this theory, spindle and astral Fibres are responsible for furrowing. The elongation of the spindle results in contraction at the centre.

4. **Astral relaxation theory :** (Wolpert 1963) According to this, the surface of the cell has uniform tension, but the tension is reduced at the poles when the asters reach the poles. However the tension (being retained) at the equator results in the contraction.

5. **Vescicle formation theory :** (Threadgold 1968) The cell elongates during anaphase causing increased electron density. With the plasma membrane also increasing its electron density, a dielectronicfibrliary plate is formed in the equator. As the furrow progresses, small membrane lined vesciles appear on either side of the plate and coalesce with one another. These seem to arise from the ER. Finally the fusing vesciles form the two daughter cells.

Thus the two daughter cells are formed from the parental cell. During division, the cell organelles like mitochondria, plastids, golgi complex, lyososomes and the cytoplasm are distributed equally to the two daughter cells. Of these organelles, the plastids and mitochondria reproduce themselves, while the behaviour of the other organelles is not known. The two daughter cells are genetically identical with each other. Soon they enter into interphase.

Mitotic apparatus : The mitotic apparatus consists of spindle, centrioles and the aster. The spindle Fibres are of three types (for details see anaphase) The centrioles lie just outside the nuclear membrane.

Frequency and Duration of mitosis : Varies according to the type of organism, tissue, nutrition, temperature and other physical and chemical forces. Bacteria divide every 20 minutes. Paramecium divides every six hours. The cells of higher plants and animals divide once or twice a day. In man about 1-2% of worn out somatic cells are replaced daily. How long does it take a cell to go through the entire process of mitotic division? This can be determined by examining a dividing cell (this is possible in cells grown under tissue culture). In human fibrocytes a duration of 18 hours is required from one interphase to another. But the actual division process from prophase to telophase requires only about 45 minutes. This means that a cell will be in prepration for cell division for over 17 hours.

In the neuroblast cells of the grass hopper embryo, the time required (in minutes) for each stage of division (at 38°C) is as follows. Prophase 102, Metaphase 13, Anaphase 9, Telophase 57, Interphase 27. In this cell, surprisingly the interphase is short compared to human cells. The duration however is dependent on temperature. Lower temperature increases the duration, while higher temperature shortens the duration, subject however to a limit.

In *Tradescantia*, the cells of the staminal hair reveal strikingly the influence of temperature on cell division. At 10°C a complete division takes about 135 minutes. Root tip cells generally take about 22 hours to complete one division at ordinary temperatures, but they can divide at 0°c also. In warm blooded animals however cell division ceases below 24°C and above 46°C.

Significance of mitosis:

1. Mitosis is an extremely regular process. It maintains the qualitative and quantitative distribution of hereditary materials to daughter cells.
2. It maintains constancy of the diploid number of chromosomes of species
3. It maintains the constancy and balance in the DNA and RNA content in the cell
4. It maintains the nuclear and cytoplasmic balance in the cell. In all the cells, the cell cannot expand indefinetly; it has to maintain a balance between the cytoplasm and nucleus called the nucleoplasmic index.
5. Mitosis balances the surface volume ratio of the cell. When the cell increases in size, the surface area in relation to increased volume becomes less. By mitotic division, cells become smaller in size and the surface/volume ratio is restored.
6. Mitosis results in the growth and development of the organism. An idea of growth brought about by mitosis can be obtained from the fact that adult human body consists of 60,000,000,000,000 cells all derived from a single cell.
7. Repair and replacement of dead cells is ensured by mitosis. In human body, dead cells of upper epidermis, gut, RBC are constantly being replaced. According to an estimate 500,000,000,000 cells are replaced daily.
8. It plays an important role in the asexual reproduction and in the formation of gametes in sexually reproducing organisms.

Role of Mitosis in cancer : The cell division in organisms is a highly controlled process and takes place at a specific rate. Usually,after every division there is a gap of time so that the cells physiologically mature, develop polarity and individuality. In other words the cells are allowed to differentiate. This differentiation is the single most important character that is responsible for the integrated and orderly growth of organs and organisms. Strictly speaking, all cells derived from a parental cell by mitosis should be alike but they will not be so due to differentiation. After two cell divisions it is not just four cells that are formed, but four cells with individual characters. If there were to be no phenomenon like differentiation, all organisms produced by zygote should have been nothing but lumps of cells.

If for some reason the mechanism of differentiation operating between cell divisions fails, cells divide at a rapid rate there being no opportunity for them to differentiate. The result will be uncontrolled, unintegrated growth thus producing a cluster of undifferentiated cells. This is called cancerous or tumorous growth. {The rate of cell division is generally uniform for a particular cell type, and cancerous cells violate this}. Also called neoplastic growth, cancerous cells bring disorder and chaos in the structure and physiology of the organ where they are produced. They consume more nutrition, more energy than is the normal case, thus depleting other parts,of the vital ingredients of life.

Mitosis is of great significance in cancer because, the latter is the result of uncontrolled division. What makes the regular rate of mitosis to alter, resulting in the hazardous growth is as yet fully not known. There are many mitotic abnormalities in cancerous cells. The chromosomes differ in their size and number. Chromatid separation is delayed resulting in multinucleate cells. There is a high quantity of DNA in the cancerous cells. The cancerous cells show loss of contact inhibition ie,, ability to grow even in contact with others indefinitely, while normal cells stop growing after they establish a contact with other cells. Much of the cytological details of cancerous cells are understood by working on a human cell culture maintained since a long time called *Hela cells*; these are cells oblained from Henrita Lacks who died of cancer.

Hypotheses on cancer : What makes mitotic division to go haywire? or why the cells become cancerous? The answer to these questions are badly needed in order to control cancer. There are many hypotheses. The following are some of the important ones.

Somatic mutation hypothesis According to this,a mutation occuring in the somatic cells (non germ ells) will alter the control mechamism or inhibit the production of repressor proteins, which in turn block the functioning of the gene.

Viral gene hypothesis : Viruses may be responsible in altering the mitotic mechanism thus producing cancerous cells. This was first discovered by Rous (1911), but people did not beleive it at that time. While there is enough evidence of Viruses causing cancer in animals, hitherto there is no concluding evidence regarding Virus induced cancer in human beings. There is however one evidence of Virus cancer relationship in Burkitt's lymphoma, a tumour occuring in certain regions of Africa.

Carcinogens : Certain compounds like coal tar, nicotine etc are known to produce cancer.

Defective immunity hypothesis : The body has its own defence mechanism or immune system which destroys cancerous cells even when they are formed. This is called immunological surveillance. If for some reason this fails, cancer cells survive and produce tumours.

In spite of the above theories, it has to be conceded that none of these satisfactorily explains what causes the cells to divide. All these merely attempt to explain the various influences (external agencies) that induce rapid cell division. But the basic question that has remained unanswered is which machinery in the cell (either at morphological or at molecular level) gets altered by mutation or immuno difficiency or whatever cause resulting in unregulated division. Possibly the ultimate answer to cancer lies in understanding the basics of interphase. It is here that the cell prepares itself for division. Even in interphase perhaps it is the S phase (as the important stage of DNA synthesis) which commits a cell for division needs to be studied in more detail to understand the rapid sequence of divisions.

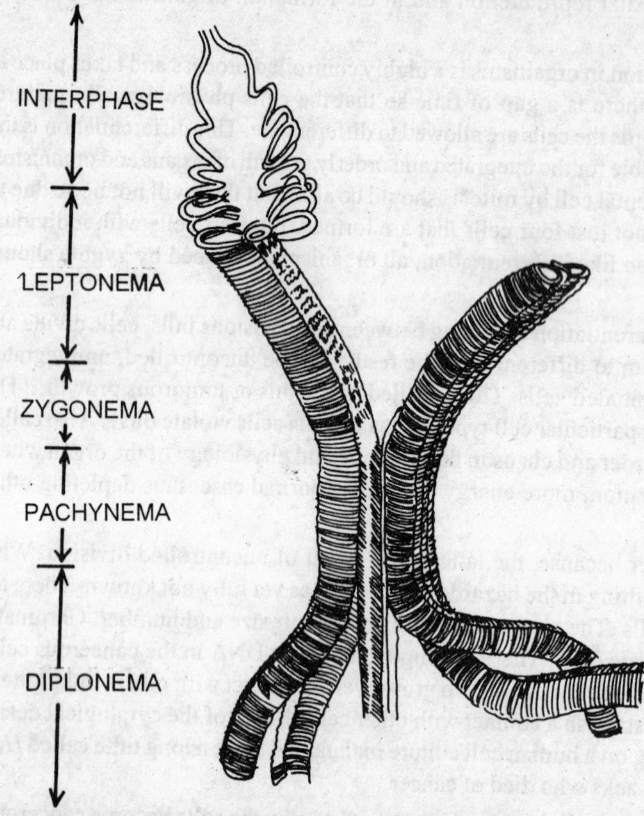

Fig. 7.5 Cell Division
Chromosomal threads during Prophase I of meiosis

MEIOSIS

In higher plants and animals, sexual reproduction takes place by the union of specialized sex cells called sperms and ova. The fusion of these sex cells that prevents the doubling of chromosomes in every generation. The answer lies in a special type of nuclear division called meiosis. It precedes the formation of gametes, so that each gamete acquires only half the number of chromosomes of the somatic cell. Thus when two gametes unite, the resulting zygote has the normal number of chromosomes. For example human body cells have 46 chromosomes in 23 pairs. But the human ovum and sperm have only 23 chromosomes, as aresult of meiosis. The reduced number of chromosomes in gametes is termed haploid, while the full complement of chromosomes in somatic cells is called diploid.

Definition of meiosis : Meiosis is a special

type of division occuring in the diploid germ cells during which the chromosomes divide once, while the nucleus and cytoplasm divide twice resulting in the formation of four haploid cells. Hence meiosis is also *reduction division* as the number of chromosomes is reduced in comparison with the parental cell. These terms were coined by J.B-Farmer and J.E.Moore in 1905.

The events of meiosis are uniform both in plants and animals. But depending on the time of occurrence, at different stages of life cycle the following types of meiosis may be identified.

1. **Sporogenetic meiosis :** In certain plants meiosis takes place in the sporophyte (2x) at the time of formation of spores. These spores develop into gametophyte.

2. **Gametic meiosis :** In most animals and some plants, reduction division takes place during the formation of gametes, so that haploid gametes are produced.

3. **Zygotic meiosis :** This is seen in lower plants where two haploid gametes produced by mitosis fuse to form a diploid zygote. The zygote undergoes meiosis to form haploid cells which develop into the haploid plant body.

Sporogenetic meiosis and gametic meiosis take place in individuals where the actual somatic body (plant or animal) is diploid, whereas zygotic meiosis takes place in individuals where the somatic body is haploid.

Factors initiating meiosis : The nature of the factors initiating meiosis is not well understood. Probably the nucleic acid balance triggers the process. If the DNA content is increased in relation to RNA, meiosis sets in. Certain hormones in the cell may bring about meiosis. However meiosis is a genetically controlled process. As meiosis takes place in the germinal cells, of the two sets of chromosomes present, one set is maternal and the other is paternal in origin. In each set, one chromosome is comparable to one other chromosome in the second set. These are called homologous chromosomes. In this way in both the sets all chromosomes have their homologues.

Meiosis is a long process and it involves the following stages.

First meiotic division

Inter phase	1	Preleptotene
Prophase I	2	Leptotene
	3	Zygotene
	4	Pachytene
	5	Diplotene
	6	Diakinesis

Prometaphase I
Metaphase I
Anaphase I
Telephase I

INTERKINESIS

Second meiotic division

Prophase II
Metaphase II
Anaphase II
Telophase II

Meiosis I is also called heterotypic division because here a diploid nucleus gives rise to two haploid nuclei. Meiosis II is called homeotypic division because here, the chromosome number is maintained. Meiosis II is actually nothing but mitosis.

INTERPHASE : The interphase consists of the same Gl phase, S phase and G_1 phase same as in mitosis. The essential events are as follows-

1. There is replication of RNA and synthesis of proteins during G_1 phase.
2. DNA is synthesized only in S phase and not during G_1 and G_2 phases.
3. DNA replication takes place by a semiconservative process. As a first step, an incision enzyme initiates a nick in one of the strands. Both the strands of DNA participate in the replication process under the influence of DNA polymerase. Together with DNA, histones are also synthesised. DNA is synthesised in short segments (Okazaki segments) which later join together to form large molecules.

PROPHASE I : This is the longest phase in meiosis. It has the following stages-proleptotene, leptotene, zygotene, pachytene, Diplotene and Diakinesis.

Proleptotene : Replication of DNA having already taken place in interphase, the chromosomes appear double stranded. The other events in proleptotene are-

1. Nucleus enlarges in size
2. Chromosomes are thin and hazy except perhaps for sex chromosomes

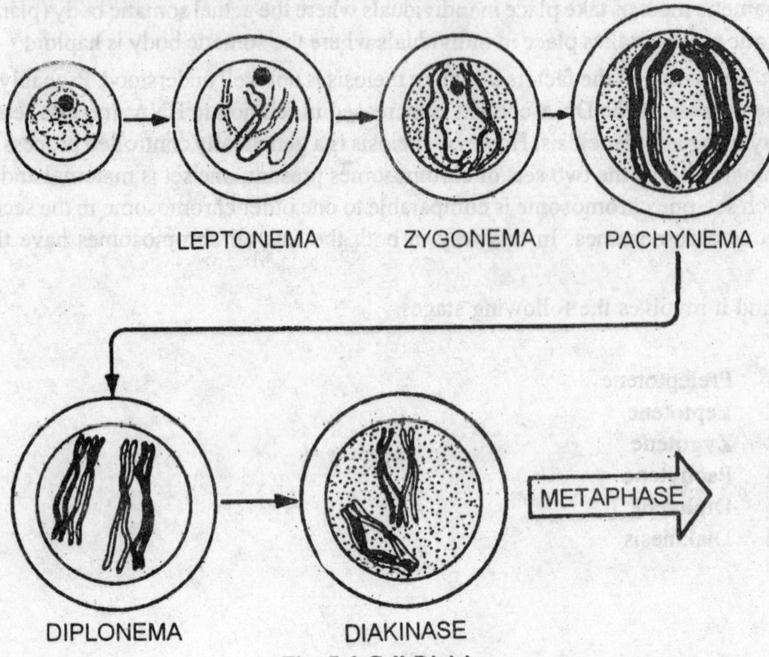

Fig. 7.6 Cell Division
Stages in Prophase I of meiosis

Leptotene (Leptonema) : This is also called the slender thread stage. Chromosomes appear distinct and the double strand (chromatid) nature can be seen clearly. The following are the other structural details of leptotene chromosomes.

1. Chromosomes are distinct,long and less coiled.
2. Chromosomes appear as threads with regularly placed beads or granules called chromomeres.
3. Electron microscopic observations have revealed that leptotene chromosomes have an axial filament with chromatin fibres attached to it in a series of lateral loops.
4. In certain areas the lateral loops cluster together to form chromomeres. The axial filament has two protein fibrils. These fibrils later get transformed into a part of the *synaptonemal complex*.
5. The homologous chromosomes are associated at their tips where they are attached to the nuclear membrane
6. Leptotene chromosomes may be uniformly distributed or localized into *bouquets*.
7. In some plants, chromosomes may form a tangled mass of fibres called *synizetic knot*.
8. Cytoplasm of a leptotene cell has many polyribosomes but vesicles are less in number. Nucleolus increases its size to the maximum due to increases in RNA synthesis.

9. The centriole duplicates and each daughter centriole migrates to the opposite poles of the cell. On reaching the pole the daughter centriole duplicates again. '

Zygotene (zygonema) : This is a very important stage where pairing of the homologous chromosomes takes place. The following are the important events with reference to zygotene chromosomes.

1. The chromosomes become shorter and thicker.
2. Lengthwise pairing of the homologous chromosomes begins as they are attracted towards each other.
3. Of the two homologues, one is maternal in origin and the other is paternal. These are called bivalents.

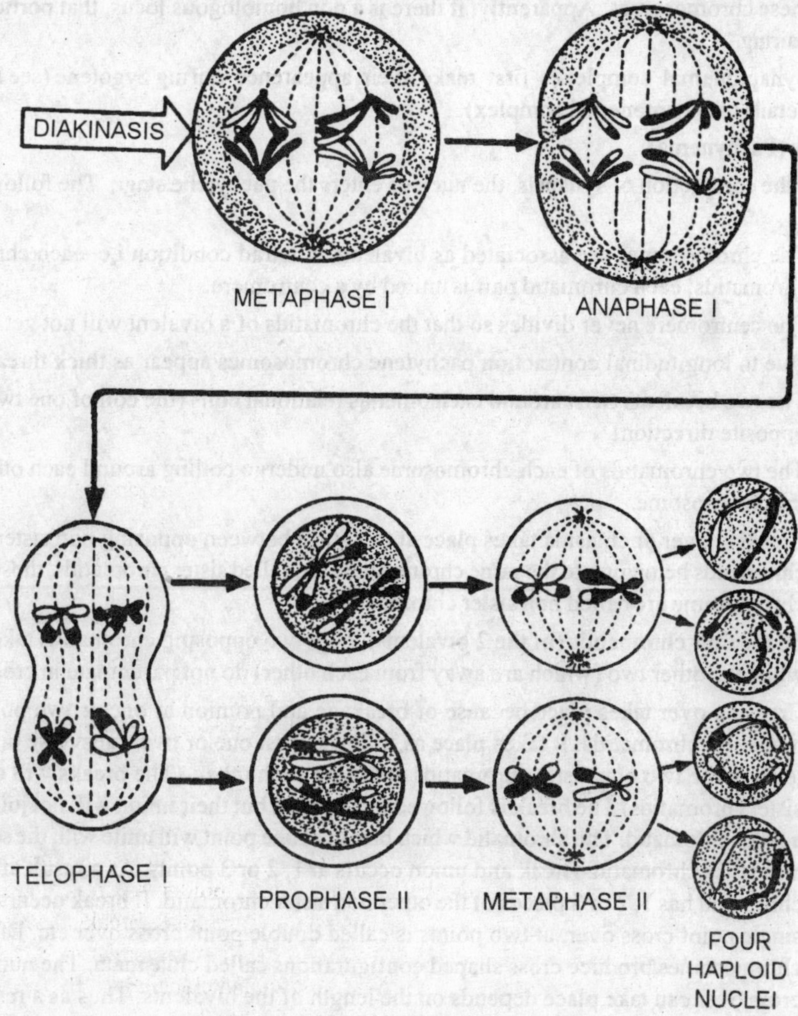

Fig. 7.7 Cell Division
Stages in meiosis (after prophase)

4. Pairing also called synopsis is precise and takes place chromomere - for chromomere. As a result,the chromosomes (homologues) align themselves parallel to each other throughout their length.
5. The bivalent has two homologous chromosomes; each homologous chromosome has two chromatids

i.e. totally there will be four chromatids (tetrad).

6. Pairing (synopsis, syndesis, synzesis) may take place in one of the following three ways.
   (a) **Proterminal synapsis :** Synapsis begins at both the ends and proceeds towards the centromere.
   (b) **Procentric synapsis :** Synapsis starts at the centromere and extends to both the ends.
   (c) **Random synapsis :** Also called localized pairing, synapsis occurs at many points apparently without any sequence or direction.
7. The pairing occurs only between homologous chromosomes but also between, homologous loci of these chromosomes. Apparently, if there is a non homologous locus, that portion will not take part in pairing.
8. Synaptonemal complexes first make their appearence during zygotene (see later in this chapter for details of synaptonemal complex).

Pachytene (Pachynema)

After the completion of synapsis, the nucleus enters the pachytene stage. The following are the important events.

1. The chromosomes are associated as bivalents in tetrad condition i.e. each chromosome having two chromatids; each chromatid pair is united by a centromere.
2. The centromere never divides so that the chromatids of a bivalent will not get separated.
3. Due to longitudinal contraction pachytene chromosomes appear as thick threads.
4. The two bivalents twist around each other as relational coils (the coil of one twining with the other in opposite direction).
5. The two chromatids of each chromosome also undergo coiling around each other creating tension on the chromosome.
6. Crossing over or chiasma takes place at this stage between opposing non sister chromatids. (The two chromatids belonging to the same chromosome are called sister chromatids, those belonging to different chromosome are called non sister chromatids)
7. Of the four chromatids (of the 2 bivalents), only two opposing chromatids take part in crossing over, while the other two (which are away from each other) do not participate in crossing over.
8. Crossing over takes place because of breakage and reunion at one or two points between opposing non sister chromatids. It takes place as follows. First one or two transverse breaks occur at the same level in the two non sister chromatids opposing each other. (The breaks will never occur in both the sister chromatids) The break is followed by reunion, but their union will not join the same segments of a sister chromatid. The chromatid which breaks at one point will unite with the segment of the opposing non sister chromatid. Break and union occurs at 1, 2 or 3 points. As a result after the cross over, each chromatid has 1, 2 or 3 pieces of the other non sister chromatid. If break occurs at one point it is called single point cross over, at two points is called double point cross over etc. Due to crossing over, the chromosomes produce cross shaped configurations called chiasmata. The number of points at which cross over can take place depends on the length of the bivalents. Thus as a result of cross over in the bivalents, there will be two unbroken or original chromatids (which are away from each other) and two broken and reunited chromatids opposing each other which have exchanged their segments.
9. Exchange of genes takes place between homologues due to exchange of segments.
10. Synaptonemal complex is very well organised at this stage. It has a central element and two lateral elements, (see later for detailed explanation)
11. Nucleolus remains unaltered; it may increase in size. It is found to be associated with the nucleolar organizing region of certain chromosomes.

Diplotene (Diplonema)

The diplotene nucleus is characterised by repulsion of the homologues, while there is coupling in pachytene. The following are the important features of diplotene stage.

1. Homologues start separating (repulsion), but are held together at one or more points called chiasmata.
2. Chiasmata are very clear in diplotene than in pachytene stage because of increased condensation of chromosomes.
3. Chiasmata are found in meiosis of almost all eukaryotic organisms. However in certain Dipteran insects including *Drosophila*, there is no chiasma in males. Meiosis here is termed non cross over meiosis or achiasmatic meiosis. Achiasmatic meiosis has also been reported in oocytes of Copepoda and Lepidoptera.
4. Chiasma may be formed at the ends (terminal chiasma) or in tshe middle (interstitial chiasma) of the chromosomes.

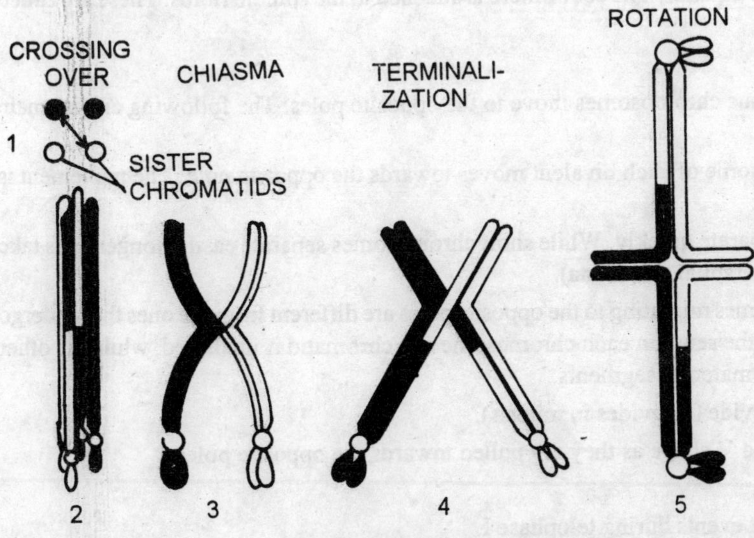

Fig. 7.8 Cell Division

5. The most important event occuring in diplotene is the teminalization of the chiasma. Terminalization or uncoupling of chiasma begins due to repulsion in homologous chromosomes. The chromatids begin to separate at the centromere and gradually progress towards the ends. As a result the terminal chiasma slip off or uncouple and progressively the chiasma towards the centromere move towards the end. When the terminalization is almost complete, the two homologues remain in contact through the terminal chiasma.

The degree of terminalization may be expressed in the form of terminalization coefficient (T).

$$T = \frac{\text{Number of terminal Chiasma}}{\text{Total number of bivalents}}$$

Chiasma frequency (Fq) is the average number of chiasma in bivalents of a nucleus. This may be calculated as follows

$$Fq = \frac{\text{Total number of chiasma}}{\text{Total number of bialents}}$$

During terminalization there is rotation of chromosomes; when there is a single chiasma, the chromosomes rotate through 180° and form a cross; when there are two chiasmata a loop is formed; when there are more than two, a number of loops are formed and rotation is brought about by the forces of repulsion.

6. Synaptonemal complexes usually disappear, but occasionally may persist towards the end of bivalents particularly where they are attached to the nuclear membrane.
7. There is increase in the endoplasmic vesicles.

Diakinesis

This is the Final stage of prophase 1. The structural details of this stage are -

1. Bivalent chromosomes are very thick and highly contracted.
2. Bivalents are distributed more towards the periphery of the nucleus.
3. Terminal chiasma may be retained maintaining contact between homologous chromosomes.
4. The nucleolous detaches from the chromosomes and finally disappears.

PROMETAPHASE I

1. The nuclear membrane disappears. The centrioles are joined by spindles which develop from tubules.
2. The chromosomes reach their maximum contraction.

METAPHASE I

At the end of diakinesis, the amphiaster (spindle) becomes clearly visible and it becomes more pronounced now. The bivalents arrange themselves on the equatorial plate. The centromeres of the bivalents lie towards the poles, wilile the arms lie towards the equator. The centromere is attached to the spindle fibres. These are called chromosomal fibres.

ANAPHASE I

It is at this stage, the homologous chromosomes move to the opposite poles. The following are the main features of anaphase I.

1. Each homologous chromosome of each bivalent moves towards the opposite pole. The movement is random unlike in mitosis.
2. The homologues do not separate quickly. While short chromosomes separate easily, longer ones take a little more time (due to interstitial chiasmata).
3. The homologous chromosomes migrating to the opposite poles are different from the ones that undergo pairing during prophase in the sense in each chromosome one chromatid is unaltered, while the other is a mixture of paternal and maternal segments.
4. The centromeré does not divide (It divides in mitosis).
5. The chromosomes assume a V shape as they are pulled towards the opposite poles.

TELOPHASE I

The following are the important events during telophase I.

1. The chromosomes reach the respective poles. The spindles persist for sometime.
2. Chromosomes undergo despiralization, becoming indistinct.
3. Nucleolus reappears, so also the nuclear membrane. In some instances nucleolus may not appear at this stage.
4. Cell membrane is formed in animal cells generally by furrowing, and in plant cells, cell wall is laid down in the form of plate. In some of the plant cells cytokinesis does not take place at this stage. It takes place when both meiotic divisions are complete. In pollen mother cells of flowering plants cell wall is laid down after meiosis I and II is called successive and if laid down only after meiosis II is called simultaneous. After Telophase I, two cells or two nuclei are formed.

INTERKINESIS

The two daughter nuclei or daughter cells undergo a resting period. This intervening phase is called interphase or Interkinesis. This phase is of a very short duration. In some instances interkinesis is absent, the incompletely formed nuclei of telophase I may enter prophase II. This is seen in plants like *Trillium*.

MEIOSIS II

The second meiotic division is essentially a mitosis in which both the chromosomes and cell divide.

But the main difference is DNA does not duplicate unlike in the interphase of mitosis. And also (as has been pointed out already) the interphase or interkinesis is short.

PROPHASE II

This is essentially similar to the mitotic prophase. The salient features of this stage are -

1. The centrioles divide into two and two pairs of centrioles are formed. They migrate to the opposite poles.
2. The microtubules get themselves arranged at right angles to the previous plane (during meiosis I).
3. The arms of the chromosomes remain separate and there is no relational coiling.
4. Nucleolus as well as nuclear envelope disappear with the formation of the achromatic figure.

METAPHASE II

1. It is of a very short duration like in mitosis.
2. Chromosomes (dyads) with their (two) chromatids arrange themselves on the equatorial plate.
3. The centromeres lie on the equator while the arms are directed towards the poles.
4. The centromere divides, thus separating the two chromatids (monads) which eventually become separate chromosomes. At this stage they can be called daughter chromosomes.
5. The centromere is attached to the chromosomal spindle fibres in the achromatic figure.

ANAPHASE II

1. The daughter chromosomes start their migration towards the opposite poles. The movement is brought about due to the contraction of chromosomal spindle fibres and stretching of continuous spindle fibres.
2. The chromosomes are not very thick and short like in anaphase, but are like the anaphase daughter chromosomes of mitosis.

TELOPHASE II

1. The chromosomes having reached the opposite poles uncoil and become less distinct.
2. Nuclear membrane and nucleolus reappear resulting in the formation of a daughter nucleus. The nucleus gets into the interphase state.
3. The nuclei are haploid; they have half the number of chromosomes in comparison with the parental cell.

CYTOKINESIS

Wall formation begins either as a furrow or as a plate and two daughter cells are formed. Thus totally at the end of meiosis four daughter cells are formed. In some instances cell wall (in some plants) would not have been laid after meiosis I. As a result at the end of Telophase II four daughter nuclei are formed.Cell walls are laid down now resulting in the formation of four cells.

SYNAPTONEMAL COMPLEX

This is a special structure visible during meiotic prophase - specially during zygotene and pachytene. Observed first with the help of electron microscope in the spermatocytes of cray fish- **Procambams darkii** (Moses 1956) and in Pigeon, Cat, Man etc (Faweett 1956), the synaptonemal complex consists of ribbon like structures associated with the chromosome core.

As these were observed during synapsis, Mosel (1958) named them synaptonemal complex (SC) subsequent to this, SCs have been found in a wide variety of organisms like Algae, Fungi, Gymnosperms, Angiosperms Protozoans, insects, molluscs, amphibians, aves, mammals etc. Hence SCs can be said to be of universal occurrence.

Structure : These are three partite ribbon like structures situated between the homologous chromosomes. Each SC is made up of a dense central axis from which arise dense lateral filamentous structures. Each lateral element on either side of the axis is attached to the inner side of the homologous chromosomes. The space between the central axis and lateral elements are traversed by a series of transverse units or LC fibres. Each lateral unit has a width ranging from 300-500 Å and the space between the lateral elements is between 900- 1200 Å. The width of the central element is betwee 150-500 Å. The SC is attached to the nuclear membrane through the lateral elements at the ends.

Comings and Okada (1971) have reported the emergence of a serie of chromatin loops from the lateral elements, which make up the chromosome material. The central axis (central element) is formed of the fusion of a series of small loops.

Each lateral element consists of two protein fibres one for each non sister chromatid. The lateral elements and LC fibres are also made up of proteins.

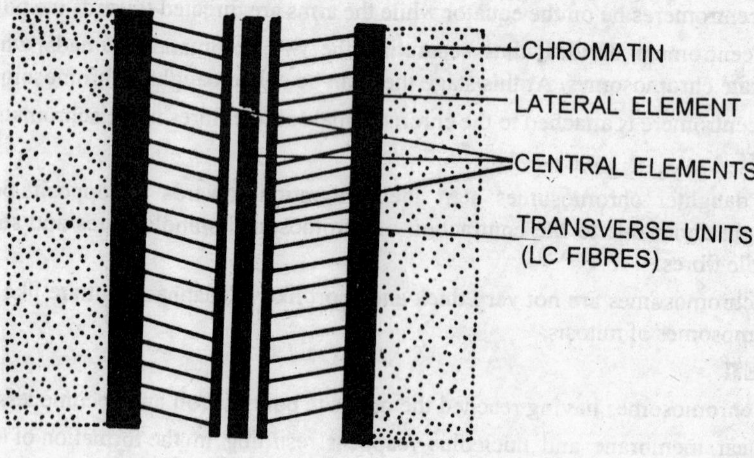

Fig. 7.9 Cell Division
Diagram of synaptonemal complex

Function of synaptonemal complex : Appearing first during zygotene, when pairing between homologous chromosomes occurs, they complete their structural development during pachytene. The SCs diasappear during diplotene, even though remnants may be seen during diakinesis. The above facts clearly attribute an important role to SCs in crossing over and exchange of segments between homologous chromosomes. According to Meyer (1964), the absence of SC in achiasmatic meiosis (as in males of dipteran insects) is proof enough of SC's role in chiasma formation. But in some instances SCs have been found in achiasmatic meiosis also.

According to Comings and Okada (1971), SCs help in chiasma not only at the morphological level but even at the molecular level also.

The SCs disintegrat gradually into cytoplasm as they disassociate from the chromosomes.

Significance of Meiosis

1. Helps to maintain constant chromosome numbers in organisms.
2. Crossing over which is the central focus in meiosis brings about recombination of genetic material leading to variation amongst individuals.

3. Meiosis in fact provides with the basic material (variation) for Evolution.
4. The random recombinations among homologous chromosomes give rise to a variety of mix up of parental characters and new characters.

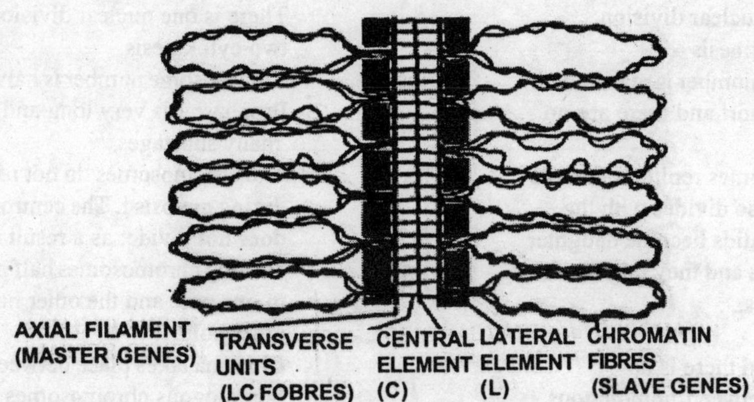

Fig. 7.10 Cell Division
Synaptonemal complex (Composite diagram showing details)

Molecular mechanism of crossing over : several models have been proposed to account for recombination of DNA molecules of the homologous chromosomes. The following is the general theme of the crossing over at the molecular level.

It is generally believed that the specific interactions between two DNA molecules of homologous chrómosomes leading to recombination are mediated by single stranded elements. Recombination begins probably by having an endonuclease enzyme make a 'nick' at one of the strands of the adjoining duplexes. As a result, a section of the single stranded DNA becomes displaced from that helix' and interacts with the complimentary strand of the other duplex. In order to facilitate this interaction, a second nick occurs in the neighbouring strand with the help of the endonuclease enzyme. Hydrogen bond formation assists in the joining of a DNA segment to another molecule. The occurrence of 'hybrid DNA' molecule (joined molecule) has been proved by observations with electron microscope.

The cuts and rejoining may not take place at precise points and at precise lengths. It is quite possible that there may be gaps and overlaps. While the overlaps are eliminated by exonucleases, the gaps in the DNA molecule are filled by the synthesis of short segments of DNA (by DNA polymerase), which are added to the gaps with the help of ligases. There is direct biochemical as well as genetic evidence to indicate the occurrence of DNA synthesis during recombination. The synthesis take place as in the mechanism of DNA repair. A very significant consequence of this DNA synthesis in order to fill the gap is that in case the segment of DNA is mispaired, the repair enzyme removes this segment and adds on a correct match, but this may alter the sequence for instance if the mismatch is GT (originally GC) instead of GC being restored it may become AT. This is the key for random recombination of genes. As a result of DNA synthesis during recombination one allele may get converted into the other. This may be called gene conversion and has been reported in *Neurospora*.

COMPARISON BETWEEN MITOSIS AND MEIOSIS

	MITOSIS	MEIOSIS
1.	Occurs in somatic cells	Occurs in germ cells
2.	There is one nuclear division and one cytokinesis	There is one nuclear division but two cytokinesis.
3.	Chromosome number is maintained.	Chromosome number is halved.
4.	Prophase is short and there are no substages	Prophase I is very long and has many substages
5.	The chromosomes replicate and the centromere also divides with the result, chromatids become daughter Chromosomes and they migrate to opposite poles.	The chromosomes do not replicate during meiosis I. The centromere does not divide; as a result among existing chromosomes half migrate to one pole and the other half to the other pole.
6.	No pairing and there is no association between homologous chromosomes	Chiasma takes place between homologous chromosomes resulting in exchange of genetic material
7.	Chromosomes duplicate during prophase.	Duplication of Chromosomes takes place during prophase II.
8.	Chromosomes occur as diads in prophase.	They occur as tetrads in metaphase - I.
9.	Centromere is directed towards the equator in the metaphase plate with the arms towards the poles.	Centromere is directed towards the poles with the arms towards the equator
10.	In anaphase the chromosomes move as monads.	Chromosomes move as dyads.
11.	During anaphase chromosomes are short and thin	During Anaphase - I Chromosomes are long and thick.
12.	Telephase always occurs.	Telephase - I is omitted sometimes.
13.	Equational division.	Reduction division.
14.	Brings about growth.	Brings about reproduction.

8

GAMETOGENESIS

One of the most remarkable abilities of living beings is their capacity of reproduction. It is by this ability only that life has maintained itself eversince it originated on this planet.

Reproduction may be defined as the ability of the organisms to produce offspring of their own kind. Reproduction generally takes place by two methods -Asexual and Sexual. In asexual reproduction there is production of offspring without involving the fusion of cells. Only one individual is involved in reproduction. In sexual reproduction generally two individuals are involved and reproduction is brought about by the fusion of two cells which are generally called gametes. These gametes are of two types - the male gamete and the female gamete. These two fuse and the resultant products viz., the zygote develops into a new individual. This is the general pattern of sexual reproduction seen in all the organisms.

Most of the higher organisms which includes higher animals reproduce generally by sexual reproduction. The sex cells namely the gametes are produced in definite structures called gonads. In the female the gonad is known as the ovary and the gamete is called the egg or the ovum. In the male the gonad is refered to as the testis and the gametes are called the sperms or spermatozoa.

The process of formation of the gametes is known as gametogenesis (Gr.*Gamos* = marriage; *genesis* = origin). In this process the reproductive structures which are diploid produce gametes by a process known as meiosis. As a result of this the gametes will have a reduced chromosome number (haploid) as against the diploid chromosome number seen in the reproductive structures.

In the body of the animals the gonads are produced by the germinal cells. These cells eventually produce the reproductive structures wherein reduction division takes place to produce the gametes.

Gametogenesis is of two types viz., spermatogenesis and oogenesis. The developmental process leading to the formation of the male reproductive cells or the sperms is called spermatogenesis. Oogenesis deals with the development of the ovum or the egg.

SPERMATOGENESIS

The process of spermatogenesis takes place in the male gonads called the testis. In most of the vertebrates the testis consists of many fine tubules called seminiferous tubules. These tubules have a wall lined by the cells of the germinal epithelium. The sperms are formed by these cells of the germinal epithelium. The seminiferous tubules also have certain cells called the nurse cells or sertoli. These are actually spermatic cells which do not undergo reduction division. Their function is to provide nourishment to the developing sperms.

The development of the sperms from the germinal epithelium, although is a continuous process may be divided into two stages. These are.

1. **Formation of spermatids and**
2. **Maturation of spermatids into sperms (spermiogenesis)**

Formation of spermatids

The germinal cells of the seminiferous tubules produce the spermatids. The germinal cells are also refered as primordial cells or primary germinal cells. To begin with, the cells are diploid and they give rise to the spermatids. This process is believed to take place in three phases - multiplication phase, growth phase and the maturation phase.

Multiplication phase : At this stage the testis consists of numerous seminiferous tubules made up of undifferentiated germinal epithelium cells. These cells have large deeply staining nuclei. The primordial germ cells increase their number by undergoing repeated mitotic divisions. As a result of this a large number of cells called sperm mother cells (spermatogonia) are formed. Each sperm mother cell is diploid and is like any other body cell in its chromosome constitution.

Growth phase : This follows the multiplication phase. In this phase there is no multiplication in the number of the cells. But the cells which are already formed absorb nutrition from the surrounding cells of germinal epithelium and grow in size. These cells have a large amount of food materials and are also rich in chromatin. These cells are spermatocytes. It has been observed that during the growth period the pairing or synapse of the chromatids becomes visible. Hence it may be said that meiosis (reduction division) is initiated in this stage.

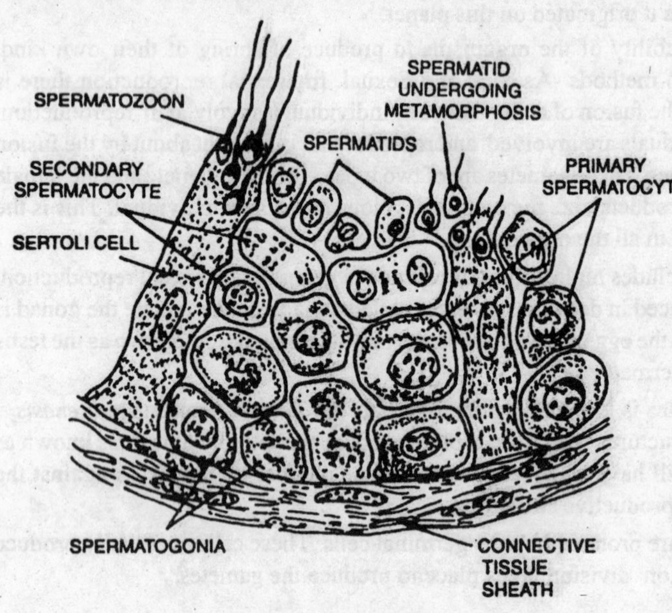

Fig. 8.1 Gametogenesis

A part of testis enlarged to show different stages of spermatogenesis and spermiogenesis

Maturation phase : During this phase the process of reduction division or meiosis which has been initiated wilt be completed. Meiosis involves two successive cell divisions - meiosis I and meiosis II. During meiosis I there is reduction division so that the two daughter cells that are formed after division will have exactly half of the number of chromosomes than present in the original cell. Meiosis II is an equational division. At the end of the division four cells are formed.

The two haploid cells that are formed after meiosis I are called secondary spermatocytes. The two secondary spermatocytes undergo meiosis II to produce four cells which are called spermatids. The spermatids do not undergo any further division. They simply undergo metamorphosis to produce the sperms.

Thus at the end of maturation phase each germinal cell would have produced four spermatids.

Spermiogenesis

During this phase the spermatids metamorphose (undergo structural ranges) themselves into sperms. The spermatid is a physiologically immature cell having nucleus mitochondira and golgi complexes. The maturation of spermatids into sperms is an extremely complex process involving many biochemical and morphological changes These changes are basically to induce a great amount of mobility to the sperm and to make it as light as possible by shedding any superfluous material present in the spermatids. The following are the major changes that take place in the spermatid on its way to maturation into a sperm.

1. **Nuclear changes :** The nucleus gradually gets reduced in size and assumes different, shapes in

different animals. In human beings the nucleus which is ovoid in the beginning becomes flattened. The RNA content of the nucleus gets reduced.

2. Formation of Acrosome : The acrosome occurs at the anterior side of the sperm nucleus. It is derived from the golgi complex and chemically it consists of protein digesting enzymes. At the time of acrosome formation the golgi complex gets associated with one or two vacuoles. These vacuoles enlarge in size and develop a dense granule known as proacrosomal granule. This granule is rich in mucopolysaccharides. Further enlargement of the vacuole makes it come very close to the sperm nucleus. Eventually the proacrosomal granule enlarges and forms the acrosome. At this stage the rest of the materials of the golgi complex gets reduced in size and are discarded from the sperm as "golgi rest"

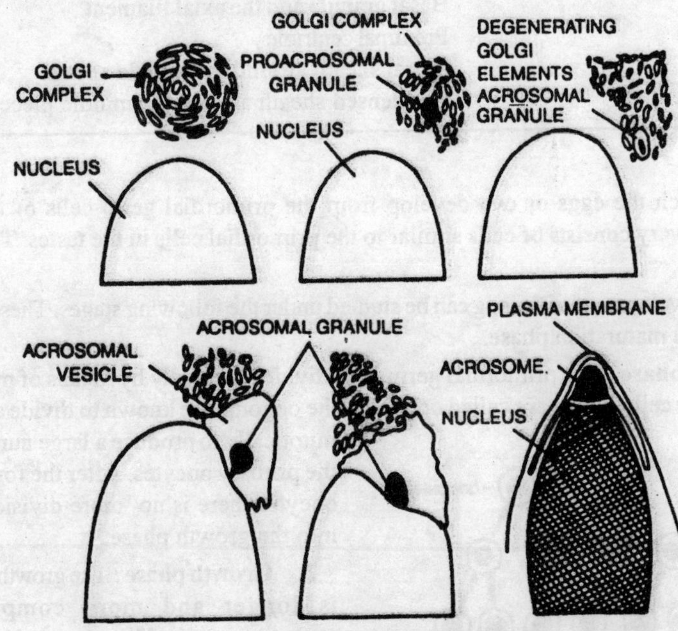

Fig. 8.2 Gametogenesis
Formation of acrosome from Golgi elements during spermiogenesis

3. Centrioles : The spermatid has two centrioles. These arrange themselves behind the nucleus The anterior one is known as the proximal centriole and the one behind is known as the distal centriole. The distal centriole functions as a basal granule and gives rise to the axial filament of the sperm. The axial filament or the flagellum of the spenn is composed of eleven fibres of which two are central and nine peripheral.

4. Changes in mitochondria : The mitochondrial mass becomes aggregated around the axial filament leading to the formation of the middle piece in the case of the development of the human sperm. The mitochondria form a spiral sheath around the axial filament. This spiral sheath is called the nebenken. The cytoplasm forms a condensed layer around the middle piece.

Structurally the maturation of the sperm from a spermatid involves the formation of two stages namely, the formation of the head and the formation of the tail. The nuclear changes and the changes that take place in the golgi bodies, eventually lead to formation of the head of the sperm. The middle piece of the sperm is formed by the cytoplasm and the spiraling mitochondria. The tail is formed by the lower portion and it involves the two centrioles.

A mature sperm has three portions, namely the head, the middle piece and the tail as in the case of human beings. The tip of the head consists of the acrosome while the remainder of the head consists of the nucleus. The middle piece has the fibres surrounded by spiral mitochondria and a condensed layer of the cytoplasm. The tail is made up of the fibers.

Derivation of parts of the sperm from a spermatid

Parts of Spermatid	Parts of the sperm
1. Nucleus	Head portion
2. Glogi	Acrosome
3. Distal centriole	Basal granule and the axial filament.
4. Proximal centriole	Proximal centriole
5. Mitochondria	Spiral sheath around the middle sheath
6. Cytoplasm	Condensed sheath around the middle piece

OOGENESIS

The process by which the eggs or ova develop from the primordial germ cells of the ovary is called oogenesis. The mature ovary consists of cells similar to the primordial cells in the testes. These however give rise to eggs.

The process of the development of the egg can be studied under the following stages. These are multiplication phase. Growth phase and maturation phase.

1. Multiplication phase : The primordial germ cells divide repeatedly by means of mitosis to produce a large number of daughter cells. These are called oogonia. The oogonia are known to divide again several times mitotically to produce a large number of cells called the primary oocytes. After the formation of primary oocytes there is no more division. The cells enter into the growth phase.

2. Growth phase : The growth phase of oogenesis is longer and more complicated than in spermatogenesis. There is considerable enlargement in the size of the primary oocyte during the growth phase. For instance the primary oocyte of frog initially has a diameter of only about 50 microns but at the end of growth phase the diameter of the matured egg well be as much as 2000 microns. Growth in the primary oocyte is brought about by the accumulation of a large amount of fats and proteins. These gel accumulated in the form of yolk. This is usually concentrated towards the lower portion of the egg called the vegetal pole. The upper side of the egg which contains cytoplasm and the egg nucleus remains often separate from the yolk-and it is called the animal pole.

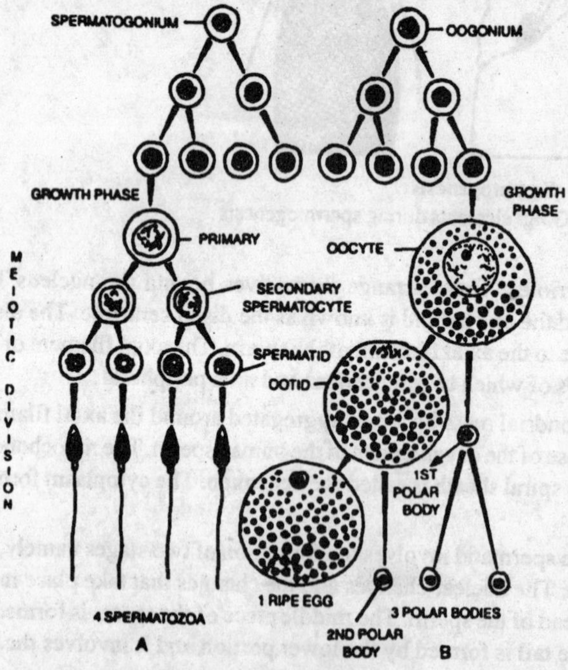

Fig. 8.3 Gametogenesis
A. Spermatogenesis, B. Oogenesis (after Balinshy, 1970)

The cytoplasm of the oocyte is very rich in RNA, DNA, enzymes, etc., other cytoplasmic organelles like mitochondria, golgi complex and ribosomes are also found in large numbers. In Oocytes of some amphibians and birds the mitochondria accumulate at certain places forming mitochondrial clouds. During this period, the nucleolar genes show increased activity of RNA synthesis and multiply their number. This is known as gene amplification or redundancy.

The nucleus enlarges in size due to the increased amount of nucleoplasm. The nucleolous also becomes large due to the synthesis of ribosomal RNA. At this stage in some of the amphibians the chromosomes change their shape and become giant lampbrush chromosomes.

Maturation phase : It is during this phase that the oocyte undergoes reduction division (meiosis) eventually to produce the egg or the ovum. The maturation division of the primary oocyte differs greatly from that of the spermatocytes. While in spermatogenesis at the end of meiosis four sperms are formed, here at the end of meiosis only one large haploid egg is formed, the remaining three cells forming three small polar bodies. This unequal division results in the formation of a single egg having a large quantity of stored food that is necessary for the development of the embryo.

The first maturation division (meiosis 1) reduces the chromosomes into half. After the homologus chromosomes undergo chiasma formation and crossing over the nuclear membrane breaks and the chromosomes move towards the opposite poles. The nuclear division is followed by the cytoplasmic division. This division (Cytokinesis) is unequal and results in the formation of a small polar body and a large secondary oocyte or ootid.

During the second maturation division (meiosis II) the haploid secondary oocyte and a small polar body undergo a mitotic division. As a result of this the polar body forms two polar bodies, while in the secondary oocytes the division results in the formation of a mature egg and a second polar body. Thus at the end of second maturation division there will be a large single egg and three polar bodies. These polar bodies eventually ooze out and degenerate, while the egg is ready for fertilization.

Differences between spermatogeneses and Oogensis

Spermatogenesis	Oogenesis
1. Occurs in male gonads (Testis)	Occurs in female gonads (Ovary)
2. Four spermatids are formed	Only one egg is formed along with three polar bodies.
3. Development involves three phases	Same

Structure of a mature egg

A mature egg is several times larger when compared with a sperm. This is due to the fact that it is full of nutrients and energy rich materials like proteins, glycogen, Yolk etc.. The size of the egg and the yolk contents vary in different animals. In general the eggs are however larger than the ordinary cells. The following are the structural details of the egg.

1. Plasma membrane : The outer envelope of the egg is the plasma membrane. It has the structure of a unit membrane.

2. Primary egg membrane : In addition to the plasma membrane the eggs of most of the animals consist of additional egg membranes known as the primary and secondary egg membranes. The primary egg membrane is secreted around the plasma membrane by the oocyte itself. In insects, molluscs, amphibians and birds the primary egg membrane is known as the vitelline membrane. The mammalian egg also has a similar membrane and it is called the zona pellucida. Chemically the vitelline membrane is composed of mucoproteins. Initially the membrane remains closely attached to the plasma membrane but at later stages a space called perivitelline space developes between them.

The secondary egg membranes are secreted by the ovarian tissues around the primary egg membrane. These are of varied nature - gelly like, chitinous, etc.

In some instances there are also tertiary egg membranes. These are formed by the oviduct or other accessory parts of the female reproductive structure. These are also varied in their nature.

3. Cytoplasm : The cytoplasm of the egg is called the ooplasm. It consists of a large amount of reserve food material stored in the form of yolk. In addition to yolk, pigment granules, water, RNA ribosomes, mitochondria and various other inclusions are also found in the cytoplasm. The outer layer of the ooplasm is microvilli and cortical granules. The microvilli are formed as outgrowths of the plasma membrane and they help in the transportation of the substances into the egg during its development. The cortical granules are spherical bodies of various size. These granules are believed to originate from the golgi complex.

Nucleus : There is a large single nucleus distributed in the cytoplasm. It has a very prominent nucleolus.

Types of eggs : The yolk content of the egg and its distribution varies in different eggs. This is used as a criterion to classify the eggs into several types.

A. Classification based on amount of yolk

1. **Alecithal :** Here there is no yolk in the egg. eg , eutherian mammals.
2. **Microlecithal :** Eggs have a small amount of yolk, e.g. Amphioxus
3. **Mesolecithal :** Eggs have a moderate amount of reserve food material, eg. Amphibians like salamander, frog, toads etc.
4. **Macrolecithal or polylecithal :** These eggs have a large amount of yolk. And also a little quantity of yolk free cytoplasm, eg. reptiles, birds and prototherian mammals.

B. Classification based on distribution of yolk

1. **Homolecithal or Isolecithal:** In this type, the quantity of yolk is very less and it is uniformly distributed all over the egg cytoplasm, eg. eggs of echinoderms.
2. **Heteropecithal :** In these types of eggs, yolk is not evenly distributed in cytoplasm. Based on the location and area it occupies, heterolecithal eggs may be of following types.
   a. **Telolecithal :** Here the yolk is concentrated in one half of the egg to form the vegetal pole and the yolk free portion is known as the animal pole. eg. amphibians, reptiles, birds and prototherian mammals.
   b. **Meiolecithal :** The quantity of the yolk is very large and it occupies almost the entire interior of the egg except for a small disc shaped portion of the cytoplasm. The cytoplasm contains the nucleus, eg. Fishes, reptiles etc.,
   c. **Centrolecithal :** Here the yolk occupies exactly the central portion of the egg while the cytoplasm is peripherial. Such eggs are found in insects.

9

FERTILIZATION

The process of the fusion of the mature male and female gamete is known as fertilization. The fused product is known as the zygote. This fusion is necessary to restore the original diploid complement. The zygote eventually develops into the embryo and then the adult individual.

In some instances the eggs without undergoing fusion may develop into a new individual. This process is known as parthenogenesis.

Types of fertilization : These are of two types-external fertilization and internal fertilization.

In external fertilization both the male and the female gametes are released into the external medium which is always aquatic. Here both the gametes are generally motile. They approach each other and fertilization is brought about. This is quite common in various invertebrates.

In internal fertilization, only the male gametes are released outside while the female gamete which is non motile usually remains inside the body of the individual. Fusion takes place internally and the aquatic medium necessary for the motility of the sperm is provided by the body fluid of the female individual.

External fertilization occurs in oviparous individuals where the egg is surrounded by a thin vitelline membrane while internal fertilization occurs in viviparous and ovoviviparous animals like mammals, certain, fishes insects, birds etc.

The process of fertilization is a highly specific one. Not any sperm can fuse with any egg. The sperms of a species can fuse with the eggs of only the same species. This specificity is very essential for the maintenance of species. It has been found out that the eggs contain a specific glycoprotein called fertilizin. The surface layer of the sperm contains a proteineous substance called antifertilizin. It is believed that there is a mutual chemical attraction between the fertilizin of the eggs and the antifertilizin of the sperms. This is again highly specific to the species.

Mechanism of fertilisation : The act of fusion of the sperm and the method of approach, entry, and eventual fusion of the male and female nuclei constitutes the mechanism of fertilisation. This is believed to take place in the following stages.

1. Movement of the sperm towards the egg
2. Capacitation and contact
3. Penetration of sperm into ovum
4. Cortical reaction
5. Activation of the ovum.
6. Fusion of male and female pronuclei (amphimixis)

Movement of the sperm towards the egg : This is the first step which brings the sperm in physical contact with the ovum. As the male and female gametes are produced in different individuals. There are various mechanisms to bring them nearer. The initial attraction of the sperms towards the egg is supposed to be chemotactic. The sperms swim towards the egg and collide with it. Usually several sperms attach themselves to the egg.

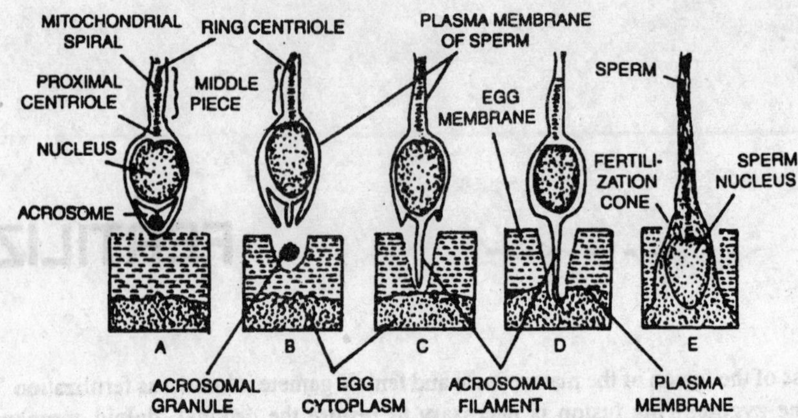

Fig. 9.1 Fertilization
Stages in the association and entry of sperm into the ovum

In individuals with external fertilization large number of eggs and sperms are released outside so that they have a favourable chance of meeting each other. In individuals with internal fertilization the sperms are discharged into the genital .tract of the females by the male individual. From here the sperms move upward and reach the egg present in the uterus.

2. Capacitatlon and contact : When a large number of sperms approach the egg for contact the fertilizin and antifertilizin interaction ensures that only a few spermatozoa are allowed to reach the ovum. The initial attachment of the sperm to the egg is believed to be due to the chemical bonding of fertilizin and antifertlizin.

3. Penetration of sperm into ovum; The acrosome portion of the sperm produces some lytic enzymes called sperm lycines which have the ability to dissolve the egg membrane allowing the entry of the sperm into the egg cytoplasm. Usually only the head and the middle piece of the sperm enter the egg while the tail is left outside.

4. Cortical reaction : The entry of the spermhead into the egg brings about several changes in the egg surface. These are the cortical changes and the development of fertilization membrane. In some echinoderm eggs some fine granules are visible in the ooplasm after the entry of the sperm head. The vitelline membrane starts lifting itself up from the -point of sperm entry and a perivitelline space is formed between it and the egg surface.

The cortical granules from the egg cortex release their contents into the perivitelline space. These contents attach themselves to the inner surface of the vitelline membrane forming what is known as a fertilization membrane. The development of the fertilization membrane prevents the entry of other sperms into the egg.

5. Activation of the ovum : The mature egg will be generally in a hibernating condition with very low rates of metabolism and inactive nucleus. The penetration of the sperm triggers the egg into activity. The metabolic rates increase allowing for entry of water and other particles. Metabolically the rate of protein synthesis goes up as these are needed for further cell divisions. At this stage the nucleus (pro nucleus) of the egg which has remained in the metaphase II stage (of the meiotic II division) becomes active completing its second division and releases the second polar body.

6. Amphimixis : In this process there is fusion of the male and female nuclei. Initially the two nuclei remain close together and at the point of contact the nuclear membranes disappear and the chromosomes come to lie on the equator. Finally the nuclear fusion is completed and it becomes a zygote nucleus. The egg is said to have been fertilized and it becomes a zygote. It is now ready to undergo cleavage to develop into the embryo.

KINDS OF FERTILIZATION

Fertilization may also be divided into several types based on the number of sperms that enter the egg apart from its classification based on the location of fertilization as to whether it is internal or external. The following are the various types of fertilization.

1. **External fertilization:** Fusion in the external media.
2. **Internal fertilization:** Fusion in the internal media.
3. **Monospermic fertilization :** Only one sperm enters the egg. This is seen in most of the animals.
4. **Polyspermic fertilization :** Here many sperms enter the egg. This is of two types -
   a. **Pathological polyspermy :** This takes place under abnormal circumstances when a monospennic egg allows the entry of many sperms. In this type the egg does not develop further.
   b. **Physiological polyspermy :** In animals with large yolky eggs many sperms enter the egg but eventually only one sperm nucleus fuses with egg nucleus and the rest degenerate. In this case the egg undergoes normal further development.
5. **Polyandry :** When two male pronuclei unite with a female pronucleus the condition is known as polyandry.eg. Rats.
6. **Polygamy :** Here two egg pronuclei unite with a single male pronucleus, eg. Sea urchins, rabbits etc.
7. **Gynogenesis :** In this only one sperm activates the egg but its pronucleus does not unite with the pronucleus of the egg. Eg. Plannarians, Nematodes etc.

10

BIOENERGETICS

All living cells are virtual biochemical factories where hundreds of reactions take place every second involving conversion, degradation and synthesis of chemical compounds that are vital for the maintenance of life. These reactions require energy or produce energy whic his either obtained from the breakdown of chemical compounds synthesis, or in the bonds of chemical compounds. In other words, these reactions are accompanied by energy changes. Thus, *bioenergetics may be defined as the study of energy changes that accompany biochemical reactions in living systems.* Bioenergetics may also be called biochemical thermodynamics, as the energy changes follow the laws of thermodynamics precisely.

In any biological system, the rectants have a higher energy level and they move **"energy down hill"** after the reaction. The energy thus liberated is wasted as heat either partially, or wholly. The energy that is liberated as heat cannot be put into any biological function and thus is a waste. The living systems, however, have developed means and mechanisms to trap the energy and store them in the form of energy rich chemical bonds. These 'energy currencies' readily donate the energy as and when required to do so, and help perform all the activities of living beings. Óne such universal energy currency is a molecule of ATP (Adenosine triphosphate). Whenever energy is required for a process, ATP is hydrolysed to ATP and the energy so released is utilized for the necessary activity. Similarly, there are a number of methods in which ATP molecules are synthesized (as during respiration).

The basic rules of bioenergetics are no different from those applicable to physico - chemical energy transformations. The only difference, perhaps, is that biomolecules involved in bioenergetics are more complex and function in a highly specialized environment. In order to understand the principles involved in bioenergetics it will be necessary first to understand the physiochemical principles involved in energy transformation reactions, and later these can be applied to biological situations.

Energy and matter

The universe is composed of only two components – *energy* and *matter* which were thought to be non interconvertible and fixed in their content. Einstein ,however, has shown that matter can be coverted into energy. What is energy? Energy may be defined as the ability or capacity of a system to carry out some work. This capacity is there in all the molecules of matter to a lesser or greater extent.

Types of energy

Basically, there are two main types of energy. These are – *potential energy* and *kinetic energy*. Potential energy is bound to the molecules. It may also be called as energy at rest i.e., not working. It can be measured in terms of how much work it is capable of doing when it is released. Potential energy is of various types – chemical, light (photic), atomic and even positional (such as that of a boulder on hill top). In biological world, potential energy source is mainly the solar radiation which is trapped, transformed and stored in the form of chemical energy during photosynthesis in greeen plants. All living beings obtain their energy requirements from the green plants directly or indirectly.

Kinetic energy is the other form of energy. It can be called as *energy in motion,* or energy released. Molecules or bodies have this energy as a result of motion. The amount of kinetic energy depends upon the mass of the body and also its velocity of motion. The following equation summaries this relationship.

$$KE = \frac{mv^2}{2} \quad \text{or} \quad KE = 1/2\, mv^2 \backslash v^2 \backslash v^2$$

In the above equation KE = Kinetic energy, *m* = mass of the body/molecule, and *v*= velocity of the body/ molecule.

Kinetic energy of molecules also depends on one other parameter viz., temperature. The amount of molecular kinetic energy in a system is proportional to the absolute temperature of that system and at absolute zero (-273.18^0 C), kinetic energy disappears, i.e., it is zero. For every rise of 10^{0}C in temperature, kinetic energy increases by the ratio of 10 to the absolute temperature.

Kinetic energy of molecules may be easily demonstrated through the Browinian movement of the particles. Brownian movement, is the to and fro movement of medium. This can be clearly seen when a beam of light is made to pass through a colloidal medium. The rate of movement of particles increases with the increase in temperature. The mechanism of Brownian movement is attributed to the random motion of molecules colliding with the visible particles, and imparting to them required energy for movement.

In addition to kinetic and potential, energy may also be of two other types. Free energy and Internal energy. Free energy may be defined as the energy that is liberated or consumed in a reaction which can be used or supplied to another system. Free energy is the most important type of energy in the biological system (see later in the chapter for a detailed discussion on free energy).

In any given substance, the internal energy is fixed under a given set of conditions. The internal energy of different substances, however, varies. Internal energy is symbolically represented by *E* or *U*.

Matter

This may be defined as something that physically occupies space and of which the entire physical universe is composed of. Originally, it was thought that matter can neither be created nor destroyed, but since the epoch making work of Albert Einstein, it has been found out that two atomic reactions *viz; nuclear fusion* and *nuclear fission* can convert matter into energy. But nuclear fusion or fission requires a very high temperature range (Nuclear fission occurs in sun and nuclear fusion is seen in atomic bombs), and they do not occur in ordinary chemical reactions.

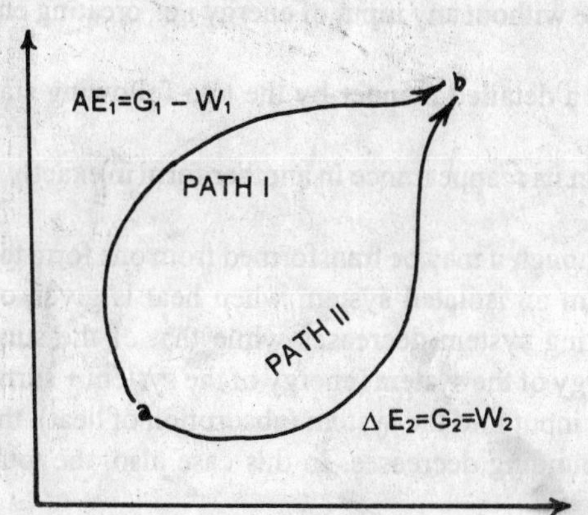

Fig. 10.1 Bioenergetics

Illustration of pathways for energy conversion of a into b

The study of matter and its energy, conversions follow the principles of thermodynamics. In order to study the principles of thermodynamics as it
relates to matter, an arbitrary position of matter, a system and its surrounding has to be imagined/considered. A system may absorb energy from the environment of liberate energy into the environment. In biological situations, *a system* may be a cell and its surroundings, or an organism and its environment or it may be an ecosystem. But the principles of thermodynamics remain the same irrespective of the size and structure of matter.

Laws of Thermodynamics

It has already been pointed out that the bioenergy conversions follow the principles of thermodynamics. Hence, an understanding of the laws of thermodynamics becomes necessary before attempting to understand the role of energy in biological reactions.

It has been mentioned above that irrespective of the size and structure of matter, the laws of thermodynamics are the same. Classic thermodynamics in reality deals with reactions in the equilibrium state, and may be regarded as *equilibrium thermodynamics.* Equilibrium thermodynamics is not concerned with the pace of the reaction, instead, it is concerned with the change in the form of energy from the initial stage (before the reaction has begun), to the final stage (when the reaction has been completed). Some of the other salient features of classic thermodynamics are –

(i) The number of steps between the initial and the final are irrelevant as long as the final state is same in different systems.

(ii) Direction of the reaction can be indicated in reversible reactions.

There are two laws of thermodynamics which are universally applicable. These are –

The first law of thermodynamics

This is known as the *Law of conservation of energy.* It states that the total content of energy is fixed and it can neither be created nor destroyed. But during reactions, energy can change form. In other words, the total energy in an isolated system will neither decrease nor increase. For instance when some amount of chemical, light or electric energy disappears (apparently), it would be transformed into equal amount of heat.

Previously, many attempts were made to disprove the principle of conservation of energy by constructing a 'perpetual motion machine', which is supposed to operate without any input of energy i.e., creating energy out of nothing. But these attempts failed.

The first law of thermodynamics may be enunciated in a detailed manner by the two following statements. These are –

(i) The disappearance of energy in one form will result in its reappearance in another form in exactly the same quality.

(ii) Energy of an isolated system will remain constant although it may be transformed from one form to another.

Elaborating the first statement, it may be said that from an isolated system when heat is given out (as in exothermic reactions), the internal energy of the reacting system decreases while that of the surrounding medium increases correspondingly. Hence, the total energy of the system (energy of the system + surrounding) remains unchanged. Alternatively, when there is energy input into the system (absorption of heat), the energy of the reacting system increases while that of the surrounding decreases. In this case also, the total energy remains unchanged.

As an illustration to elaborate the second statement, the burning of coal in a steam engine may be cited. Coal has a lot of internal energy. When coal is burnt, a lot of energy is released as heat, which is converted into mechanical energy of the engine i.e., it is allowed to boil water and release steam to rotate the wheels. In this, the chemical energy of coal gets transformed into mechanical energy of the engine.

Energy exchanges in chemical systems are studied with the help of a 'bomb calorimeter'. In the working principle, a calorimeter is nothing but a slightly modified thermos or Dewar flask. Basically, it has an insulation mechanism to isolate a system with internal energy as in a thermos flask.

A bomb calorimeter consists of an outer jacket (made up of insulating material not to allow heat to escape). This jacket surrounds a chamber which is filled with water. The centre of the calorimeter has a central chamber (combustion chamber) containing a sample which releases heat. The heat released by the sample increases the temperature of the water and its temperature is noted with the help of a thermometer. Since a calorimeter may not have a 100% insulation, corrections must be made for possible thermal leaks. With the help of this, it will be possible to show that the rise in temperature of water corresponds to the quantity of heat released by the sample.

Calculations of total energy (as per the first law)

Calculations of total energy in a reacting system can be made by finding out the quantities of the following :

(i) The work done

(ii) The heat that is exchanged, and

(iii) The energy stored in the system

Mathematically the above may be expressed as follows.

$$\blacktriangle E = q - w$$

Where ▲ *E* is change in the internal energy of the system, q is the heat absorbed by the system and *w* is the work done.

The values of *q* and *w* may be positive or negative, depending on the nature of reaction. If heat is absorbed, *q* has a positive value. If heat is given out, *q* has negative value. In the same way if the system performs work, *w* will be positive and if the medium performs work on the system, *w* will be negative.

The change in total energy does not depend whether *q* and *w* are negative or positive. For instance if molecules of *b, the change in energy is* ▲ E_1, and the other case ▲ E_2 then ▲ E_1 = ▲ E_2 The energy change her may also be represented as –

$$\blacktriangle E = Ea - Eb$$

Where Ea and Eb are energy levels of molecules of *a* and *b* respectively. The first law of thermodynamics needed some modification when Albert Einstein discovered that matter is convertible into energy. Accordingly, it is now agreed that matter (mass) can be converted into energy according to the following equation.

$$E = mc^2$$

Where E = energy; m = mass and c is velocity of light in cm/sec.

The reverse of this system i.e., conversion of energy into mass is also possible. Based on these observations, the first law of thermodynamics may be restated as *matter and energy can neither be created nor destroyed but can be inter converted.*

The interconversion of matter and energy however, does not occur in biological systems.

Second law of Thermodynamics

As per the first law of thermodynamics, the total energy in a reaction remains unchanged, but the transformation of energy from one to another proceeds in such a fashion that the capacity of the total energy to do work decreases. This is because all natural processes have a tendency to move towards equilibrium. The second law is actually related to the equilibrium of the process. Stated in very simple terms, it means during every energy change the capacity of total energy to do work decreases.

The second law of thermodynamics states that spontaneous reactions have a direction for the flow of energy. Once a reaction has reached an equilibrium, the reactants cannot return spontaneously to the initial stage.

Potential energy of molecules is often referred to as energy with a high degree of orderliness, whereas the kinetic energy is said to be random or disorderly. When once potential energy changes into kinetic energy, its randomness or disorder increases. This increase in randomness is called entropy. The stage of entropy is energy in that state where energy is incapable of doing any work because of its disorderliness. In otherwords when energy is transformed from one state to another state, the entropy goes on increasing untill it reaches an equilibrium. At equilibrium, entropy is maximum.

In nature, randomness or disorder is more a rule than exception. This is the reason why orderly arrangement which is special, tends to always move towards randomness. Entropy can be zero if a system is fully orderly. For instance, at absolute zero, the entropy of a crystalline substance is zero because there is no randomness.

The second law also stipulates that in any transformation of one type of potential energy into another type, the efficiency of the process is not perfect because some quantity of potential energy gets transformed into kinetic energy. Although theoritically, total kinetic energy can be transformed into potential energy, practically it is not possible because of inefficiency of potential energy conversions. Some transformation does occur as in the case of steam turbine or internal combustion engine, but in these instances only a small fraction of the kinetic energy becomes potential mechanical energy.

Ultimately, all potential energy of the system (whether it be a molecule, body, or even the entire universe) gets converted into kinetic energy or heat and in course of time (in finite), the entire potential energy may get exhausted i.e. Total entropy.

The fundamental principles of the second law of thermodynamics also operates in a food chain, in an ecosystem where starting from producers with a high degree of order (more number of individuals), succeedingly there will be less and less consumersin number of the primary, secondary and tertiary variety respectively. This is due to the fact that the whole of the potential energy of a producer cannot be transformed into the potential energy of a consumer. This is because of the intervening kinetic energy which introduces some amount of randomness (entropy).

Operation of the principles of second law of thermodynamics can be seen in our day to day life. For example water runs downhill and not uphill and when it has flown down its capacity to work gets decreased.

Green plants in this respect can be regarded as the most efficient mechanism to store energy for their own use and and for the use of other life forms. They trap the radiant energy and store an appreciable amount of it in the form of chemical energy. It is this potential energy which forms the fulcrum on which revolve all forms of life.

The concept of free energy

One of, the most important thermodynamic concepts that is of great relevance to biological systems is that of **free energy.** It is the energy that is liberated or consumed in a reaction and can be used or supplied to a system. It is denoted by the letter *G.* It may be regarded as a measure of the potential energy of a substance but it cannot be measured directly. However, change in free energy content that accompanies a reaction can be measured. This change is denoted by ΔG, may be defined as that quantity of total energy change in a system that is available to do the work. Change in free energy (ΔG) of a reaction is calculated by considering the difference between the sum of free energies of the reactants and products. This may be represented as follows.

$A \longrightarrow B$

In the above, *A* is reactant and *B* is in the product. Here, free energy is the maximum energy made available, as *A* is getting converted to *B*. If the free energy content of the product B *(Gb)* is less than the free energy content of the reactant A *(Ga)*, then ▲G will be negative (Here, reactant has more energy than the product). This may be shown as follows :

▲G = *Gb-Ga*

Here, *Ga* is greater than *Gb*, hence ▲G is negative. There is decrease in energy level as A gets converted to *B*. Similarly, when the reaction is reversible, *B* gets converted to *A*. In this reaction the level of free energy is increased (as B has less free energy).

In this case ▲G will be positive. Based on the fact whether ▲G is positive or negative (whether it gets reduced or increased), reactions are categorized into two types **endergonic** and **exergonic.**

Exergonic reactions

These reactions do not require the input of external energy. Normally, reactants have more energy than products. Products are *downhill* in energy level while reactants are *uphill* in energy level. Hence, no energy input from the medium is required to carry out the reaction. The reaction is driven by the inherent potential energy of the reactant molecules. Hence, the reaction can take place without any external energy supply. During exergonic reactions energy is released.

Endergonic reactions

Here, the reactant molecules have less free energy than the products. That is, product is *uphill,* and reactant *downhill.* Naturally, the direction of the reaction cannot be from down hill to uphill. The reactant molecules cannot negotiate on their own the uphill migration. Hence, energy from outside has to be supplied to the reactants to make them participate in the process. Here ▲G is positive *(GB–GA* and *GB* is greater than *GA,* where *GB* is free energy of reactant, and GA is free energy of the product).

The difference between endergonic and exergonic (energy requiring and energy releasing) processes can be explained with an example. Suppose a huge boulder is on top of hill, no extra effort is required to push it down hill; just a shove would do it. It has greater potential energy because of its position on hill top. When it reaches foot hill, its free energy is less because of its different position (location). This can be called exergonic process. Suppose the same boulder has to be now pushed back (from foot hill) uphill it requires an enormous effort (input of energy), which should come from outside and comparable to input of energy into a system. The boulder that is pushed uphill will again have more amount of free energy than the boulder at foot hill. This process is called endergonic, as there is energy requirement.

Photosynthesis and respiration serve as the best examples of endergonic and exergonic processes. For instant the overall formula of the processes.

1. $C_6H_{12}O_6 + 6O_2 + 6H_2O \longrightarrow 6CO_2 + 12H_2O +$ energy (respiration)
2. $6CO_2 + 12h_2O \xrightarrow{\text{light energy}} C_6 12_{12}O_6 + 6O_2\ 6H_2O$ (photosynthesis)

In the above equation for respiration, reactant (glucose) has more free energy than products. Hence, along with products energy also will be released. Reaction takes place without any energy input into the system as the reactants are *energy uphill,* and products *energy downhill.* The reaction is said to exergonic.

Equation for photosynthesis on the other hand shows that product viz., glucose has more free energy than the reactants viz CO_2 and H_2O. Hence, to drive the reaction *uphill,* there is requirement of energy input into the system which is provided by light energy. The reaction here is said to be endergonic.

Calculation of standard free energy change

In an exergonic reaction where ▲ G is negative, the rate of reaction however, is determined by the surrounding conditions like temperature (heat) etc. For instance suppose 10 molecules of reactants have to become products, not all the 10 molecules may participate in the process as they may have lost some energy because of collision among molecules. Hence, only some molecules which have the required high energy participate in a reaction. This is known as *activation energy.* Activation energy for a process is the higher potential energy that is required to start a reaction. Acvitation energy may sometimes be increased depending on the environment, like a medium having higher temperature. This means then the reactant molecules have varying speeds of the reaction even in exergonic reactions depending on the medium. For example, glucose can be oxidized to CO_2 and H_2O in different surroundings. In a bomb calorimeter, oxidation takes place in a few seconds; in living organisms it takes minutes to several hours. Similarly, even in the presence of O_2 (oxidizing agent), glucose can be kept in a bottle for years without large scale oxidation. In a reaction where A and B are reactant and product respectively.

s ▲G= ▲G⁰+RT B/A▲

Here ▲ G⁰ is standard free energy. R is the gas constant and T is absolute temperature in Kelvins. This equation shows that when conditions are ideal ▲ G and ▲ G⁰ are equal (▲ G = free energy and ▲ G⁰ is standard free energy). *Standard free energy change may be defined as the free energy change of chemical reaction when reactants and products are in their standard state.* In otherwords, standard free energy change of a substance is that value which it can ideally release when all conditions are optimum. Standard free energy change of a process can be used to find out whether a reaction is proceeding along the ideal path of not. If there is any difference between the standard free energy change and the actual free energy change that is going on in a reaction, it indicates that the conditions are not ideal necessiating corrective measures.

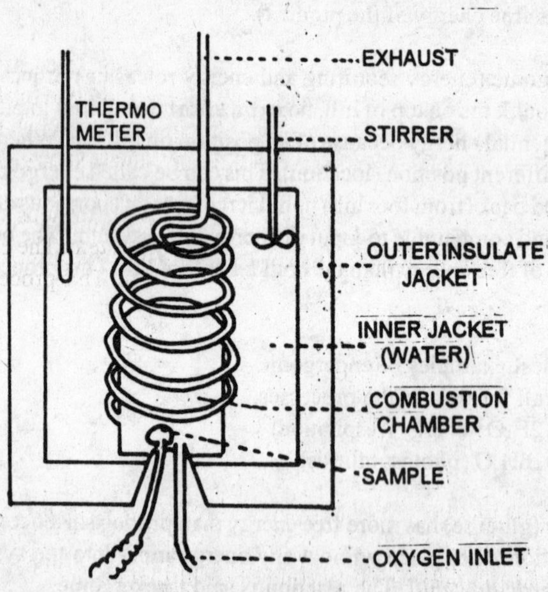

Fig. 10.2 Bioenergetics
A bomb calorimeter for determining heat of combustion of compounds

To summarize, the standard free energy change can be calculated as follows :

A+B ___ C+D (1)

Where A, B, C and D are molecules of substances participating in a reaction. The free energy change ▲ G, at constant temperature and pressure is given by

$$\blacktriangle G = \blacktriangle G^0 + RT \ln \frac{C+D}{A+B} \quad (2)$$

In the above, ▲ G = is free energy change; ▲ G⁰ = standard free energy change;
R=gas constant and T= absolute temperature.

In the equation (2), above ▲ G⁰ has a

If the reaction at equation (1) is allowed to proceed towards equilibrium, as the free energy decreases, the reaction will be able to carry out work at constant temperature and pressure. When equilibrium is reached, ▲G will be at its minimum or zero, and the system halts as it cannot do any more work. The equation at equilibrium is

$$O = \blacktriangle G^0 + RT \ In \frac{C+D}{A=B} \quad (3)$$

If we rearrange the above equation

$$\blacktriangle G^0 = -RT \ In \frac{C+D}{A+B} \quad (4)$$

Since the equilibrium constant K for the reaction would be

$$K = \frac{C+D}{A+B} \quad (5)$$

K can be substitued in the equation (4) to state as follows

$G^0 = -RT \ln K$ or $G^0 = 2.303\ RT \log K$

The above equation shows that if we can calculate the equilibrium constant for any reaction, the standard free energy change can also be calculated. In the same way, if ▲G^0 is known, ▲G is calculated either in calories/mole or joules/mole depending on whether gas constant is given in calories (R=1.98 mol^{-1} K-1) or joules (R=8.31 J mol^{-1} k-1)

As a conclusion to the discussion, the following summarized statements may be made about standard free energy change

(a) "The standard free energy change of a given chemical reaction is the difference between the sum of the free energies of the products and the sum of the free energies of the reactants, each reactant and product being present in its standard state" (Lehninger, 1970). The standard condition on the various components in a reaction system are –
   (i) Cencentration of 1.0m
   (ii) Temperature 25°C or 298°k
   (iii) Pressure 1.0 atm

Each compound has a particular architectural pattern that decides its inherent free energy. Hence, the standard free energy ▲G^0 of a chemical reaction may be expressed as follows

$$\blacktriangle G^0 = \underset{(product)}{EG^0} - \underset{(reactant)}{EG^0}$$

(b) The standard free energy change may also be expressed in terms of the equation A+B_____C+D. Here, ▲G^0 is that quantum of free energy absorbed or lost per mole when molecules of A and B get converted into molecules of C and D, when the conditions of all the components are standard. The following table gives the free energy changes of some chemical reactions.

Standard free energy changes at pH7 and at 25°C of some chemical reactions
(Adopted from Arthur Giese, 1976)

Oxidation	▲**F**
Glucose + $6O_2$ ----- $6CO_2$ + $6H_2$	-6,86,000
Lactic acid + $3O_2$ ----- $3CO_2$ + $3H_2O$	-3,26,000
Palmitic acid + $23O_2$ ----- $16CO_2$ + $16H_2O$	-2,338,000
Hydrolysis	

Sucrose + H_2O ----- glucose + fructose	-5,500
Glucose 6 phosphate + H_2O ----- glucose + H_3PO_4	-3,300
Glycyglocine + H_2O ----- 2 glycine	-4,600
Rearrangement	
Glucose 1, Phosphate ----- glucose 6. Phosphate	-1,745
Fructose 6, Phosphate ----- glucose 6. Phosphate	-400
Ionization	
CH_3COOH + H_2O ----- H_3O_3+ + CH_3COO -	+6,310
Elimination	
Malate ----- fumarate + H_2O	+750

A sample calculation of ▲ G^0

In order to elucidate the calculation of ▲ G^0 from the data obtained from equilibrium constant, the reaction in which the enzyme phosphoglucomutase can be considered. This enzyme catalyses the reversible reaction of glucose 1. Phosphate to glucose 6. Phosphate. We can take 0.020m glucose 1. Phosphate add the enzyme allowing forward direction change or take 0.020m glucose 6. Phosphate and allow the reaction to go in reverse direction. In either case, at equilibrium the medium contains 0.001m glucose. 1. Phosphate and 0.019m glucose 6. Phosphate at 25°C and at photosynthesis 7.0. The equilibrium constant can be calculated as follows

$$k = \frac{\text{glucose 6. Phosphate}}{\text{glucose 1. Phosphate}} = \frac{0.019}{0.001} = 19$$

From the value of k ▲ G^0 can be calculated as follows

▲ G^0 = RT in K

= - 1.987 x 298 In 19

= - 1.987 x 298 x 2.303 log 19

= - 1.745 cal mol^{-1} or - 7301 KJ mol^{-1}

Energy rich compounds

Living systems must get free energy from their environment in order to conduct life processes. Autotropic organisms obtain energy from sun light and their metabolism is linked to the exergonic process (respiration) to obtain energy. On the other hand, heterotrophic organisms couple their metabolism to the breakdown of complex to say that an endergonic process is always linked to an exergonic process.

In the living systems, a number of organic compounds are present which are energy rich (high energy compounds) because they undergo a large decrease in free energy change during their breakdown. These are also called energy currencies because they readily give out energy by undergoing hydrolysis. Some of these compounds are –

1. Pyrophosphate compounds
2. Acyl phosphates
3. Enolic phosphates
4. Thiol esters
5. Guanidine phosphates

Pyrophosphate compounds

Among the pyrophosphate compounds, the most important ones are ATP (Adenosine triphosphate) and ADP (Adenosine diphosphate), from the point of view of energy changes.

ATP: Historically, ATP was first discovered by C.F. Fiske and Y. Subbarow, in USA, and by K. Lohmann in Germany in 1929. After its discovery, it was thought that ATP was primarily involved in only muscle contraction reactions. Subsequent findings of Otto Warburg and Otto Meyerhof, indicated that ATP is generated from ADP during the anaerobic breakdown of glucose to lactic acid. Later, Kalckar of Denmark, and Belister of Russia, showed the generation of ATP (from ADP) during aerobic respiration also. Hydrolysis of ATP to ADP was demonstrated by Engelhardt and Lyubjmova during energy requiring reactions.

Finally, it was Fritz Lipman in 1941, who organized all the known facts of ATP into a general hypothesis concerning energy transformation reaction in living systems. Lipman opined that – ATP functions in a cyclic manner as a carrier ATP is used for various cellular activities. When fuel molecules undergo degradation, ATP is generated by the coupled phosphorylation of ADP. The ATP so generated can donate its third phosphate group for

energy requiring processes such as biosynthesis of macromolecules, transport of substances across the membrane against the concentration gradient etc. As the energy from ATP is released, it undergoes cleavage to form ADP. ADP will be rephosphorylated to ADP through the energy obtained by degradation of fuel molecules thus completing the energy cycle. The terminal (third) phosphate group is the key substance that undergoes continuous cleaving and regeneration.

For Adenosine nucleoside also, there are three types. These are –

(a) Adenosine monophosphate (AMP)
(b) Adenosine diphosphate (ADP)
(c) Adenosine triphosphate (ATP)

The phosphate groups of these are designated as α, β, γ. AMP has only phosphate group, ADP has α, and β, while ATP has α, β, and γ phosphate groups. The γ phosphate group is energy rich and undergoes cleaving and regeneration.

The structure of ATP was first deduced by Lohmann (1930), and subsequently confirmed through total chemical synthesis by Alexander Todd *et al,* (1948). The three compounds (ATP, ADP and AMP), are of universal occurrence in all living beings.

Fig.10.3 Bioenergetics
space filling model of ATP

The sun total of these three is constant (between 2 to 10mm) in a given species. In actively dividing cells, the amount of ATP exceeds those of ADP and AMP.

At pH 7.0 in aqueous solution both, ATP and ADP are highly charged ions. ATP has four anions while ADP has three protons. Three of the four protons of ATP get fully ionized at pH 7.0 while the fourth one will be dissociated about 75% as it has a pH of 6.95. The high concentration of closely placed negative charge around the triphosphate group in ATP has a significant bearing on its ability to carry out energy transformations.

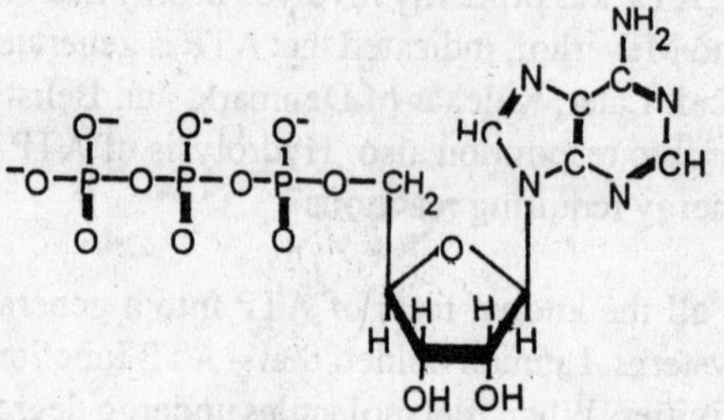

Fig.10.4 Bioenergetics
Molecular structure of ATP

Hydrolysis of ATP and Standard free energy

Theoretically, ▲ G^0 for ATP can be calculated by determining the equilibrium constant at pH 7.0. However there are several practical difficulties in this method as it is not easy to find out when exactly equilibrium has reached. One of the easier methods is to find out of the additive values ▲ G^0 when ATP participants are in consecutive reactions. For instance, reaction of ATP with glucose and its subsequent breakdown into glucose + phosphate will help us to calculate the total standard free energy of hydrolysis of ATP.

1. ATP + glucose ———— flucose 6. Phosphate + ADP
   Here k = 661 therefore ▲ G^0 = 4.0 cal mol⁻
2. Here k = 171 ▲ G^0 = 3.3 k cal/mol

Adding 1 and 2 of the above (-4+3.3) we get -7.3 k cal/mol as the total standard free energy available during the hydrolysisn of ATP.

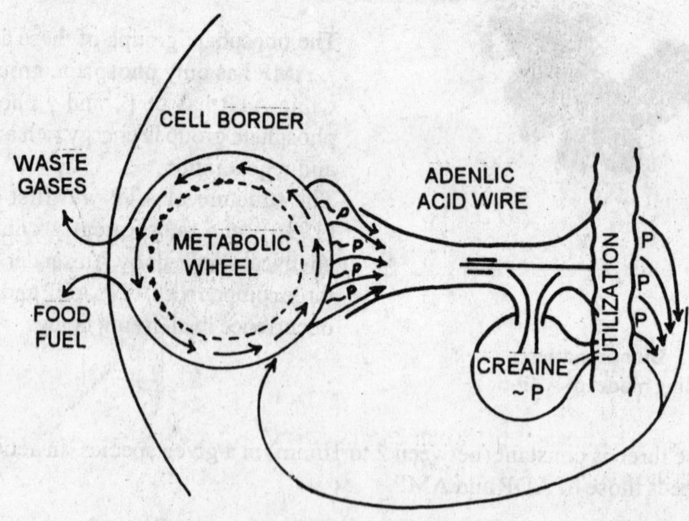

Fig. 10.5 Bioenergetics
The ATP cycle as depicted by Lipmann (1941)

The second phosphate group of ADP also has a similar value (-7.3 k cal mol.), while the single phosphate group of AMP has G^0 value of - 3.40 k cal/ mol. This is due to the fact that the first phosphate group is linked to the nucleoside with an ester linkage, while the other two are anhydride linkages. It is well known that anhydride linkages have a great G^0 than ester linkages.

In the above discussion on energy rich pyrophosphates, attention is given to only ATP and ADP, as these participate in almost all energy transformation reactions, hence called universal energy currency. However, there are other energy rich pyrophosphate compounds such as GTP, GDP, CTP, CDP, UTP, UDP, dATP, dGTP,dTTP and dCTP (These correspond to the nucleotides of RNA and DNA). These, however, participate only in specific reactions. Thus, UTP is used in the biosynthesis of carbohydrates - polysaccharides); GTP is used in protein synthesis and CTP is utilized in lipid metabolism. These three along with ATP are involved in RNA

synthesis, while dATP, dGTP, dCTP and dTTP are involved in DNA synthesis.
Another pyrophosphate compound that is energy rich is cyclic AMP (CAMP). Eventhough on hydrolysis, CAMP has a large negative free energy, it is not known to function due its high negative energy and unstable anhydride ring, instead, it acts as an allosteric effector and second messenger. Cyclic AMP is known to stimulate inactive enzymes (phosphorylase) into the active state.

2. Acyl phosphates

1.3 diphosphogylceric acid is an example of acyl phosphate. Its standard free energy on hydrolysis is – 11.8 k cal/mol. The strain of the C = 0 bond is responsible for a very high ▲ G^0 in this kind of compounds. High energy is required (during synthesis) in these bonds to overcome the repulsive changes of carbon and phosphorous atoms and consequently on hydrolysis a large amount of energy is released.

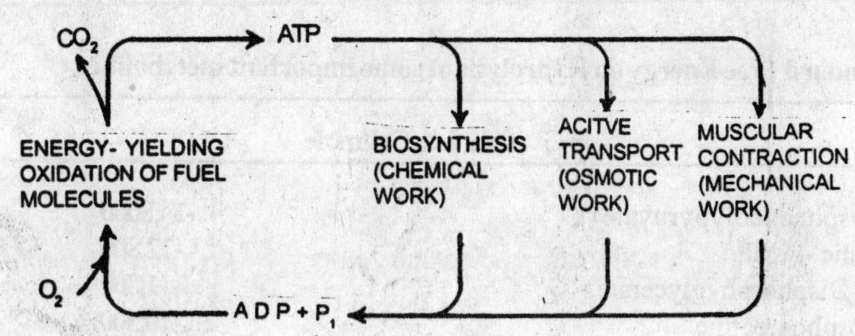

Fig. 10.6 Bioenergetics
The major uses of ATP energy

3. Enolic phosphate

An example of enolic phosphate is phosphoenol pyruvic and PEP. This is formed during the glycolysis when glucose gets converted intp pyruvate. On hydrolysis, it has a standard free energy of - 14800 cal/mole at pH 7.0

The large negative ▲ G^0 observed on hydrolysis of this compound, is due to stabilization of enolic form of pyruvic acid in PEP by the phosphate ester group. On hydrolysis, the unstable enol will instantly isomerize into the much more stable keto structure. It has been in ▲ G of about 8000 cal/mol thus bringing the total upto - 14800 cal/mole. This tautomerization of PEP makes it one of the most 'energy rich' compounds in biological systems.

4. Thiol esters

An example of thio ester is Acetyl CoA. This is also an energy rich compound which can be utilized to generate ATP from ADP. The ▲ G^0 of this compound on hydrolysis is approximately - 75v00 cal/mole. The higher negative value for ▲ G^0 is due to the fact that thioesters do not exhibit the resonance forms that is exhibited by ordinary O_2 atom. The sulphur in thio esters do not readily release the electrons for double bond formation and hence, resonance *forms possible* for oxygen ester. Usually, oxygen es atom is knocked to the carboxyl acid carbon by means of a double bond. As the thio ester linkage is unstable (to S), the ▲ G is higher on hydrolysis.

5. Guanidine compounds

These represent a fourth type of energy rich compounds. A typical example of guanidine compound is guanidine phosphate that plays an important role in energy transfer and storage. Also known as *phosphagous,* these compounds are formed by the phosophorylation of creatine to arginine with ATP in the presence of appropriate enzyme. Hydrolysis of these compounds has a standard free energy of – 1–300 cal/mole i.e., about – 3000 more than that of ATP.

The guanidine phosphates are not inherently less stable because of bond strain as in ATP and ADP. There are no ionization or tautomerization processes which account for greater stability of reactants than their products.

In addition to the above mentioned 5 compounds which are energy rich, there are other compounds'. Some of these are, glucose 6. Phosphate, fructose 6. Phosphate etc. These have a standard free energy less than that of ATP (i.e., less than - 7300 cal/mole). The table below gives the standard free energy (on hydrolysis) of some important compounds.

Standard Free Energy on Hydrolysis of some important metabolities

	▲ G' at pH 7.0 (Cal/mole
Phospheoneolypyruvate	-14,8000
Cyclic - AMP	-12,800
1,3-Disphosphyglycerate	-11,800
Phosphocreatine	-10,300
Acetyl phosphate	-10,100
S - adenosylmethionine	-10,000
Pyrophosphate	-8,000
Acetly - CoA	-7,500
ATP to ADP and Pi	-7,300
ATP to AMP and pyrophosphate	-8,600
ADP	-6,500
UDP - glucose to UDP and glucose	8,000
Glucose - 1 - Phosphate	-5,000
Fructose - 7 - Phosphate	-3,800

Glucose - 6 - Phosphate	-3,300
sn - Glycerol - 3 - Phosphate	-2,200

Coupling reactions

In biological systems, ATP is the source of energy for most of the endergonic reactions. ATP itself is synthesized in exergonic reactions by the process of phosphorylation. The endergonic and exergonic reactions are always coupled so that the energy synthesized from one (exergonic) is used to drive the other (endergonic). The net effect of this coupling is the transfer of kinetic energy from one system to the other. This coupling is mediated by a number of specific enzymes. Coupling reactions or energy coupling for efficient energy transfer is a common phenomenon. Let us consider an example in energy coupling.

The conversion of phosphoenol pyruvate to pyruvate is an exergonic process (energy releasing).

The liberated energy is used for the formation of ATP from ADP. This ATP can be hydrolysed to provide energy for other reactions. For instance, formation of glucose 6. Phosphate from glucose and phosphoric acid requires energy to thc tunc of – 3k cal/mole. Phosphorylation of ADP to ATP however, has free energy of –4k cal/mole. Participation of ATP in this reaction provides energy for the formation of glucose-phosphate bonds and the extra energy (of about -1k cal/ mole) is liberated into the system (waste). The above reaction may be shown as follows :

1. Glucose + Pi ———————— glucose 6, Phosphate + H_2O

(This is an endergonic reaction requiring energy of about –3k cal/mole)

2. ATP ———————— ADP + Pi

(This is an exergonic reaction releasing energy of about –4 k cal/mole)

Here reaction 1 cannot take place unless there is energy input. In order to do this reaction 1 gets coupled with 2 mediated by the enzyme hexokinase as follows :

Glucose + ATP hexokinase glucose 6. Phosphate + ADP

Energy coupling reactions produce complex compounds apparently against the second law of thermodynamics. But it has to be noted that the algebraic sum of all the components of coupled reactions is negative i.e., the sum of

energy liberated is more than the sum of energy consumed (-3 k cal/mole is used while the energy liberated -4k cal/mole). In biological systems, hydrolysis of energy rich compounds is always associated with an endergonic reaction in order not to waste the energy. This requires effective energy management which is provided.

Redox potential

The ability of any particular atom, or molecule, or ion to eject an electron (getting oxidized), is measured by its oxidation – reduction potential (ability to lose an electron - oxidation, and also receive an electron - reduction). This is known as redox potential and may be defined as *"The quantitiative measure of the affinity of a compound to lose or gain electrons"*.

During oxidative phosphorylation in mitochondria ATP is produced when electron are tossed from one acceptor to another. This flow occurs according to the relative affinities of the substrate to gain or lose electrons. This is accompanied by the alternate oxidation - reduction of the system. A compound losing an electron is oxidized, while the one accepting an electron is reduced. For instance, ferric ion is oxidized to ferrous ion by losing an electron, and when ferrous ion accepts an electron, it gets reduced to ferric ion.

Fe^{+++} oxidation $Fe^{++} + e$

reduction

A redox system can be compared to a dry electric cell. In the cell, the electrons are then transferred through a wire from one electrode to another, generating electric current. In such a system the capacity to lose or gain electrons is called the electrode potential. This can be measured through a standard hydrogen electrode. The electrode potential of a reducing oxidizing system also can be measured is a similar way. The electrode potential is here called the redox potential.

11

ENZYMES

A plant cell is verily a microchemical factory in which hundreds of chemical reactions are taking place at amazingly high speeds and that too at normal temperature ranges. These biochemical reactions provide sustenance for a plant to absorb, assimilate, synthesize, degrade, grow and reproduce which are essential for survival. Many of these reactions are simple, but most of them are complex and involve several sequential steps in order to produce a particular product. Conversion of one compound to another has to be precise and exact, there being no excuse for mistakes. For a long time biologists and biochemists wondered as to how biological reactions occur in *vivo* in such precise manner and at such rapid rates without involving undue increase of temperature. Very soon they got an answer, when Buchner (1897) accidentally discovered that a juice extracted from a yeast cell can perform fermentation as the living yeast cells can do. This means then that the juice contained some substance which can bring about fermentation. The chemical substance bringing about fermentation was soon isolated and given the name *zymase.* Thus started the glorious story of enzymes. Enzymes can be described as *Chemical middlemen* mediating and accelerating a variety of biological reactions. It may not be an exaggeration to state that virtually no metabolic activity in living organisms is possible without the participation of enzymes. Each reaction requires the participation one, two or some times many specific enzymes. What are these enzymes? What is their role in plant metabolism? How do they accelerate chemical reactions? We shall try to find answers to these in this chapter.

The word *enzyme* is derived from Greek meaning -*En* = in ; *zyme* = leaven or living. The term was first coined by Kuhne (1878) while working on fermentation, even though it was Buchner who first made an extract of enzyme from yeast cells. Robert Sumner (1926) was the first scientist to purify and crystallize an enzyme revealing the proteinaceous nature. Eventhough the scientific discovery of enzymes may be comparatively recent, their use for practical purposes was known to early Greeks and Indians. The *Arthasastra* of Kautilya (4th century BC) mentions the use of extracts of barks of certain trees helpful in the process of fermentation.

Definition of enzymes : a. Biocatalysts; b. Organic catalysts; c. specific proteins, simple or compound in structure acting as specific catalysts; d. Biological middlemen.

Structure and chemical composition : All the known enzymes are proteinaceous. These proteins called functional proteins have a very complex structure and have very high molecular weights sometimes ranging up to several millions. Even some of the simple enzymes like *pepsin* (35,000) *urease* (4,83,000) etc., also have high molecular weights. While all enzymes are made up of proteins, they also have a non protein part. The bulk of the enzyme however is proteinaceous.

The total enzyme is given the name *holoenzyme* (Protein + non protein), while the protein part is called the *apoenzyme* and the non protein part is given the name *prosthetic* group. When the prosthetic group is not tightly bound to the apoenzyme it is called a *coenzyme* or *cofactor.* The presence of a nonprotein part in the enzyme is a must as otherwise the enzyme will be inactive. The following are some of the structural attributes of the enzymes.

(i) Apoenzyme : The protein part constitutes the major part of an enzyme molecule. The complex proteins that make up the apoenzyme have specific sequences of aminoacids contributing to the specific reactivity of the enzymes. As the proteins are colloids, they offer a large area in relation to their volume.

Very few enzymes are wholly made up of proteins. E.g. proteolytic enzymes.

Active site : All enzymes possess an area in their molecular organization where substrate materials bind themselves in order to undergo chemical change. This binding site is called the *active site of* an enzyme. An enzyme may have one or more active sites. In addition to the active sites an enzyme may also have *regulatory sites,* to which regulatory substances bind and regulate the activity of the enzyme.

Prosthetic group : Most of the enzymes possess a non protein part tightly bound to the protein part knows as the prosthetic group. Many metallic ions like Cu, Zn, Mn, Mo etc., constitute the prosthetic groups. Organic compounds such as cytochromes, flavoproteins, pyridoxal phosphate, biotin etc., are also known to be prosthetic groups of certain enzymes.

Coenzymes and Cofactors : Strictly speaking there is no distinction between a prosthetic group and a coenzyme except that the latter is loosely bound to the enzyme molecule. The term cofactor is employed for inorganic ions such as CV, Mg^{++} etc., while organic molecules are called coenzymes.

Distribution of enzymes in plant cells

Enzymes are found distributed all over the plant cell. But there is "a qualitative distribution i.e., not all enzymes are found in all the areas. Enzymes have a localized distribution with reference to the functions they have to perform. The distribution of some important enzymes are listed below.

1. **Nucleus :** *Polymerases, Transferases* etc.
2. **Chloroplast :** *Carboxydismutases, Phosphatases, Kinases, Dehydrogenases* etc.
3. **Mitochondria :** Enzymes of respiratory pathway - *Dehydrogenases, Cytochrome oxidases* etc.
4. **Cytoplasm :** *Aldolases, Isomerases, ATPases, Phosphorylases, Transphospnolylases* etc.

(The list is not comprehensive)

Nomenclature and Classification

The name of the enzyme is usually descriptive and consists of two parts - the first part denotes the substrate on which the enzyme acts, while the second part indicates the type of action of the enzyme. For example *Isocitric dehydrogenase* indicates the enzyme acts on Isocitric acid and removes , hydrogen from it. In some of the older nomenclature, this is not followed and the enzymes are named arbitarily. For E.g. *Pepsin, Trypsin, Chymotrypsin* etc.

According to the commission on Enzymology of the International union of biochemistry, all enzyme names should end with *ase.* Enzymes are classified into six major divisions. These are -

1. Oxidoreductases
2. Transferases
3. Lyases
4. Isomerases
5. Ligases (= synthetases) and
6. Hydrolases.

Oxidoreductases : Enzymes which bring about oxidation reduction reactions are included under this group. This group is further subdivided into the following.

(i) *Oxidases* : Reactions involving oxidation of the substrate with molecular oxygen serving as the electron acceptor belong to this category.

(ii) *Peroxidases* : These catalyze the removal of hydrogen from the substrate which combines with hydrogen peroxide (H_2O_2)

$$SH_2 + H_2O_2 \longrightarrow 2H_2O + S$$

(iii) *Dehydrogenases:Enzymes* which bring about oxidation of the substrate by removing hydrogen from it belong to this category. The hydrogen acceptor however is not molecular oxygen.

$$SH_2 + R \longrightarrow S + RH_2$$

(iv) *Reductases* : These remove only electrons but not hydrogen ions from the substrate.

(v) *Oxygenases*: These catalyze the incorporation of both the atoms of oxygen into the substrate.

(vi) *Hydroxylases* : These enzymes catalyze the reaction in which only one atom of oxygen is added to the substrate.

(vii) *Catalases* : These liberate molecular oxygen from hydrogen peroxide.

2. Transferases : These enzymes transfer a particular group from one substrate to another. These are further divided as follows based on the type of group transfer which they catalyze.

(i) *Transketolases* : Transfer a ketone group

(ii) *Transaminase* : *Traasfer* an amino group

(iii) *Transaldolases* : Transfer an aldose group

(iv) *Transphosphorylases* : Transfer an energy rich phosphate group

(v) *Transcarboxylases* : Transfer a carboxyl group

(vi) *Hexokinases* : Transfer an ordinary phosphate group

(vii) *Transglycosylases* : Transfer of a glycosyl group

3. Hydrolases : These enzymes catalyze the hydrolysis of complex substrates into simpler ones (starch into glucose). These include *Carbohydrases* (hydrolysis of carbohydrates), *Lipases* (hydrolysis of fats), *Phosphatases* (removal of phosphate groups) etc.

4. Lyases : These catalyze the removal of groups from the substrate without the addition of water (non hydrolytic).

Fructose 1,6 diphosphate ⟶ Dihydroxy acetone phosphate + phosphoglyceraldenyde

Other examples of lyases are - *Aconitase, Isocitrase, Fumarase* etc.

5. Isomerases : They bring about isomerization of one substrate to other (intramolecular shifting of atoms)

Glucose 6 phosphate $\xrightarrow{\textit{Isomerase}}$ Fructose 6 phosphate

6. Ligases : Also called synthetases, they help in the synthesis of a new compound generally with the clevage of ATP or other nucleoside phosphates.

Isoenzymes : Enzymes which perform similar functions but having different molecular structures are called *Isoenzymes* or *Isozymes*. In other words *different molecular species of an enzyme can be called Isozymes.* Isozymes can be identified based on the appearance of different bands on a polyacrylamide gel. For example the enzyme *lactic dehydrogenase* has five isozymes.

Zymogens : These are enzyme precursors or inactivated forms of enzymes secreted by cells. These can be activated to the normal enzyme state by chemical modification.

Constitutive and adaptive enzymes : Not all enzymes are present in the site of action always. Some are synthesized at the time of reaction and soon get degraded after the reaction. Such enzymes are known as *adaptive enzymes.* The other category - *constitutive enzymes* are present always and participate in a reaction whenever required.

Properties of enzymes (Physical and chemical) : Enzymes are characterized by the following properties.

1. Catalytic properties : All enzymes are catalysts and as such they accelerate the pace of a reaction.

Being true catalysts they do not undergo any chemical change and do not figure in the end products of a reaction. They also do not disturb the equilibrium of a reaction. Being catalysts they are required in very small quantities. The number of molecules of substrate converted into product per minute is called the *turnover number* of the enzyme indicating its efficiency. Turnover number for different enzymes varies from 100 - 3,00,000.

The velocity of enzyme action like that of a true catalyst is proportional to the concentration of the substrate/concentration of the enzyme.

2. Reversibility of action : Like all catalysts enzymes can accelerate the pace of a reaction in both directions. The direction depends on the factors present at that particular time. It should be specified here that an enzyme by itself does not initiate a reaction but merely speeds up the rate. Thus an enzyme by itself cannot change the direction of a reaction. For example, the enzyme *starch phosphorylase* catalyzes the hydrolysis of starch to glucose during day, but brings about the synthesis of starch during darkness. So here it is not the enzyme that decides the direction of a reaction, but the presence or absence of light, pH etc., which do so. The enzyme merely speeds up the rate of the reaction in whichever direction it is proceeding.

3. Specificity : Enzymes are highly specific in their action. A particular enzyme catalyzes only a particular kind of a reaction. Sometimes if a reaction involves several steps, an enzyme will be specific for each step. Enzyme specificity is of the following types.

(i) *Absolute specific* : Acts on only one kind of a substrate. For example *malic dehydrogenase* acts only on malic acid.

(ii) *Absolute group specific* : Acts on a given organic group. For e.g. alcoholic dehydrogenase acts only on alcoholic group.

(iii) *Relative group specific* : Enzyme catalyzes more than one organic group. E.g. *Trypsin* acting on ester bonds involving Arginine or lysine residues.

(iv) *Optical specificity* : Certain enzymes can distinguish between optical isomers in the substrate and catalyze any one only. For example *l-amino acid oxidase* will not act on D- amino acids and *vice versa.*

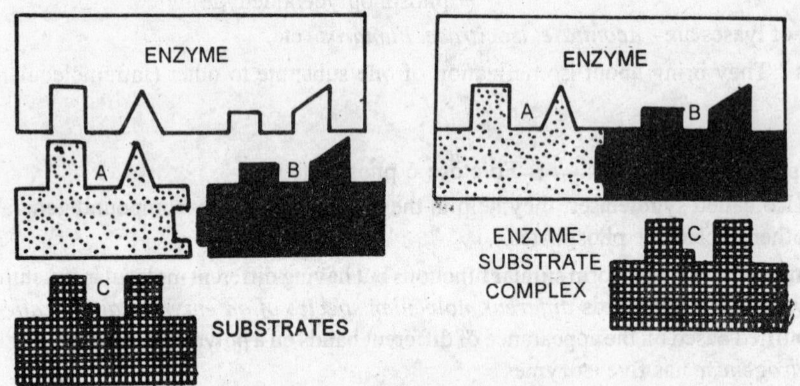

Fig. 11.1 Enzymes
Diagram to illustrate enzyme substrate specificity

4. Heat sensitivity : Enzymes remain active under normal temperature and are denatured under high temperatures (60 - 70°C). Under dehydrated conditions however as in seeds they can withstand temperature up to 120°C. Hence enzymes may be termed *thermolabile.*

5. pH sensitivity : Most of the enzymes are active only under a particular range of pH. Any alteration in pH would render the enzymes inactive. Enzymes are denatured by strong acids or alkali. Optimum pH for each enzyme varies- depending on the nature of the buffer system (if present, absorbed ions, presence of activators/ inhibitors etc.

6. Inhibition of enzyme action : Most of the enzymes have active sites to which the substrate binds to undergo conformational changes. In some instances it has been noticed that certain other substances which are partially identical to the substrates bind themselves to the enzyme and inactivate it. These substances arc called inhibitors and their action is called inhibition. There are two types of inhibition *viz.* competetive inhibition and non competetive inhibition.

(i) *Competetive inhibition* : In this, the inhibitor molecule is structurally analogous with the substrate molecule. It competes with the substrate molecule to occupy the active site of the enzyme. Due to the occupation of the active site by the inhibitor molecule, the enzyme gets inactivated i.e., it cannot accept a substrate molecule. For example *succinic dehydrogenase* is inhibited by molecules of malonic acid.

Competetive inhibition can be overcome by the addition of more molecules of substrate to the system which can now favourably compete with the inhibitor molecules for the active site of the enzyme.

(ii) *Non competetive inhibition* : The inhibitor molecules are not structural analogs of the substrate, hence they do not bind to the active site of the enzyme, instead they bind themselves to the enzyme in some other area, all the same the enzyme gets inactivated due to conformational changes. Heavy metallic ions such as Mg^{++} Ag^{++}, Pb^{++} etc combine with the sulfhydryl groups of the enzyme rendering it inactive.

Non competetive inhibition is irreversible. Addition of extra molecules of substrate is of no consequence since the active sites are anyway available to the substrate.

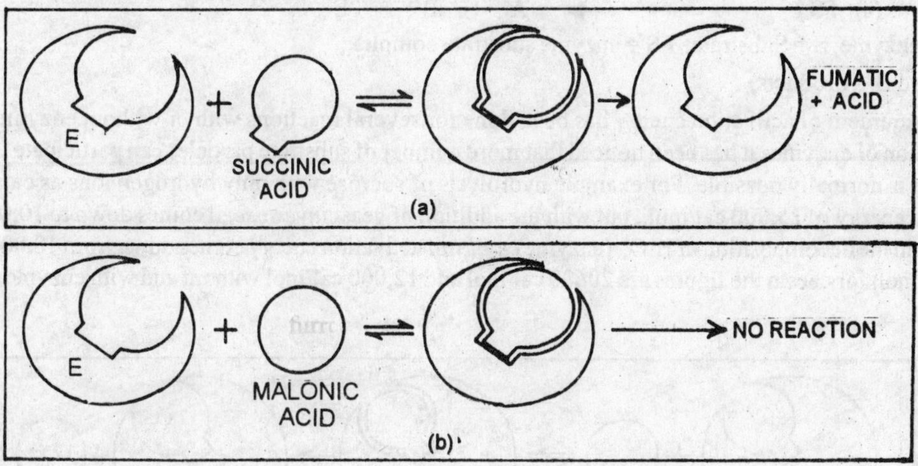

Fig. 11.2 Enzymes
Competetive inhibition of enzyme action

7. Colloidal condition : The enzymes are large molecules of proteins and exhibit all the colloidal properties. They cannot pass through a membrane *of collodion.* The enzymes have the capacity of adsorption.

8. Redox potential value : Certain enzymatic reactions which are reversible, depend on 'the redox potential value for the direction of the reaction.

MECHANISM OF ENZYME ACTION

How does an enzyme catalyze or speed up a reaction? The following explanation may be offered by considering the conditions of a biochemical reaction.

Normally chemical reactions take place when molecules of any substance possessing on an average a higher kinetic energy change into a substance whose molecules have a lower average of kinetic energy. Although on an average all molecules of reactants (substrate) possess equal amount of kinetic energy, a few of them possess higher kinetic energy and a few of them possess lower than the average 'due to collisions among molecules. Thus among the molecules of a substrate there are *energy rich* and *energy poor* molecules. If we for instance take a reaction where molecules of A get converted into B, at any given time only few molecules of A can get converted into B because of their higher kinetic energy. This extra energy required for the participation in the reaction is called activation energy. The activation energy is always above the average kinetic energy of the molecules. In a non enzymatic reaction, the average kinetic energy of molecules may be raised to activation energy by increasing the temperature of the medium. This speeds up the rate of reaction. Theoretically continuous increase of temperature should indefinitely increase the rate of reaction. But this is not possible actually because, a higher temperature beyond a certain point would kill the whole system. The enzymes in living cells do the same thing *in vivo* what the temperature does *in vitro* (they speed up the rate). But the enzymes do not increase the kinetic energy of the molecules. Instead they combine with the molecules and lower the activation energy necessary for reaction. The combined product of enzyme and substrate known as *enzyme substrate complex* crosses the activation energy barrier much easily than what the plain substrate molecules can do. All enzymatic reactions are believed to take place as follows.

(i) E + S ⟶ ES

(ii) ES ⟶ A + B + E

(E = enzyme; S = Substrate; ES = enzyme substrate complex;

'A and B ~ products)

Measurement of activation energy has been done for several reactions with or without enzymes. With the participation of enzymes it has been noticed that more number of substrate particles can participate in a reaction than what is normally possible. For example hydrolysis of sucrose with only hydrogen ions as catalyst has an activation energy of 25,000 cal/mol., but with the addition of yeast *invertase,* it comes down to 10,000 cal/mol. Similarly in the decomposition of H_2O_2 (enzyme *catalase)* activation energy comes down from 18,000 cal/mol to 2000 cal/mol; for caesin the figures are 20600 cal/mol and 12,000 cal/mol without and with enzyme *(trypsin).*

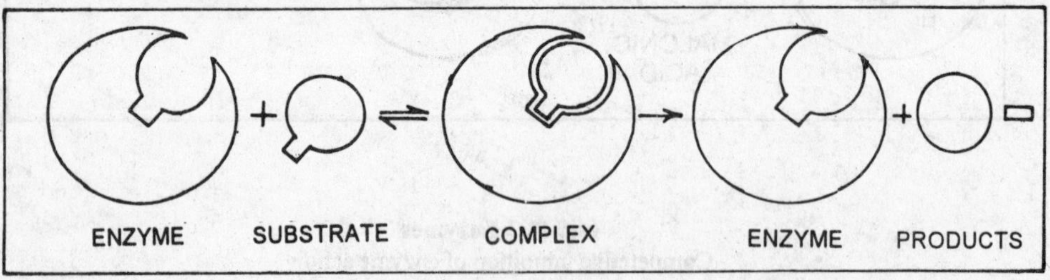

Fig. 11.3 Enzymes

Graph to demonstrate lowering of activation energy due to enzyme catalysis

Enzyme kinetics (Michaelis-Menton equation): Michaelis and Menton (1913) proposed a theory for enzyme action, in which they believed that the substrate and enzyme form a loose complex, and at the end of reaction the complex breaks releasing the enzyme and products as explained above. In order to explain the rate of substrate conversion, they applied the laws of kinetics which may be represented as follows.

$$E + S \underset{K_2}{\overset{K_1}{\longrightarrow}} ES$$

(E = enzyme; S = substrate; ES = enzyme substrate complex; K_1 velocity constant of the formation of ES and K_2 velocity constant of dissociation of ES)

OR

$$E + S \xrightarrow{K_1} ES \xrightarrow{K_2} P + E$$

(K_2 = velocity constant for decomposition of ES. P = end product)

The equation can be further simplified as $\dfrac{K_1 + K_2}{} = K_M$

K_M is known as the velocity constant or Michaelis and Menton's constant. Its value is characteristic for each enzyme and substrate and by comparing *KM* values of an enzyme with different substrates, the affinity of different substrates to an enzyme can be calculated. When the velocity of the reaction is half of the maximum velocity, *KM* is equal to the concentration of the substrate. However Michaelis Menton equation is applicable to only single substrate reactions.

MODELS FOR ENZYME ACTION

Two models have been proposed to account for the specificity and mode of enzyme action. These are (i) the lock and key theory and (ii) the induced fit theory.

Lock and key model : This theory originally proposed by Fischer (1898) believes that the enzyme specificity can be accounted for by assuming that the enzyme and the substrate fit into each other structurally like the two pieces of a jigsaw puzzle or the lock and key. Just as only a particular key can fit into a particular lock because of the alignment of levers, enzymes have specific active sites, to the contours of which the substrate molecules can fit in.

No other molecule can fit in to this slot. When a reaction mixture consists of several substrates, like A, B, C, D etc., only A and B can fit into the active sites and undergo conformational change resulting in the product P (This is an example for a synthetase enzyme; if the reaction involved is clevage of one substrate molecule into two products only one active site will be present on the enzyme and only one substrate A or B can fit into the active site)

The lock and key mechanism assumes a rigid structural configuration for the enzyme molecule and does not allow even minor adjustments between the enzyme and the substrate.

(ii) **Induced fit model :** Many experimental findings questioned the rigid unchangeable configuration for the enzyme molecule as proposed by the lock and key model. It has been noticed that the enzyme molecule does allow for minor adjustments in its geometry to permit the substrate to fit into the active site. Secondly it has also been observed that the substrates induce proper orientation of the catalytic groups in the enzyme in order to activate enzyme action.

In order to make provision for the above findings, Koshland (1959) proposed a theory called the induced fit theory. According to the induced fit model a substrate after linking itself with the enzyme alters the geometry of the enzyme protein in such a way as to achieve perfect alignment.

Allosteric regulation : In reactions which involve several sequential steps many enzymes participate, each one catalyzing a particular step. Such enzymes constitute the multienzyme complexes. In multienzyme complexes the product of one step is a substrate for the next step. The intermediate products are transfered from one enzyme to the other. In these enzyme complexes, there is end product inhibition of all the enzymes involved. In otherwords the end product of the pathway blocks the functioning of the first enzyme (of the complex) by binding itself to a site called the *allosteric site* which is different from the active site. The end product which binds to the allosteric site is called the *modulator* and the enzyme itself is known as the allosteric enzyme (allosterism means another space). The name allosteric enzyme was first proposed by J.Monod, J.P. Changeux and F. Jacob. Allosteric regulation is different from non competetive inhibition in that it is the *end product* of the pathway which blocks the enzyme action of the first in the series of the enzyme complex. Allosteric regulation need not always be inhibition of enzyme action *(negative modulation)*. In some cases the end product may even activate the first enzyme of the series *(positive modulation)*.

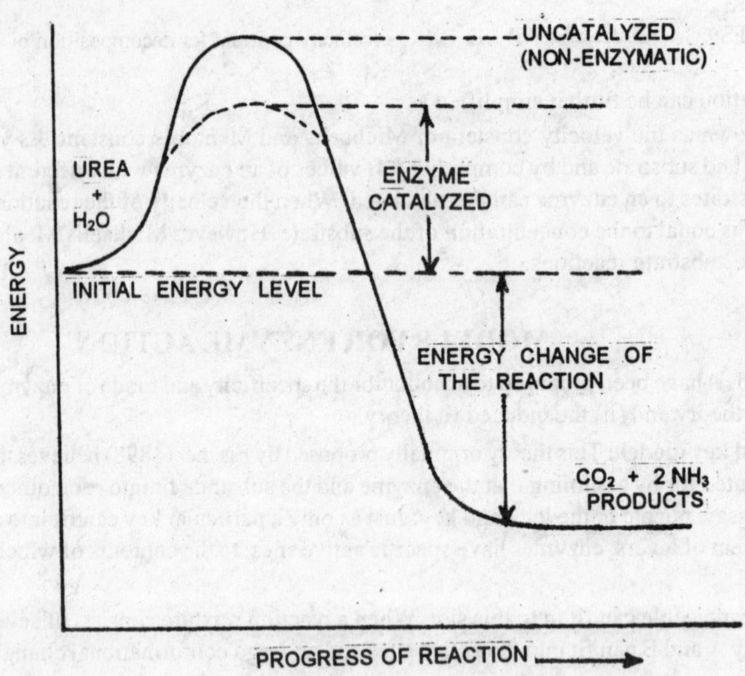

Fig. 11.4 Enzymes

Lock and Key mechanism of enzyme action

Double displacement (ping pong) reactions : In enzyme reactions involving a single substrate, it is well known that the substrate concentration influences the rate of the reactions. In many biological reactions however, a single reaction with the enzyme involves two or more substrates participating simultaneously. For example in the following reaction involving *hexokinase,* ATP and glucose are substrates with ADP and glucose - 6 - phosphate being the products.

$$\text{Glucose + ATP} \xrightarrow{\text{hexokinase}} \text{Glucose 6 phosphate + ADP}$$

In a situation as mentioned above what will be the influence of the substrate concentration on the enzyme activity? In several bisubstrate reactions it has been observed that the enzyme has a separate velocity constant (KM) for each of the substrates involved.

Bisubstrate reactions usually involve the transfer of a functional atom or group from one substrate to other.

This may proceed in one of the two following ways - a) *single displacement method* and b) *double displacement method.* In the first type, both the substrates bind to the enzyme simultaneously to form the enzyme substrate complex.

$$A + B + E \rightarrow EAB$$

(A – substrate EAB – Enzyme substrate complex
B – substrate E – enzyme)

EAB then breaks up to form the products C and D

$$EAB \longrightarrow C + D + E$$

In the second category of bisubstrate reaction called the *double displacement type,* only one substrate can combine with the active site of the enzyme at any given time. After the substrate transfers its functional group to the enzyme molecule, it (substrate) leaves the enzyme as the product. The second substrate now binds to the enzyme (which has the functional group of first substrate) accepts the fanctional group donated by the first substrate and leaves the enzyme as product.

$$E + A \rightarrow EA$$
$$EA \rightarrow E^{I} + C$$
$$E^{I} + B \rightarrow E^{I}B$$
$$E^{I}B \rightarrow E + D$$

(A and B = substrates, C and D = products, E = enzyme EA = enzyme substrate complex, E' = enzyme with the functional group of substrate A, E'B = enzyme substrate complex; enzyme has the functional group of A.

The enzyme which has the functional group of one substrate after it (substrate) has been released as the product, may be called *a partial enzyme substrate complex*

This type of a reaction in which the substrates alternately bind and get released to the enzyme is called **ping pong reaction or double displacement reaction.**

Transamination reactions involving the enzyme *transaminase* provide a typical example of ping pong reactions. In a reaction involving glutamic acid and oxaloacetic acid, the enzyme glutamic transaminase has pyridoxal phosphate as the prosthetic group. In the first step of the reaction, glutamic acid binds to the prosthetic group of the enzyme (pyridoxal phosphate) and donates the amino group and is released as *a* keto glutaric acid (product). In the second step, oxaloacetic acid binds to the pyridoxal phosphate (which has the amino group bound to it) accepts the amino group and is released as Aspartic acid (product).

```
 COOH                            COO
  |                               |         H
 CH2            E-C-H            CH2        |
  |                 ‖             |    + E-C-H
H-C-NH3             O            CH2        |
  |                               |        NH2
 COOH                            C=O      Enzyme
                Enzyme            |     with amino group
Glutamic acid  (pyridoxal phosphate)  COOH
                                 β ketoglutaric acid

 COOH                            COOH
  |             H                 |
 CH2            |                CH2
  |       +  E-C-H                |      +  E-C-H
 C=O            |              H-C-NH3          ‖
  |            NH3                |             O
```

COOH COOH Enzyme
Oxalo acetic acid Aspartic acid *(Pyridoxal phosphate)*

FACTORS AFFECTING ENZYME ACTION : The following factors influence the enzyme action.

1. Enzyme concentration : At specific concentrations and in the presence of large number of substrate molecules, the reaction will proceed at a particular speed. Any addition of substrate at this stage is not of any consequence as all the anzyme active sites are already occupied. In such a situation increasing the concentration of the enzyme will increase the rate of the reaction. This increase stops at a point when substrate molecules become a limiting factor.

Substrate concentration

As the enzyme molecules are very large in size when compared with the substrate molecules, relatively large amount of substrate is required to engage all the enzyme molecules. At a particular concentration of the enzyme, if the substrate molecules are at a low concentration, the rate of the reaction will not proceed at the required pace since many active sites of the enzyme are unoccupied. At this stage addition of more substrate to the system would accelerate the pace of the reaction as many enzyme molecules can be engaged. Indefinite increase of substrate concentration however would not increase the pace of reaction indefinitely as enzyme concentration would become a limiting factor then.

Temperature

Like every other biological reaction, enzyme catalysis is also influenced by temperature. Since enzymes are proteinacious, they are very sensitive to temperature changes. They can function at a very narrow range of temperature since enzymes are functional even at as low a temperature as 0ºC but the rate of the reaction will be at its minimum. A rise in temperature at this stage would bring in a geometric progress in the rate of the reaction. Very soon this reaches a peak and this temperature level is known as the optimum. Further increase beyond the optimum temperature would actually retard the rate of reaction. A higher temperature of 50ºC and above would incapacitate the enzyme.

The Q_{10} for all enzymatic reactions is about 2.0. The higher value of Q_{10} clearly shows that temperature has a marked influene on enzymatic reactions.

Hydrogen ion concentration (pH)

Most of the enzymes are catalytically active only at a particular pH. Sometimes they may have a range of pH at which their activity will be sustained. Any increase or decrease from this pH would render the enzymatic activity impossible. The different enzymes however would vary in their pH requirement. While some enzymes are active at a low pH, others prefer an alkaline pH; still others require a neutral pH for their optimum activity.

Inhibitors

The presence of certain inhibitor molecules in the system would render the enzyme inactive. Various substances are known to inhibit enzyme action. These are of 3 kinds.

1. competitive inhibitors
2. Non - competitive inhibitors
3. End product accumulation

In many cases if the end products of a reaction are not removed from the system they acumulate and bring about inhibition of enzyme action known as feed back inhibition. The feed back inhibition many a time may be due to allosteric effect as can be seen in reactions involving multienzyme complexes. The end product inhibition of enzyme action may also be due to a difference in the pH value of the substrate (at which enzyme is active) and of the end product.

12

PHOTOSYNTHESIS

Eversince its origin, sometime in the remote past, life has been able to maintain itself in myriads of forms mainly because of its capacity of reproduction. An organism however has to exist, maintain itself in the healthy condition and only then can it reproduce. The existence of all forms of life on this planet has been made possible mainly by means of energy conversions. Organisms trap energy from environment which they utilize for their activities and after a series of interconversions it (energy) is returned to the environment.

The phenomenon of energy input from the abiological environment to the biological system is always through the plants-green plants. .Green plants thus can be regarded as the basic fulcrum on which revolve all forms of life. In fact it is impossible to imagine what course the evolution of life would have taken but for green plants. All forms of life are dependent on green plants (directly or indirectly) for their sustenance.

This unique phenomenon of energy input into the biological system is known as photosynthesis. The energy input has to be mediated only through the green plants for they alone are capable of absorbing the energy from solar radiation .

Source of energy : There are only two sources of energy on this planet. One is atomic energy and the other is solar energy. Of these two-atomic energy is not easily negotiable and available. Solar energy however is easily available in large quantities and living organisms have developed in the course of evolution mechanisms to trap and utilize this unique source of energy.

All forms of life however cannot directly harness solar energy. The capacity to trap solar energy is unique only for green plants. Green plants possess the green pigment *viz.,* chlorophyll which can capture, transform, translocate and store energy which will be readily available for all activities of life

Photosynthesis may thus be explained as the process in which the light energy will be converted into chemical energy. Except for green plants all other organisms cannot directly utilize the solar energy; hence they are dependent on green plants for their nutritional requirements. Green plants are called autotrophic (they can prepare their own food) while all other organisms are called **heterotrophic** (they cannot prepare their own food). There is however one group of exceptional organisms which are autotrophic in spite of being non green. These are the **Chemosynthetic bacteria,** which obtain energy by the oxidation of inorganic substrates such as ferrous ions and sulphur dissolved from the earth's crust or H_2S released from volcanic action. But considering the overall energy budget of lifeforms the energy produced by chemosynthesis is very meagre and is of least significance quantitatively.

Definition : Photosynthesis (photo = light) may be defined as the process in which CO_2 and H_2O are put together (by green plants) into an organic molecule(sugar) utilizing the .solar energy.

$$CO_2 + H_2O \xrightarrow[\textit{Green Plants}]{\textit{Light}} \text{Sugar (Carbohydrates)}$$

Photosynthesis may also be defined as the reduction of CO_2 to glucose ($C_6H_{12}O_6$)with the H necessary for reduction coming from water. Photosynthesis involves both oxidation as well as reduction ultimately to synthesize energy rich compounds. Water is oxidized (electrons are removed) and the H released from water is used to reduce CO_2 to carbohydrates. Oxidation of water and the units of light energy (absorbed) produce energy intermediary compounds called ATP (Adenosine triphosphate) and reduced coenzymes called NADPH + H (Nicotinamide adenosine diphosphate). These two compounds help in the fixation (reduction) of carbon dioxide into carbohydrates. These carbohydrates in turn constitute the starting point for the synthesis of various other metabolites necessary for the survival of the organism.

The definition and brief explanation of photosynthesis given above sums up the significance of the process. A few other definitions (different only in the usage of the words) of photosynthesis are given below:

1. "Biosynthesis of simple carbohydrates from CO_2 and H_2O in the presence of sunlight inside chlorophyll containing cells"
2. "Production of carbon containing compounds from CO_2 and hydrogen donor by illuminated green cells"
3. "Photosynthesis is a redox process"
4. "Photosynthesis is a sensitized photochemical oxidation and reduction mechanism between water (H_2 donor) and CO_2 taking place at biological temperature"
5. "Photosynthesis is the formation of carbon containing compounds from CO_2 and water by illuminated green cells, water and oxygen being the by products".
6. Photosynthesis deals with the capturing and transforming of light energy into chemical energy"

The overall reaction of photosynthesis may be represented as follows

$$6CO_2 + 6H_2O \xrightarrow[\text{Chlorophyll}]{\text{Light}} C_6H_{12}O_6 + 6O_2$$

Oxygen which is released as a byproduct originates from water (Ruben and Kamen, 1941). In order to account for this,the equation should be changed as follows

$$6CO_2 + 12H_2O \xrightarrow[\text{Chlorophyll}]{\text{Light}} C_6H_{12}O_6 + 6O_2 + 6H_2O$$

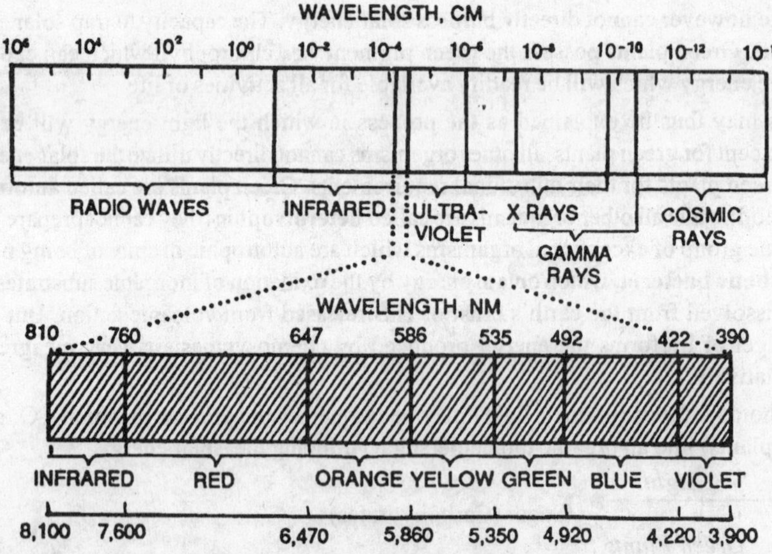

Fig. 12.1 Photosynthesis
Electromagnetic Spectrum

MAGNITUDE OF PHOTOSYNTHESIS

As has already been pointed out, photosynthesis is the most important biological process that supports all forms of life. Either directly or indirectly all forms consume photosynthetic energy. From the point of view of human welfare the contribution of photosynthesis is immense. Coal,gas,oil etc. represent the photosynthetic products of the plants belonging to early geological periods.

The significance of photosynthesis as a vital process contributing to the survival of life can be understood when we take into consideration the overall rate of photosynthesis.

According to an estimate, the overall area under green vegetation on this earth is approximately 510 million sq.km. Of this oceans alone account for 361 million sq.km.

Regarding the availability of light, solar radiation which is emitted from Sun in the form of electromagnetic radiation reaches the outer atmosphere. In terms of energy, the outer atmosphere receives about 1300 X 10^{21} cal/ year. Much of this is shielded by the atmosphere and the surface of the Earth receives light equivalent to 650 X 10^{21} cal/year. Considering the distribution of water and land masses, marine plants receive about 90 x 10^{21} cal/ year and terrestrial green vegetation receives about 25 x 10^{21} cal/year. If we assume photosynthetic efficiency is 2% of this energy, it amounts to 50 x 10^9 tons of organic carbon for terrestrial green plants and 180 x 10^9 tons for marine plants.

Rabinowitch (1951) calculates that photosynthesis produces approximately 155 x 10^9 tons of organic carbon of which 90% is contributed by the oceans. However some of the recent estimates (Rhyer and Wood Well, 1970) suggest a ratio of 1:3 fixation of carbon between terrestrial and marine plants. Of the total amount of light received on Earth's surface (650 X 10^{21} cal/year) about 60% is in the form of very short and very long wavelength (far and infra red) radiation and does not participate in photosynthesis. This leaves a total of 180 x 10^{21} cal/year for oceans and 80 x 10^{21} cal/year for terrestrial area. Added to this is the fact that more than half of the land area is devoid of vegetation. As a result (efficiency of photosynthesis being 2%) this represents only 0.5 X 10^{21} cal/year. Sheӧder estimated the production of organic carbon by land plants to be about 16 X 10^9 tons, while Leibeg and Ebermayer estimate the carbon production by land plants at 30 X 10^9 tons and 24 X 10^9 tons respectively. In oceans where planktons and other plants absorb about 50% of the incident light, the estimates come to 1.8 x 10^{21} cal/year or 180 X 10^9 tons of organic carbon.

HISTORICAL ACCOUNT

Eventhough scientific study of photosynthesis is very recent, from time immemorial man has looked upon Sun as the source of all light and sustenance. Early writings of Indians, Romans and Greeks bring out this idea very clearly. The Rig veda says that the 'Sun is the soul of the entire world'. Bhagavadgita also mentions that all' forms of life are dependent on Sun. Apart from these early writings, the following may be mentioned as significant names in the development of photosynthetic knowledge.

Aristotle (257 - 80 B.C), Theoprastus (370 - 285 B.C) and others up to the 17 th century believed that plants obtained their nutrition from the soil, and that Sun had nothing to do with plant nutrition.

Van Helmont (1648) : In his experiments with willow seedlings, he observed that water and not the soil was the principle source of nutrition.

Wood Ward (1699) : Opined that plantbody is made up of not water but a peculiar soil substance which is absorbed along with water.

Stephen Hales (1727) who is regarded as the father of plant physiology proposed that green plants obtained their nutrition partly from the air through the leaves and sunlight may have a role in it. He however was not aware of the part played by carbon dioxide.

Joseph Priestley (1772) who conducted several experiments on mint plants staled that plants can purify air. He also observed that impure air (CO_2) would get purified (vital air - O_2) if kept in contact with the green mint plants. Pri'elstley also knew nothing about the role of either CO_2 or light in photosynthesis.

Jan Ingen Housz (1771) : A Dutch physician demonstrated that plants purified air only in the presence of light. He also stated that plants too contributed to 'bad air' during darkness.

Jean Senebier (1782), a Swiss minister agreed with the findings of Ingen Housz and commented that plants absorb CO_2 from the atmosphere during daytime *and* exhale it (CO_2) during darkness. The role of two gases (O_2 and CO_2) had been implicated in plant nutrition by 1782. Works of Lavoisier and others identified that these gases were O_2 and CO_2.

N.T. de Saussure (1804) identified that water is an active participant in photosynthesis together with 002. He also made the first quantitative measurements of photosynthesis and opined that approximately equal volumes of 002 and 02 are exchanged during photosynthesis.

Mayer (1842) who formulated the law of conservation of energy, observed that during photosynthesis solar energy is stored in the form of chemical energy.

In 1817, two French Chemists **Joseph Bienaime Caventon** and **Piere Pelletier** isolated the chlorophyll pigment from 'green plants. **Richard Willstatler** obtained chlorophyll in the pure form. He also identified two forms - **Chlorophyll a** and **Chlorophyll b** which differed in the manner of their light absorption.

Leibeg (1843), stated that the source of all carbon inplants is the CO_2 of the atmosphere.

Julius Sachs (1864), a German plant physiologist,observed the growth of starch grains in chloroplasts, only in the area of the leaf exposed to light. Thus he was the first one to note the increase in organic content due to photosynthesis.

Boussingault (1864), calculated the ratio of CO_2 intake and O_2 release and came to the conclusion that it is 1: 1.

Baeyer (1870), proposed the formaldehyde hypothesis. He argued That CO_2 .gets reduced and forms CH_2O which on polymerization forms the glucose molecule. The formaldehyde hypothesis however has been abandoned not only because it has not been detected in plants, but it is also poisonous (to the plants) if it is present.

Blackmann (1905), identified that photosynthesis involves two distinct phases *viz.* light phase and dark phase. Van Niel (1931) showed that photosynthesis is a biological oxidation reduction reaction.

Robin Hill (1937), showed that photosynthesis is localized in the chloroplasts. In the presence of certain electron acceptors (oxidizing - reducing agents), isolated chloroplasts accept light energy and release O_2. This phenomenon of illuminated (isolated) chloroplasts splitting water and releasing O_2 has come to be known as **Hill's reaction.**

Samuel Ruben, Randall Kamen and Hyde (1941) showed (using radioisotopes of oxygen) that the oxygen released during photosynthesis comes only from water and not from CO_2.

Consden, Gordon and Martin (1944) developed paper chromatography technique for the separation of amino acids. This technique later was employed to detect the various intermediate compounds during the dark fixation of CO_2.

Arnon (1951 - 60) developed the concept of photophosphorylation and showed that coenzyme II picks up the H released by splitting up of water.

Benson, Bassham, Melvin Calvin and others have worked out in detail the carbon reduction cycle with the help of techniques like chromatography and radioautography.

RAW MATERIALS FOR PHOTOSYNTHESIS

There are some basic requirements (raw materials) for the process of photosynthesis. These are CO_2, a hydrogen donor (water) and light energy, besides of course the structural framework of the green plant in the form of chloroplasts.

Carbon dioxide : The only source of carbon for plants for the purpose of assimilation is the atmospheric carbon dioxide. Approximately about 20 X 10^{11} tons of carbon dioxide is used for the production of 50 x 10^{9} tons of organic carbon. Besides atmosphere, other sources of carbon are biological respiration, decaying organic matter etc. Carbon dioxide is in the gaseous form in the atmosphere, while it is in dissolved condition in water.

Carbon dioxide is present in trace quantities in the atmosphere (0.035%). There has been a gradual increase in the concentration of the atmospheric carbon dioxide since the middle of 19th century perhaps due to increased pollution (Wood Well, 1970).

Hydrogen donor : In almost all photosynthetic organisms, water is the hydrogen donor. Water splits into H and OH ions in the presence of light. H is picked up by the enzyme NADPH for reduction of CO_2, while OH recombines to form water and oxygen

Certain chemosynthetic bacteria however use compounds of Iron and Sulphur as hydrogen donors.

Light : Solar radiation is the only source of light for all forms of life. The radiation from Sun is emitted in the form of electromagnetic radiation with radiowaves at one end and cosmic rays at the other. Much of this radiation however is shielded by the outer reaches of the atmosphere and only visible light together with some amount of U.V rays reaches the surface of the Earth. Light may be defined as the visible (to human eye) part of the electromagnetic spectrum. The visible spectrum is confined between wavelengths of 390 mμ and 760 mμ. Spectral analysis of the visible light shows the following rays with their respective wavelengths given in parenthesis.

Violet	(390 - 430 mμ)
Blue	(430-470 mμ)
Bluegreen	(470 - 500 mμ)
Green	(500 - 560 mμ)
Yellow	(560 - 600 mμ)
Orange	(600 - 650 mμ)
Red	(650 - 760 mμ)

Rays of the radiation with wavelengths shorter than violet are called U.V (Ultra violet -100 Å-390 Å), X rays (0.1 Å - 100 Å), Gamma rays (0.001 Å-0-.1 Å) and cosmic rays (less than 0.001 Å). Radiation with wave lengths longer than red are called infra red (7600 Å - 1000,000 Å). Beyond this are electric and radio waves, which are measured (in terms of wavelength) in kilometeres.

Radiation travels in the form of discrete corpuscles or particles called **Quanta or Photons.** Each quantum represents a unit of energy. Energy quantity of each ray depends on its wavelength; shorter the wavelength higher is the energy. The wavelengths of radiation are so small that they are measured not in terms of millimeteres or centimeteres, but in terms of microns (μ), millimicrons (mμ) or nanometers (nm) or Angstrom (Å) units.

1μ	=	10^{-6} meters or 10^{-4} cm.
Imμ	=	10^{-9} meters por 10^{-7} cm.
Inmμ	=	10^{-9} meters or 10^{-7} cm.
1μ	=	10^{-6} meters or 10^{-4} cm.
1 A^0	=	10^{-10} meters or 10^{-8} cm.
1 A^0	=	10^{-10} meters or 10^{-8} cms

Each photon has an energy constant E = hc/λ where h is planck's constant (= 6.625 X 10^{-27} erg/see) c is velocity of light (3 x 10^{10} cms/sec) and λ is the wavelength as given by Einstein.

The energy value of different light rays is given in the following table.

Energy Value of Different Wavelengths of Light

Sl. No.	Wavelength in mm	Approximate colour	Energy in cal
1.	300	Ultra Violet	95,200
2.	420	Violet	68,000
3.	470	Blue	60,760
4.	530	Green	53,890
5.	620	Orange	46,060
6.	700	Red	40,800
7.	683	Red	41,810
8.	672	Red	42,490
9.	650	Red	43,940
10.	850	Infra Red	33,590
11.	1000	Infra Red	28,560

THE PHOTOSYNTHETIC APPARATUS

Within the cell of the green plants are present certain unique organelles called plastids. These perform many vital functions in cell physiology. Among the plastids, the most important from the point of view of photosynthesis are **chloroplasts**. All the physiological reactions, starting from the absorption of light energy to the fixation of CO_2 take place in the chloroplasts. No wonder then chloroplast is aptly called the photosynthetic apparatus.

Chloroplasts : These are by far, lhe commonest and the most plaxlids. As lhe primary;sites for 'rapping and converting solar very vital for the existolice of not only the green plants, but for world.

Shape : Chloroplasts have varied shapes. The algal chloroplasts exhibit an array of shapes. They may be cup shaped *(Chlamydornonas)*, girdle shaped *(Utothrix)*, reticulate *(Hydrodictyon)*, stellate *(Zygnema)*, ribbon shaped *(Spirogyra)* or discoid. They are generally spherical or ovoid in higher plants.

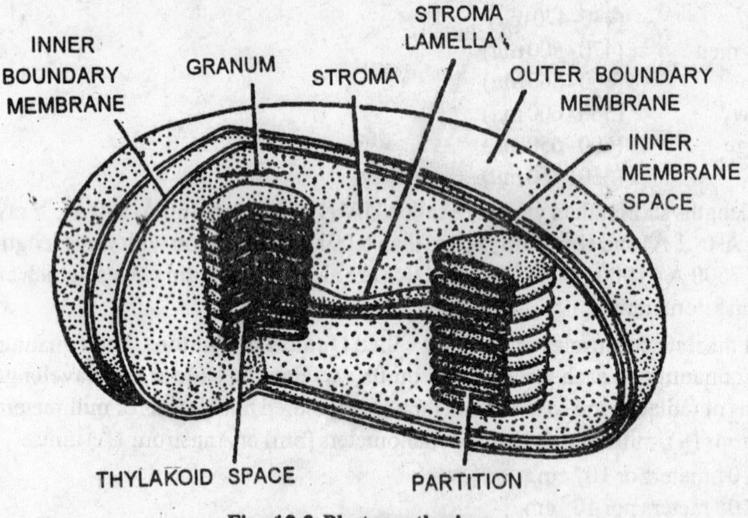

Fig. 12.2 Photosynthesis
Model of Ultrastructure of chloroplasts

Size : There is a great variety in size. Normally they are about 1μm thick and 4-6μm in length. Chloroplasis ofpolyploid cells are generally larger than in the diploid cells. Cells in the shaded area have longer chloroplasts but with less intensity in colouration.

Distribition : They arc linifonniy distributed all over cytoplasm, but in some instiince they cluster towards the nucleus. The concentration of chloroplasts will also depend on light intensity.

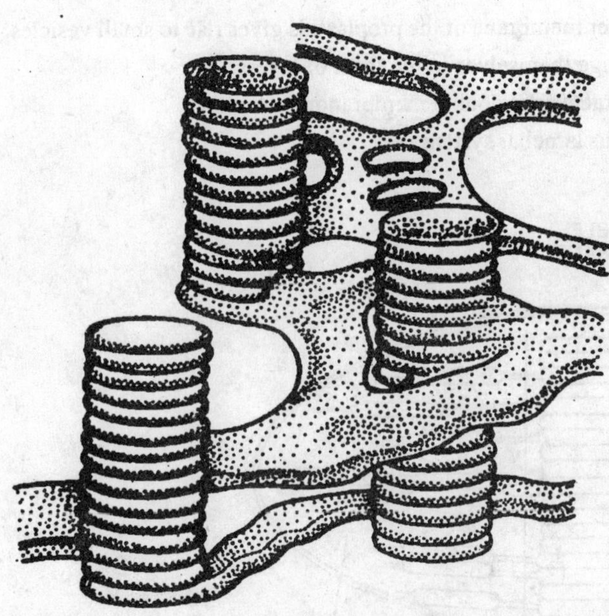

Fig. 12.3 Photosynthesis
Three dimensional view of Grana (enlarged view)

Number : The number of choloroplasts per cell varies. In some algae lik *Chkmydomonas,* there is only one chloroplast per cell or two as in *Zygnema.* In higher plants, cells have a number of chloroplasts. The leaf of *Ricinus communis* has nearly lour lakh chloropliists per square millimeter area.

Clemical composition : The chloroplasis contain carbohydrates, lipids, protein! chlorophyll, ca.rotenoid and xanthophylls. The protein lipid ratio is 40:30. Besides the above, nucicic acids have also been demonstrated in the chloroplasts.

Ultra stnucture of the chloroplasts : The chloroplast has a covering of two membranes with an inner membrane space. There membranes are smooth and there arc no perforations or particles. The membranes are differentially permeable (Mudrak A section of the chloroplast reveals an intricate system of membranes enclosed in a granular matrix. These membranes are called lamellae and the surrounding matrix-the *stroma.* In a sectional view, the lamellae can be seen packed and these stacks are called *thylakoids.* In the chloroplasts of higher plants, the thylakoids themselves form highly compact bundles called *grana.* Some thylakoids of granum extend into the stroma and maintain contact with other grana. These are *caUed-Stroma thylakoid* or *stroma lamellae* or *inte~grana.*

The chloroplasts of algae lack the granum arrangement and in cyanobacteria (blue green algae), the thylakoids lie naked in cytoplasm without any envelope. In such instances, pigments are uniformly distributed on or in the lamellae.

The stroma has in addition to the lamellae - granules (globuli), lipid droplets, starch grains and vesicles.

Structure of the lamellae : All the photosynthetic pigments are concentrated in the chloroplast lamellae. The lamellae as well as the outer membranes are composed of lipoprotein sub units, each sub unit according to Weir and Benson has a protein core surrounded by a lipid sheath. The chloroplast membranes have a single layer of sub units, whereas in thylakoids it is double due to the juxtaposition of the membranes. The appressed areas are called partitions. These are hydrophobic in nature The spa~e between the two membranes of thylakoid (fret membrane) is called *fret channel.* The space between two thylakoids is called *loculus* while the ends of disc shaped thylakoids are called *margins.* These areas (loculus, fret channel are hydrophilic. The chlorophyll molecules are concentrated in the fret channel ie., in the space between the two membranes of thylakoids. The chlorophyll molecules may be entirely in the space or the *heads* may be in the fret channel while the *tails* are buried in the sub units, occasionally the entire chlorophyll molecule may be found in the fret membrane. Other pigments like cytochromes, carotenoids, etc., are also found in the fret membranes.

Four sub units of the partition constitute a photosynthetic unit or a *quantosome.*

Development and origin of chloroplasts : The development from the early stages has been followed by Wettstein (1959). There are five stages -

1. Proplastids arise in the cytoplasm. The inner membrane of the proplastids gives rise to small vesicles.
2. The vesicles attach to each other and arrange themselves in the form of layers.
3. The layers grow and fuse forming the characteristic double membranous structures.
4. The lamellae multiply and form a continuous lamellar system.
5. Finally grana region get differentiated.

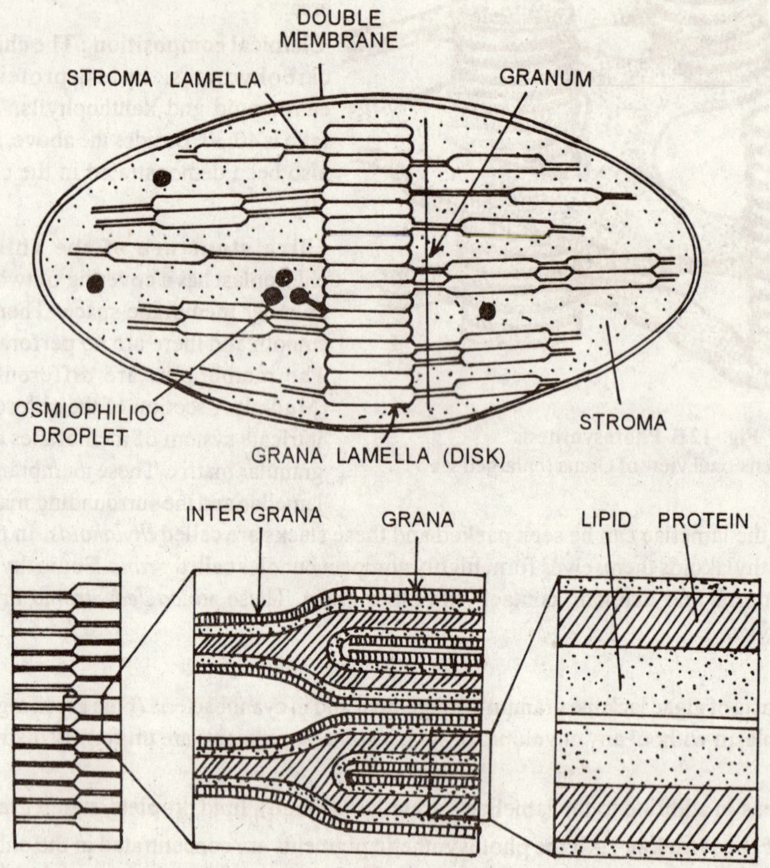

Fig. 12.4 Photosynthesis

Diagrammatic representation of Ultrastructure of chloroplast showing the lamellae.

A. Entire chloroplast, **B-D.** Show successive enlargement of a part of the granum

The origin of chloroplasts is as debatable as that of mitochondria. According to some physiologists, they arise *de novo,* while some argue for self replication of chloroplasts. The division of chloroplasts has been observed in some algae and fern gametophytes.

Nucleic acids in chloroplasts : Ris and Plant (1962) have reported DNA in the chloroplast of *Chlamydomonas.* This has later been demonstrated in many other plants. Woodcock and Fernandez (1968) have isolated DNA fragments as long as 150 μ from the matrix of chloroplasts. Goffean and Branchet (1965) have demonstrated DNA in the chloroplasts of *Acetabularia* from which the nucleus had been removed. Chloroplast DNA is different from the nuclear DNA and similar to bacterial DNA. The chloroplast DNA is also capable of replication.

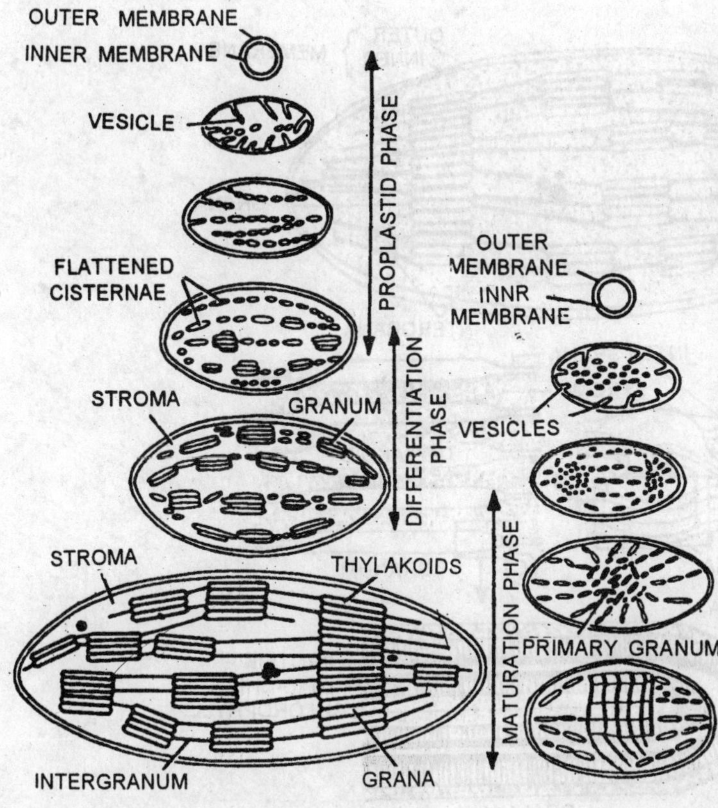

Fig. 12.5 Photosynthesis
Stages in the development of chloroplast

Ribosomes and RNA have also been isolated from the chloroplasts indicating a machinery for protein synthesis.

Since the chloroplasts, have their own RNA, DNA ribosomes and protein synthesis, according to Deviin (1975) "One might be inclined to view the chloroplast as a cell within a cell". De Robertis (1980) suggests "It has been suggested that chloroplasts may have resulted from a symbiotic relationship between an autotrophic microorganism, able to tmasform energy from height, and a heterotrophic host cell" But Bogorad (1977) opines that the chloroplast components depend to a great extent on the nuclear genes for their assembly into an integrated organelle.

Pigments in chloroplast : Some of the important pigments present in chloroplasts are chlorophylls, carotenoids, cytochromes etc. In algal cells the thylakoids also have phycobilins.

Chlorophylls : These are the most important pigments involved in the basic photochemical reaction that sustains life. Atleast nine types of chlorophylls have been flistinoilishefl so far - chinronhvils *a*.. *h*. *c*. *d* and *e*. hacterio Chloronhvils a and b and the chlorobium chlorophylls 650 and 660mm.

Among the chlorophylls, chlorophyll *a* and *b* are best known and most widely distributed. These are absent in pigmented bacteria. Chlorophylls *c, d* and e are found only in algae in combination with chlorophyll a. Bacteriochlorophylls and chlorobium chlorophylls are found only photosynthetic bacteria.

The chlorophyll molecule is a complex structure basically made up of a 'head' and 'tail' resembling a tennis racquet. The 'head' is a *porphyrin* structure made of four *Pyrrole rings* attached to each other at the centre by an isocyclic ring containing Mg atOm at the centre. Extending from one of the pyrrole rings is the 'tail' - the alcoholic chain *(phytol)*. The emperical formula for the chlorophyll molecule is CH, 0,N Mg. The phytol chain is esterified on the C atom of one of the pyrrole rings and has only one double bond.

Chlorophyll *a* and *b* differ structurally in having different atomic groupings at the C, atom in the pyrrole ring. In chlorophyll a C, atom has a methyl group, while chlorophyll *b* has an aldehyde group; besides the two pigments have different absorption spectra. The peaks are as follows

Chlorophyll *a* 429, 410 and 660nm

Chlorophyll *b* 430,453 and 442nm

(The above spectra are for chlorophyll in *vitro)*

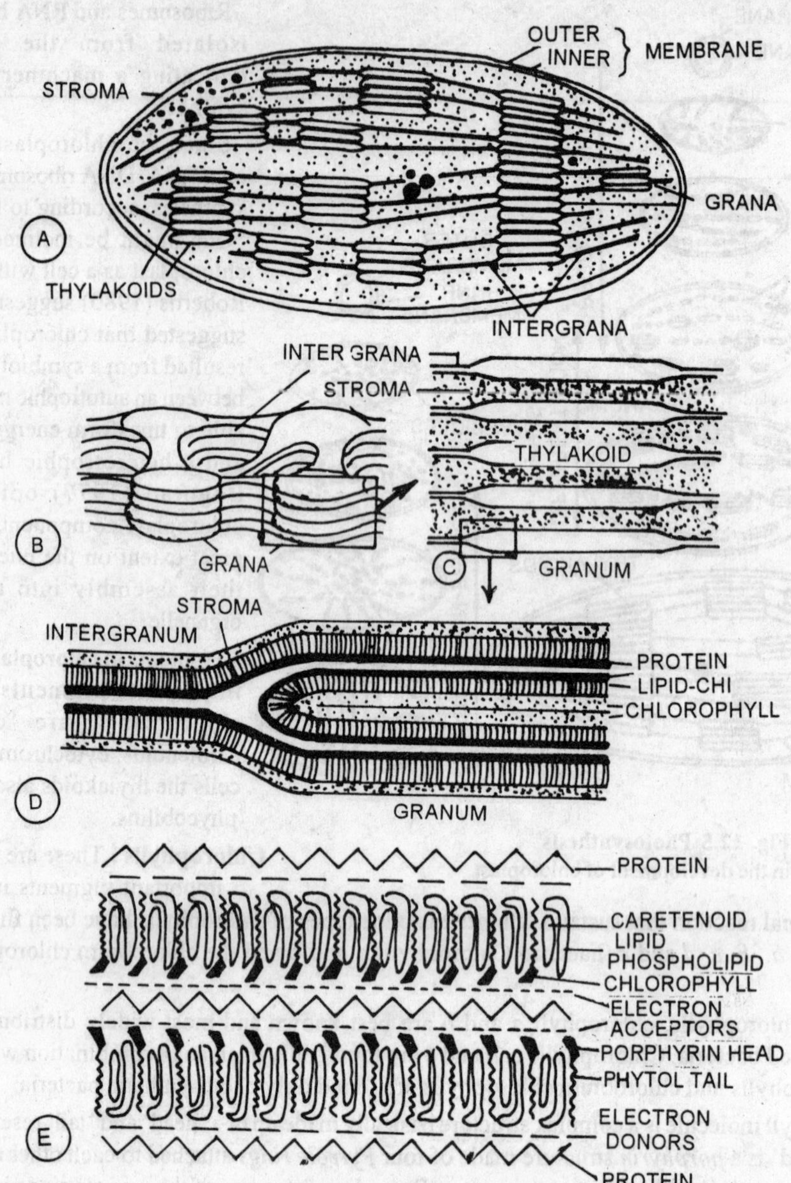

Fig. 12.6 Photosynthesis

A. Entire chloroplast, **B.** Grana and Intergrana (diagrammatic), **C.** Section through a granum, **D.** Lamellar structure (in detail), **E.** Model for the arrangement of pigments in a granum

Carotenoid pigments : These are lipid compounds ranging in colour from yellow to purple. They are widely distributed in both plants and animals. They are also present in microorganisms including red alage, cyanobacteria, photosynthetic bacteria, fungi, etc. Wackenroder (1931) isolated the first carotenoid - carotene from tissue.

Carotenoids are derivatives of *lycopene* a red pigment found in many plants. They (carotenoids) are known to have eight isoprene ($CH_2 = C[CH_3] - CH = CH_2$) like residues.

The main carotenoid found in plants is the orange yellow coloured carotene with some quantity of a carotene.

The carotenoids are also located in the chloroplast. Accrding to Goodwin (1960), carotenoids and chlorophylls may be combined with the same protein to form a complex known as *photosynthein.*

Carotenoids protect the chorophyll form photooxidation and transfer the light energy they absorb to chlorophyll *a*.

Phycobilins: These are found only in algae. There are two types of pycobilins - the *Phycoreythrin* (red) and *Phycocyanin* (blue) These pigments are strongly associated with a protein and consequently it is difficult to isolate the pigments in the pure state.

Like carotenoids phycobilins are also involved in the transfer of light energy (they absorb) to the chlorophyll.

The absorption spectra of phycobilins presents an intersting study considering the fact that they act as accessory pigments in photosynthesis. R - phycoerythrin has peaks at 495,540 and 545nm, while R - phycocyanin has peaks at 550 and 615nm.

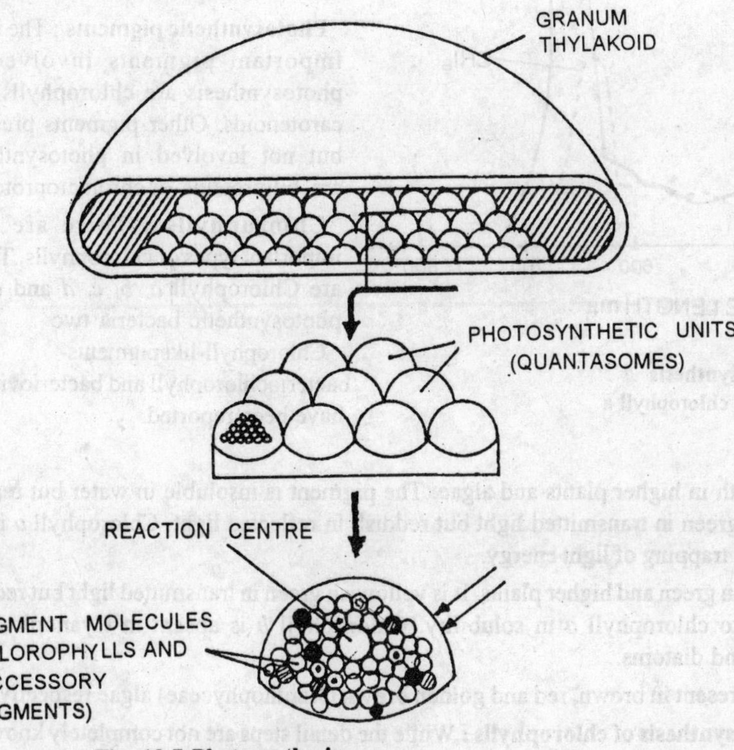

Fig. 12.7 Photosynthesis
Diagrammatic representation of a thylakoid, quantosomes and the reaction centre

Photosynthetic units : A photosynthetic unit may be defined as "the smallest group of coordinating pigment molecules necessary to effect a photochemical act, *i.e* absorption and transportation of a light quantum to a trapping center where it causes excitation and release of an electron".

Emerson and Arnold (1932) who experimented in *Chlorella* plants, observed that about 2500 molecules of the chlorophyll pigment are necessary to fix a molecule of CO_2 and they termed this number (2500) as a photosynthetic unit They also observed that to release one molecule of O_2, 10 quanta of light are necessary. Some of the recent works however have shown that on the basis of absorption of 10 quanta, a photosynthetic unit should consist of 230 molecules of the chlorophyll pigment. Park and Beggins(1964) working on the chloroplasts opined that photosynthetic units are distinct morphological identities and called them quantasomes. The quantasomes are 180 Å broad and 100 Å thick (18 X 16 x 10nm). Each quantasome has 230 molecules of chlorophyll (160 chlorophyll *a* and 70 chlorophyll *b),* 48 carotenoids, 46 quinone compounds, 116 phospholipids, 144 digalactosyl diglycerides, 346 monogalactosyidiglycerides, 48 sulfolipids unidentified lipids and many sterols. Quantasomes have a molecular weight of about two millions and are arranged in a monolayer on the thylakoid bodies. Electron microscopic studies have revealed that the individual quantasome may have four or more sub units. This middle region of the quantasome is called the reaction centre.

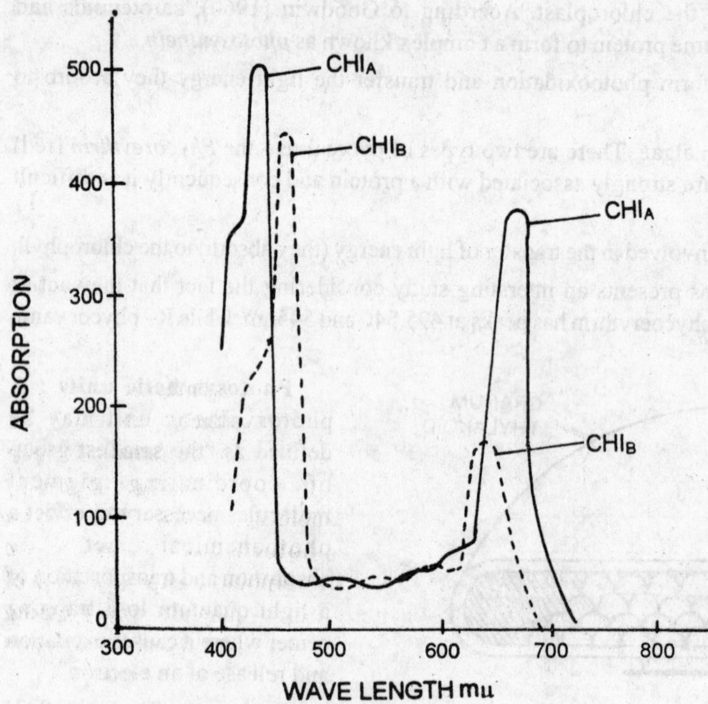

Fig. 12.8 Photosynthesis
Absorption spectra of chlorophyll a and b

Robert Huber, Johann Deisenhofer and Hartmutt Michel were awarded the 1988 chemistry Nobel prize for unravelling the secret of the reaction centre. Working at the Max Planck Institute in Martinsried, Germany, these scientists, made their breakthrough by extracting the reaction centre from the membranes of certain photosynthetic bacteria. They then crystallized it and analysed the structure using x - ray diffraction.

Photosynthetic pigments : The most important pigments involved in photosynthesis are chlorophylls and carotenoids. Other pigments present, but not involved in photosynthesis are biliproteins or chromatoproteins.

Chlorophylls : There are five important types of chlorophylls. These are Chlorophyll *a, b, c, d* and *e*. In photosynthetic bacteria two Chlorophyll-likepigments-bacteriochlorophyll and bacterioviridin have been reported.

Chlorophyll a is present both in higher plants and algae. The pigment is insoluble in water but readily soluble in ether. It appears blue green in transmitted light but reddish in reflected light. Chlorophyll *a* is the principal pigment involved in the trapping of light energy.

Chlorophyll b is also found in green and higher plants. It is yellowish green in transmitted light but reddish in reflected light. It is similar to chlorophyll *a* in solubility. Chlorophyll *b* is absent in Cyanophyceae, Rhodophyceae, Phaeophyceae and diatoms.

Chlorophyll c, d and e are present in brown, red and golden brown (Xanthophyceae) algae respectively.

Chemical structure and biosynthesis of chlorophylls : While the detail steps are not completely known, it is evident that the synthesis of chlorophyll is linked to one of the Krebs cycle intermediaries namely succinyl CoA. Succinnyl CoA together with amino acid glycine is known to initiate the synthesis of chlorophyll. These two combine to form an unstable α amino β ketoadipic acid which on decarboxylation yields σ aminolevulinic acid and is catalyzed by the enzyme σ aminolevulinic acid synthetase. This reaction is known to be light dependent (Grassman and Bogorad, 1967). Two molecules of H_2O are used in the reaction. Four molecules of porphobilinogen combine to form Unroporphyrinogen III. This reaction is mediated by uroporphyrinogen III. This reaction is mediated by unroporphyrinogen III synthetase. Decarboxylation in the acetic acid substitutes of uroporphyrinogen III. This reaction is mediated by urophyrinogen III synthetase. Decarboxylation in the acetic acid substitutes of uroporphyrinogen III by the enzyme decarboxylase results in the formation of Coproporphyrinogen III. This, later undergoes decarboxylation to give rise to protoporphyrinogen IX. This, on oxidation gives rise to protoporphyrin IX. Mg is incorporated into this compound resulting in the formation of Mg^- protoporphyrin IX. A methyl group is added to this to form Mg^- protoporphyrin IX monomethyl ester. This ester gives rise to Protochlorophyllide. Protochlorophyllide is reduced to form Chlorophyllide. This is a photo reduction process with H being added to carbon atoms 7 chlorophyllase results in the formation of chlorophyll a. Chlorophyll b is known to be synthesised from chlorophyll a (Bogorad 1966 - The biosynthesis of chlorophyll).

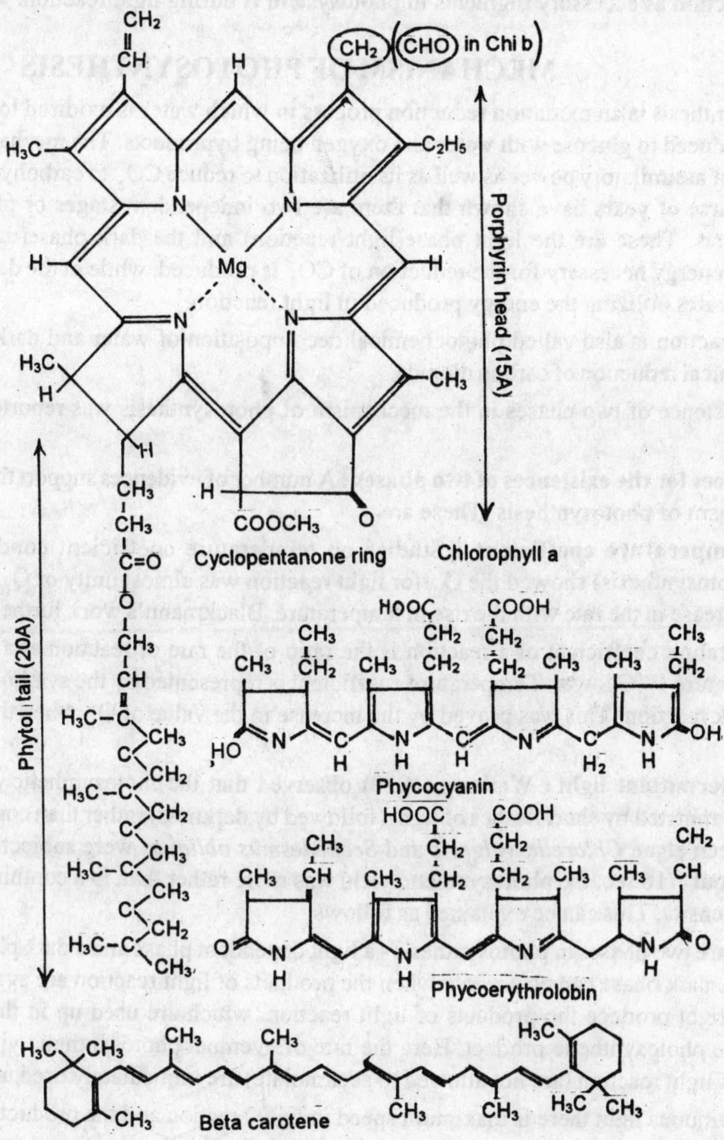

Fig. 12.9 Photosynthesis

Molecular structure of photosynthetic pigments

Carotenoids : These are yellow, brown, orange or red in color. Carotenoids are of two categories *viz.* Carotenes and Xanthophylls. They are insoluble in water but soluble in organic solvents Structurally carotenoid pigments are related to the phytol tail of the chlorophyll molecule. The pigments help in the release of oxygen during light-induced ionization of water.

Carotenoids perform a variety of functions in relation to photosynthesis. These are

1. Absorb light energy in the middle of the visible spectrum and transfer it to chlorophyll *a*
2. Function as accessory pigments in photosystem II during light reaction.

MECHANISM OF PHOTOSYNTHESIS

Photosynthesis is an oxidation reduction process in which water is oxidised to H^+ and OH^-, while carbon dioxide is reduced to glucose with water and oxygen being byproducts. The mechanism has to account for the production of assimilatory power as well as its utilization to reduce CO_2 to carbohydrates. Evidences collected over the course of years have shown that there are two independent stages or phases in the mechanism of photosynthesis. These are the light phase(light reaction) and the dark phase(dark reaction). During light reaction the energy necessary for the reduction of CO_2 is produced, while in the dark reaction CO_2 is reduced to carbohydrates utilizing the energy produced in light reaction.

Light reaction is also called photochemical decomposition of water and dark reaction is also called the thermochemical reduction of carbon dioxide.

The existence of two phases in the mechanism of photosynthesis was reported as early as 1905 by F.F. Blackmann.

Evidences for the existences of two phases : A number of evidences support the existence of two phases in the mechanism of photosynthesis. These are-

a. **Temperature coefficient :** Studies on temperature coefficient conducted by Blackmann (for photosynthesis) showed the Q_{10} for light reaction was almost unity or Q_{10} == 1. It does not show any increase in the rate with the rise in temperature. Blackmann's work further showed that Qio is

1. Temperature coefficient of a reaction is the ratio of the rate of reaction at a given temperature and at a temperature 10°C lower. Temperature coefficient is represented by the symbol that Q_{10} is always 2 or 3 for the dark reaction. This was proved by the increase in the value of Q_{10} when the system was supplied with CO_2.

b. **Intermittent light :** Warburg (1920) observed that the photosynthetic yield is more when plants are illuminated by short flashes of light (followed by darkness) rather than continuous illumination. When green algae *Chlorella vulgaris* and *Scenedesmus obliquas* were subjected to intermittent flashes of about 1/16 sec, the photosynthetic yield was more rather than in a continuous supply of light of same intensity. This can be explained as follows:

There are two phases in photosynthesis - a light dependent phase and a dark phase. The two are interlinked in the sense dark phase can occur only when the products of light reaction are available. In intermittent light flashes of light produce the products of light reaction, which are used up in the intervening dark phase to produce the photosynthetic product. Here the rate of eventual photosynthetic yield will be maximum as the products of light reaction (are not allowed to accumulate) are immediately used up in dark reaction.

In continuous light there is maximum speed in light reaction and the products accumulate as they can not be used up in dark reaction as fast as they are produced. Accumulation of products of light reaction retards the rate of dark reaction (i.e production of carbohydrates)

c. **Dark reaction of CO_2 :** By far the best evidence for the existence of two phases comes from tracer experiments. Pre- illuminated leaves kept in darkness subsequently and supplied with C^{14} have been known to Fix CO_2 eventually producing carbohydrates.

LIGHT REACTION

The light reaction or photochemical decomposition of water takes place in the grana region of the chloroplast. According to the current understanding, light reaction (photochemical excitation) takes place in about 10^{-9} seconds. Within this short span of time there will be the synthesis of molecule(s) of ATP and a molecule of NADPH. These two are utilized in the dark fixation of CO_2. In this section we will study the following aspects in detail with respect to light reaction.

(a) Source of oxygen released during light reaction
(b) Hill's reaction
(c) Arnon's work
(d) Quantum requirement
(e) Emerson effect and two pigment systems
(f) Primary photochemical reaction(mechanism of pigment excitation)
(g) Non-cyclic Photophosphorylation
(h) Cyclic photophosphorylation
(i) Pseudocyclic photophosphorylation and
(j) Mechanism of ATP formation

SOURCE OF OXYGEN

The overall reaction of photosynthesis was until 1930 believed to be the exact opposite of respiration and oxygen evolved during photosynthesis came from carbon dioxide.

$$6CO_2 + 6H_{2O} \rightarrow C_6H_{12}O_6 + 6O_2.$$

The discovery by Van Niel(1931) that in certain purple photosynthetic bacteria photosynthesis takes place in the presence of 002 but without releasing oxygen. Surely, if the source of O_2 were to be CO_2 even in the absence of water O_2 should have been released. These photosynthetic bacteria used H_2S as hydrogen donor instead of water. According to Van Niel (1931), in purple bacteria H_2S, splits releasing H to reduce CO_2 and in the process elemental sulphur is being formed as follows.

$$6O_2 + 2H_2S \rightarrow (CH_2O) + H_2O + 2S$$

Van Niel (1931) further argued that, if elemental sulphur comes from H_2S in purple bacteria, in green plants which use water, 02 should come from water only and not from 002. In order to account for the release of O_2 from water the equation should be modified as follows.

$$6CO_2 + 12H_2O \rightarrow C_6H_{12}O_6 + 6O_2. + 6H_2O$$

According to Van Niel (1931), water is split up by light (photolysis of water) into H^+ and OH^- ions. The H^+ ions are used to reduce CO_2, while the OH^- ions react with one another to form water and oxygen as follows

$$4H_2O \rightarrow 4H^+ + 4(OH)$$

$$4(OH) \rightarrow 2H_2O + O_2$$

$$CO_2 + 4H^+ \rightarrow (CH_2O) + H_2O$$

Here (CH_2O)represents not the actual photosynthetic product but the empirical formula of a bigger molecule (glucose).

The work of Ruben, Kamen and Randall (1941) using radio active oxygen (O^{18})clearly showed that O_2 came only from water and not from CO_2. When the experimental plant was supplied with labelled water (H_2O^{18}), the released oxygen was of the O^{18} type, and when the plant was supplied with labelled $CO_2{}^{18}$ the O_2 released was of the normal type (O^{16})

HILL'S REACTION : Hill (1937) and Hill and Scaribrick (1940) showed that the initial reaction of photosynthesis can take place even in isolated chloroplasts in the cell-free environment. When isolated chloroplasts (obtained from ground leaves) are supplied with light, water and a suitable hydrogen acceptor, O_2 will be released and the H_2 acceptor gets reduced. A similar release of O_2 may be observed, when grana from disintegrated chloroplasts are suspended in water with ferric salts and benzoquinones acting as hydrogen acceptors. Besides these, chromates, indophenols etc also can act as H_2 acceptors. The various H_2 acceptors used by Hill are called Hill oxidants. Hill attributed the release of oxygen to the splitting of water into H and OH by light. Hill's reaction proves the following:

(a) The entire photosynthetic activity including the reduction of CO_2 takes place within the chloroplasts and

(b) O_2 released during light reaction comes entirely from water.

ARNON'S WORK (production of assimilatory power) :

The products of light reaction provide the necessary inputs for the reduction or assimilation of CO_2. That the assimilatory power (input) is utilized for the dark fixation of CO_2 was demonstrated clearly by the work of Arnon(1951).

According to Arnon (1951), the hydrogen released by water was accepted by coenzyme II (NADP). He found that in isolated chloroplasts, the release of O_2 takes place at a faster rate in the presence of NADP. Arnon also opined that ATP molecules will also be synthesized and he termed this as photophosphorylation

$$\text{ADP} + \underset{(inorganic\ phosphate)}{\text{Pi}} \xrightarrow{\ \ energy\ \ } \text{ATP}$$

Arnon also showed that CO_2 fixation takes place in the pigment free portion of the chloroplast (stroma), while the assimilatory power (ATP molecules and NADPH) is produced in the grana portion.

QUANTUM REQUIREMENT

The solar radiation that reaches the Earth consists of two portions - the visible (400nm-800 nm) and UV and infrared (less than 400 nm and more than 800 nm). Whether the light consists mainly of waves or particles, as far as photosynthesis is concerned, it is absorbed in units of energy called photons or quanta.

The quantum yield may be explained on the basis of molecules of CO_2 reduced and amount of O_2 released. Consequently a minimum quantum is the one required to fix one molecule of CO_2 and release one molecule of O_2. According to Warburg and Negelein (1922), the minimum quantum requirement is four i.e. 4 quanta are required to fix one mol of CO_2 and release one mol of O_2. This (requirement) concurs with the fact that every CO_2 molecule requires $2H_2O$ molecules for reduction i.e. 4 hydrogen atoms. Emerson and his co-workers (1938) however opine that eight light quanta is the minimum requirement. This has been further confirmed by the work of Govindjee (1968) on *Chlorella* plants. Thus two quanta of light are required for the transfer of every atom of hydrogen during photosynthesis.

EMERSON EFFECT AND TWO PIGMENT SYSTEMS

Emerson and Lewis(1943) worked on different wavelengths of light (monochromatic light) as to their quantum yield and came to the conclusion that 8 quanta of light are required for the reduction of one molecule of CO_2 (or to release one molecule of O_2). Accordingly the quantum yield is 1/8 or 12%. Since 4 electrons are required in the release of one molecule of O_2, Emerson opined that 2 light quanta are required to excite one electron. Proceeding further, Emerson *et al* (1943) determined the quantum yield of different beams of monochromatic light and found that the average quantum yield (12%) suddenly dropped near the far red end of the spectrum. This decline in quantum yield is called **red drop.** Red drop begins at wavelengths greater than 680 nm in green plants and greater than 650nm in red algae. The phenomenon of red drop was also confirmed by the works of Haxo and Blinks (1950) on the red alga *Porphyra nereogtis*.

CONTINUOUS LIGHT
YIELD LESS
L D L D L D L D L D L D L D
YIELD ENHANCED
FLASHING LIGHT

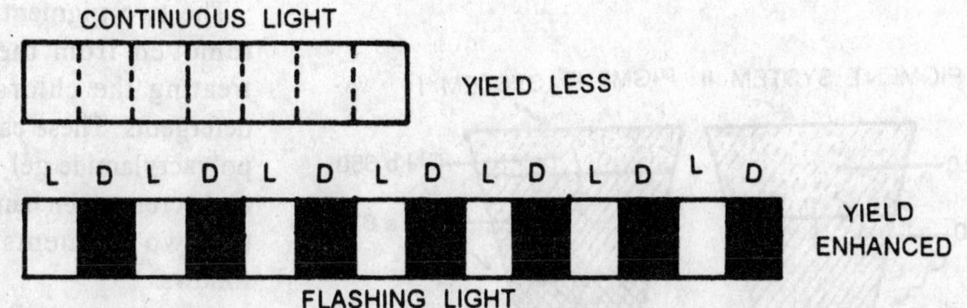

Fig. 12.10 Photosynthesis
Red drop and Emerson effect in Chlorella

The phenomenon of red drop was studied further by Emerson and Chalmers (1958), who found out that the reduction in the rate of photosynthesis due to red drop can be rectified by providing a beam of monochromatic light of shorter wavelength. In other words,by providing shorter wavelengths of light either simultaneously or quickly alternating with a longer wavelength of light, the rate of photosynthesis can be enhanced (effect of red drop is reversed).

This enhancement of photosynthetic rate under the influence of two wavelengths (long and short) of light is called **Emerson effect.** This may be represented as follows

$$E = \frac{\text{Rate of evolution of } O_2 \text{ in double beam} - \text{Rate of Evolution } O_2 \text{ in red beam (short)}}{\text{Rate of evolution of } O_2 \text{ in far red beam (long)}}$$

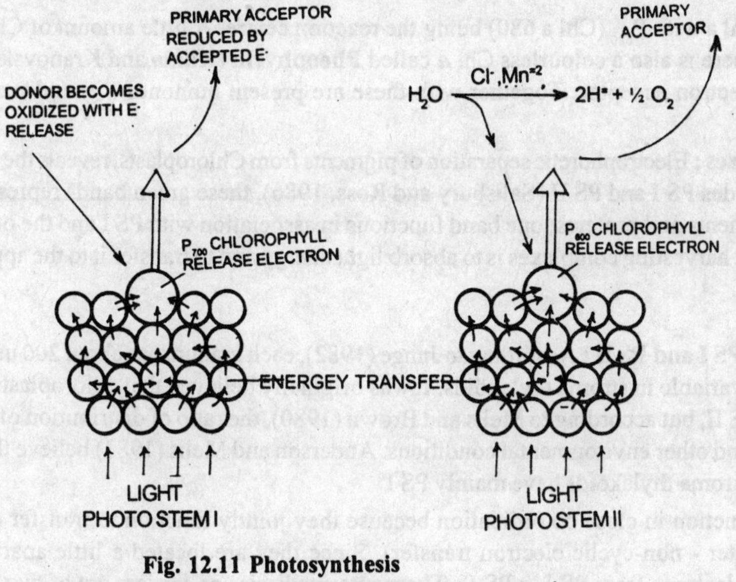

PRIMARY ACCEPTOR REDUCED BY ACCEPTED E⁻
PRIMARY ACCEPTOR
DONOR BECOMES OXIDIZED WITH E⁻ RELEASE
Cl⁻,Mn⁺²
H₂O → 2H⁺ + ½ O₂
P700 CHLOROPHYLL RELEASE ELECTRON
P680 CHLOROPHYLL RELEASE ELECTRON
ENERGEY TRANSFER
LIGHT
PHOTO STEM I
LIGHT
PHOTO STEM II

Fig. 12.11 Photosynthesis
Energy transfer in Photosystem I and Photosystem II in green plants

Reason for Emerson effect

The discovery of the Emerson effect led to the idea that there must be two different pigment systems in the plants which differed in their absorption peak(of light). While one pigment system was more efficient in longer wavelength, the other was efficient in shorter wavelength. Obviously under a specific wavelength(either short or long) the rate of photosynthesis will not be maximum as one or the other pigment system will be ineffective. Butler(1966) discovered two forms of chlorophyll *a viz* chl *a* 673 and chl *a* 683 (with absorption peaks at 673nm and 683nm respectively). Clayton (1966) discovered another form of chl *a* with absorption peak at 700nm which he named P_{700}.

It has now been confirmed by various researches that two pigment systems PS I and PS II are involved in photosynthesis, each pigment system having different absorption peaks and consisting of chl *a* and other accessory pigments.

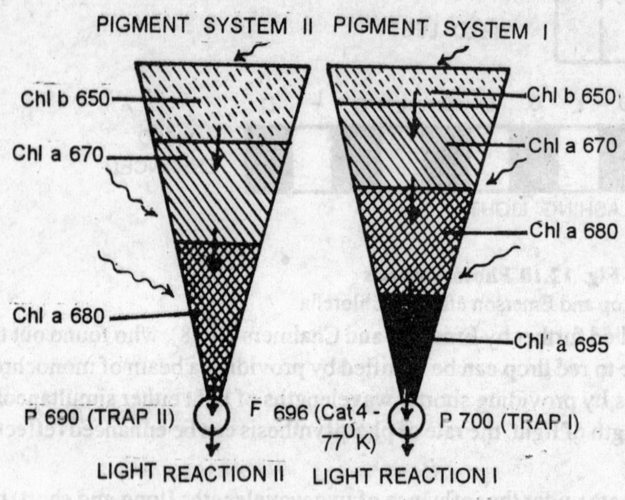

Fig. 12.12 Photosynthesis
Distribution of different pigments in PS I and PS II

The two pigment systems may be removed from the thylakoids, by treating the chloroplast with mild detergents. These can be separated on polyacrylamide gel when they appear as discrete green bands. The details of the two pigments systems are as follows.

Pigment system I (PSI)

According to Malkin (1982), Anderson(1982) and Barber(1983), PSI consists of chl *a* 670, chl *a* 680 and chl *a* 710(P 700), chl b and some β carotenes. P_{700} is the reaction centre in PS I into which light energy is transferred by all other pigments. Besides, PS I also consists of iron containing proteins Fe-S proteins and Cytochromes. PS I is weakly fluorescent.

Pigment system II (PSII)

It has several forms of Chl a with P_{680} (Chl a 680) being the reaction centre. A little amount of Ch *b* and *b* carotenes are also present. There is also a colourless Chl *a* called **Pheophyrin** (Klima and Kranovskii, 1981) which acts as the primary electron acceptor. Together with these are present quinones (Q) and manganese proteins.

Light harvesting complexes : Electrophoretic separation of pigments from Chloroplasts, reveals the presence of two other green bands besides PS I and PS II (Salisbury and Ross, 1986), these green bands represent light harvesting complexes of pigments and proteins; one band functions in association with PS I and the other with PS II. The function of the light harvesting complexes is to absorb light energy and to transfer it to the appropriate pigment system.

Co-ordination between PS I and PS II : According to Junge (1982), each granum has about 200 units of PS I and II, while the number is variable in stroma thylakoids. It was originally believed that chloroplasts contain equal quantities of PS I and PS II, but according to Melis and Brown (1980), the ratio of distribution of PS I and PS II varies with the species and other environmental conditions. Anderson and Melis (1983) believe that grana consists mainly PS II, while stroma thylakoids have mainly PS I

PS I and PS II should function in close co ordination because they jointly function to transfer electrons from water to NADP (see later - non-cyclic electron transfer). Since they are located a little apart, certain intermediaries help to carry electrons from PS I to PS II. These intermediaries or carriers are of two types (a) copper containing proteins called *plastocyanins* bound loosely to the inside of the thylakoid membranes and (b) a group of quinones called *plastoquinones (PQ)* I. Plastoquinones carry two electrons and two H^+ from PS II to PS I.

Besides the above, additional electron transport components in the form of cytochromes also occur in thylakoids. *Cytochrome* b_6 and *Cytochrome f* lie physically in between PS I and PS II and help in electron transport between them. Another cytochrome - cyt b_3 and an Fe-S protein (Ferredoxin) also participate in the light mediated electron transport.

Also present in thylakoids and necessary for light mediated ATP synthesis is a group of complex proteins called *coupling factor (CF)* complexes or ATP *ases* (see later under mechanism of ATP formation for more details). The CF complex consists of a spherical head attached to the thylakoid membranes towards the stroma and a stalk that extends across the lipid bilayer.

MECHANISM OF PIGMENT EXCITATION

Normally the chlorophyll molecule will be in its stable state (no excitation), in the sense the two electrons of its atoms will be spinning in opposite directions and the molecule has no magnetic movement.

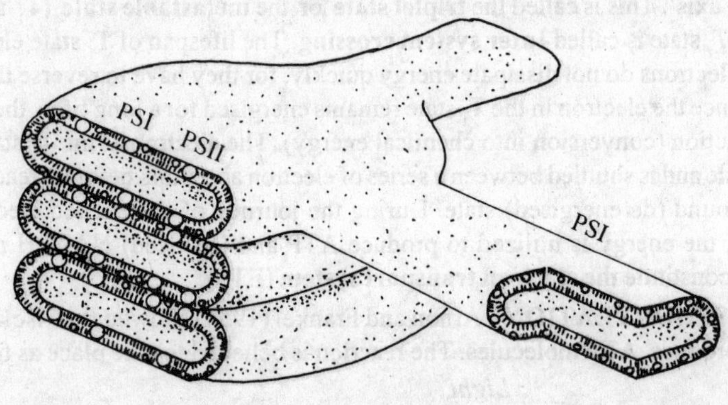

Fig. 12.13 Photosynthesis

Model of the structure of Granum thylakoid and a stroma thylakoid showing the distribution of PS I and PS II

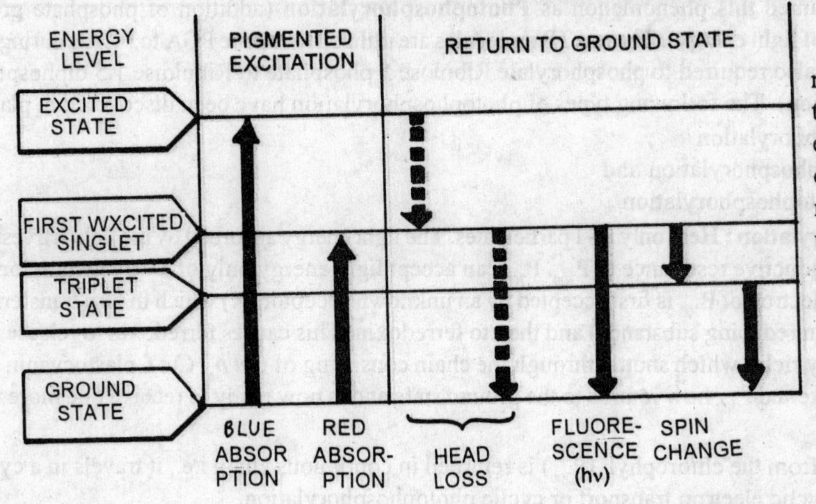

Fig. 12.14 Photosynthesis

Diagram illustrating the absorption of light and changes in the electronic configurations in chlorophyll

When the chlorophyll molecule absorbs light energy in the form of photon, one of the electrons is raised to a higher orbit (higher energy) and hence no magnetic movement. The uniexcited state of the chlorphyll molecule (before the absorption of light) is called the *ground state* or the *singlet state.* Here the energy level is zero hence it is also called the E_0 state. After the excitation, the stage is known as the **first excited state** or the S_1 state with the energy being E_1. The life span of the S_1 state is very short and very soon (10^{-9} sec) the excited electron returns to the ground state by losing energy in the form of heat, light, or by the transfer of energy. The chlorphyll molecule reaches the S_1 state by absorbing the light at the red level (660 nm). If the light is absorbed at the blue level (400nm), the chlorophyll molecule is known to reach the S_2 level (with energy level being E_2) because the 'blue photon' has more energy. However even here the magnetic movement is nil as the spin of the electrons is still in the opposite directions. The electron of S_2 state reaches the ground state via the S_1 state by losing energy either as heat or light (fluorescence or phosphorescence). Since the electron reaches the ground state very soon (10^{-8} or 10^{-9} sec) the energy cannot be put into use (it cannot be converted into chemical form).

Under certain circumstances, an electron at the S_1 state may lose some energy in reversing the spin and drops into a low orbit. At this stage there will be two electrons in two different orbits, but spinning in the same direction. The molecule is now said to possess magnetic movement with electrons oriented in three different ways relative to the axis'. This is called the **triplet state** or the **metastable state** (T_1 state). The conversion of *Si* state electron to T_1 state is called **Inter system crossing**. The lifespan of T_1 state electron is longer (10^{-2} *or* 10^{-3} sec). The electrons do not dissipate energy quickly, for they have to reverse their spin before reaching the ground state. Since the electron in the T_1 state remains energised for a long time, the energy can be made use of for chemical reaction (conversion into chemical energy). The electron at the T_1 state is dislodged from the chlorophyll molecule and is shuttled between a series of electron acceptors before it reaches back the chlorophyll molecule in the ground (de energised) state. During the journey of the excited electron through a series of electron acceptors, the energy is utilized to produce ATP and (reduced) NADPH molecules. The series of electron acceptors constitute the **electron transport system** (ETS).

PHOTOPHOSPHORYLATION : Arnon and Frankel (1954) showed that isolated chloroplasts, under illumination can produce ATP molecules. The reaction is believed to take place as follows

$$ADP + Pi \xrightarrow[\text{Chlorophyll}]{\text{Light}} ATP$$

(ADP = Adenosine diphosphate
ATP = Adenosine triphosphate
Pi = Inorganic phosphate)

Arnon *et al* (1954) named this phenomenon as **Photophosphorylation** (addition of phosphate group toADPunder the influence of light energy). These ATP molecules are utilized to reduce PGA to PGAL during the fixation of CO_2. They are also required to phosphorylate Ribulose 5 phosphate to Ribuloise 1.5 diphosphate (For details, see dark reaction). The following types of photophosphorylation have been discovered in plants.

1. Cyclic photophosphorylation
2. Non-cyclic photophosphorylation and
3. Pseudocyclic photophosphorylation.

Cyclic Photophosphorylatlon : Here only PS I participates. The light energy absorbed by the light harvesting complex is transfered by inductive resonance to P_{700}. P_{700} can accept light energy only of a wavelength longer than 680 nm. The excited electron of P_{700} is first accepted by an unknown acceptor (X) which in turn transfers its electron to FRS (Ferredoxin reducing substance) and then to ferredoxin. This causes ferredoxin to release one of its own electrons (energy rich), which shuttle through the chain consisting of *Cyt* b_6, *Cyt f*, plastocyanin and Finally back to P_{700} - The excited P_{700} now returns to the ground state and is now ready to receive one more unit of radiant energy.

The electron released from the chlorophyll (P_{700}) is returned in continuous chain i.e., it travels in a cyclic fashion, hence the name cyclic electron transport or cyclic photophosphorylation.

Synthesis of ATP during cyclic photophosphorylation : The electron ejected from P_{700} has an energy of 0.42 v which gets reduced to 0.41 when it reaches ferredoxin. Here the electron passes through 3 down hill migration

(*Cyt* b_6, *Cyt f and P)* steps before it is returned to P_{700} - The potential gap between Ferredoxin to P_{700} (ground state) is about one volt which is sufficient to produce at least two molecules of ATP. ATP molecules are formed at two stages - a) between ferredoxin and *Cyt* b_6 and *b*) between *Cyt* and *Cyt f.*

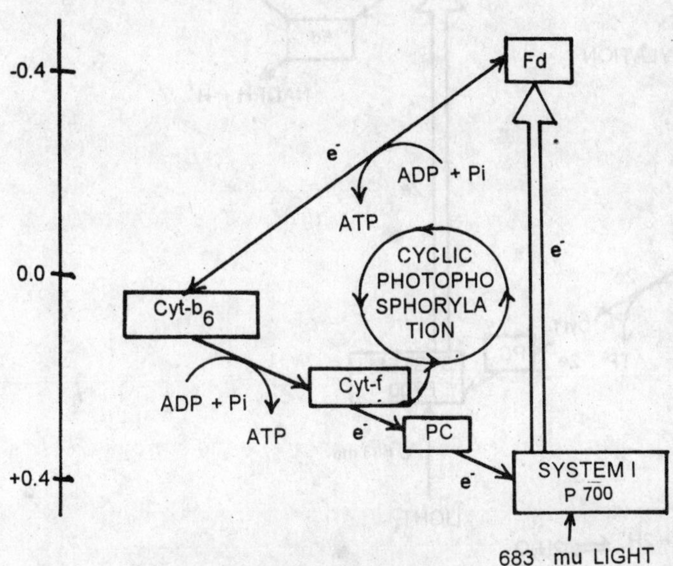

Fig. 12.15 Photosynthesis
Systematic representation of cyclic photophosphorylation

In cyclic photophosphorylation water does not participate; hence there is neither evolution of O_2 nor is there any formation of reduced NADPH molecule (for there is no hydrogen donor). Further PS II is not involved in any stage.

According to Park and Sane (1970) there are two types ofPS I, PS I_s and PS I_g. PS I_s is present in the stroma thylakoids and PS I_g is present in the granum. PS Is participates in cyclic photophosphorylation, while PS I_g participates in non-cyclic photophosphorylation with PS II.

Ramriez *et al* (1968) opine that cyclic photophosphorylation is of limited occurrence in plants and a high rate of cyclic reaction may even retard the CO_2 fixation as NADPH will not be available. They argue that cyclic reaction is not involved in the main path of carbohydrate synthesis as it can only provide part of ATP requirement for CO_2 fixation.

Non-Cyclic Photophosphorylation

This is the major pathway of light reaction involving both PS I and PS II.

Non-cyclic photophosphorylation is also called the 'Z' scheme electron transport because of the zig zag fashion of electron travel. It takes place as follows.

(a) The radiant energy is trapped by PS I (P700) which ejects one of its electrons and is trapped by X, FRS and finally goes to ferredoxin. The reduced ferredoxin now transfers its electrons to NADPH. NADPH gets reduced. PS I however remains in the excited state as it has not got back its electron.

b) Meanwhile PS II absorbs light energy (680nm) and ejects one of its own electrons which is trapped by an unknown quinone acceptor (Q). From Q- the electron moves downhill to cyt *b* and , plastoquinine *cyt f*, Plastocyanin (PC) and finally to PS 1. PS I which is in an excited state, now returns to the ground state as it has got an electron from PS II.

How does PS II come back to the ground state now that it has sent its electron to fill up the 'hole' in PS I ?. Because unless the PS II returns to the normal state, the system cannot continue functioning. The answer lies in the dissociation of water molecules. Water splits it into H^+ and OH^- ions. What is the mechanism of this splitting of water? Some physiologists call it photolysis of water, others are not sure of the mechanism but state that Mn^+ and Cl^- ions are essential for the breakdown of water. In any case the excited PS II returns to the normal state by getting an electron from the OH' ions of water. According to Salisbury and Ross (1986), the excited PS II returns to the ground state by attracting an electron from an adjacent Mn' protein, which in turn gets an electron from the OH^- of water. The H^+ ions released by water are accepted by NADPH to become NADPH + H^+. Thus a reduced NADPH is formed.

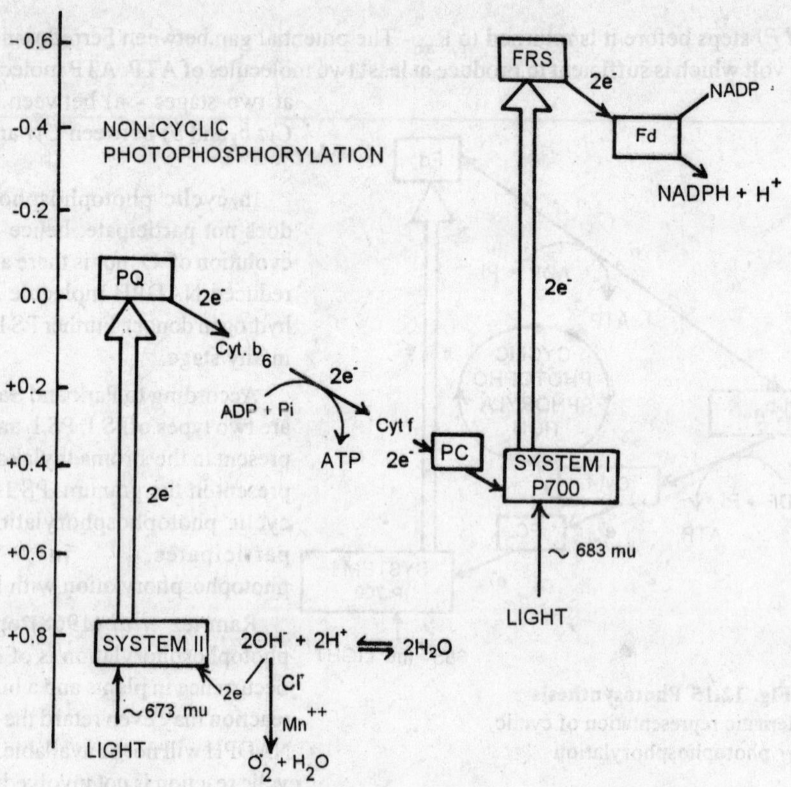

Fig. 12.16 Photosynthesis
Systematic representation of non-cyclic photophosphorylation

The electron transport system or photophosphorylation is called non cyclic because the excited PS I returns to the normal state by getting an electron from PS II, while the excited PS II returns to the ground state by receiving an electron from water. Hence the movement of electron is not in a cyclic fashion.

ATP Synthesis during non-cyclic photophosphorylation : The electron (+ 0.8 v) released by PS II is accepted by a quinone from where the downhill migration begins to reach PS I *via Cyt* b_6, PQ, *Cyt f* and PC. One molecule of ATP is synthesized when the electron shuttles between PQ and *Cytf*.

Release of oxygen and formation of NADPH⁺ + H⁺ : Four molecules of water split up for every turn of non-cyclic photophosphorylation. This is necessary (see fig) to provide four electrons to the excited PS II to bring it back to the ground state. Similarly, four H ions are required to reduce 2 molecules of NADP. These reactions take place as follows.

$$4H_2O \rightarrow 4H^+ + 4OH^-$$
$$4OH^- \rightarrow 2H_2O + O_2$$

(4 electrons from OH reach the excited PS II)

(four electrons are required to reduce 2 molecules ofNADP and these come from the excited PS I *via* ferredoxin)

Thus the end products of non-cyclic photophosphorylation are :

(a) One molecule of ATP
(b) Two molecules of reduced NADPH
(c) 2 molecules of water and
(d) one molecule of O_2.

While a and b are used for the dark fixation of CO_2, c and d are byproducts.

Comparison between cyclic and non-cyclic photophosphorylation

	Cyclic	Non-cyclic
1.	Only PS I participates	PS I and PS II participate.
2.	PS I gets back the electron in a cyclic fashion.	PS I gets back the electron from PS II and PS II from wate
3.	Only ATP molecules are formed.	Both ATP and NADPH molecules are produced.
4.	Water does not participate hence no evolution of O_2.	O_2 evolved as water participates
5.	Found predominantly in bacteria.	Found predominantly in green plants.

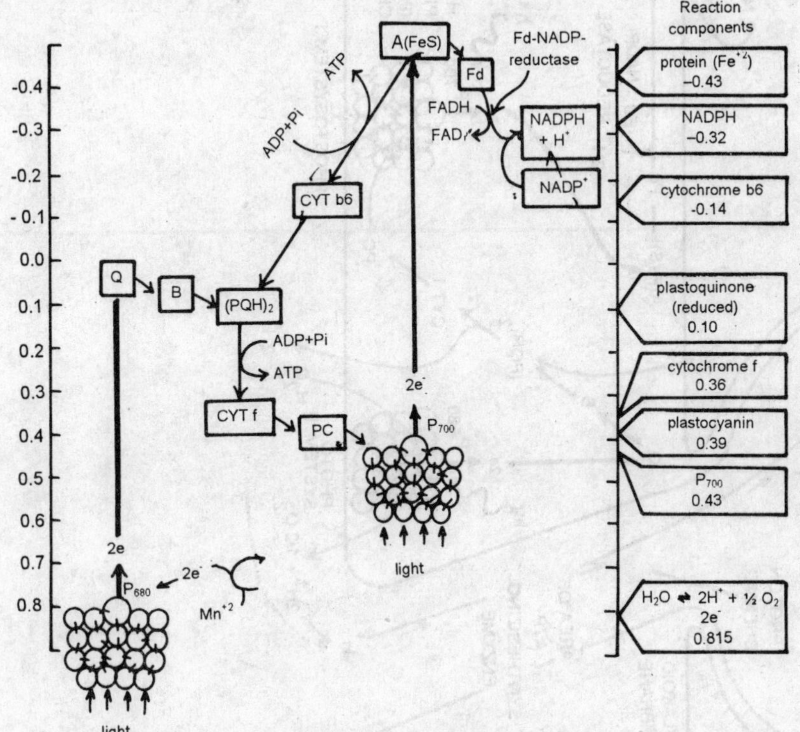

Fig. 12.17 Photosynthesis
Systematic representation of overall electron flow in light reaction combining both cyclic and non-cyclic photophosphorylation

Pseudocyclic Photophosphorylation : Arnon and his co-workers demonstrated another kind of Photophosphorylation, when illuminated chloroplasts produced ATP molecules in the absence of CO_2 and NADP but in the presence of FMN or vit K and oxygen. This involves the reduction of FMN by water with the evolution of oxygen. The reduced FMN is reoxidizable by oxygen. Hence Arnon (1954) called it oxygen **dependent** FMN catalysed photophosphorylation or pseudocyclic photophosphorylation.

MECHANISM OF ATP FORMATION

The source of energy for the synthesis of ATP molecules from ADP comes obviously from the electron transport chain. There are many evidences which point out to this relationship brought about by some common agents. For instance, if some uncoupling agents are introduced into the system, electron transport continues, while the ATP synthesis is inhibited. If the uncoupling agent is removed, ATP synthesis is resumed. Inhibition of electron transport also inhibiting ATP synthesis is another evidence to show that both are coupled.

While there does not seem to be any doubt about the coupling of electron flow and phosphorylation, the mechanism ,is as yet not fully understood (according to some physiologists). The following are the three hypotheses proposed for the mechanism of ATP synthesis.

a. Conformational coupling
b. Chemical coupling
c. Chemiosmotic coupling

(The mechanism of ATP synthesis is same for both photophosphorylation during photosynthesis as well as for oxidative phosphorylation during respiration).

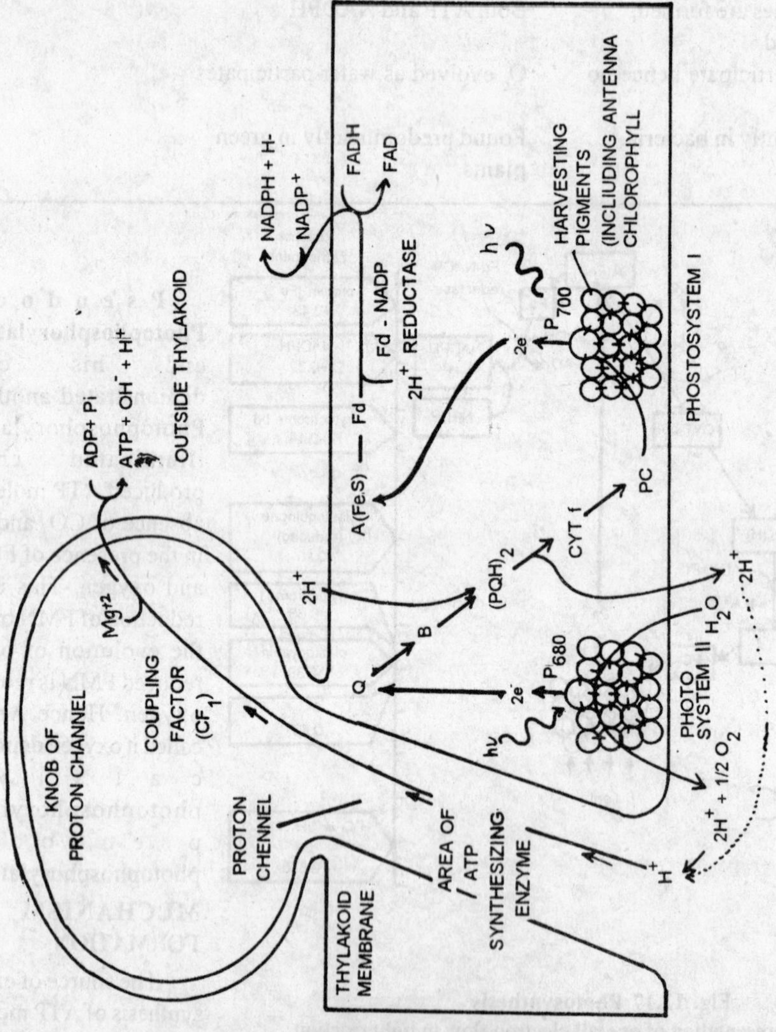

Fig. 12.18 Photosynthesis
Photophosphorylation and ATP synthesis in granum thylakoid according to Chemiosmotic hypothesis

Conformational coupling : During electron transport, both in mitochondria as well as in chloroplasts, some conformational (structural) changes occur which facilitate the release of energy resulting in the production of ATPase mediated ATP synthesis. ATPase is a reversible action enzyme which decomposes ATP into ADP and Pi under less energy conditions, while it facilitates synthesis of ATP during high energy conditions. According to Devlin and Witham (1983), electron micrographs of membranes indicate structural variations during peak activity. But there does not seem to be any definite relationship between activated membrane and ATP synthesis.

Chemical coupling : This hypothesis believes in the presence of a coupling factor (most probably a protein) mediating the energy transfer between electron transport and ATP synthesis. Initially the coupling factor (CF) forms a high energy CF complex with one of the electron carriers. Later an inorganic phosphate replaces the electron carrier in the CF complex. This results in the formation of high energy phosphorylated complex (CFP). The CFP complex releases its energy rich phosphate to ADP resulting in the formation of ATP.

Chemiosmotic coupling : Among the hypothesis proposed for the mechanism of ATP synthesis, this is by far the most acceptable. Originally proposed by Michell during 1960s for oxidative phosphorylation in mitochondria it has recently been applied to photophosphorylation in chloroplasts also (Jagendorf; 1975).

As it relates to oxidative phosphorylation, the flow of electrons from NADH to atmospheric oxygen through the respiratory electron transport, influences the extrusion of protons from the mitochondrial matrix. As-a result, a proton concentration gradient is produced across the mitochondrial membrane. In effect, the energy of electron flow in the respiratory chain is partially conserved as the proton concentration gradient or proton motive force across the inner mitochondrial membrane. The protons return to the mitochondrial matrix down the electrochemical gradient through ATPase resulting in the formation of ATP from ADP and Pi.

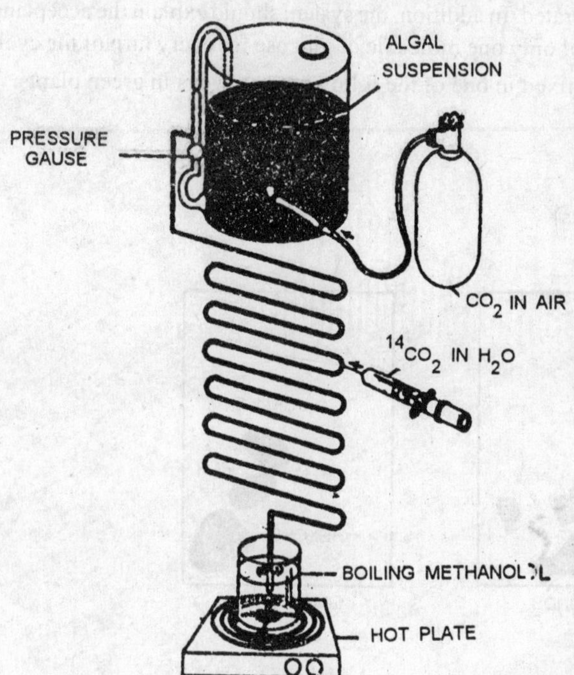

Fig. 12.19 Photosynthesis
Apparatus used by Calvin et al for short time exposure to C^{14}

Ubiquinone, one of the components of the respiratory chain is known to transfer protons across the mitochondrial membrane through alternate oxidation and reduction. The flow of protons across the membrane is due to a force called proticity (Noggle and Fritz, 1983) which is compatable to electricity. According to Noggle and Fritz (19sj), the word chemiosmotic was proposed to emphasize the fact that chemical energy (oxidation energy) is converted to osmotic energy (difference in the proton concentration gradient between the inner and outer faces of the membrane).

Oxidative phosphorylation may be halted by treating the membrane with **uncouplers.** These are weak acids, dissolve in the protoplasmic membrane and disturb the proton concentration gradient by allowing the free flow of protons. Thus the proton uncouplers (protonophores) disturb the ATP synthesis.

The Chemiosmotic theory also explains ATP formation in chloroplasts. Jagendorf (1975) demonstrated that a pH gradient established across the thylakoid membranes stimulates the ATP synthesis. The electrons are transported across the membrane to the interior through the electron transport chain. Light energy provides the necessary drive for the flow of electrons from water to NADP and for transport of protons across the membrane. Plastoquinone, an electron acceptor transfers electrons to *Cytf and* picks up H^+ ions on the outside of the thylakoid membrane and releases protons to the interior of the channel. This results in the building up of a pH gradient (the interior of the thylakoid becomes acidic due to accumulation of protons) across the membrane. The accumulated protons flow from the inside to the stroma side of the membrane through pathways of coupling factor (which include stalks and knobs) which are the sites of photophosphorylation. The downhill migration of protons provides the necessary energy for the addition of Pi to ADP to form ATP, with ATPase acting as a mediator.

DARK REACTION

The phenomenon of thermo-chemical reduction of CO_2 was first established by Blackmann in 1905. He opined that light is not necessary for the reduction of CO_2. In dark reaction the products of light reaction *viz.* ATP and NADPH + H are used to reduce CO_2 into glucose. But the precise manner in which CO_2 is absorbed, and reduced resulting in the formation of a molecule of glucose was understood fairly recently thanks to the availability of radioactive isotopes. Utilizing the isotopes of carbon dioxide, one can trace the different stages of entry of CO_2 into the plant system untill it gets converted into glucose. Many physiologists, biochemists etc have contributed a great lot towards the understanding of the path of carbon in Photosynthesis. The names of Bassham (1957 - 1964), Melvin Calvin (1961), Benson (1951) etc are worth mentioning in connection with the unravelling of the secrets of CO_2 fixation. Melvin Calvin was awarded the Nobel Prize in 1961 for tracing the CO_2 reduction cyde, which is also known as the Calvin cycle.

In CO_2 reduction cycle what we have to precisely understand is how the CO_2 enters the plant system, which is the compound that accepts the CO_2 molecule and how a molecule of glucose is synthesized and how the compound (that originally accepts the CO_2) is regenerated. In addition, the system should explain the acceptance of 6 CO_2 molecules, their reduction and formation of only one molecule of glucose for every turn of the cycle.

Modern researches have indicated that CO_2 is fixed in one of the following pathways in green plants.

1. O_3 pathway or Calvin Cycle
2. Hatch and Slack cycle or C^4 pathway
3. Glycollate pathway (O_2 plants)
4. Crassulacean acid metabolism (CAM plants)

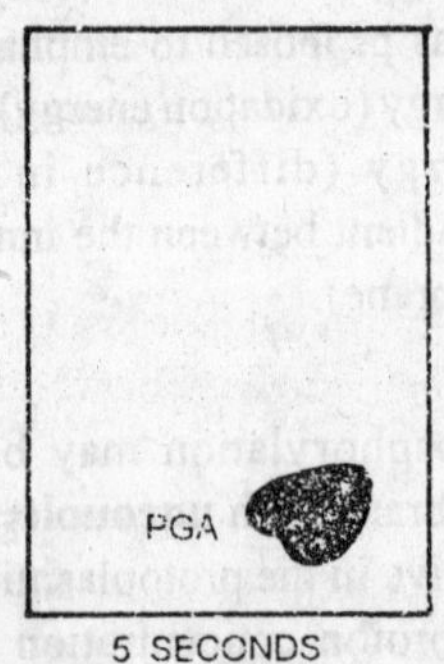

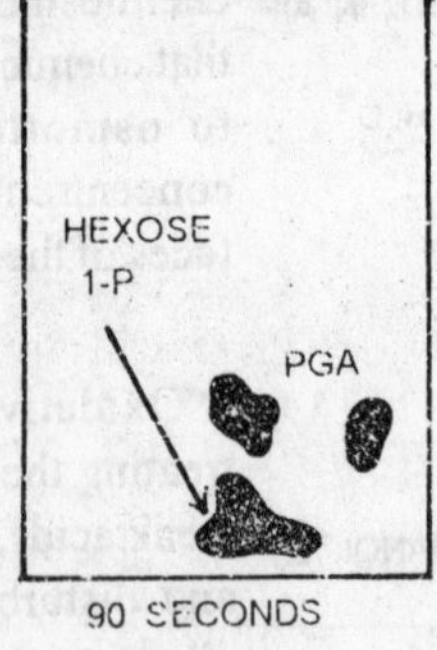

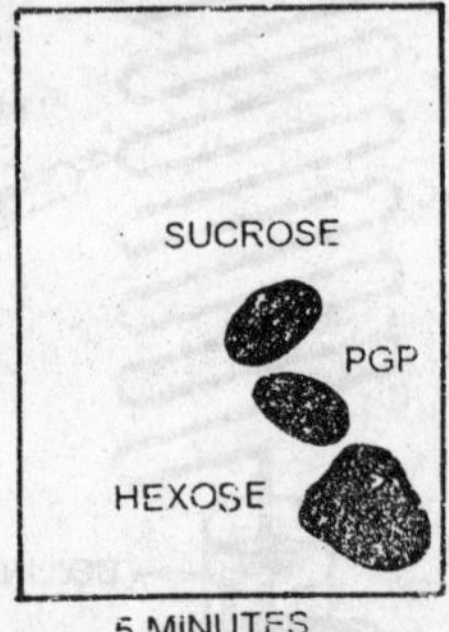

Fig. 12.20 Photosynthesis

Auto radiograph (Diagrammatic representation of chromatograms showing the fixation of C^{14} after different periods of exposure to light

CALVIN CYCLE

This is the major pathway for the fixation of CO_2 in green plants. It is called Calvin Cycle owing to its discovery by Melvin Calvin of the university of California. It is also called the O_3 pathway because, the first stable compound formed after the entry of CO_2 into the plant system is a triose sugar (O_3) viz phosphoglyceric acid (PGA).

Some experimental details of Calvin's work

In order to understand the significance of the discovery of CO_2 reduction cycle, we should understand the experimental methods followed by Calvin. To trace the various steps of conversion of CO_2 after its entry into the plant system untill the formation of glucose, Calvin allowed the experimental plants to photosynthesize for various periods of time and analysed the various compounds formed after different periods of photosynthesis.

Calvin chose for his experimentation two green algae - *Chlorella pyrenoidosa* and *Senedesmus obliquas.* The plants have the same pigmentation as higher green plants and can be easily cultured and harvested. To make sure that the compounds to be analysed can be easily traced and not mistaken for the compounds already present in the .plant, the experimental plants were starved of CO_2 and later supplied with a radio active isotope of CO_2 (C^{14}). Only those compounds in which C^{14} was incorporated were taken into consideration because they were produced after the experiment began. By varying the time of photosynthesis (1/16 sec to 5 sec) and then killing the plant tc stop photosynthesis and analysing the intermediate compounds formed, Calvin and his co-workers were able to map the whole of the CO_2 reduction cycle.

Two, very simple but efficient tools namely Chromatography and Radioautography helped Calvin to trace the path of carbon in Photosynthesis.

Chromatography, in particular paper chromatography is an ideal technique for the separation of compounds present in a small mixture. The extract (mixture) of the plant to be analysed is placed as a spot on a Filter paper and dipped in a suitable solvent. The solvent travels on the paper due to capillary action and carries with it the different substances in the extract. The rate of migration of the different substances depends upon their solubility and specific gravity.

The filter paper (Chromatogram) should be removed from the solvent before it reaches the edge of the paper and dried. By spraying a suitable reagent (For e.g ninhydrin for amino acids) the compounds can be detected as several spots widely separated on the paper. The substances may be identified by their relative positions on the Chromatogram or by comparing with a standard Chromatogram. (In a standard Chromatogram compounds whose identity are known are spotted and their position is marked after the solvent run).

Radioautography is a technique to find out, if a particular source consists of a particular type of isotope. Different radioactive isotopes emit different types of radiation. Chiefly there are 3 types. These are: a) Alpha rays, b) Beta rays and c) Gamma rays.

When we use a particular isotope we already know what is the type of emission. For instance, C^{14} emits β rays. When these rays are directed to a photographic film they form characteristic markings and by these one can identify the type of radiation and thereby the isotope.

In the present experiment, Melvin Calvin and associates allowed the algae to photosynthesise for varying periods (using C^{14}) after which they (plants) were killed in boiling methanol. The plant extract was then spotted on the Chromatogram to identify the compounds. In order to make sure that these contain C^{14} the Chromatogram was held against a photographic film of equal size. Only those compounds which bad incorporated C^{14} leave markings on the film and these were taken into consideration (as produced during experimentation). In order to vary the time of photosynthesis, flashes of light from 1/16 sec to 5 sec were given to identify the earliest compound formed. In this way Calvin and his associates found out all the intermediaries in the CO_2 reduction cycle.

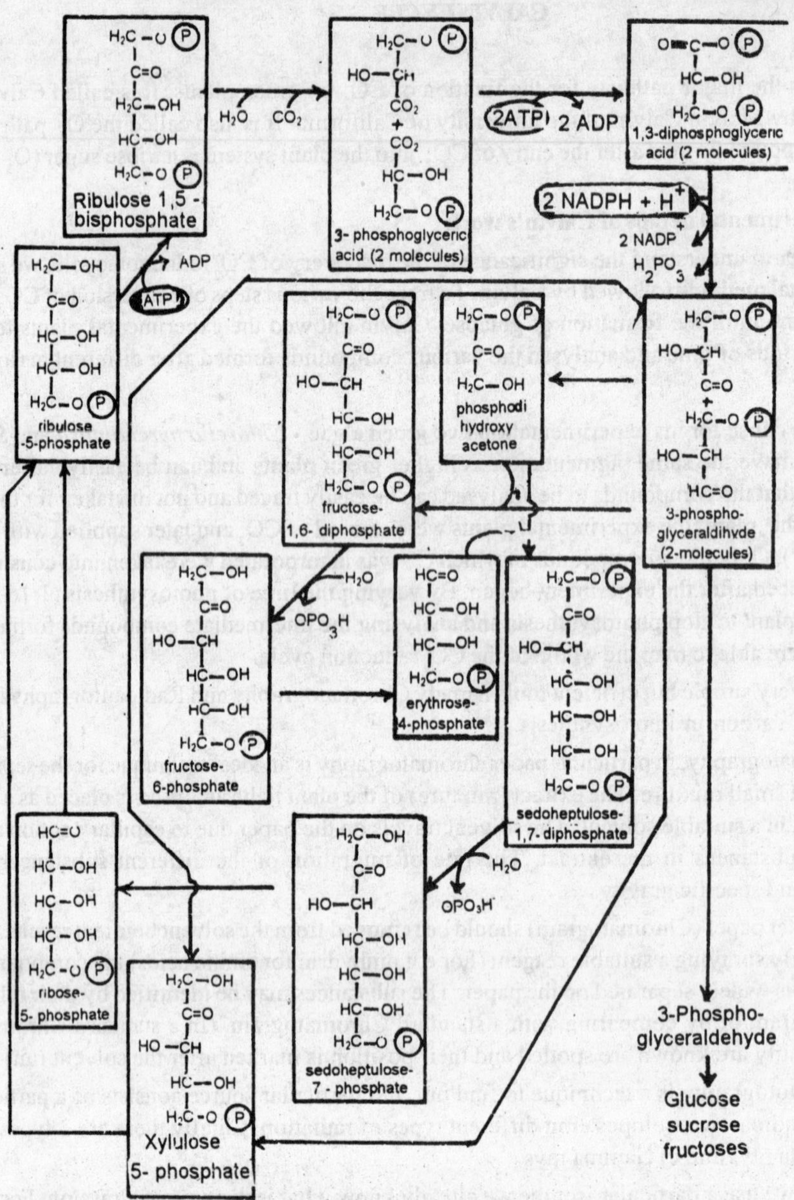

Fig. 12.21 Photosynthesis
Systematic representation of C3 pathway (Calvin Cycle)

The first stable compound formed after the initial entry of CO_2 is a triose (PGA). The triose cannot be the acceptor molecule for it has to be initially a 2C compound to accept one CO_2 molecule to become triose. Benson, however soon found out that a 5 carbon compound is the first acceptor molecule (Ribulose 1.5 diphosphate). After accepting a molecule of CO_2 an unstable 6C compound is formed, which immediately breaks up into 2 molecules of PGA (triose) the first stable compound.

The essentials of the Calvin cycle are as follows. The cycle should be self inclusive in that it should remain unaltered, but provide for the continuous entry of 6 molecules of CO_2, formation of one molecule of glucose and regeneration of six molecules of Ribulose 1.5 diphosphate (RUDP)[1] ,the initial acceptor molecule. The important events of CO_2 cycle can be detailed as under

(a) Fixation of CO_2,
(b) Reduction of PGA,
(c) Formation of sugars, and
(d) Regeneration of RUDP.

1. Rudp is also called Rubp (Ribulose 1,5- biphosphate)

Fixation of CO_2 : CO_2 is first accepted by molecules of a pentose (5C) sugar Ribulose 1.5 diphosphate and an unstable 6 carbon compound is formed. This compound (tentatively called *carboxy dismutase system)* immediately breaks up into 2 molecules of a triose (3C) called phosphoglyceric acid (PGA). The reaction is catalyzed by the enzyme carboxy dismutase or Rudp carboxylase (Rubisco). For every molecule of CO_2, 2 mols of PGA are formed; hence 12 PGA moles for 6 CO_2 mols.

Reduction of PGA

The molecules of PGA are reduced to phosphoglyceraldehyde (PGAL) by NADPH + H. The reaction requires energy and hence ATP participates. The enzyme triose phosphate dehydrogenase catalyses this reaction.

Rudp + CO_2 + H_2O → PGA (2 molecules)

PGA + NADPH + H $\xrightarrow{\text{ATP}}$ PGAL

The above reaction takes place at two stages. In the first stage

PGA + ATP $\xrightarrow{\text{Phosphoglycerokinase}}$ 1.3 - diphosphoglyceric acid + ADP

In the second stage

1.3 diphosphoglyceric acid + NADPH + H → 3 PGAL + NADP

Since 6 molecules of CO_2 are taken into the system (by 6 mols of Rudp) 12 PGAL mols are formed after reduction of the 12 molecules of PGA. Only two are utilized for the formation of sugars, while the rest are channeled back for the regeneration of Rudp molecules. Several intermediate compounds are formed in the process.

Formation of sugars : The PGAL molecule (one out of two) undergoes isomerization to form a mol of dihydroxyacetone phosphate. The enzyme triose phosphate isomerase catalyses this reaction. Ultimately sugars are formed as follows.

1.3 PGAL $\xrightarrow[\textit{isomerase}]{\textit{Triosephosphate}}$ Dihydroxyacetone phosphate

2.3 PGAL + dihydroxyacetonephosphate $\xrightarrow{\text{Aldolase}}$ Fructose 1,6-diphosphate

Fructose 1.6-diphosphate now gets converted into fructose 6 monophosphate due to the loss of one phosphate group under the influence of the enzyme phosphotase

3. Fructose 1.6-dip + H_2O $\xrightarrow{\textit{Phosphatase}}$ Fructose 6. monophosphate (FMP)

Fructose 6 monophosphate isomerises into glucose monophosphate under the influence of the enzyme

isomerase.

4. Isomerase $\xrightarrow{\text{Isomerase}}$ Glucose 6.monophosphate (GMP)

Now FMP or GMP may undergo dephosphorylation to form either fructose or glucose. These monosaccharides unile to form sucrose and various other carbohydrates.

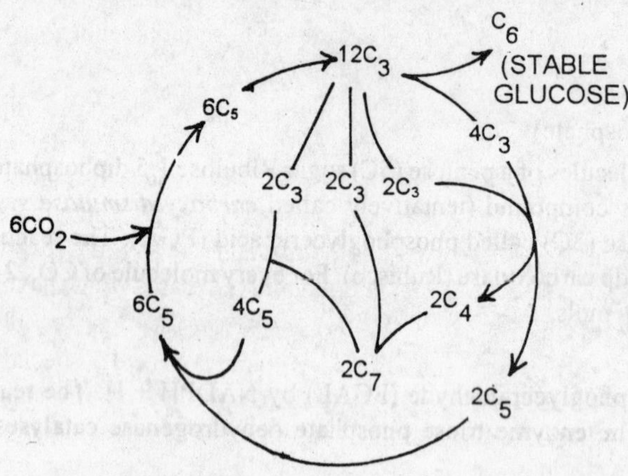

Fig. 12.22 Photosynthesis
Balance Sheet of carbon Calvin's Cycle

Regeneration of Rudp : In the above explanation of formation of sugars, it has been pointed out that Rudp accepts a molecule of CO_2 to ultimately form a molecule of sugar. But the overall formula for photosynthesis accounts for the entry of 6 molecules of CO_2, hence there must be 6 molecules of Rudp to accept the CO_2 molecules and for every turn of the cycle, this should result in the formation of 6 molecules of glucose. But the output of the cycle for every 6 CO_2 molecules is only one molecule of glucose. Then what happens to the remaining 6 molecules of hexose (5 moles of 6C). Obviously they undergo transformation to regenerate 6 molecules of Rudp. This is necessary because, the cycle should be self-sufficient and there must always be 6 molecules of Rudp ready to accept 6 molecules of CO_2. If all the six molecules of Rudp combine with 6 molecules of CO_2 and form 6 molecules ofglucose,the cycle will cease to operate; for the next turn of the cycle there will be no Rudp to accept CO_2 molecules. Thus for the entry of every 6 CO_2 molecules, the output is one mole of hexose and the rest go for the regeneration of Rudp.

The balance sheet of the molecules of carbon from the entry of CO_2 to the regeneration of Rudp is given in the accompanying figure.

In the following equations, regeneration of one mole of Rudp is explained and this may he multinlied to obtain 6 molecules of Rudp).

1. Fructose 6 phosphate + PGAL $\xrightarrow{\text{Transketolase}}$ Erythrose 4. phosphate + Xylulose 5. phosphate
2. Erythrose 4. phosphate + Dihydroxyacetone phosphate $\xrightarrow{\text{Aldolase}}$ Sedoheptulose 1.7 diphosphate
3. Sedoheptulose 1.7 diphosphate $\xrightarrow{\text{Phosphatase}}$ Sedoheptulose 7. phosphate + H_3PO_4.
4. Sedoheptulose 7. phosphate · PGAL $\xrightarrow{\text{Transketolase}}$ Xylulose 5 phosphate + Ribose 5 phosphate

Xylulose 5 phosphate and Ribose 5 phosphate get converted into Ribulose 5 phosphate under the influence of the enzyme **phosphoplentose isomerase**

5. Ribulose 5 phosphate $\xrightarrow[\textit{– kinase}]{\textit{Phosphopento}}$ Rudp.

HATCH AND SLACK CYCLE (C_4 pathway)

For a long time it was thought that the only method of carbon fixation in green plants is by Calvin cycle (C_3 pathway). Indeed a large majority of plants (from algae to higher) investigated seemed to confirm this view. The findings of Kortschah and Hartt (1964) in sugarcane plants however doubted the universality of occurrence of the O_3 pathway being the only mode of carbon assimilation.

On feeding the sugar cane plants with labelled CO_2 (C^{14}), they found that the first stable products formed are certain dicarboxylic acids and not PGA as expected. Malic acid and aspartic acid were found to be the first labelled products on the plants being exposed to shortest period of photosynthesis. Subsequently, extensive researches of Australian physiologists Hatch and Slack (1966) confirmed C_4 dicarboxylic acids to be the first stable products of photosynthesis in many plants. The pathway of carbon fixation in sugarcane then came to be known as **C_4 path way** or **Hatch and Slack pathway** or **β carboxylation pathway.**

Originally discovered in sugarcane, the C_4 pathway has since been found to occur in a good majority of grasses like maize, *Panicum maximum, Chloris guyana, Atriplex spongosa* etc. It has also been reported in the guard cells of *Tulipa gesneriana* and *Commeliana communis.* Besides *Poaceae,* C_4 pathway is also known to occur in plants belonging to families Cyperaceae, Amaranthaceae, Chenopodiaceae, Asteraceae, Euphorbiaceae, Portulacaceae etc. However not all graminaceous plants have a C_4 pathway. Wheat and rice have a C_3 pathway. The plants which have a C_4 pathway of carbon fixation are called C_4 plants as against C_3 plants which have a C_3 pathway.

Structral peculiarities of C_4 plants : Most of the plants exhibiting a C_4 pathway have a special leaf anatomy called Kranz anatomy ("Wreath" in German). The anatomy is characterized by the lack of differentiation of the mesophyll into palisade and spongy parenchyma, the most important feature being the vascular bundles (veins) surrounded by a layer of distinct parenchyma cells arranged radially. These constitute the **bundle sheath.** In a transverse section of the vascular bundle, the sheath appears like a "Wreath", hence the name Kranz anatomy. In addition to this, C_4 plants are known to possess dimorphic chloroplasts. The chloroplasts in the bundle sheath cells lack grana, while those of mesophyll cells possess them. Dimorphic chloroplasts however need not occur in all C_4 plants (e.g Bermuda grass). Another characteristic feature of 'Kranz' leaves is the close packing of mesophyll cells (less intercellular spaces) which helps in the easy diffusion of CO_2 to carboxylation sites.

The C_4 pathway

Phosphoenolpyruvate (PEP) is the initial acceptor of CO_2. Phosphoenolpyruvate carboxylase found abundantly in mesophyll cells catalyzes the fixation of CO_2 with PEP, which gets converted into oxaloacetate. This is the first stable compound in Photosynthesis. The reaction requires the participation of a molecule of water; phosphoric acid is a byproduct

$$PEP + CO_2 + H_2O \xrightarrow[\textit{Carboxylase}]{\textit{PEP}} \text{oxaloacetic acid}$$

The function of PEP carboxylase is similar to that of Rubisco in the C_3 pathway. Oxaloacetic acid is relatively unstable and readily gets converted into either malate or aspartate. Malate is produced by the activity of malic dehydrogenase in the presence of NADPH + H^+. The malate thus produced is transferred to the chloroplast of bundle sheath cells. Here malate is decarboxylated (under the influence of the enzyme *malate dehydrogenase)* to produce pyruvate, CO_2 and NADPH. NADPH travels back to the mesophyll cells to regenerate malate, while pyruvate also travels back to the mesophyll cells, where it utilizes the light generated ATP to produce PEP again.

CO_2 released by the decarboxylation of malate is fixed in the bundle sheath cells by the typical C_3 pathway (it is accepted by Rudp).

In the C_4 plants there are two carboxylations one by the atmospheric CO_2 which forms the dicarboxylic acids and second by the internally generated CO_2 entering the Rudp. In effect, in C_4 plants, there are both C_4 arid 03 pathways. CO_2, for fixation by the C_3 pathway comes from the C_4 pathway. Further C_3 and C_4 pathways are delimited to bundle sheath cells and mesophyll cells respectively. Because there is carboxylation at two sites, the pathway is also known as Dicarboxylation pathway.

Chollet and Ogren(1975) have recognised three categories of C_4 plants. This categorization is based on the kind of stable product formed initially in photosynthesis. These are :

1. Fixation of CO_2 by PEP results in the formation of oxaloacetate, which gets converted into malale .e.g. maize, sugarcane etc.,
2. Oxaloacetate gets converted into asparatate in the mesophyll cells and it is transported to bundle sheath cells. In the bundle sheath cells, asparatate is reconverted to oxaloacetate. Oxaloacetate then gets converted lo pyruvate and CO_2 by PEP carboxykinase. E.g. *Panicum maximum, Chloris guyana.*
3. Asparatate produced in mesophyll cells is transported to bundle sheath cells where it gets transaminated to oxaloacetate first and then gets reduced to malale in the mitochondria (using NADH).The malale is decarboxylated to produce pyruyate and CO_2 e.g. *Atriplex spongosa*

Significance of C_4 pathway : The discussion of 04 pathway above shows that it involves cytoplasmic as well as chloroplast and mitochondrial enzymes. Thus the old concept that photosynthesis exclusively takes place in the chloroplast is strictly not true (Bidwell, 1979). Taxonomically C_4 plants are found in several groups of tropical monocotyledons and dicotyledons. In many families both C_3 and C_4 plants are found. It will be too much to interrelate the various members taxonom'cally on the basis of C_4 pathway. This leads to the conclusion that C_4 pathway must have arisen independently several times during Evolution.

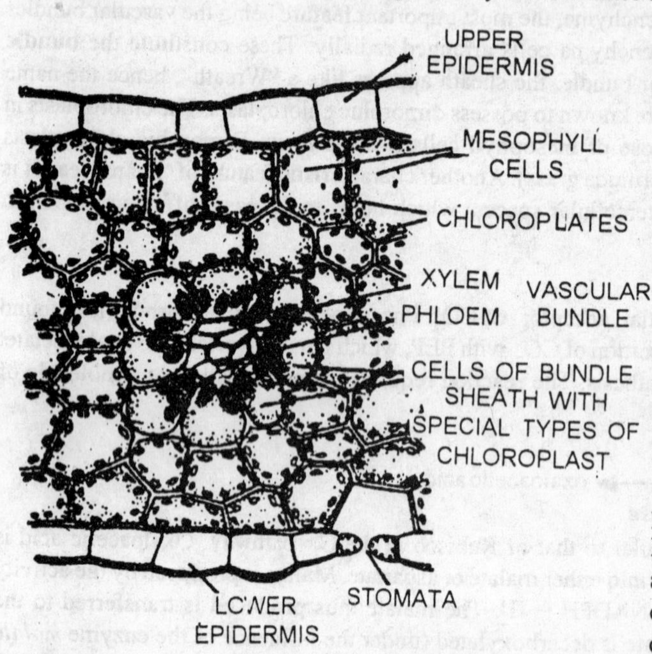

Fig. 12.23 Photosynthesis
TS of a monocat leaf (grass) showing Kranz anatomy

What is the significance of C_4 pathway for the plants which possess it? It is widely believed that C_4 plants are capable of higher rates of photosynthesis due to the high affinity of PEP carboxylase to CO_2. C_4 plants can carry on photosynthesis even under very low CO_2 concentrations(10 ppm). Even when the stomata are nearly closed, C_4 plants can continue to photosynthesize. This is of very great advantage in plants growing under xeric conditions (hence found in many tropical plants). There is no photorespiration in C_4 plants, hence no loss of CO_2.

Many of the weeds and some of the high yeilding crops are C_4 plants (Bidwell, 1979). But it is not always true that C_4 plants are always better than C_3 plants. Some C_3 plants compare favourably well with C_4 plants in terms of productivity.

In a way, C_4 pathway is less efficient than C_3 because it uses more light energy to fix CO_2. In other words, the excess light available in tropical plants is used to run the C_4 cycle to concentrate CO_2 in bundle sheath cells where it can be fixed at a rapid rate by the C_3 cycle. Here C_4 pathway can be regarded as a guiding factor to provide increased CO_2.

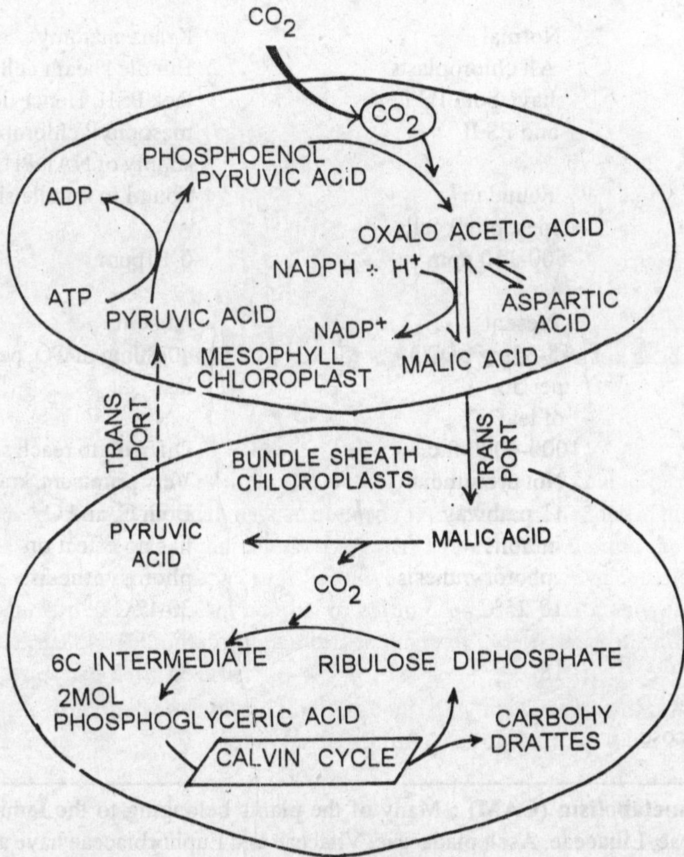

Fig. 12.24 Photosynthesis
Systematic representation of Hatch and Slack pathway

The real significance of C_4 pathway therefore is to regard it as an adaptation to a tropical climate where the interpolation C_4 pathway leads to a better utilization of available CO_2 in the C_3 fixation. Another important aspect of C_4 pathway is the fact that such a fundamental phenomenon like photosynthesis also need not have one absolute pathway. Even here nature seems to have experimented with variations depending on the environmental factors. But an inescapable conclusion is that C_4 pathway by itself does not produce carbohydrates. It is only a contributory pathway for the C_3 cycle. Thus the pre-eminence of C_3 as the chief (or sole) pathway of carbohydrate production still stands.

The following table lists the differences between C_3 and C_4 plants.

No.	Character	C_3 Plants	C_4 Plants
1.	CO_2 acceptor	Rudp	PEP
2.	First stable product	PGA	Oxaloacetate
3.	Type of Chloroplast	One type	Dimorphic, bundle sheath chloroplasts lack grana. Mesophyll cells have normal chloroplasts.

4.	Leaf anatomy	Normal	Kranz anatomy
5.	Pigment systems	All chloroplasts have both PS I and PS II	Bundle sheath cell chloroplasts lack PSII. Hence dependent on mesophyll chloroplasts for supply of NADPH.
6.	Enzyme of C_3 pathway	Found in mesophyll cells	Found in bundle sheath cells.
7.	CO_2 compensation point	500-150 ppm	0-10 ppm
8.	Photorespiration	Present	Absent
9.	Net rate of photo - synthesis in full sunlight	15-35mg of CO_2 per dm^2 of leaf	40-80mg of CO_2 per per dm^2 of leaf
10.	Saturation intensity	1000-4000 ft.c.	Difficult to reach saturation.
11.	Bundle sheath Cells	Not prominent	Very priminent; kranz like
12.	Co_2 fixation	C_3 pathway	Both C_3 and C_4
13.	High rate of O_2	inhibits photosynthesis	has no effect on photosynthesis.
14.	Temperature (optimum)	10-25°C	30-45°C
15.	ATP molecules required to synthesize one molecule of glucose	18	30

Crassulacean acid metabolism (CAM) : Many of the plants belonging to the families Crassulaceae, Orchidaceae, Bromeliaceae, Liliaceae, Asclepiadaceae, Vitaceae and Euphorbiaceae have a diurnal pattern of organic acid formation. Many of these plants like *Kalanchoe, Sedum, Agave* etc., which grow in acidic habitat have very low rates of transpiration with succulent leaves and stems.=

In CAM plants the total content of the organic acids decreases during day and the pH of the leaf cell sap increases, whereas during night the organic acid content increases and the pH decreases. Accompanying these changes in the leaf is the increase of storage of carbohydrates during day and their decrease during night.

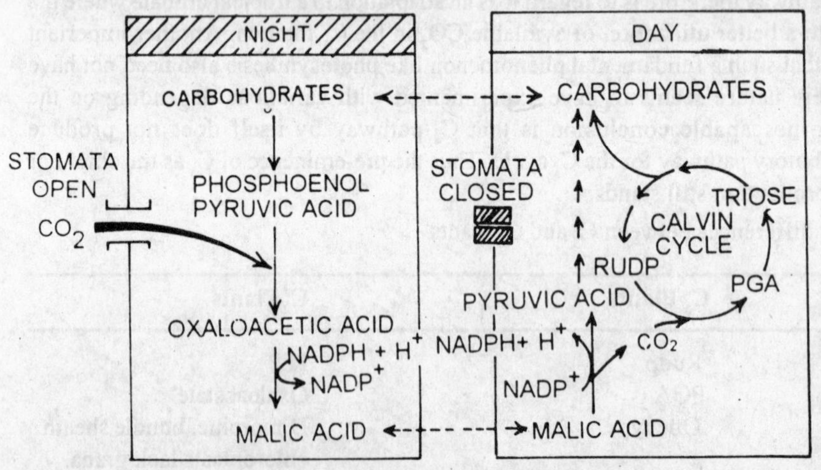

Fig. 12.25 Photosynthesis
Systematic representation of CO_2 fixation in CAM Plants

The diurnal pattern of organic acid content in CAM plants is associated with a diurnal stomatal opening and closing. In CAM plants stomata open during night a.nd close during day (the reversal of what is seen in other plants). Since most of the CAM plants inhabit arid areas, this type of stomatal behaviour (closure during day) helps in water conservation by checking the rate of transpiration.

During night when stomata are open, CO_2 is fixed by phosphoenol pyruvic acid (PEP) under the influence of the enzyme *PEP carboxylase (PEPcase)*. PEP is first converted to oxaloacetic acid with the help of *PEPcase*. Later, oxaloacetic acid gets converted to malic acid with the help of malic dehydrogenase. Malic acid eventually gets converted into starch, glucose etc., in the presence of light. It undergoes decarboxylation to form pyruvate and CO_2. CO_2 enters the C_3 cycle, while pyruvate goes for the regeneration of PEP.

The above discussion underscores the point that the metabolic fixation of CO_2 is essentially same in both C_4 plants and CAM plants. In CAM plants however both C_3 and C_4 pathways occur in mesophyll cells,while in C_3 plants they [C_3 and C_4 pathways] are separated between mesophyll and bundle sheath cells. Again C_4 and C_3 pathway occur simultaneously in C_4 plants,while in CAM plants they occur during night and day respectively. *Thus the C_3 and C_4 pathways are separated in space in C_4 plants but separated in time in CAM plants.*

It may be pertinent to analyse as to what made the CAM plants to seperate C_4 and C_3 in time. CAM plants are essentially C_3 plants, which had to resort to nocturnal photosynthetic activity due to the closure of stomata during day time.Maintaining only the C_3 pathway would have been difficult for them, for there could be little or no photosynthetic activity; as there would be no supply of atmostpheric CO_2 which is necessary for the C_3 pathway. Again, the stomata could not be kept open during daytime to facilitate CO_2 supply for the C_3 pathway as it would not be in the best interest of water economy, given the arid conditions in which the CAM plants grow. Hence the adaptation of both C_4 and C_3 pathways separated in time but not in space [there is no Kranz anatomy] is the best the CAM plants could have bargained for.

FACTORS AFFECTING THE RATE OF PHOTOSYNTHESIS

Like any other biological process,photosynthesis is influenced by a number of environmental conditions. These factors or environmental conditions occur at varying concentrations and consequently the rate of the process also varies. Sachs[1860], who made a study of the influence of concentration of factors on a biological process proposed the concept of three cordinal points.These are a **minimum, an optimum** and a **maximum.** Minimum point represents the lowest concentration of the factor when the process barely begins,and the optimum represents the highest rate of the process and a maximum [of the factor] represents the limit beyond which the process is impossible. The rate of the process beyond the maximum of the factor actually comes down.

This concept of three cardinal points is based on the isolated study of a factor as to its influence on a process to the exclusion of other factors. But in actual fact when a process is conditioned by several factors, a particular factor not only influences the process but will influence other factors also mutually. For instance an optimum of a factor will fluctuate depending on the influence of other factors. Optimum concentration of CO_2 for example is not fixed, it is greater with increased light intensity and *vice versa*, same is true for light also. Hence the factors that influence a process are all interrelated and they condition a process not in isolation but in totality.

The problem of influence of various factors and their optimum intensity was sought to be solved by Liebig[1843] who proposed the Law of the minimum which states that when a physiological process is conditioned by several factors, the slowest factor controls the rate of the process. Blackmann [1905] slightly elaborated this and proposed the Law of limiting factors which states *"when a physiological process is conditioned as to its rapidity by a number off actors, the rate of the process is governed by the pace of the slowest factor."*

According to this principle, the magnitude of photosynthesis is controlled by only one of the factors at a time. This does not mean that other factors are not necessary. When they are available in abundance, the one that is available in the least governs the rate of the process. The rate of photosynthesis is proportional to the quantity of this factor. But this is not fixed; the rate of photosynthesis alters when the availability of another factor becomes least. Then this factor [which has become least] becomes limiting.

The law of limiting factors can be graphically demonstrated as seen in the Figure.

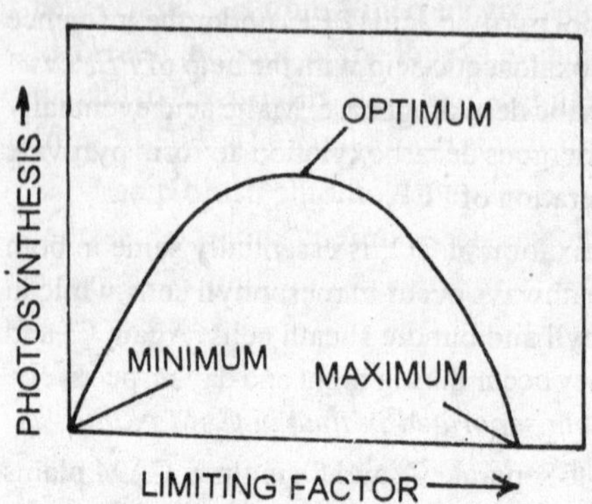

Fig. 12.26 Photosynthesis
Sach's concept of 3 cardinal points

Suppose a leaf is exposed to light to utilize a maximum of 5mg of CO_2 per hour but only one mg is available, CO_2 will become the limiting factor. At this stage, increase of CO_2 supply to 2mg per hour will increase the rate of photosynthesis. But when the CO_2 supply is increased to 6mg it ceases to become the limiting factor [becuase supply is more than saturation]and light becomes the limiting factor. At this stage, the rate of photosynthesis can be increased by increasing the concentration of light.

In the graph representing the law of limiting factors, the rate of photosynthesis increases from A to B_1 when the supply of CO_2 is increased to 5mg per hour. Thereafter the rate of the process remains constant along B_1-B_2 and further supply of CO_2 has no effect, because light now becomes the limiting factor. Increasing the supply of light at this stage increases the rate of photosynthesis from B_1-C_1. The rate again becomes constant along C_1-C_2 unless there is further increase in light supply. With the increased supply of light the rate increases along the line C_1-D_1, and then becomes constant along C_1-C_2. Any further increase of light at this stage may not increase the rate of the process becauase some other factor would have become limiting.

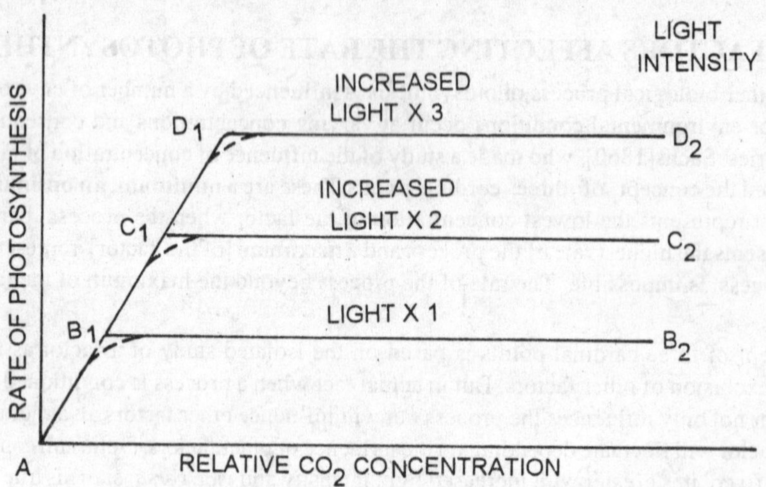

Fig. 12.27 Photosynthesis
Graph representing Law of Limiting Factors

Demerits of the law of limiting factors : The principle of the law of limiting factors as proposed by Blackmann [1905] was criticised by Boysen-Jensen [1918), Warburg 11918], Hander 11921], James [1928] and Hoover [1933]. These workers felt that if there is only one limiting factor functioning at any given time as assumed by Blackmann, any change in the concentration of the factor should result in abrupt change or transition in the rate of the process. But in actual practice, change in the rate of the process is always gradual so that in a graphic representation a smooth curve is obtained and not a skewed one. This is due to the fact that at any given time, not one but many factors may be limiting. Besides, the millions of chloroplasts which are photosynthesising may not be exposed to the conditions [factors] uniformly. For example, for certain chloroplasts light may be a limiting factor,while for others at the same time CO_2 may be a limiting factor.

In order to satisfy the possibility of more than one factor being "limiting" simultaneously, the law was modified and termed "relatively limiting factor" to denote that a factor quantitatively smallest need not be limiting by itself, but any factor whose quantity is less than the actual amount required will act as "limiting".

Another possibility not considered by Blackmann is that a factor present far in excess may also inhibit the process. However this does not invalidate the principle of "limiting" influence of the factors. Under experimental conditions, where one can strictly control the factors, the limiting factor influence does operate (Rabinowitch, 1951). While the law of limiting factors provides an approximate evaluation of the influence of factors, its strict application may not be possible (Devlin -Baker, 1971).

The factors that influence the rate of photosynthesis may be classified into External and Internal. The external factors are - Light, CO_2, Temperature, Water and Oxygen. The internal factors are - Chlorophyll content, Protoplasmic factors, accumulation of end products and availability of nutrients. We will discuss each one of these factors in detail.

LIGHT

The solar radiation is the sourse of light, although photosynthesis can be carried out in the presence of artificial light also. The radiation from the Sun is available in the form of electromagnetic radiation much of which is shielded by the Earth's atmostphere. Although some bacteria can carry on photosynthesis under infra red. much of photosynthesis is carried out under the visible light. Broadly speaking, three aspects of light affect photosynthesis. These are - light intensity,light quality and light duration.

Light Intensity : The morphology, architecture and internal structure of the leaf is best suited to utilize maximum amount of incident light. However leaves can utilize only a fraction of the light that falls on them. Of the total quantity of incident light, a part is reflected, a part is transmitted and only a little is absorbed. According to Seybold [1932] about 80% of the visible light falling on the leaves is absorbed. However most of the absorbed light is used up in raising the temperature of the leaf and is lost as heat. Only about 2% of the light perhaps is made use of in photosynthesis.

According to Wolkoff (1866), the rate of photosynthesis is directly proportional to the intensity of light. But this is true only when other factors are not limiting. It has been observed that a maximum rate of photosynthesis can be achieved by leaves under a light intensity, less than the maximum.

Light acts as a limiting factor for plants growing in shade (this may not apply to *Psciophytes*, which are genetically shade loving) or when the envirnoment is cloudy or foggy. The rate of of photosynthesis seems to be at the maximum under diffused light while in bright light of noon or afternoon, the rate is slow.

During daytime, the rate of photosynthesis is 10 times faster than the rate of respiration. At a certain intensity of light, the amount of CO_2 used up during photosynthesis and the amount of CO_2 released during respiration is volumetrically equal. This point is known as the compensation point. In shade plants, compensation point is retained for a longer time than in light plants (plants growing in Sunny area). Light compensation point is of the order of 100-200 fc for heliophilic leaves, while only about 10 fc for psciophilic (heliophobic) leaves

While very low intensities of light lower the rate of photosynthesis by causing stomatal closure and blocking the entry of CO_2, an increase in the light intensity initially brings about an increase in the rate of photosynthesis. But this is subject to a certain limit. Beyond this point, the rate of photosynthesis actually is retarded. This point is known as the saturation point. High intensity of light not only retards photosynthesis, but it is also injurious to the plant. Besides causing excessive transpiration and reducing the water content of the plant, high light intensity also causes a phenomenon called Solarization. Solarization causes photooxidation of the Chlorophyll molecule and is dependent on O_2. Solarization can be observed when shade plants are kept in direct sunlight. Leaves become chlorotic and develop a burnt-like appearance. Soarization is explained on the basis of excitation of more molecules of Chlorophyll (under the influence of high intensity of light) than can be utilized in photosynthesis and are oxidized in the presence of O_2. According to Thomas(1955) and Stanier(1959), the electron transport chain cannot support the large number of excited electrons that are ejected out of the chlorophyll molecules. Thomas(1955) has also reported decreased rates of protein and carbohydrate synthesis during high light intensity. Carotenoids are supposed to play a protective role for Chlorophyll molecules under high light intensity. They absorb high light energy and divert it from Chlorophyll dissipating it as heat.

Light quality : Normally the rays of white light between the range of 390nm and 760nm are useful in photosynthesis. Studies of Hoover have indicated maximum absorption for Chlorophyll in the range of blue (440nm) and Red (655nm). Photosynthetic bacteria however can carry on the process even at 900nm also. Most of the light in the green region is reflected indicating least absorption.

Some physiologists believe that the rate of photosynthesis actually depends on the energy absorbed irrespective of the wavelength absorbed. But quality of lieht is also important as seen from the fact, that the rate of photosynthesis is not high in the undergrowth of a thick forest as they get mostly the green light filtered through the leaves of tall trees.

Duration of Light : When light intensity and other factors are favourable, the rate of photosynthesis is directly proportional to the duration of light. It has been found out that the quantum yield is higher when plants are exposed to continuous daylight of 10-12 hours. Findings ofMitchell(1936), Boehring(1949) and others have indicated that plants continue photosynthesizing for long periods without any ill effect on them. Devlin and Baker(1971) however have suggested that continuous light is damaging to photosynthesis. The rate of photosynthesis has been noted to be high under diffused light for longer duration, than under high intensity of light for shorter duration.

CARBON DI OXIDE

The quantity of CO_2 in the air is extremely small. It is about 3 parts in 10,000 or 0.3% by volume. This amount is constant and its concentration affects the rate of photosynthesis markedly.

The sources of CO_2 are plant and animal respiration, bacteria in the soil, water and ocean, decay of organic substances and combustion of various types of fuels. It is beleived that the "golden age" for green plants as for as photosynthesis is concerned was the carboniferous age (300 million years ago). This is due to the fact that the humid and warm atmosphere of the carboniferous age contained far more CO_2 than it is today and this supported a higher rate of photosynthesis. Much of the fossil fuel we have today represents the stored photosynthetic products of the carboniferous period.

Godlewski (1873) found that there is considerable increase in the rate of photosynthesis with the increased supply of CO_2. But this is subject to a certain limit. A higher 002 concentration may actually retard the process possibly by affecting other processes (Steeman-Nielson 1955).

Optimum level of CO_2 for maximum photosynthesis varies with plants. In *Triticum sativum*, maximum photosynthesis is achieved at 0.15% of CO_2. In water plants even 1.1% of CO_2 is conducive for a higher rate of photosynthesis.

TEMPERATURE

Like any other physiological process photosynthesis too is affected by fluctuations in temperature. It has a range when the process is active, a minimum as well as a maximum. This range however is not fixed. It varies with plants. While it is true that the light phase of photosynthesis is unaffected by temperature/a high temperature however not only retards photosynthesis, but might also even prove harmful to the plant itself.

It is generally believed that photosynthesis is possible at a temperature range tolerated by protein compounds [0°C 60°C]. Between 6°C and 35°C the rate of photosynthesis increases with increase in temperature. At this range it follows Van Hoff's rule that for every rise of 10°C, the rate of photosynthesis doubles [$Q_{10} = 2$]. Many of the desert plants can photosynthesize comfortably at 55°C. In majority of the plants an optimum of 30-35°C is required for the maximum rate of photosynthesis.

Extremes of temperature cause injury or even permanent damage to the plants. A very cold temperature inactivates photosynthesis by blocking the activity of enzymes. Besides, in cold temperature *ice* formation takes place resulting in the change of colloidal nature of cytoplasm. Ice formation also disturbs the permeability of the membranes.

WATER

It is difficult to analyze the effect of water scarcity on photosynthesis largely because, plants absorb a large quantity of water of which only a minute quantity is utilized for photosynthesis. Further, scarcity of water makes its deleterious effect felt on the general survival of the plant before it influences the rate of photosynthesis. Thus due to water scarcity various other forces make the functioning of the photosynthesis impossible,before the effect [of water scarcity] on photosynthesis could be analysed.

Schneider and Childers [1941] noted almost 50% reduction in photosynthetic efficiency in dry soil conditions which favour high transpiration. Elimination of water results in general dehydration, closure of stomata, inactivation of enzymes etc. Besides, dehydration may cause irrepairable damage to the micromolecular structure of the chloroplast membranes.

OXYGEN

It is one of the byproducts of photosynthesis. While many plant physiologists have opined that O_2 is necessary for photosynthesis, there does not seem to be any experimental evidence in support of this assumption The accumulation of oxygen infact retards the rate of photosynthesis. Findings of Warburg [1920] have proved the inhibitory effect of O_2 on photosynthesis. Photosynthetic inhibition by O_2 is called Warburg effect. Working on *Chlorella* plants, Warburg [1920] demonstrated such an effect. In Warburg effect, both O_2 evolution and CO_2 assimilation will be inhibited in the presence of atmostpheric O_2. Gibbs [1970] has demonstrated that high concentration of oxygen brings about competetive inhibition of Ribulose phosphate carboxylase. High concentration of oxygen also inhibits photosynthesis by encouraging photorespiration.

INTERNAL FACTORS

Chrophyll content : Chlorophyll is indispensible for photosynthesis. Earlier Willslatter and Stoll [1918] thought that no direct relationship exists between the rate of photosynthesis and chlorophyl content. In a study of green and yellow leaved varieties of the same species they found that the total quantity ofthe chlorophyll had little bearing on the rate of photosynthesis. They expressed the relationship between the quantity of chlorophyll [in grams] and the quantity of CO_2 absorbed [in grams] during one hour in terms of **photosyntlietic number** or **assimilation number**. Working with the fresh water green alga *Chlorella,* Emerson [1929] demonstrated that the chlorophyll content has a direct bearing on the rate of photosynthesis. According to Thomas [1955], the rate of photosynthesis is slow in young leaves and optimum in mature leaves. However mature leaves of old age show reduced rate of photosynthesis due to "ageing factor". According to Clendenning and Gorham [1950] ageing changes the chloroplast structure in that grana might get disintegrated. This is also supported by the observations of Shaw and Nanocha (1965).

Protoplasmic factors : Chloroplasts are completely independent units as for as photosynthesis is concerned. The works of Hill and Arnon have clearly demonstrated that chloroplasts are capable of photosynthesis even when taken out of the cell. Therefore the opinion of earlier physiologists that some cytoplasmic factors influence photosynthesis may not be true. However enzymes which are cytoplasmic in origin are necessary for photosynthesis and they remain active only in the hydrated state of cytoplasm.

Nutrition : Macro and micro elements play a vital role in plant nutrition. Hence any deficiency of mineral elements is bound to affect photosynthesis adversely. Many elements like Mg and Fe influence the functioning of Chlorophyll and Cytochrome respectively. Many other elements like copper which constitute the cofactors of many photosynthetic enzymes reduce the rate of photosynthesis when they are in deficient supply.

Osmotic relations : Osmotic relationship, between different cells has an indirect effect on photosynthesis as it plays a vital role in the availability of water.

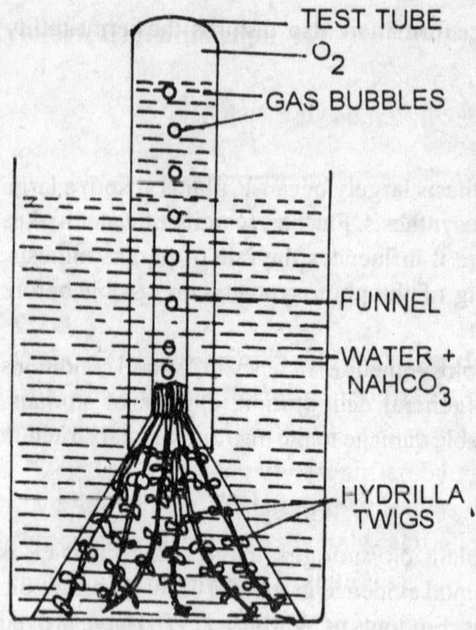

Fig. 12.28 Photosynthesis
Experiment to demonstrate the release.

Acumulation of end products : End product inhibition is noticed in photosynthesis also. Generally, food is translocated as soon as it is formed. This helps the process [photosynthesis] to go on. If for some reason, food accumulates at the site of its synthesis [chloroplasts], it has an adverse effect on the rate of photosynthesis. This is called "end product inhibition". A slow rate of translocation slows down the rate of photosynthesis. In fact translocation keeps pace with photosynthesis.

Experiments

1. **Release of O_2 during photosynthesis :** A glass trough half filled with water is taken. In this are kept a few twigs of the water plant *Elodea*. A funnel is inverted over this, to keep the plants intact. A test tube filled with water is inverted over the stem of the funnel. The apparatus is kept in sunlight. After some time it may be observed that small bubbles come out of the plant, get into the test tube and displace the water there. In this way ultimately the entire water in the test tube is displaced by the gas released by the plant. On testing (introduce a burning splinter), the gas may be identified as oxygen.

2. **Necessity of CO_2 for photosynthesis :** (Mohl's half leaf experiment). The apparatus consists of a wide mouthed tube half filled with KOH solution. A leaf is introduced into the tube with the help of a split cork so that half of the leaf is outside the bottle. The apparatus is kept in sunlight and the leaf is allowed to photosynthesise. After a few hours the leaf is removed and tested for starch (the leaf should be boiled in alcohol to remove chlorophyll). Iodine solution should be poured over the leaf.

   The portions which have starch turn blue. The half of the leaf inside the bottle- does not turn blue indicating the absence of starch. The portion of the leaf inside the bottle could not photosynthesise for want of carbon dioxide because whatever carbon dioxide that was there in the tube would have been absorbed by KOH solution.

3. **Necessity of light for photosynthesis :** (Ganong's light screen experiment). The apparatus consists of the two flaps of a metal piece covered by black paper. This can be fixed to a leaf with the help of a clip. A potted plant kept in darkness for over 24 hours is selected. Light screen apparatus is attached to a part of the leaf so that it does not receive any light. The plant is kept in sunlight and allowed to photosynthesise.

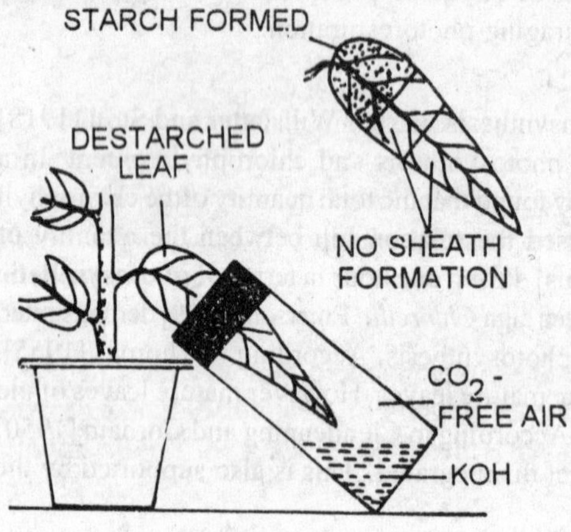

Fig. 12.29 Photosynthesis
Experiment to demonstrate the necessity of light for photosynthesis

After a few hours the light screen is removed from the leaf and the leaf is tested for starch. The result shows the absence of starch in covered portion. This shows *that* the covered portion of leaf does not photosynthesise for want of light.

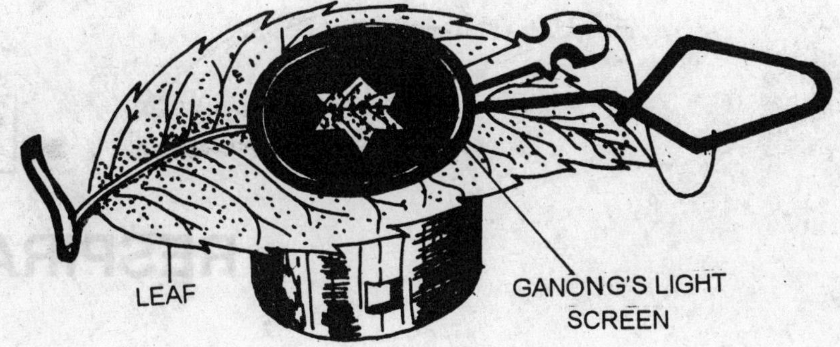

Fig. 12.30 Photosynthesis
Experiment to demonstrate the necessity of light for photosynthesis

13

RESPIRATION

The most basic requirement of all living organisms is *energy*. Without energy no life activity is possible. The solar energy that is trapped by green plants will be stored in cells in the form of carbohydrates and other nutrients. These are oxidized and the energy so released will be utilized to carry on the vital processes of living beings.

The process of energy release by the (oxidation of) substrate material is called respiration. The term respiration which bas been in usage since a long time is derived from the latin word *respirare* (= breathe). Breathing as a phenomenon of life has been noticed in animals long before its relation to respiration and its universal occurrence in all forms of life.

Breathing and respiration : Breathing and respiration are often used as synonymous. But it is not correct. Breathing as a separate process is noticed only in few organisms (animals), while respiration occurs in all organisms. Further, breathing is a physical process, while respiration is a chemical process taking place in all the cells. Breathing nlay be regarded as the last step found only in aerobic respiration, where atmospheric oxygen participates. There can be respiration without breathing, but the reverse is not possible.

Respiration and combustion : Respiration is often compared to the process of combustion, because in both the instances the substrate material gets burnt (oxidized) up. But the process of respiration differs from mere combustion in many respects. In combustion the burning up or oxidation of the substrate is sudden while in respiration it is stepwise. In combustion, a large amount of energy is wasted as heat, while in respiration it is channelised and stored up for further usage. Hence respiration may be termed as an orderly combustion found in living organisms.

Cellular respiration : In a multicellular organism where there are thousands of cells, how does energy reach each cell ? Each cell obtains its own energy by the oxidation of substrate material present within. Hence respiration takes place in all the cells wherever they may be present in animals and plants. This is called cellular respiration. Cellular respiration is a complex process which includes various aspects.

These are the following.

(a) Absorption of oxygen (not in all)

(b) Degradation of substrates such as carbohydrates to CO_2 and H_2O (oxidation)

(c) Energy is released in the process of oxidation; some amount will also be lost in the form of heat.

(d) Formation of many intermediate products which play different roles in metabolism

(e) There will be some loss of weight due to the process of oxidation.

From the point of view of plant metabolism, cellular respiration is of paramount significance as it is the basic biological process that provides energy for all the vital activities. In short, the following aspects may be mentioned as indicating the significance of respiration.

(i) Energy released in respiration is used for various metabolic processes. In fact most of the physiological activities are driven by respiratory energy.

(ii) Many intermediate compounds are formed which play a key role in the intermediate metabolism.

(iii) Insoluble food is converted into soluble form.

(iv) Potential energy is converted into kinetic energy.

(v) CO_2 which is released during respiration is an important step in maintaining the carbon balance of nature.

The overall formula for respiration is as follows

$$C_6H_{12}O_6 + H_2O + 6\,O_2 \rightarrow 6\,CO_2 + 2\,H_2O + 686\text{ K cal energy.}$$

Respiratory substrate : The energy rich material (food substrate) that is used for the purpose of oxidation is called the *respiratory substrate.* In other words a respiratory substrate is defined as "any organic plant constituent oxidized partially or completely with the concomitant release of energy together with CO_2 and H_2O".

Plants use a variety of organic constituents as their respiratory substrates. These may be a variety of carbohydrates (hexose sugars, disaccharides, polysaccharides), organic acids, fats etc. Under certain circumstances like starvation, even proteins may be used as respiratory substrates.

Among the carbohydrates, sucrose and starch are the principal respiratory substrates. When simple monosaccharides like glucose and fructose are available, they are directly utilized as respiratory substrates, while, sucrose and starch are first hydrolysed to simple sugars before being channelled into respiratory metabolism.

In certain seeds like castor bean, fats are reserve food materials. They are first broken down into fatty acids and glycerol. As already mentioned proteins will be channeled into respiratory metabolism under special circumstances such as starvation. Blackmann, who has studied respiration in detail, observed that in leaves kept in darkness, the respiratory rate which will be normal initially, declines but does not stop altogether. Blackmann termed the normal respiration (when carbohydrates are utilized) as *floating respiration* and the slowed down rate of respiration as *protoplasmic respiration,* when the essential proteins are oxidised to keep the cells alive. In the long run however this is harmful to the cells. When proteins are the respiratory substrates, they are first broken down into amino acids and these by deamination get converted into keto acids and-enter the respiratory chain.

Respiratory Quotient (RQ)

RQ may be defined as *"the ratio between the volume of carbon di oxide given out and oxygen taken in simultaneously by a given weight of the tissue in a given period of time at standard temperature and pressure".*

RQ may be calculated as follows

$$RQ = \frac{\text{Volume of } CO_2 \text{ evolved}}{\text{Volume of } O_2 \text{ consumed}}$$

The value of RQ depends upon the nature of the respiratory substrate, amount of oxygen present (in the substrate), and the extent to which the substrate gets broken. Normally the value of RQ should be unity(l), but many deviations are noticed due to the alterations in the levels of oxidation and reduction. The value of RQ also depends upon whether all the oxygen consumed is used up only for respiration or some of it is utilized for other purposes also. It is for these reasons that RQ may be unity, less than unity or even more than unity. RQ value also gives an idea of the respiratory substrate that is being used. The following are some RQ values when different respiratory substrates are involved.

RQ for Carbohydrates : Carbohydrates are the principal respiratory substances in a large majority of the organisms. These include a variety of monosaccharides (glucose, fructose etc), disaccharides (sucrose) and polysaccharides(starch, inulin etc). When simple sugars are available they are directly, channelised into the respiration stream, but when poly and disaccharides are involved, they are first hydrolysed into monosaccharides.

When hexose sugars are the respiratory substrate, volume of CO_2 evolved equals volume of O_2 used and hence the value will be unity. This will be illustrated by the following equation.

$$C_6H_{12}O_6 \rightarrow 6O_2 + 6H_2O$$

$$RQ = \frac{CO_2}{O_2} = \frac{6CO_2}{6O_2} = 1 \text{ or unity}$$

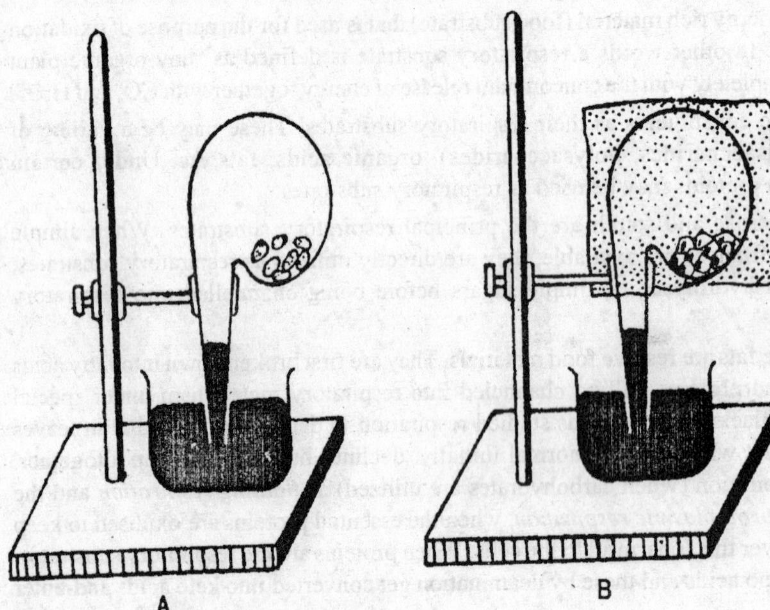

Fig. 13.1 Respiration

Demonstration of value of RQ for carbohydrates, **A.** Initial level, **B.** After 24 hours

It has to be noted however that not in all cases when sugar is the respiratory substrate, the RQ will be unity. There will be many deviations due to the following reasons.

(i) Incomplete oxidation of sugars (as in anaerobic respiration)

(ii) Involvement of many reductive events such as sulphate and nitrate reduction.

(iii) Oxidation and decarboxylation processes unrelated to respiration.

(iv) Retention of CO_2 within the cell fluids instead of releasing it to the exterior

(v) Non photosynthetic carboxylation reactions resulting in accumulation of theseproducts as seen in succulent plants

(vi) Metabolic utilization of CO_2

RQ for Fats : Fats are not generally found in vegetative parts, but they are found as storage products in many of the angiosperm seeds like sunflower, castor, niger seed, groundnut etc. At the time of the germination of seeds, a major amount of fat is converted into carbohydrates, while the rest is used in respiration.

Fats are poor in oxygen content, and the ratio of oxygen to carbon is always less in fats when compared with carbohydrates. Fats are rich in hydrogen and as a result they require more O_2 for complete oxidation. In addition to this, fats are not directly oxidised. They are first broken down *into fatty acids* and *glycerol* which then act as respiratory substrates. Oxygen is required for the breakdown of fats into fatty acids and glycerol. As a result, for the overall respiratory process (when fats are involved) more oxygen is required. Hence the value of RQ will be less than unity.

(a) $C_{18}H_{36}O_2 + 2_6O_2 \rightarrow 18CO_2 + 18H_2O$
(stearic acid)

$$RQ = \frac{26O_2}{18CO_2} = 0.69$$

(b) $C_{57}H_{164}O_6 + 8O_2 \rightarrow 57CO_2 + 52H_2O$
(Triolein fat)

$$RQ = \frac{57}{80} = 0.7$$

When simple fatty acids like acetic acids are involved RQ will be unity as seen below

(c) $CH_3COOOH + 2O_2 \rightarrow 2CO_2 + 2H_2O$
(Acetic acid)

$$RQ = \frac{2O_2}{2CO_2} = 1 \text{ (unity)}$$

(d) $2C_{51}H_{98}O_6 + 145O_2 \rightarrow 102CO_2 + 98H_2O$
(Tripalmitin)

$$RQ = \frac{102}{145} = 0.7$$

Interestingly fats liberate approximately three times more energy when compared with Carbohydrates. While a gram of carbohydrate yields about 3.8 K cal, the same amount of fat produces approximately 9.1 K cal of energy. This may be the reason as to why the seeds prefer to store reserve food material in the form of oil, more amount of energy can be obtained by less amount of food (it requires less space).

RQ for proteins : Proteins and amino acids serve as respiratory substrates only under certain situations like starvation etc. Like fats, proteins also have less oxygen and the volume of oxygen to carbon is low. When proteins are hydrolysed, they require more oxygen for complete oxidation as a result of which RQ value will be less than unity. The RQ value for proteins and their derivatives varies between 0.5 - 0.8; very often it will be 0.79.

The value of RQ may be 0.99 when ammonia is released or it may be 0.79 when amides are formed.

RQ for organic acids : The value of RQ is more than unity when organic acids are used as respiratory substrates. This is due to the fact that in organic acids, the proportion of oxygen in relation to carbon is always more, or less of O_2 is required for oxidation. The following are some of the examples.

1. $C_4H_6O_5 + 3O_2 \rightarrow 4CO_2 + 3H_2O$
(malic acid)

2. $2(COOH)_2 + O_2 \rightarrow 4CO_2 + 2H_2O$
(oxalic acid)

3. $2C_4H_6O_6 + 5O_2 \rightarrow 8CO_2 + 6H_2O$
(Tartaric acid)

4. $2C_6H_8O_7 + 9O_2 \rightarrow 12CO_2 + 8H_2O$
(Citric acid)

$$RQ = \frac{12}{9} = 1.33$$

RQ for succulent plants

In succulent xerophytes such as *Opuntia,* and many other plants there is no evolution of CO_2 . As only O_2 is taken in and there is no release of CO_2, RQ is zero. In succulent plants stomata are open during night, O_2 is taken in with the result carbohydrates are incompletely oxidized, while during day time, the organic acids are completely oxidized releasing CO_2. But this CO_2 is not released as it is used up for photosynthesis. Hence with no evolution of CO_2, RQ will be zero.

$$2\,C_6H_{12}O_6 \quad + \quad 3O_2 \rightarrow 3\,C_4H_6O_5 \quad + \quad 3\,H_2O$$

$$RQ = \frac{CO_2}{O_2} = \frac{0}{3} = 0.0$$

RQ when O_2 is used for other activities : O_2 is also utilized for various other processes like conversion of fats to carbohydrates, synthesis of anthocyanins etc. In such cases the amount of CO_2 evolved will not correspond to the amount of O_2 consumed. Here RQ falls below unity.

RQ for anaerobic respiration : Certain tissues respire anaerobically (no participation of atmospheric O_2) in which case only CO_2 is evolved without the absorption of O_2. In such circumstances RQ will be more than unity.

$$C_6H_{12}O_6 \quad \rightarrow 2\,C_2H_5OH_2 \quad + \quad 2CO_2$$

$$RQ = \frac{CO_2}{O_2} = \frac{2}{0} = 2$$

The following is a list of RQ values for some of the substrates occurringo in various plant parts.

No.	Plant part	Substrate	RQ Value
1.	Leaves	Carbohydrates	1.0
2.	*Opuntia* shoots	Organic acids	.03
3.	Seeds	Starch	1.0
4.	Linseed	Fat	0.64
5.	Buckwheat seed	Protein	0.5
6.	Pea seeds	Carbohydrates	1.5-2.4

Measurement of RQ : RQ can be measured with the help of *Ganong's respiroscope.* The apparatus essentially consists of a graduated tube with a bulb at the upper end and a levelling tube. The two tubes are connected at their base with a long rubber tube which assumes a 'U' shape. The bulb in the graduated tube has a glass stopper fitted into its neck. Both the neck and the stopper have holes which can be brought into a straight line by twisting the stopper. The apparatus is fitted into an iron or wooden stand and filled with mercury (Instead of Hg saline solution also can be used). Respirable material such as plant parts or germinating seeds are kept in the bulb and the stopper is rotated to allow the air enter into the bulb. Mercury level in both the tubes is brought at the same level. Now the stopper is rotated to cut off the communication.

The initial level of Hg is noted. If the respiratory substrate is carbohydrate CO_2 released will be equal to O_2 absorbed so that there is neither rise nor decrease in the level of Hg. If the substrate is fat or protein, CO_2 released will be less than O_2 consumed and therefore a vaccum will be created in the chamber. This results in the increase of the level of Hg. If the CO_2 released is more than O_2 used up, the initial level of Hg will fall after the experiment. RQ in this case will be more than one. RQ may be calculated with the help of the following formulae.

$$RQ = \frac{V_2}{V_2 + V_2}$$

V_1 = Initial rise of Hg; V2 Further rise in Hg level after introduction of causuc poiasn. *i* ins is imrotilicea io ima out inc excess release ot CO_2 it any.

$$RQ = \frac{V_2}{V_2 + V_1}$$

V_1 = Fall in the initial level of Hg. V_2 = Rise in level after introduction of caustic potash.

TYPES OF RESPIRATION

The involvement of atmospheric oxygen during respiration constitutes the basis for its categorization into **aerobic** and **anaerobic** types

Aerobic respiration : The respiratory substrate is completely broken down into CO_2 and H_2O with the participation of atmospheric CO_2.

$$C_6H_{12}O_6 + 6O_2 \longrightarrow 6CO_2 + 6H_2O + 686\ K\ Cal.$$

Anaerobic respiration : Here there is no participation of atmospheric oxygen. Food substances are incompletely oxidized. Hence all the carbon in the substrate will not be released in the form of CO_2. The end products are not only CO_2, but intermediate products like ethyl alcohol, acids etc.

$$C_6H_{12}O_6 \longrightarrow 2C_2H_5OH + 2CO_2 + 56\ K\ Cal.$$

Besides the above, there are some more differences between aerobic and anaerobic respiration. They are listed below.

	Aerobic	Anaerobic
1.	O_2 necessary	O_2 not necessary
2.	Common to most of the organisms	Restricted to certain organisms, and certain tissues.
3.	Occurs always	Occurs temporarily in certain tissues, besides microorganisms.
4.	Oxidation complete	Oxidation incomplete
5.	Large amount of energy is liberated.	Less energy is liberated.
6.	End products are CO_2 and water	End products are ethyl alcohol, organic acids etc and CO_2
7.	The entire carbon content of the of the substrate comes out in the form of CO_2 (inorganic)	Due to incomplete oxidation substrate, carbon will still be left the organic compounds (ethyl alcohol)
8.	Process non toxic to plants	Toxic to plants
9.	Respiratory enzymes present in mitochondria and cytoplasm.	Enzymes present only in cytoplasm.
10.	Output is 38 molecules of ATP	Output is only 2 molecules of ATP

Fermentation : Fermentation is essentially anaerobic respiration found in microorganisms like certain Fungi and bacteria. In fermentation, the substrate for oxidation is extracellular (not present within the cell) since the organisms are basically saprophytic.

Fermentation as a process to produce alcohol was known to human beings since time immemorial, eventhough its scientific understanding dates back to only the middle of the 19th century. Experiments conducted by **Louis Pasteur** established that fermentation is carried out with the participation of living cells.

Buchner in 1897 found out that the extract of yeast cells (which does not contain living cells) can bring about fermentation of sugar solution. Subsequently it was discovered that the enzyme *Zymase* present in the yeast extract is capable of fermenting activity (break down of carbohydrates together with the formation of alcohol and CO_2).

The enzyme *Zymase* will be active only in the presence of phosphate. *Sucrase* and *maltase* are the other carbohydrate fermenting enzymes secreted by yeasts. In fermentation also, as in anaerobic respiration atmospheric oxygen does not participate. Oxidation is accomplished by the intermolecular atomic shifts, with the result the product will have less energy than the original substrate. Much of the energy released during fermentation is wasted as heat, while the remainder is used for the conduct of metabolic activities.

The end product of fermentation need not always be alcohol, it may be various types of organic acids such as butyric acid, lactic acid etc. Depending on the end product, fermentation may be termed as alcoholic fermentation, lactic acid fermentation etc. The following are the equations for these fermentations.

(i) $C_6H_{12}O_6 \longrightarrow 2C_2H_5OH + 2CO_2$ (ethyl alcohol)

(ii) $C_6H_{12}O_6 \longrightarrow C_4H_8O_2 + 2H_2 + 2CO_2$ (butyric acid)

(iii) $C_6H_{12}O_6 \longrightarrow 23H_6O_3$ (Lactic acid)

(iv) $C_2H_5OH + O_2 \longrightarrow CH_3COOH + H_2O$ + energy (acetic acid)

In the equation (iv) given above, for the formation of acetic acid from ethyl alcohol, O_2 is necessary. Hence acetic acid fermentation is different from other types of fermentation.

Fermentation is carried out by microorganisms, a point which has been emphasized earlier. The following is a list of organisms which carry out the various types of fermentation.

	Organism	Type of fermentation (based on end product)
1.	*Bacillus butyricus* and *Clostridium butyricum*	Butyric acid
2.	*Bacterium lactic acidi*	Lactic acid
3.	*Acetobacter aceti*	Acetic acid

MECHANISM OF RESPIRATION

As has already been pointed out, during respiration the substrate material is broken down with the release of energy, CO_2 and H_2O.

In the mechanism, we will be studying the various stages of the breakdown of the substrate material, enzymes involved etc., and account for the release of CO_2 and energy released in the form of molecules ofATP. In the foregoing paragraphs various steps involved in the aerobic breakdown of carbohydrates are given.

Broadly categorizing, the following steps are identifiable in the oxidative breakdown of carbohydrates.

1. Hydrolysis. 2. Glycolysis 3. Pyruvic acid oxidation, 4. Krebs cycle or citric acid cycle and 5. Electron transport system or oxidative phosphorylation.

The mechanism of anaerobic respiration consists of the following steps

1. Glycolysis and 2. anaerobic oxidation of pyruvic acid.

From the above account it can be understood that the step glycolysis is common for both aerobic and anaerobic respiration.

HYDROLYSIS

When starch is the reserve food material, it should first be hydrolysed into simple sugars before it can enter into the respiratory cycle. The process of breakdown of starch into simple sugars with the addition of water is called hydrolysis.

GLYCOLYSIS

During glycolysis one molecule of glucose will be broken into two molecules of pyruvic acid. Originally, the name glycolysis was meant for the breakdown of glycogen *(lysis* or breakdown of glycogen) found in muscle cells. Enzymes responsible for glycolysis are present in the cytoplasm. As will be seen later, no participation of atmospheric O_2 is required for the formation of pyruvic acid. Glycolysis is the common pathway for both aerobic and anaerobic respiration.

Glycolytic breakdown of glucose into pyruvic acid is also known as the **Embden - Meyerhof - Parnas** pathway (EMP pathway) after three scientists who discovered the various steps of glycolysis. EMP pathway is also called cytoplasmic respiration as the whole process of glycolysis takes place only in cytosol(cytoplasm minus organelles). EMP pathway was first worked out in muscle cells and yeasts and later it was observed that it is common for all the organisms. The overall raction of glycolysis may be summed up as follows

$$C_6H_{12}O_6 \longrightarrow 2CH_3COCOOH + 4H.$$

The details of the above reaction are given below.

Phosphorylation : Any sugar that has to enter into the glycolysis pathway must first undergo phosphorylation (i.e. a phosphate group has to be added). In this process ATP participates and donates a phosphate group (there will be utilization of a molecule of ATP).

(i) Glucose + ATP — Glucose 6. phosphate + ADP

The reaction is catalyzed by the enzyme *hexokinase* with Mg^{++} as the cofactor. This reaction was first worked out by Meyerhof in 1927.

Glucose 6. phosphate may be produced by another means also if starch is the original food material. It is as follows .

$$Starch + H_3PO_4 \longrightarrow \text{Glucose 1 phosphate}$$

The enzyme phosphorylase catalyses this reaction. In the next step

Glucose 1. phosphate — Glucose 6. phosphate

In the above reaction Glucose 1. phosphate changes into Glucose 6 phosphate i.e. the phosphate group attached to the first carbon atom is shifted to the sixth carbon atom. The reaction is catalyzed by the enzyme *Phosphoglucomutase*

Isomerization : In this step, glucose 6. phosphate isomerises into Fructose 6 phosphate under the influence of the enzyme *Phosphohexoisomerase*

Glucose 6. phosphate ⟶ Fructose 6 phosphate

Fructose 6. phosphate may be produced from fructose directly when the latter is available

$$\text{Fructose} + \text{ATP} \xrightarrow[Mg^{++}]{\textit{Hexokinase}} \text{Fructose 6. phosphate}$$

Formation of Fructose 1.6. diphosphate : This is a key reaction in the initial break down of glucose. Fructose 6. phosphate utilizes a molecule of ATP to get converted to Fructose 1.6 diphosphate under the influence of the enzyme *Phosphohexokinase*

Fructose 6. Phosphate + ATP ⟶ Fructosee 1.6. diphosphate + ADP

The reactions of phosphorylation leading to the formation of Fructose 1.6. diphosphate may be summarized as follows

(i) Glucose + ATP —*Hexokinase* / Mg^{++}→ Glucose 6. phosphate + ADP

(ii) Glucose 6 phosphate —*Hexokinase – Isomerase*→ Fructose 6 phosphate

(iii) Fructose 6. phosphate + ATP —*Phosphohexo kinase* / Mg^{++}→ Fructose 1.6 diphosphate + ADP

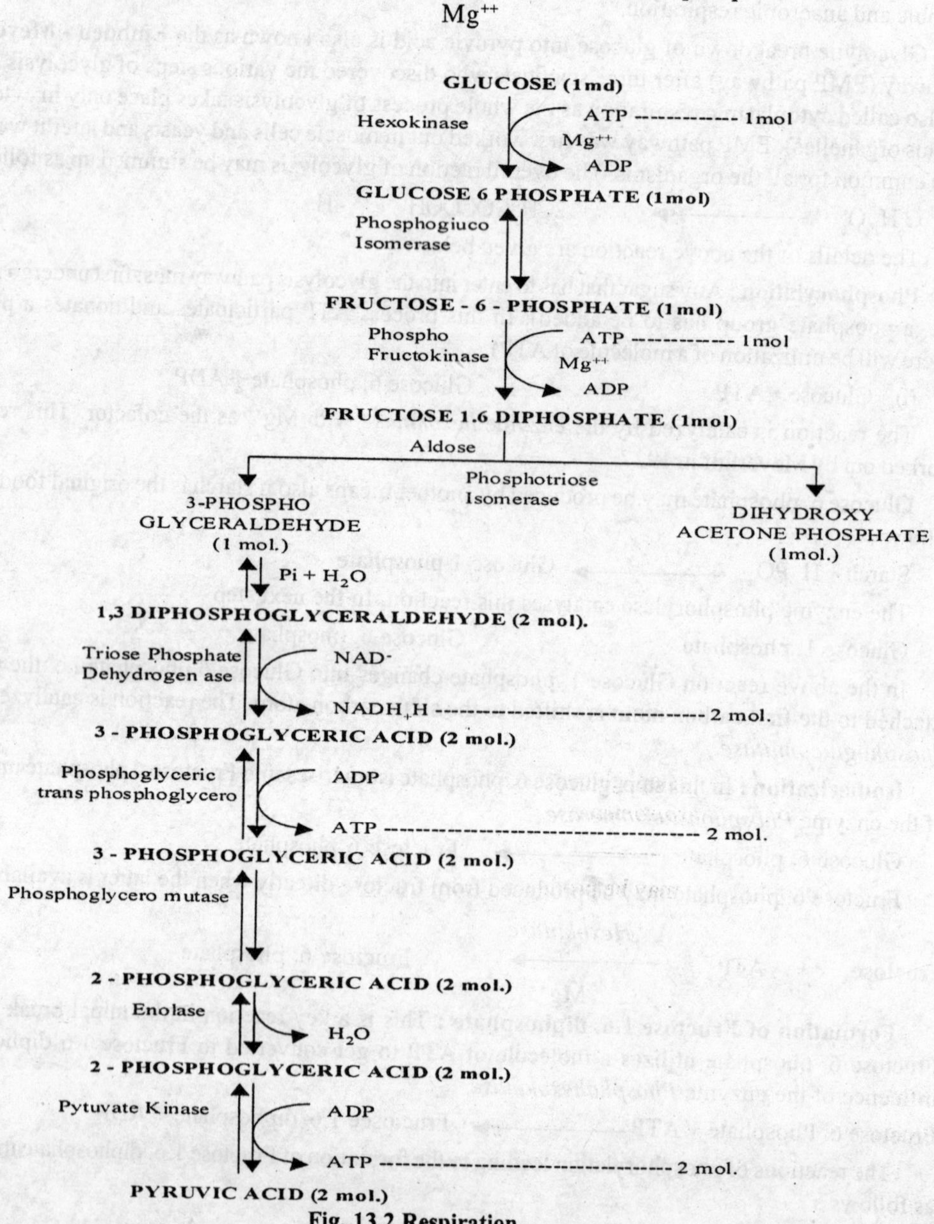

GLUCOSE (1md)

Hexokinase — ATP ------------ 1mol; Mg^{++}; ADP

GLUCOSE 6 PHOSPHATE (1mol)

Phosphogiuco Isomerase

FRUCTOSE - 6 - PHOSPHATE (1mol)

Phospho Fructokinase — ATP ------------ 1mol; Mg^{++}; ADP

FRUCTOSE 1.6 DIPHOSPHATE (1mol)

Aldose

Phosphotriose Isomerase

3-PHOSPHO GLYCERALDEHYDE (1 mol.) — DIHYDROXY ACETONE PHOSPHATE (1mol.)

$Pi + H_2O$

1,3 DIPHOSPHOGLYCERALDEHYDE (2 mol).

Triose Phosphate Dehydrogen ase — NAD- ; NADH,H ------------ 2 mol.

3 - PHOSPHOGLYCERIC ACID (2 mol.)

Phosphoglyceric trans phosphoglycero — ADP ; ATP ------------ 2 mol.

3 - PHOSPHOGLYCERIC ACID (2 mol.)

Phosphoglycero mutase

2 - PHOSPHOGLYCERIC ACID (2 mol.)

Enolase — H_2O

2 - PHOSPHOGLYCERIC ACID (2 mol.)

Pytuvate Kinase — ADP ; ATP ------------ 2 mol.

PYRUVIC ACID (2 mol.)

Fig. 13.2 Respiration

Glycolysis - Various reactions to the formation of Pyruvic acid

Cleavage of Fructose 1.6 dip : Further reactions in glycolysis begin with the breaking up of Fructose 1.6 diphosphate into two molecules of triose under the influence of the enzyme *Aldolase* The presence of this enzyme in plant tissues was first discovered by Tewfik and Stumf (1949).

1. Fructose 1.6 diphosphate ——— Glyceraldehyde 3 phosphate + Dihydroxyacetone phosphate

97% of Fructose 1.6 diphosphate gets converted into glyceraldehyde 3 phosphate while about 3% will change into dihydroxyacetone phosphate. Dihydroxyacetone phosphate isomerises into glyceraldehyde 3 phosphate under the influence of the enzyme *phosphotriose isomerase.* The reaction may be represented as follows.

```
   CH2O (P)
      |
      C =  O
      |
HO – C – OH                  CHO                          CH2O–
      |         Aldolase      |      Phosphotriose         |
 H – C – OH ————————————→ H – C – OH ——————————————→ C=O
      |                       |         Isomerase          |
 H – C – OH              CH2O – P                       CH2OH
      |
   CH2O (P)
Fructose 1.6            Glyceraldehyde          Dihydroxyacetone
diphosphate             3- Phosphate                phosphate
```

To sum up, Fructose 1.6 diphosphate (6C) breaks up into 2 molecules of glyceraldehyde 3 phosphate (3C). From now on the reactions are explained for a molecule of triose (glyceraldehyde 3 phosphate). While accounting for hexose each reaction has to be doubled.

2. In the next step glyceraldehyde 3 phosphate is oxidized (hydrogen removed) with the help of the enzyme *phosphoglyceraldehyde dehydrogenase*. Inorganic phosphate and NAD participate in this reaction. 2 electrons and lwo protons (H) are released from glyceraldehyde 3. phosphate and the energy is utilized to link the inorganic phosphate to the oxidized glyceraldehyde 3 phosphate to form 1,3 - diphosphoglyceric acid.

Glyceraldehyde 3 phosphate + H_3PO4 + NAD
$\longrightarrow$ 1.3 diphosphoglyceric acid
+ NADH + H^+

As may be noticed, while the electrons provide the energy necessary for the linking of inorganic phosphate, the H goes for the reduction of NAD to NADH + H^+

```
  CHO                                    O
   |                                     ||
H – C – OH  +  H3PO4  +  NAD            C – O – P
   |                                     |
  CH2O - P                         H – C – OH  +  NADH  +  H+
                                        CH2O – P
Glyceraldehyde - 3 Phosphate      1, 3 - diphosphoglyceric acid
```

3. 1,3 - disphosphoglyceric acid is transformed into 3 phosphoglyceric acid, when one of the phosphate groups is released, which is accepted by ADP to synthesize a molecule of ATP. *Phosphoglyceric kinase* enzyme catalyzes this reaction.

The enzyme has Mg^{++} as the cofactor.

1.3 disphosphoglyceric acid + ADP $\rightarrow$ 3 Phosphoglyceric acid + ATP

$$\begin{array}{l} O \\ \| \\ C-O-P \\ | \\ H-C-OH+ADP \\ | \\ CH_2O-P \end{array} \xrightarrow[\textit{Kinase} + Mg^{++}]{\textit{Phosphogyceric}} \begin{array}{l} COOH \\ | \\ H-C-OH+ATP \\ | \\ CH_2O-P \end{array}$$

4. 3 phosphogyceric acid gets transformed into 2 phosphologyceric acid (phosphate group is shifted from the 3rd carbon position to the 2nd carbon position)under the influence of the enzyme *phosphoglyceromutase*

$$\begin{array}{l} COOH \\ | \\ H-C-OH \\ | \\ CH_2O-P \end{array} \xrightarrow[\textit{- mutase}]{\textit{Phosphoglycero}} \begin{array}{l} COOH \\ | \\ H-C-O-P \\ | \\ CH_2O-OH \end{array}$$

3 - phosphoglyceric acid

5. 2-phosphoglyceric acid (2PGA) gets converted into phosphoenolpyruvic acid (PEP) under the influence of the enzyme *enolase* with Mg^{++} ions as cofactor. A molecule of water is eliminated from 2PGA in the process with the result an energy rich (P) centre is created

$$\begin{array}{l} COOH \\ | \\ H-C-O-P \\ | \\ CH_2OH \\ (2\,PGA) \end{array} \xrightarrow[Mg^{++}]{\textit{Enolase}} \begin{array}{l} COOH \\ | \\ H-C-O-P+H_2O \\ | \\ CH_2 \\ (2\,PEP) \end{array}$$

6. In the molecule of PEP, much of the energy is centred round the phosphate group. This energy rich phosphate group is removed from PEP, which is accepted by ADP to produce a molecule of ATP. In the process the dephosphorylated PEP becomes converted to pyruvic acid. This reaction is catalyzed by the enzyme *pyruvate kinase* with Mg^{++} and K^{+} ions as cofactors.

$$\begin{array}{l} COOH \\ | \\ H-C-O \quad P+ADP \\ | \\ CH_2 \\ (PEP) \end{array} \xrightarrow[Mg^{++}\ K^{+}]{\textit{pyruvate kinase}} \begin{array}{l} COOH \\ | \\ C=O+ATP \\ | \\ CH_3 \\ (\text{Pyruvic acid}) \end{array}$$

With the formation of pyruvic acid the reactions of glycolysis come to an end. One molecule of pyruvic acid is produced for every triose, hence for hexose, it should be two molecules of pyruvate.

ATP account during glycolysis : The production of pyruvic acid terminates glycolysis. When we briefly glance through the glycolytic reactions we realize that there are two important stages - (i)phosphorylation and (ii) oxidation of Fructose 1.6 diphosphate to pyruvic acid.

During phosphorylation of glucose to form Fructose 1.6 diphosphate, two molecules of ATP are used up. Thus during phosphorylation of glucose 2 molecules of ATP are used.

In the second stage, at two stages ATP molecules are synthesized-once during the conversion of 1,3 - diphosphoglyceric acid to 3 phosphoglyceric acid (2 molecules - i.e., one for every triose and two for every hexose) and again during the conversion of phosphoenol pyruvic acid to pyruvic acid (2 molecules of ATP, reactions are to be doubled while accounting for hexose). Thus in all, towards the end of glycolysis 4 molecules of ATP are produced and two molecules of ATP are consumed (during phosphorylation). Hence the net gain of ATP during glycolysis is only two molecules.

But the statement that the net gain of ATP during glycolysis is only two molecules is not completely correct, because during the oxidation of glyceraldehyde 3 phosphate to phosphoglyceric acid 2 molecules of NADH + H are produced. These two molecules pass through the electron transport chain producing a total of 6(3+3) ATP molecules (Details explained later). Hence the net gain of ATP during aerobic glycolysis is 6 + 2 = 8 molecules of ATP.

Breakdown of pyruvic acid

Further breakdown of pyruvic acid depends upon the nature of the organism, type of respiratory pathway and the presence or absence of O_2. Broadly categorizing, the following options are available to the pyruvic acid for breakdown. These are -

(i) Alcoholic fermentation
(ii) Lactic acid fermentation
(iii) Amino acid synthesis and
(iv) Aerobic breakdown to produce CO_2 and H_2O.

Options (i) and (ii) can be carried out in the absence of O_2 and (iv) is possible in the presence of O_2 which happens to be the major pathway. Option (iii) leads to the biosynthesis of alanine by transaminase reaction with glutamate. This is not discussed here as it is not relevant lo the respiratory pathway.

Alcoholic fermentation : This is very common in yeasts, certain types of bacteria and some tissues of higher plants (tissues of fruit during maturation) In this process pyruvic acid is converted into ethyl alcohol and CO_2 in two stages.

(i) $$\underset{\text{pyruvic acid}}{CH_3COCOOH} \xrightarrow[M^{++}]{\textit{pyruvic decarboxylase}} \underset{\text{Acetaldehyde}}{CH_3CHO} + CO_2$$

(ii) $$CH_3CHO + NADH_2 \xrightarrow[\textit{dehydrogenase}]{\textit{Alcohol}} \underset{\text{Ethyl alcohol}}{C_2H_5OH} + NAD$$

Lactic acid fermentation : In bacteria like *Lactobacilli, Clostridium* etc., and in animal tissues (muscle) pyruvic acid is broken down to lactic acid under the influence of the enzyme *lactic dehydrogenase*

$$\underset{\text{Pyruvic acid}}{CH_3COCOOH} + NADH_2 \xrightarrow{\textit{Dehydrogenase}} \underset{\text{Lactic acid}}{CH_3CHOH\,COOH} + NAD$$

A question that naturally arises here is from where do the $NADH_2$ molecules come to participate in the reaction during alcoholic and lactic acid fermentation. If we remember the reactions of glycolysis we realize that the two molecules of $NADH_2$ are produced when 3 phosphoglyceraldehyde gets oxidized to 1.3 diphosphoglyceric acid. Hence in the case of anaerobic respiration the $NADH_2$ will not enter the electron transport chain and the net gain of ATP during glycolysis will be only (4-2) two molecules.

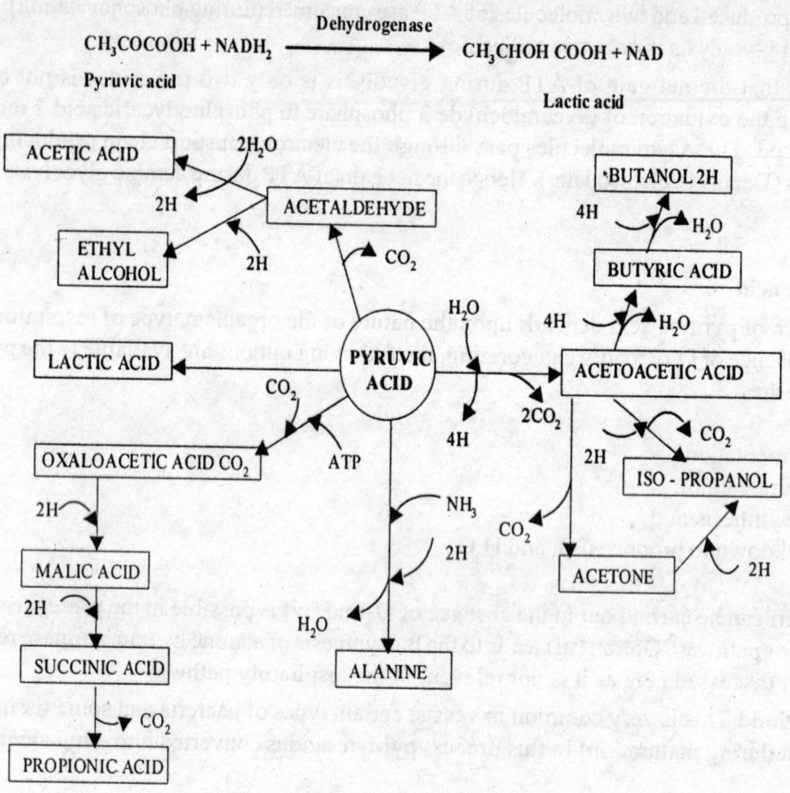

Fig. 13.3 Respiration
Anaerobic degradation of Pyruvic acid

Aerobic oxidation ofpyruvic acid

This is the most important phase of aerobic respiration in which pyruvic acid gets converted into a two carbon compound called acetaldehyde, before entering the Krebs cycle for complete decarboxylation and oxidation. The reaction is mediated by a complex enzyme with atleast five essential cofactors (Gunsalnus 1954). The acetaldehyde combines with co enzyme A to form acetyl CoA which enters the Krebs cycle.

Formation of Acetyl Co A : As already pointed out, the enzyme which catalyzes the decarboxylation of pyruvic acid is a complex one which includes 3 enzymes - *pyruvic acid decarboxylase, dihydroxylipoyl transacetylase* and *dihydrolipoyl dehydrogenase.*
The enzyme complex also requires five cofactors-Thiamine pyrophosphate (TPP), Mg ions, NAD , Coenzyme A (CoA) and lipoic acid.

The first step includes the formation of a complex between TPP and pyruvate, followed by the .decarboxylation of the latter. In the second step the acetaldehyde fragment reacts with the cofactor lipoic acid to form an *acetyl lipoic acid complex.* In this reaction acetaldehyde is oxidized to the acid and lipoic acid is reduced. In the third step acetyl group is removed from lipoic acid and added on to CoA, resulting in the formation of Acetyl CoA and reduced lipoic acid. In the final step, oxidized lipoic acid is regenerated by transforming the electrons from reduced lipoic acid to NAD. This is necessary because, there should be a continuous supply of oxidized lipoic acid to accept pyruvate and produce acetyl CoA. The NAD which gets reduced to form NADH + H eventually enters into the electron transport system to produce 3 molecules of ATP. The following reactions summarize the formation of Acetyl CoA from pyruvate.

(i) Pyruvate + TPP $\xrightarrow{Mg^{++}}$ TPP Complex + CO_2

(ii) TPP Complex + Lipoic acid (oxidized) $\longrightarrow$ Acetyl lipoic acid Complex + TPP

(iii) Acetyl lipoic acid + CoA Complex $\longrightarrow$ Acetyl CoA + Lipoic acid (reduced form)

(iv) Lipoic acid (reduced form) + NAD $\longrightarrow$ Lipoic acid (oxidized form) + NADH + H^+

The net result may be given by the following equation.

Pyruvate + CoA + NAD $\rightarrow$ Acetyl CoA + CO_2 + NADH + H^+

Finally the two carbon compound Acetyl CoA formed in cytoplasm enters the mitochondria to react with the acids of the Krebs cycle for the removal of remaining 2 carbons and to produce energy to be stored in the molecules of ATP.

KREBS CYCLE

Acetyl CoA, the two carbon compound undergoes further decarboxylation when it enters a cycle of interconverting carboxylic acids present in mitochondria. Together with decarboxylation, there is oxidation also (dehydrogenation) resulting in the formation of molecules of ATP (to trap the released energy).

Kerbs Cycle (named after its discoverer) was actually worked out by Wood *et at* (1942) and Krebs (1943). The cycle has several names - Tricarboxylic Cycle (TCA), Citric acid Cycle, Mitochondrial respiration etc. Basically the cycle consists of several 6 carbon, 5 carbon and 4 carbon acids. When these are getting interconverted, Acetyl CoA enters facilitating the interconversion, and in the process undergoes decarboxylation at two steps. The details of individual reactions are given below.

1. Formation of citric acid : Acetyl CoA, the two carbon compound, reacts with a 4 carbon carboxylic acid - *oxaloacetic acid* in the presence of water to form a six carbon carboxylic acid *citric acid.* The reaction is mediated by the enzyme *citrate sy'nthetase.* In the process CoA is released which goes back to form some more Acetyl CoA.

CO COOH
|
$CH_2COOH + CH_3CO\ S.CoA + H_2O \xrightarrow[\textit{synthetase}]{\textit{Citrate}}$

CH_2 COOH
|
C (OH) COOH + CoA + HS
|
CH_2COOH

Oxaloacetic acid, Acetyl CoA → Citric acid

2. Citric acid loses a molecule of water to form *cis aconitic acid* under the influence of the enzyme *Aconitase* with Fe^{++} as cofactor.

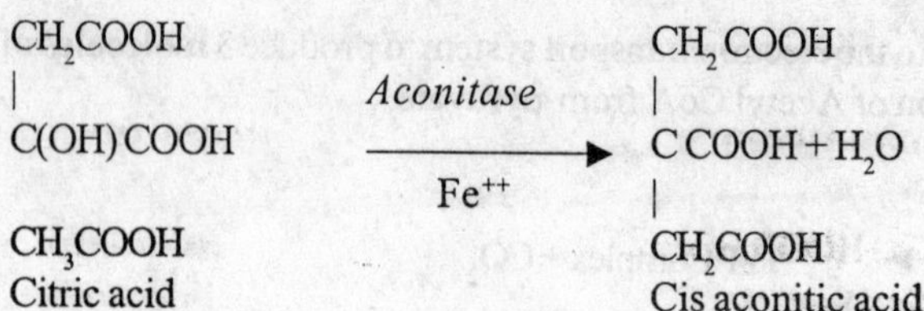

3. Cis aconitic acid now gets converted into *Isocitricacid* (isomer of citric acid) by the addition of a molecule of water. The *enzyme Aconitase* catalyzes this reaction also

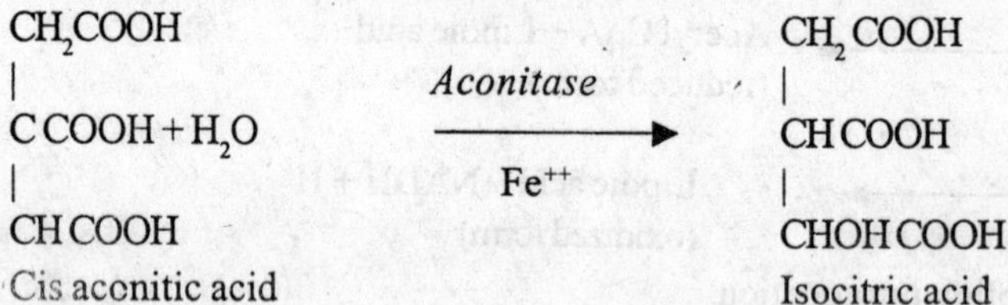

The addition of water molecule to cis aconitic does not produce Citric acid, but its isomer isocitric because H_2O combines in a different position in the molecule of cis aconitic acid.

4. Isocitric acid undergoes oxidation (removal of hydrogen) under the influence of the enzyme Isocitric dehydrogenase to form *oxalosuccinic acid* The released H atoms are accepted by NAD to become NADH+ H^+

The NADH H^+ gets reoxidised after entering into the electron transport chain (see later).

CH_2COOH | $CH\,COOH + NAD$ | $CO\,COOH$ (Isocitric acid) —*Isocitric dehydrogenase*→ CH_2COOH | $CH\,COOH + NADH + H^+$ | $CO\,COOH$ (oxalosuccinic acid)

5. Oxalo succinic undergoes decarboxylation to form a 5 carbon acid α *ketoglutaric acid.* The reaction is catalysed by the enzyme *decarboxylase* with Mn^{++} as the cofactor.

CH_2COOH | $CH\,COOH$ | $CO\,COOH$ —*Decarboxylase*, Mn^{++}→ CH_2COOH | $CH_2CO\,COOH + CO_2$

6. The next step is the oxidation of *a* ketoglutaric acid which is achieved by the removal of H. Simultaneously it also undergoes decarboxylation leading to the formation of Succinyl CoA. The enzymes involved are α *ketoglutaric dehydrogenase* and α *ketodecarboxylase,* together with coenzymes NAD and CoA. NAD gets reduced to NADH + H^+ while a molecule of CO_2 is liberated. Together with CoA, TPP, Mg^{++}, FAD and lipoic acid also participate in the reaction.

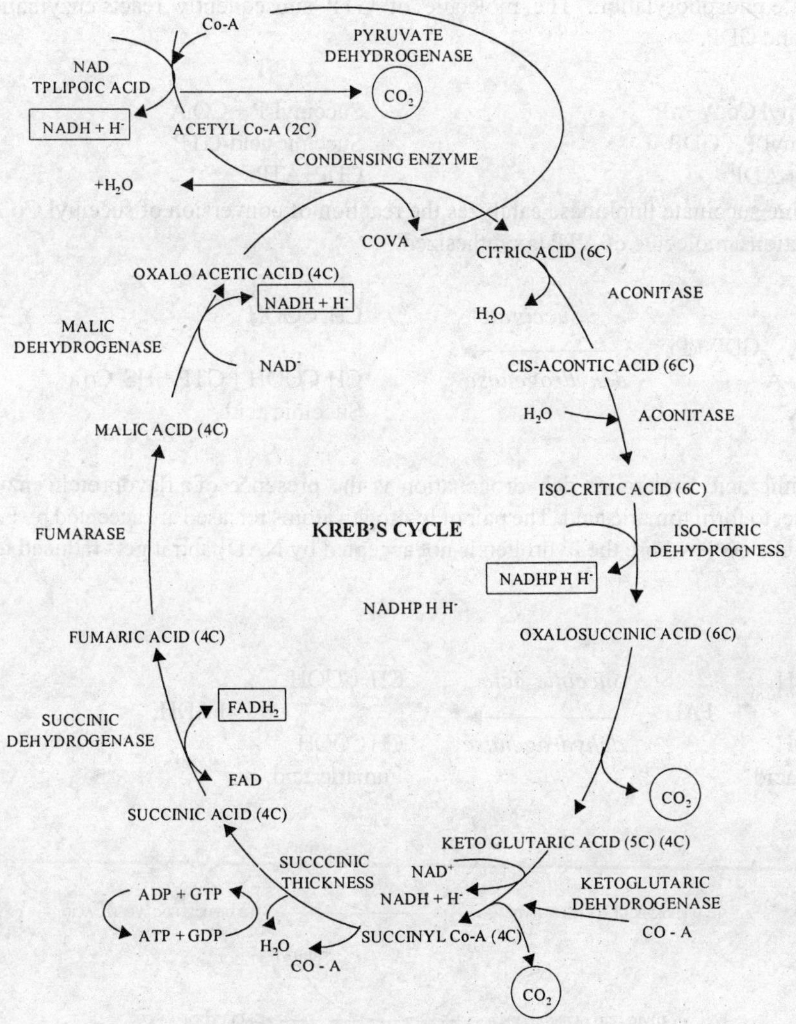

Fig. 13.4 Respiration
Reaction of Kerb' s cycle

$CH_2\,COOH$ | + HS CoA + NAD $CH_2\,CO\,COOH$ α Ketoglutaric acid	*dehydrogenase* ⟶ Mg^{++} FAD TPP, Lipoic acid, FAD	$CH_2\,COOH$ | CH_2 Co S.Co.A+ NADH + H^+ + CO_2 Succinyl Co A

7. Succinyl CoA is hydrolysed with a molecule of H_2O to *succinic acid,* regenerating CoA for continued participation in the cycle. Inorganic phosphate and a phosphorylating enzyme are involved in the formation of succinic acid. The release of CoA accompanies the release of energy, which will be used to synthesize a molecule of Guanosine triphosphate (GTP) .by the addition of energy rich phosphate to Guanosine disphosphate (GDP). This is the only place where GTP instead of ATP is synthesized. This type of phosphorylation is called substrate phosphorylation. The molecule of GTP subsequently reacts enzymaticallv with ADP to produce ATP and GDP.

(i) Succinyl Co.A + iP Succinyl-P + CO.A
(ii) Sucemyl-P + GDP Succinic acid-GTP
(iii) GTP + ADP GDP+ATP

The enzyme succinate thiokinase catalyzes the reaction of conversion of succinyl Co.A to succinic acid. From this reaction a molecule of ATP is synthesized.

$$\begin{matrix} CH_2COOH \\ | \\ CH_2CO^-S.CoA \\ \text{Succinyl Co.A} \end{matrix} + GDP + Pi \xrightarrow[\textit{dehydrogenase}]{\textit{Succinate}} \begin{matrix} CH_2COOH \\ | \\ CH_2COOH + GTP + HS.Co\,a \\ \text{Succinic acid.} \end{matrix}$$

8. Succinic acid undergoes dehydrogenation is the presence of a flavoprotein enzyme - succinic acid dehydrogenase, to form fumaric acid. The pair of hydrogen atoms released are accepted by FAD (this is the only reaction in Krebs cycle where the hydrogen is not accepted by NAD) and it gets reduced to $FADH_2$.

$$\begin{matrix} CH_2COOH \\ | \\ CH_2COOH \\ \text{Succinic acid} \end{matrix} + FAD \xrightarrow[\textit{dehydrogenase}]{\textit{Succinic acid}} \begin{matrix} CH_2COOH \\ | \\ CHCOOH \\ \text{Fumaric acid} \end{matrix} + FADH_2$$

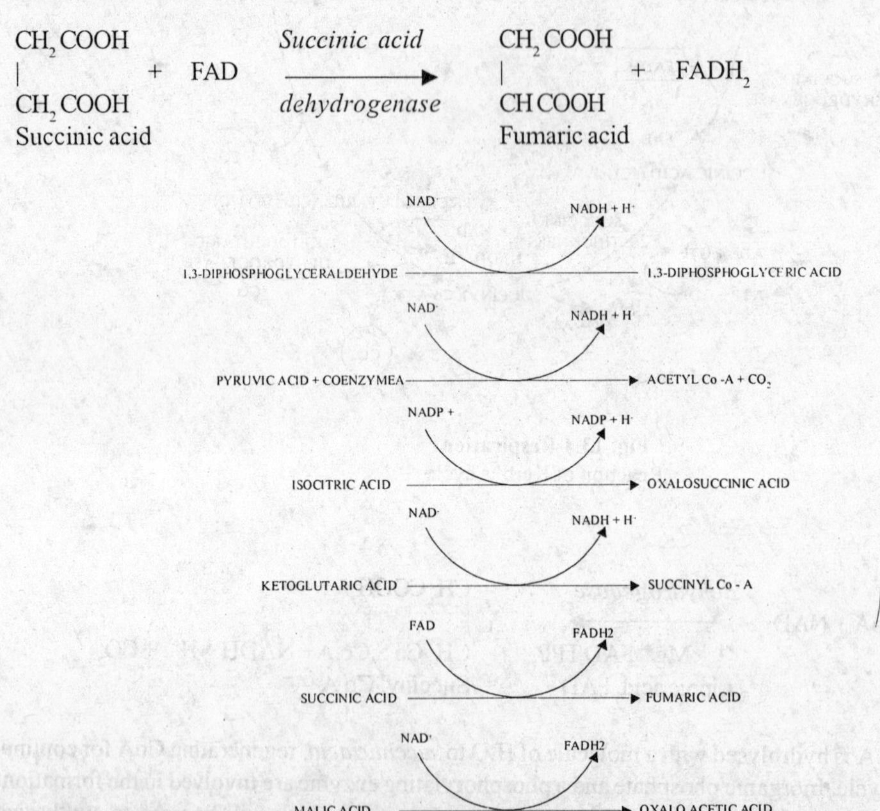

Fig. 13.5 Respiration
Dhydrogenation reactions in aerobic respirations (various stages)

9. One molecule of water is added to Fumaric acid enzymatically (Fumarase) to produce *malic acid*

CH COOH | CH COOH Fumaric acid	+ H_2O	*Fumarase* ⟶	CHOH COOH | CH_2 COOH Malic acid

10. In the last oxidation reaction in Krebs cycle malic acid undergoes dehydrogenation under the influence of *malate dehydrogenase* to produce *oxalo acetic acid,* which combines with acetylCoA to produce citric acid and continues the cycle. The hydrogen released from malic acid is accepted by NAD to produce NADH + H^+

CHOH COOH | CH_2 COOH Malic acid	+ NAD	*Malate* ⟶ *dehydrogenase*	CO COOH | CH_2 COOH Oxaloacetic acid

Summary of Krebs Cycle : In the Krebs cycle there are four 6 carbon acids - *citric, cisaconitic, isocitric* and *oxalosuccinic* acids, one 5 carbon acid *a ketoglutaric acid* and four 4 carbon acids *succinic, fumaric, malic* and *oxaloacetic* acids. When the 2 carbon acetyl CoA enters the cycle decarboxylation takes place at two stages - once to form the *a* ketoglutaric acid and again to form succinic acid. With this, all the carbon atoms of the original glucose molecule have been removed.

The reactions of the Krebs cycle are for one pyruvate; since two molecules of pyruvate (3C) are formed at the end of glycolysis, the reactions of the Krebs cycle are to be doubled to account for complete glucose. The removal of CO_2 from glucose may be summarized as follows

Glycolysis

$$\underset{\text{Glucose}}{C_6H_{12}O_6} \longrightarrow \underset{\text{Pyruvic acid}}{2CH_3COCOOH}$$

In the above, there is no decarboxylation as all six carbon atoms are found in the organic molecule of pyruvate

$$\underset{\text{(pyruvic acid)}}{2CH_3COCOOH} \qquad \underset{\text{(acetaldehyde)}}{2CH_3CHO + 2CO_2}$$

In the above, there is removal of two carbon atoms. Four more carbon atoms are left. They are removed as follows in the Krebs Cycle.

(i) 2 molecules of oxalosuccinic acid (6C) → 2 molecules of α ketoglutaric acid (5C) + $2CO_2$

(ii) 2 molecules of *a* ketoglutaric acid (4C) → 2 molecules of succinic acid (4C) + $2CO_2$

Thus in the aerobic respiration which involves Krebs Cycle there is complete removal of all the organic carbon atoms to inorganic CO_2. Another important aspect of Krebs Cycle is the dehydrogenation. The energy rich electrons released from the carboxylic acids at the following stages enter into the electron transport system (ETS) to produce molecules ofATP.

Stages of dehydrogenation in the Krebs Cycle

(i) Conversion of Isocitric acid to oxalosuccinic acid
(ii) Conversion of α ketoglutaric acid to succinyl Co.A
(iii) Conversion of succinic acid to Fumaric acid
(iv) Conversion of malic acid to oxaloacetic acid

These are the stages at which the energy is trapped for the synthesis of ATP molecules. Besides the above, during the conversion of succinyl CoA to succinic acid there is substrate phosphorylation (not through the ETS). All these are explained below in detail.

ELECTRON TRANSPORT SYSTEM (ETS)

Electron transport system or oxidative phosphorylation is the final stage in aerobic respiration. Oxidative phosphorylation takes place in the F particles or oxisomes of the cristae in mitochondria.

In various steps in the respiratory process, the substrates are dehydrogenated (oxidised) and a pair of hydrogen atoms are removed. These H atoms are accepted by a series of acceptors which in turn get reduced and oxidised alternately. As hydrogen is shuttled between acceptors it ultimately reaches the last acceptor and then combines with 1/2 molecule of O_2 (O) to form water. The series of acceptors constitute the ETS.

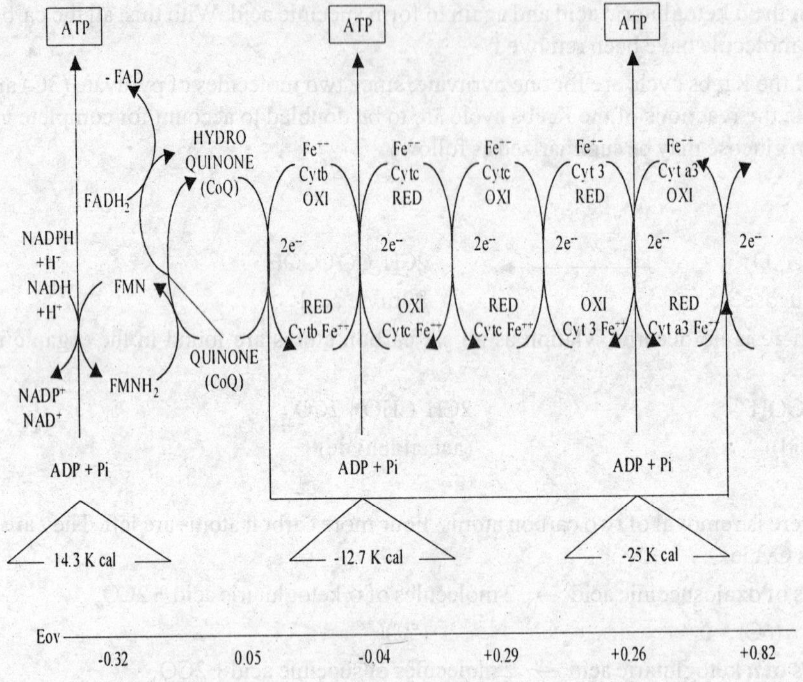

Fig. 13.6 Respiration
Schematic representation of electron transport system in oxidative phosphorylation

The bonded hydrogen cannot combine with atmospheric oxygen to form water. In ETS hydrogen, is released from its bonded state. The H atoms get disassociated into two protons and two electrons ($2H^{+}+ 2e^{-}$) The two protons released in the medium as H ion? are available for combination with oxygen to form water. At each dehydrogenation, two atoms of hydrogen are released with the result at every step, after diasssociation four protons ($4H^{+}$) and four electrons ($4e^{-}$) are released. While the H^{+} ions are used for the formation of water, electrons, more specifically the energy in them is utilized for the formation of molecules of ATP.

Components of ETS

The pairs of hydrogen ($2H^{+} + 2e^{-}$) removed from the substrate at various steps are eventually transported to the oxygen through an assembly of enzymes which constitute the respiratory chain. The enzymes of the respiratory chain mediate in the transport of the hydrogen through their prosthetic groups or coenzymes. The prosthetic group of an enzyme in the respiratory chain undergoes reduction when it accepts hydrogen or electrons from the preceeding enzyme and passes it on to the next enzyme in the chain getting oxidized (coming back to normal) in the process.

It is beleived that most of these enzymes which are involved in dehydrogenating the substrate are loosely bound to the mitochondial inner membrane, while the others (enzymes) constitute the integral part of the lipoprotein matrix of the inner membrane.

The ETS has the following enzyme complexes and two mobile carriers.

Complex 1 : This consists of a flavoprotein with attached $NADH_2$ dehydrogenase enzyme having FMN (Flavin mononucleolide) as the prosthetic group 2-4 atoms of nonheme iron with iron-sulphur (Fe-S) centre is found in the enzyme.

Complex II : This also consists of a flavoprotein called *succinic dehydrogenase* with FAD (Flavin adenine dinucleotide) as its prosthetic group. In addition there are two Fe-S centres and an Fe-S protein. Between Complex I and II, is present a mobile carrier ubiquinone (CoQ). CoQ is reduced by $NADH_2$ as well as by succinic dehydrogenase.

Complex III : This consists of Cytochrome *b* and C_1 with a non heme iron protein with only one Fe-S centre. However the presence Cytochrome C_1, is doubtful in plant ETS.

Complex IV : This is also called cytochrome oxidase or terminal oxidase. The complex consists of cytochromes *a* and a_3 having one or two atoms of firmly attached copper. Another mobile carrier is found between Complex III and IV which has a prosthetic group made up of iron porphyrin IX. This is identical with the heme of hemoglobin.

Complex V : This is the enzyme system called ATPase and is mainly concerned with the synthesis of ATP from ADP and inorganic phosphate.

The plant electron transport system consists of these complexes and their location and sequence may be decided by the evidence provided by redox potential and respiratory inhibitors.

Evidence from Redox potential : In the biological system, certain compounds more readily get oxidized (give out electrons), while others get reduced (accept electrons). This phenomenen of release or acceptance of electrons is called **Redox potential** (E_0) It is usually defined as the ratio between the reduced and oxidized forms of a compound. The standard electrode potential (E_0) may be defined as the potential at 50% oxidation for an oxidation reduction system at pH 7. The standard electrode potential measurement (measured in milivolts) of the various components of the ETS gives the following sequence (in the order of decreasing potential) with exception of CoQ. The sequence is NAD, FMN, FAD, *Cyt b, Cyt c, Cyt e, Cyt a, and Cyt* C_3. The standard redox potential of $NADH_2$ is 0.32 volts while that of water is + 0.82 volts. It is for this reason that electrons from hydrogen to oxygen will yield energy.

Evidence from respiratory inhibitors : The correct sequence of the components of ETS can also be ascertained by the use of specific inhibitors, which block the function of one or the other components. For instance when dilute cyanides are used which block the functioning of Cyt *a* and a_3, all other components get reduced i.e function normally indicating Cyta and a_3 are at the end. If on the other hand antimycin A is used, it inhibits the functioning of NAD, all other components remain in the oxidized state (will not get reduced), thus indicating that NAD must be the first electron acceptor.

Stages of dehydrogenation : Total oxidation of the substrate material (glucose) into CO_2 and water involves several dehydrogenations at various steps (in glycolysis and Krebs cycle). The released hydrogen is picked up by the NAD or FAD and passed on to the next electron acceptor.

1. 2 H is removed in glycolysis when a molecule of glyceraldehyde 3. phosphate is converted into 1.3. diphosphoglyceric acid. NAD picks up the hydrogen and gets reduced to $NADH_2$.
2. During the decarboxylation and dehydrogenation of pyruvic acid into acetyl CoA, 2H is removed. Hydrogen is accepted by NAD to form $NADH_2$.
3. 2H are removed from isocitric acid to form oxalosuccinic acid. NAD accepts hydrogen and is reduced to $NADH_2$.
4. During the oxidation of *a* ketogulatric acid to succinic acid 2 H are removed. This is again accepted by NAD to become $NADH_2$
5. During the formation of fumaric acid from succinic acid, dehydrogenation occurs which reduces FAD to $FADH_2$ (Note: This is the only instance where FAD and not NAD is the first electron acceptor from the substrate).
6. During the dehydrogenation of malic acid to oxaloacetic acid 2 H atoms are removed which are picked up by NAD to get reduced to $NADH_2$.

Thus in the above 6 steps (one in glycolysis, one in pyruvate and four in Krebs cycle) a total of 12 hydrogen atoms arc removed. Since this number is for triose only, it has to be doubled for a hexose and it gives a number of 24 hydrogen atoms. These hydrogen atoms travel down their electrical gradient along the various electron acceptors, which alternately get reduced and oxidized as the electron is shuttled through. At specific stages the energy in the electron is made use of for the synthesis of ATP (oxidative phosphorylation).

The scheme of electron movement

The ETS begins with the acceptance of 2 atoms of hydrogen from the substrate by NAD which gets reduced to $NADH_2$. $NADH_2$ is oxidized when the 2H are transferred to the prosthetic group of lipoproteins of the FMN which in turn gets reduced to $FMNH_2$.

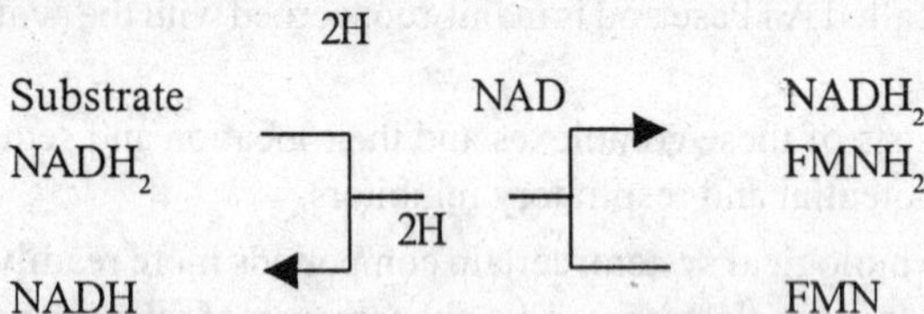

The 2H atoms are now transferred from $FMNH_2$ to the mobile carrier CoQ (Coenzvme 0 or ubiauinone). CoQ is chemically related to vitamin K and E and similar to plastoquinone of chloroplasts. This CoQ links the flavoproteins of the FAD to Cytochrome.

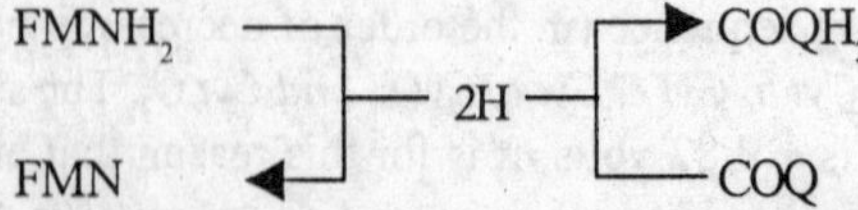

For hydrogen removed in all instances NAD is the initial acceptor except in the case of succinic acid (Krebs Cycle). Here the hydrogen is picked up by FAD (Flavin adenine dinucleotide) which is reduced as follows

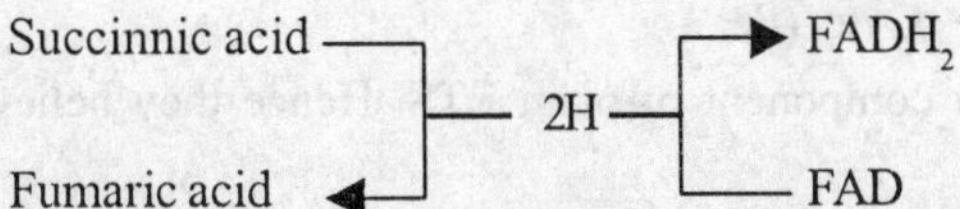

The reduced FAD($FADH_2$) then passes on its hydrogen to CoQ or *Cyt b*. The actual point of hydrogen entry (either to CoQ or *Cyt b* is not yet definitely established).

Some of the current researches have indicated that at two places (pyruvate and α sketoglutarate) hydrogen released from the substrate does not go direct to NAD but through lipoate and FAD and then to NAD. This is however subject to further proof.

According to Kligenberg and Kroger (1967) CoQ not only picks up hydrogen from succinate *via* FAD but also from *a* glycerophosphate and Acetyl CoA. The FAD of acetyl CO.A, an additional flavoprotein ETF (electron transporting flavoprotein) is present.

The reduced CoQ is oxidized by the release of 2 hydrogen atoms. Soon after release, hydrogen undergoes disassociation into two protons (2 H^+) and two electrons ($2e^-$). At this stage the protons (2 H^+) are released into cytoplasm, while only the electrons are further transported through the chain (i.e., through *Cyt b, Cyt e, Cyt a* and *Cyt* a_3). The two electrons finally combine with oxygen which in turn combines with protons (H^+) to form water as follows

$$O_2 + 4e^- \longrightarrow 2(O_2)$$
$$2(O_2) + 4H^+ \longrightarrow 2H_2O$$

(In the above equation 4e and 4H+ are shown instead of2H and $2e^-$. We have to remember that the reaction sequences ofglycolysis (from 3 PGA onwards) and Krebs Cycle are for only half of glucose (one triose), hence they have to be doubled while accounting for a complete glucose molecule).

The passage of electrons are two at a time from NAD upto CoQ. But when electrons are transported from CoQ to *Cyt b,* each molecule of *Cyt b* can accept only one electron. But if we account for the passage of only one electron at a time, it will not be sufficient to reduce molecular oxygen. How the two electron chain is changed to a single electron chain (at the same time delivering two electrons towards the end) is as yet completely not known. The following are some of the views

(a) Some unknown intermediate compound transfers the electrons on by one into the cytochromes

(b) The heme prosthetic group of the enzyme system Cytochromes brings about electron transfer one at a time

(c) The electron chain is forked at the point of CoQ so that two electron can pass through the two branches (one in each branch). Each brand has the same enzyme system as the other. Finally the two electrons released from the branches combine with molecular oxygen.

The third view seems to be possible as it satisfactorily accounts for the smooth flow of both the electrons in separate but parallel channels.

The electrons given out by CoQ are first accepted by Cytochrome *b* which gets reduced. *Cyt b* is reoxidised by transferring its electron to *Cyt c*. From these, the electron passes to *Cyt c, Cyt a,* and finally to *Cyt* a_3. *Cyt* a_3 gets oxidized by transferring its electron to the atmospheric oxygen. How is it released? Each enzyme consists of a heme prosthetic group called *protoheme*. This iron alternately *gets.* reduced and oxidized bu accepting *protoheme*. This iron alternately gets reduced and oxidized and donating an electron at a time as follows.

Cyt b CFe^{2+} → Cyt*c* (Fe^{2+})

e

Cyt b (Fe^{3+}) ← $Cytc_1$ (Fe^{3+})

Some physiogists argue that *Cyt* C_1 is not a component of plant ETS. Hence they believe that *Cyt b* transfers the electrons to *Cyt c*.

ATP synthesis (oxidative phosphorylation)

All biological reactions require energy and this depends on the formation and utilization of 'energy rich' phosphate bonds. A high energy bond does not actually mean that the bond is strong, on the contrary the bond is labile and when it is broken in a reaction a large amount of energy is released. The energy requiring reactions always obtain energy by using certain compounds which readily release energy. These are called energy currencies. Similarly during the storing of energy that is released in a reaction, these intermediate molecules are synthesized. Several molecules which act as energy intermediaries are known in biological systems. Of these, the most important are ATP (Adenosine triphosphate) and ADP (Adenosine diphosphate). Both ATP and ADP are derived from AMP (Adenosine monophosphate). AMP consists of a purine (adenine), a pentose sugar (ribose) and phosphoric acid. ADP and ATP are similar to AMP except in the number of phosphate groups. While ADP has two phosphate groups, ATP has three. The second and third phosphate bonds are energy rich in ATP. On hydrolysis of the phosphate bond, energy equivalent to 7600 cal per mol is released.

To sum up wherever energy is released ATP molecules are synthesized and whenver energy is required ATP molecules are utilized. This may be illustrated by the following equations.

Synthesis

$$\text{ADP} + \text{Pi} \xrightarrow{\text{energy}} \text{ATP}$$

(inorganic phosphate)

Utilization

$$\text{ATP} + \text{X} \longrightarrow \text{X-(P)} + \text{ADP}$$

ATP molecules are synthesized in living systems by two methods

(i) Photophosphorylation and

(ii) Oxidative phosphorylation

In photophosphorylation (seen in green plants) radiant energy is used to add iP to ADP to produce ATP. In oxidative phosphorylation ATP is synthesized using the energy of oxidation.

Oxidative phosphorylation is the most significant function of ETS. The number of ATPs produced may be explained on the basis of P/O index (phosphorylation/Oxygen) P/O index is the ratio of ATP produced for each atom of oxygen used.

Observations of Lehninger point out that the P/O ratio in the case of NAD linked ETS is three *i.e.,* at three sites (between NAD and molecular oxygen) ATP may be produced. Hence three molecules of ATP are produced. For a question as to why there are only three energy producing sites the answer may be provided as follows.

When the electron passes through the chain from one acceptor to the other, at every step some energy is released. The energy so released will be equivalent to the difference in energy level between the electron donor and acceptor. At certain steps the energy released is so less that it is insufficient to bind Pi to ADP (to produce an ATP) and hence wasted in the form of heat. But at certain steps, the amount of energy released is sufficient to bind Pi to ADP and hence a molecule of ATP is synthesized. As per some calculations, it has been noted that 0.28 volts of energy is sufficient to synthesize a molecule of ATP from ADP and Pi, and in the ETS, whereever the energy difference between the donor and acceptor is 0.28 volts or more a molecule of ATP is synthesized. As has already been pointed out, in substrate systems linked to NAD there are three sites, where a molecule of

ATP is synthesized.

These are

(i) NADH	FAD
(ii) Cyt *b*	Cyt *c* or c_1 and
(iii) Cyt *a*	Cyt a_3

In the oxidition ofsuccinilic acid which is linked to FAD directly, obviously there are only two sites for the synthesis of ATP hence only two molecules of ATP are synthesized.

The scheme of ETS and the sites of ATP molecule synthesis proposed by Bidwell (1979) are slightly different, though he also agrees that in NAD linked substrate systems there are three sites of ATP synthesis and in FAD linked systems there are only two. But he places the CoQ (Ubiqinone) between *Cyt b, Cyt c* and not earlier to the cytochrome systems (between FAD and Cytochrome b). This poses a problem for the movement ofelectronsasubiquinonerequires2H^+ atoms for reduction and not just electrons. But before the 2 H reaches the Cytochrome b, it would have undergone disassociation with two protons (2 H^+) getting released into cytoplasm. Hence Bidwell (1979) believes that once again (after ionization before they reach *Cyt b)* 2 H^+ and 2e' unite to form 2 H, and this reduces the ubiquinone. From ubiquinone when 2 H has to reach *Cyt c,* there is ionization, 2 H^+ is released into water while 2e continues in the ETS. Hence in this chain, there is ionization of hydrogen twice and not once as believed by other physiologists.

Mechanism of ATP formation : Same as in photophosphorylation.

Balance sheet of energy in Respiration (Account of ATP molecules synthesized). The following is a total account of ATP molecules synthesized during complete breakdown of glucose to CO_2 and H_2O.

Glycolysis

Reaction sequence	**ATP used**	**ATP Synthesized**
1. Glucose to Glucose 6 phosphate	1	–
2. Fructose 6 phosphate to Fructose- 1,6-diphosphate	1	–
3. 13 diphosphoglyceric acid to 3 phosphogyceric acid (1 X 2)	–	2
4. Phosphoenol pyruvic acid to pyruvic acid (1 X 2)	–	2
5. Dehydrogenation ofglyceraldehyde-3 phosphate to 1.3 diphosphoglyceric acid (through ETS with participation of NAD) 3 X 2	–	6
Total	**2**	**10**

Net gain of ATP during Glycolysis = 8 molecules.

(Note: reactions 1 and 2 are for entire glucose molecule, while 3,4 and 5 are only for a triose hence they are doubled)

Pyruvic acid oxidation

Reaction sequence	ATP used	ATP Synthesized
Conversion of pyruvic acid to Acetyl Co A; dehydrogenation through ETS with initial acceptor being NAD (3 x 2)		6
Krebs Cycle (There is no ATP loss in Krebs Cycle)		
		ATP Synthesized
1. Isocitric acid to oxalosuccimc acid dehydrogenation through ETS with NAD participation (3 X 2)		6
2. Conversion of *a* ketoglutaric acid to succinyl CoA, ETS with NAD as the initial accepter (3 X 2)		6
3. Conversion of succinyl CoA to succinic acid; substrate level phosphorylation. No ETS (1 X 2)		2
4. Conversion of Succinic acid to Fumaric acid. ETS beginning with FAD and not NAD. Hence only two sites ofATP synthesis (2 X 2)		4
5. Conversion of malic acid to Fumaric acid. ETS beginning with NAD (3 X 2)		6
Total		**24**

Total ATP used up during entire aerobic respiration 2

Total ATP produced during glycolysis, pyruvic acid oxidation and Krebs Cycle.

$$10+6+24=40$$

Net gain of ATP molecule for aerobic oxidation of one molecule of glucose

$$40-2=38$$

Differences between photophosphorylation and oxidative phosphorylation

No.	Character	Photo phosphorylation	Oxidative Phosphorylation
1.	Occurrence	Photosynthesis	Respiration
2.	Cell organelle involved	Chloroplast	Mitochondrion
3.	Occurrence within the organelle	Thylakoid membranes	Inner membrane of cristae
4.	Need for molecular oxygen	Not needed	Needed
5.	Source of energy	Light	Reduction - oxidation reactions.

6.	Output of ATP molecules	Less	More
7.	Fate of ATP	Used for the reduction of CO_2	Released into cytoplasm and molecules available for all metabolic reactions.

Efficiency of Respiration : A molecule of glucose consists of a total amount of 6,86,000 (A F) calories of energy in the form of chemical bond. During the complete oxidation of a glucose molecule 6,74,000 (673.6 kcal) calories of energy is released. Out of this, only a part is trapped in the energy rich phosphate bonds of ATP molecules, while the rest is released as heat. Each ATP molecule has about 7600 calories of energy. The total amount of energy trapped during the breakdown of glucose depends on the number of ATP molecules formed.

In case of oxidative phosphorylation the total number of ATP molecules produced will be 38. This means a total of 38 x 7600 X- 100/686,000 or about 40% of the total energy will be negotiable while the rest (60%) is wasted. In the alternative pathways of glucose breakdown (pentose phosphate shunt) a similar amount of energy will be made available in the form of molecules of ATP. This gives an effieicncy of 40% for aerobic breakdown of glucose.

In anaerobic respiration only two molecules of ATP are formed per glucose molecule. The efficiency will be

$$C_6H_{12}O_6 \longrightarrow 2C_3CH_4O_3 + OH$$

(F = 52 K Cal)

$$= (2 \times 7.6/52) \times 100 = \mathbf{29.2\%}$$

The energy efficiency for alcoholic fermentation would be (here also only two ATP molecules are formed)

$$C_6H_{12}O_6 \longrightarrow 2C_2H_5OH_3 + 2Co_2$$

(F = 52 K Cal)

$$= (2 \times 7.6/52) \times 100 = \mathbf{27\%}$$

PASTEUR'S EFFECT

The main difference between aerobic and anaerobic respiration lies in the complete decarboxylation in the former and partial decarboxylation in the latter. The following equations show that

Anaerobic respiration

$$C_6H_{12}O_6 \longrightarrow 2C_2H_5OH + 2CO_2$$

Aerobic respiration

$$C_6H_{12}O_6 + 6O_2 \longrightarrow 6H_2O + 6CO_2$$

a ratio of 1:3 is maintained in the release of carbon dioxide between anaerobic and aerobic respiration. The noted scientist Pasteur however noticed that supply of O_2 to an anaerobic system does not increase the output of CO_2. This means that in anaerobic respiration the ratio 1:3 is not maintained always. The amount of CO_2 released many a time in anaerobic respiration is far less when compared with the equations given above. This is due to the presence of excess O_2 in the system. This decreased rate of glucose breakdown in anaerobic systems caused by the presence of O_2 is called Pasteur's effect. In other words inhibition of sugar breakdown occurs due to the presence of oxygen under aerobic conditions (in anaerobic systems).

According to Dixon (1937), Pasteur's effect prevents excessive conversion of carbohydrates into products of fermentation and thereby maintains the level of carbohydrates. Pasteur's effect may be due to any one or all of the following reasons.

(a) Prevention of functioning of glycolytic enzymes
(b) Increased rate ofglycolysis with decreased oxygen tension
(c) Release of excess CO_2 from degradation of compounds other than the respiratory substrate
(d) Formation of partial glycolytic products

FACTORS AFFECTING THE RATE OF RESPIRATION

Like any physiological process respiration is also influenced by a number of external and internal factors. These are

Internal factors - protoplasmic factors and concentration of respiratory substrate

External factors - Temperature, light, O_2 concentration, CO_2 concentration, water, injury, chemical, mechanical effects etc.

INTERNAL FACTORS

Protoplasmic factors : The rate of respiration depends on the metabolic state of protoplasm. Young and meristematic cells which are actively dividing have dense non vacuolated cytoplasm. Besides, the presence of respiratory enzymes is another protoplasmic factor influencing the rate of respiration. In general the rate of respiration is more in undifferentiated cells than in mature cells.

Concentration of respiratory substrate : The availability of the respiratory substrate is very important for the proper rate of respiration, poor supply of respiratory substrate retards the rate of respiration. When other conditions are favourable, the rate of respiration increases with the supply of respirable substance. Rate of respiration increases in a leaf after it is transfered from darkness to light (due to adequate supply of substrate as a result of photosynthesis). Etiolated leaves have a low rate of respiration.

EXTERNAL FACTORS

Temperature : The influence of temperature on respiration is the same as on other metabolic processes. Plants generally are capable of respiring in the temperature range between 0-45°C. Within this range any increase or decrease of temperature would have a similiar effect on respiration.

In several experiments it has been noticed that the increase in the rate of respiration follows the Vant Hoff's law that the rate of reaction is doubled for every 10°C increase of temperature, if no other factor is limiting. In general an optimum tempeature of 30°C is most suitable for respiration. Beyond this range there is no further increase but the rate flattens until 145°C when it starts declining.

At higher ranges of temperature the rate of respiration depends on the time factor. Greater duration at higher temperature retards the rate of respiration possibly due to inactivation of respiratory enzymes. According to Platenius (1942) Q_{10} for respiration between 0°C-30°C is usually 2.0 or 2.5. Mayer and Anderson (1952) have reported that the following causes may retard the rate of respiration at higher ranges of temperature.

(i) At higher ranges of temperature, O_2 may not get proper access to the cells.
(ii) CO_2 accumulates in the cells at higher temperature retarding the rate of respiration.
(iii) Supply of respiratory substrate is affected at higher temperature.

At very low temperature the .rate of respiration is very low. This has been made use of in cold preservation of fresh fruits and vetetables, when oxidation of substrate is retarded. The cold temperature again is believed to be responsible for bigger size of the tubers in potato plants grown in hilly regions. Low rates of respiration in hilly regions increases the quantity of storage food materials.

Comparison of temperature effect on photosynthesis and respiration is a factor contributing to the longevity of plants. While photosynthesis has a temperature range of 0-25°C, that of respiration is 0-35°C. This means that maintenance of temperature beyond 25°C increases the metabolic processes thereby affecting plant health. The ratio between photosynthesis and respiration (P/R ratio) based on temperature gives a metabolic index. A higher metabolic index favours plant growth, while a lower index retards the growth. Besides temperature, drought, and high and low intensities of light have an influence on the metabolic index. It is this higher metabolic index, that is responsible for the bigger size of the tubers in potatoes grown in mountainous regions.

Light : It is generally believed that the influence of light on respiration is indirect even though physiologists like Emerson and Lewis (1943) have reported, that illuminated *Chlorella* cells have a higher rate of respiration than those kept in darkness. A similar effect on barley seedlings has been observed by Johnson (1944). But there is no stoppage of respiration in darkness. This indicates that light affects respiration indirectly by governing the supply of respirable materials *via* photosynthesis. Another influence of light, is that by increasing temperature, it increases the rate of respiration.

Oxygen concentration : Oxygen is a must for aerobic respiration, though it is not necessary for anaerobic respiration. Since the concentration of O_2 in the atmosphere is constant, slight variations in O_2 concentration do not seem to have a lasting influence on respiration. But if the concentration of O_2 goes below 2.0% it will retard the rate of respiration as there is rise in CO_2 level due to fermentation.

Carbon dioxide concentration : The atmosphere has a carbon dioxide concentration of 0.03% and it is almost constant. Hence generally its influence on respiratory rate is negligible. However any increase in the carbon dioxide concentration has a retarding influence on respiration. Soil has a higher concentration of CO_2 than atmosphere due to the activity of microbes. Hence seeds sown in such a soil generally fail to germinate. Ploughing the soil increases the aeration and enhances the rate of germination.

Decrease in respiration due to increased CO_2 concentration may be used to preserve vegetables and fruits (in a CO_2 rich atmosphere) by keeping the rate of respiration at minimum. This is not however practised as a higher CO_2 concentration may prove to be poisonous. Higher CO_2 concentration also affects all other vital activities which require biological energy.

Water : Hydration of the medium is a basic requirement for all the metabolic activities and respiration is no exception. The rate of respiration is minimum in dry seeds, and as soon as they come in contact with water the rate of respiration registers a marked increase. Respiratory enzymes become active only in a hydrated medium. Water maintains proper turgidity of the cells and provides a suitable environment for respiratory reactions.

A slight decrease or increase of water however does not influence the rate of respiration. Wilting plants on the other hand show a higher rate of respiration due to higher rate of conversion of starch into sugars. Another effect of water is that it is the medium through which O_2 diffuses into the respiring cells.

Injury : The works of Hopkins (1927) and others have demonstrated, that injury enhances the rate of respiration. It has been observed that sugar content of cells near the wounded region increases preparatory to increased rate of respiration. In the wound healing process new cells have to be added, with the result there is increased metabolic activity, thereby accounting for increased respiratory rate. Potato tubers cut into pieces show enhanced rate of respiration due to the conversion of starch into sugars.

Chemicals : Many of the chemicals like ether, alcohol, chloroform, formaldehyde, cyanides, carbon-monoxide, alkaloids, azides, iodoacetate etc., have a retarding influence on respiration. Many of these chemicals (azides, cyanides etc) are enzymatic inhibitors. At a low concentration however some of the above mentioned chemicals may even enhance the rate of respiration. But at higher concentrations, they are certainly injurious to the plant. The initial enhancement of respiration is possibly a defensive action of plant cells, as the plants require more energy to counteract the influence of these chemicals.

Mechanical effects : Mechanical stimulation of respiratory rate has been demonstrated in many plants. Audus (1939) has shown that rubbing or twisting the lamina will increase the rate of respiration by as much as 20-183% over the normal rate. This increased rate according to Audus (1939) is maintained for several days and seems to influence the aerobic phase of respiration. However repeated application of mechanical effects at periodic intervals seems to stop the enhancing influence on respiration. The reason for such a behaviour has not been explained till now.

EXPERIMENTS

1. Release of Carbon dioxide during aerobic respiration

This may be demonstrated with the help of the appa.atus called *Respiroscope.* This consists of a vertical tube which is bent on one side. To the bent end is attached a bulb with a stopper. The other end of the tube is dipped in a beaker containing a solution of KOH (potassium hydroxide). Before the beginning of the experiment note the level of KOH in the respiroscope tube. Put some germinating seeds into the bulb of the respiroscope. After sometime it maybe noticed that the level of KOH solution in the tube has risen.

The rising in the level of the KOH solution is due to absorption of CO_2 which is released by the activity of respiting seeds.

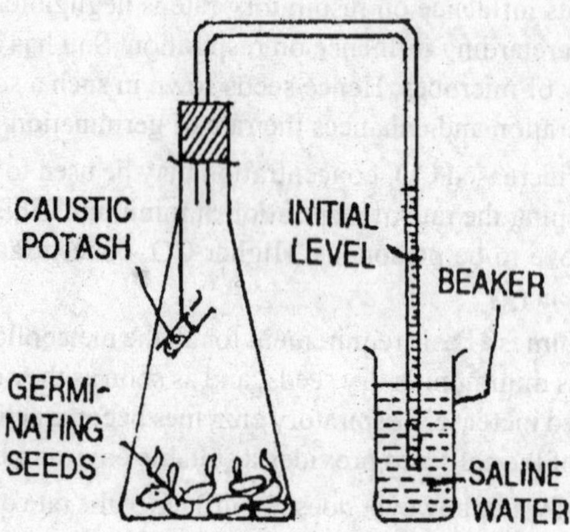

Fig. 13.7. Respiration
Experiment to demonstrate the release of CO_2 during aerobic respiration

2. Evolution of heat during respiration

Keep some germinating seeds in a thermos flask. Insert a thermometer into this through a split cork. The mercury bulb of the thermometer must be dipped in germinating seeds. Note the initial temperature. After sometime it may be noticed that the temperature has risen as recorded by the thermometer. This is due to the heat evolved by the rapid respiration on the germinating seeds.

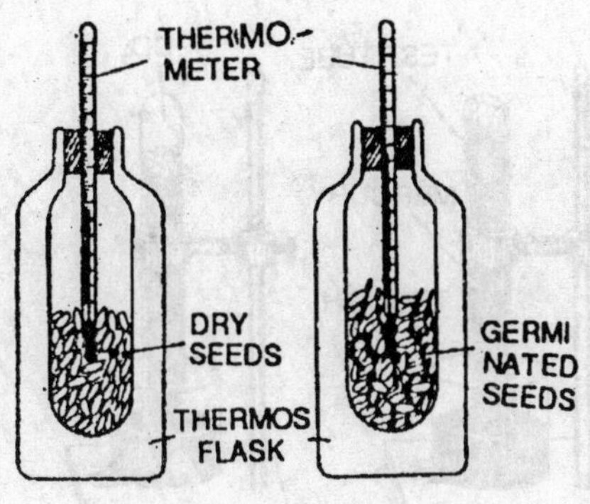

Fig. 13.8 Respiration
Experiment to demonstrate the release of heat during respiration

3. Demonstration of Fermentation

Fermentation may be demonstrated with the help of Kuhne's Fermentation flask. The apparatus consists of an upright tube to which is attached a bulb at the side. The flask is filled with 10% sugar solution. A small quantity of baker's yeast or toddy (contains yeasts) is added to this. The opening of the bulb is plugged with cotton. After sometime it may be noticed that small bubbles rise up in the vertical tube and the level of solution comes down. The solution is replaced by a gas. If we introduce a piece of KOH (Caustic potash) the gas is absorbed and the level of solution once again goes up. Alcohol is also produced.

This experiment demonstrates that saprophytic microorganisms respire and produce CO_2 and alcohol.

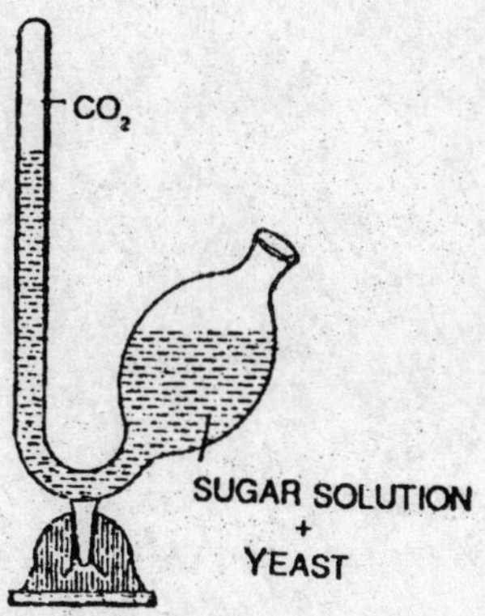

Fig. 13.9 Respiration
Kuhne's Fermentation flask

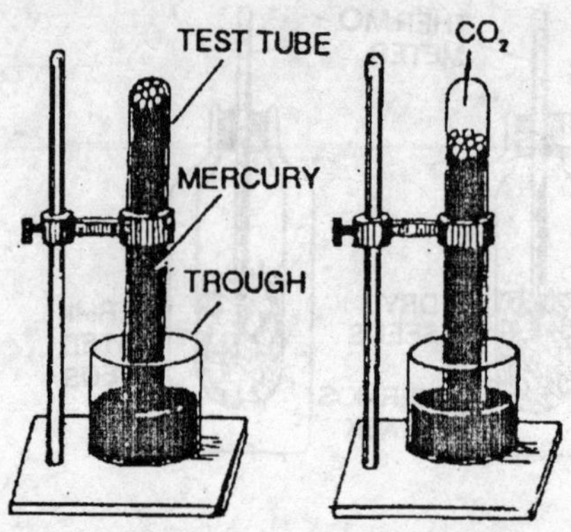

Fig. 13.10 Respiration
Evolution of CO_2 during anaerobic respiration

4.Demonstration of anaerobic respiration

Take a small petri dish and fill it with mercury. Carefully invert a test tube also filled with mercury over this. Introduce a few seeds into the test tube. The seeds rise up because of their light weight. Initially the mercury completely fills the test tube. After some time it may be noticed that the level of mercury has come down. This is due to the evolution of gas by the germinating seeds which press down the mercury. If a solid piece of KOH is introduced into the test tube it absorbs the gas and the mercury level goes up. Since KOH absorbs CO_2 the gas must be CO_2 evolved during anaerobic respiration. This respiration is anaerobic because the seeds in the test tube (filled with mercury) did not have any oxygen.

14

PHOTORESPIRATION

It is widely beleived that while Photosynthesis takes place during day time, respiration takes place during all the 24 hrs and its rate is uniform both in darkness as well as in light. But some of the current researches have indicated that the rate of respiration varies between day and night and the daytime respiratory rate is at least three to five times higher than during night. Obviously this higher rate of respiration must be due to light, as it is the major factor of difference between day and night. **This phenomenon of increased rate of respiration induced by light is called photorespiration**. Photorespiration however is seen only in green parts of the plant. Further there are different pathways for dark respiration and photorespiration. In green leaves and algae photorespiration replaces dark respiration during daytime.

Site of Photorespiration : Normal respiration takes place in Cytosol (cytoplasm without organelles) and in mitochondria,while photorespiration takes place in a co-operative way between peroxisomes, chloroplasts and mitochondria.

The fact that photorespiration is encountered only in green leaves is an indication that CO_2 evolution in light must be interlinked with the photosynthetic apparatus. Originally it was thought that the entire process of photorespiration is limited to chloroplasts and that some early products of photosynthesis act as respiratory substrates. But recent discoveries of peroxisomes found very close to chloroplasts and the presence of enzymes ofglycolate metabolism (glycolate catalase) in them (peroxisomes) indicate a correlation between .peroxisomes and photorespiration (Tolbert, 1981; Huang et al, 1983). Further, studies of Kişaki and Tolbert (1981) have indicated that peroxisomes produce the necessary substrate for photorespiration, which actually has its origin in chloroplasts. According to Tolbert (1971), CO_2 evolution during photorespiration occurs in-mitochondria. In view of these findings it is concluded that the coordinated functioning of chloroplasts, peroxisomes and mitochondria bring about photorespiration.

Photorespiration has been reported in the green cells of *Beta, Phaseolus, Gossypium, Pisum, Capsiciim,Petunia, Antirrhinum, Glycine, Helianthus, Chlorella,Nitetla, Oryza* etc.,

Conditions necessary for photorespiration : Light is indispensible for photorespiration. Temperature also plays an important role. A high rate of photorespiration has been reported between 25-35°C. Another important factor is the concentration of oxygen, which increases the rate of photorespiration in geometric progression (even up to 100%).

Evidences for photorespirartion : A number of evidences have been adduced to prove the occurrence of photorespiration. The following are some of them.

1. Experiments on illuminated tobacco leaves using labelled carbon have indicated that the First carbon atom of glycolate is liberated as CO_2. The work of Zelitch(1966) has proved this.
2. The works of Tregunna (1966), Hew (1968) etc have revealed that, effect of light on CO_2 production can be offset by the elimination of chlorophyll. Non-chlorphyll cells of the leaves do not exhibit photorespiration.

3. The observation of Eisharkawy (1967) that illuminated corn leaves which do not release CO_2 in the free air, begin to do so when treated with photosynthetic inhibitors like DEMU (3,4- dichlorophenyl-1 dimethyl urea) is another evidence for the occurrence of photorespiration.
4. Photorespiralion is inhibited by glycolate oxidase inhibitor.

Mechanism of Photorespiration : The respiratory substrate in normal respiration is sucrose or glucose. For photorespiration however the substrate is a 2C compound called glycollic acid. Until 1971 physiologists were not sure as to the mode of biosynthesis of glycollic acid. Bowes, Orgren and Hageman in the year 1971 discovered the dual role **Rubisco** (Ribulose 1.5 biphosphate carboxylase). The type of action of **Rubisco** depends on the conditions of the surrounding air. Under normal conditions Rubisco catalyzes the fixation of -CO_2 resulting in the formation of two molecules of Phosphoglyceric acid. But under conditions of high O_2 concentration (high O_2\ low CO_2), *Rubisco* does not fix CO_2, but resorts to oxygenase activity (O_2 competes with CO_2 for Rubisco) resulting in the cleaving of the Rudp molecule into a 2C phosphoglycollic acid and PGA-3C phospho - glyceric acid. The two reactions may be represented as follows.

(a) **Carboxylase activity :** *Rudp* $\xrightarrow[CO_2 \; + \; H_2O]{\textit{Rubisco}}$ 2 moles of 3 PGA.

(b) **Oxygenase activity :** *Rudp* $\xrightarrow[+ \; O_2]{\textit{Rubisco}}$ phosphoglycollic acid + 3PGA
(2C) (3C)

In the next step phosphoglycollic acid undergoes dephosphorylation to form glycollic acid. The reaction is catalyzed by the enzyme phosphotase

Phosphoglycollic acid + H_2O $\xrightarrow{\textit{phosphatase}}$ Glycollic acid + phosphoric acid.

The ability of Rudp carboxylase either to catalyse CO_2 fixation or bring about oxygenation reaction solely depends on the relative concentrations of $CO_2 : O_2$ in the atmosphere. In effect the $CO_2 : O_2$ concentration of the atmosphere determines the relative rates of photosynthesis and photorespiration.

The glycollic acid thus formed in chloroplasts is transported out into the peroxisomes. As has already been pointed out peroxisomes have a spatial relationship with both mitochondria and chloroplasts.

In the peroxisomes, glycollic acid is converted into glyoxylic acid in the presence of the enzyme glycolate oxidase as reported by Zeiitch (1966). O_2 is one of the reactants. Together with glyoxylic acid, hydrogen peroxide is also formed which is readily destroyed by the enzyme catalase into H_2O and O_2. If H_2O_2 is not destroyed by catalase, it (H_2O_2) oxidizes glyoxylic acid into formic acid and CO_2. But in most instances catalase destroys H_2O_2.

In the next step of the reaction glyoxylic acid is convered into glycine under the influence of the enzyme glutamate glyoxylate **aminotransferase**. This is a transamination reaction and takes place as follows.

Glyoxylic acid + Glutamic acid $\rightarrow$ Glycine + α ketoglutaric acid

Glycine which is formed in the peroxisomes is transported to mitochondia *via* cytoplasm. Since the three organelles (chloroplast, peroxisomes and mitochondria) are situated close by, transportation between them is easy.

Glyoxylic acid may also form glycine by transference of amino group to it from serine instead of glutamic acid. The reaction takes place as follows.

Glyoxylic acid + Serine $\xrightarrow[\textit{aminotransferase}]{\textit{Serine glyoxylate}}$ Glycine + Hydroxypyruvic acid

In the mitochondria two molecules of glycine join together to form a molecule of serine. CO_2 and NH_3 are

also released in this reaction. This oxidative decarboxylation reaction is highly complex and possibly involves more than one enzyme (Tolbert, 1980 Journet *el al,* 1981).

2Glycine + H_2O + NA → Serinc + CO_2 + NH_3 + NADH

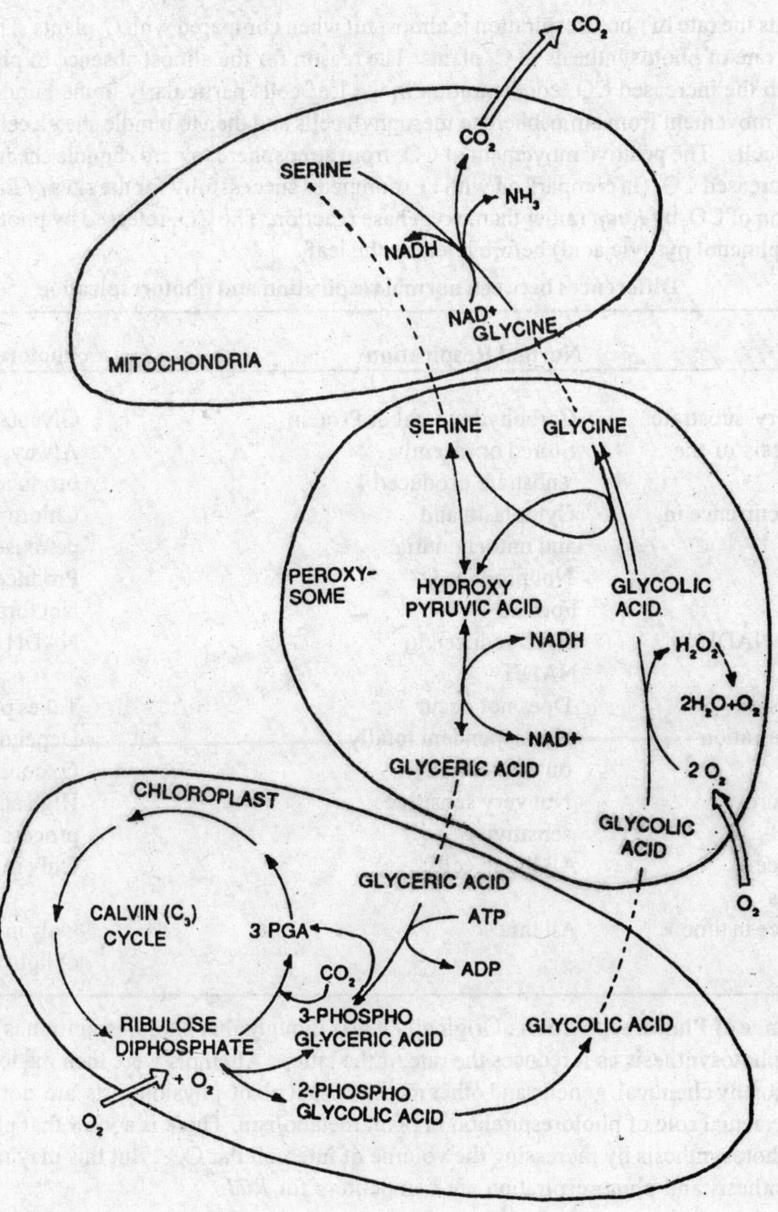

Fig. 14.1 Photorespiration
Various steps of photorespiration in chloroplasts, peroxisomes and mitochondria

It is at the stage of formation of serine that light induced liberation of CO_2takes place. Serine is then transported out of mitochondria and into peroxisomes, where it is converted successively into hydroxypyruvic acid and then to glyceric acid. Finally glyceric acid is transported to chloroplast where it undergoes phosphorylation by ATP to form 3PGA. The molecules of 3PGA join the pool in the CO_2 reduction cycle.

In C_4 plants the rate of photorespiration is almost nil when compared with C_3plants. This perhaps accounts for the higher rate of photosynthesis in C_4 plants. The reason for the almost absence of photorespiration in C_4 plants is due to the increased CO_2 concentration in the leaf cells particularly in the bundle sheath cells. The pattern of CO_2movement from atmosphere to mesophyll cells and then to bundle sheath cells, concentrates CO_2 bundle sheath cells. The positive movement of CO_2 from atmosphere towards bundle sheath cells is called **CO_2 pump**. The increased CO_2 (in comparison with O_2) competes successfully for the sites *of Rudp carboxylase* and favours fixation of $CO_2$by *Rudp* rather than oxygenase reaction. The CO_2 released by photorespiration is fixed by PEP (phosphoenol pyruvic acid) before it leaves the leaf.

Differences between normal respiration and photorespiration

	Feature	Normal Respiration	Photorespiration
1.	Respiratory substrate	Carbohydrate, fat or Protein	Glycotate
2.	Biosynthesis of the	Stored or recently substrate produced	Always recently produced
3.	Site of occurrence in the cell	Cytoplasm and and mitochondria	Chloroplasts, peroxisomes
4.	H_2O_2	Not produced	Produced
5.	ATP	Formed	Not formed
6.	NAD and NADH	NAD reduced to NADH	NADH oxidised to NAD
7.	Transmission	Does not occur	Takes place
8.	O_2 concentration	Not dependent totally on concentration	Dependent totally on O_2 O_2 concentration
9.	Temperature	Not very sensitive sensitivity	High acceleration of the process between 25⁰-35⁰C
10.	Occurrence in organisms	All living cells	Only in green cells
11.	Occurrence in time	All times	Only in daytime (in presence of light)

Significance of Photorespiration : Originally it was thought that photorespiration is a redundant process in relation to photosynthesis as it reduces the rate of the latter. Attempts were then made to check the rate of photorespiration by chemical, genetic and other methods. But plant physiologists are not totally aware even now about the actual role of photorespiralion in plant metabolism. There is a view that photorespiration may indeed help photosynthesis by increasing the volume of intercellular CO_2. But this may not be true always as both photosynthesis and photorespiration are competitors for *Rudp.*

One of the current hypothesis is actually to think that photrespiration provides the protective mechanism against light destruction of chloroplasts in C_3 plants. Photorespiration consumes O_2 and thereby helps to prevent the build up of O_2 accumulation iA chloroplasts. As reduced components of PSI react with photosynthetically produced O_2, free radicals of the gas (O_2) are produced which have the potential to destroy the membranes.

Another opinion regarding the advantage of photorespiration is that it is a mechanism for the utilization of excess ofATP and reducing power (NADPH) produced at higher levels of light. In photorespiration both these are used up without fixing CO_2. Utilization of excess ATP and NADPH will prevent solarization.

Another intercsting aspect is that while photorespiration is seen in many of the C_3 plants it is essentially absent in C_4 plants because bundle sheath cells where Rubisco is present, a very high concentration of CO_2 is always maintained, not allowing the glycollic acid synthesis.

Is Photorespiration a necessary evil ? Photorespiration seems to be a process working as an antithesis to photosynthesis, because it essentially competes for the same substrate *Rudp*. Photosynthesis being such a vital process, why photorespiration has not been eliminated due to selection pressure during the course of Evolution? Some physiologists argue that photorespiration is a necessary consequence of the architectural makeup of **Rubisco** whose active site cannot distinguish between CO_2 and O_2 **Rubisco** was evolved certainly to fix CO_2 in ancient times when atmospheric CO_2 level was high. But as O_2 started accumulating by the light splitting of water by algae, **Rubisco** started playing host to O_2 also. Hence photorespiration is a necessary consequence of the makeup of **Rubisco** and cannot be avoided. It is like the anatomical makeup of the leaf meant essentially for gaseous exchange also promoting water loss (transpiration). Hence can photoresipration also be termed a necessary evil? Unless we know more about the process it is difficult to answer.

Another aspect that perhaps merits serious research is the nature of **Rubisco** itself. Since photorespiration is not present in all plants, is there any difference in the nature of Rubisco between photorespiring and non-photorespiring plants? If so, it may be possible to modify or alter **Rubisco** in photorespiring plants so that it (Rubisco) favours only CO_2 fixation. But before attempting this, the actual role of photorespiration in overall plant metabolism has lo be identified.

Measurement of Photorespiration : There are several methods of which two are mentioned here. In one method air deviod of CO_2 is made to pass through an illuminated leaf placed in a glass chamber. The air then passes through an infra red gas analyser which measures the amount of CO_2 in air after its passage over an illuminated leaf. This, in camparison with control (air passed through non-illuminated leaf) gives an estimate of light induced CO_2 evolution. But this may not give an accurate data of CO_2 evolution, because some amount of CO_2 released may immedialtcly be fixed in photosynthesis, before it gets out of the leaf.

In another method regarded as more accurate, and described by Ludwigand Canvin (1971) photosynthetic refixation of internally evolved CO_2 is avoided. Here after photosynthesising for some time in a glass chamber, the leaf is fed with C^{14}. The disappearance of C^{14} and C^{12} are measured simultaneously within seconds after the introduction of C^{14}. From these measurements the rate of photorespiration maybe calculated (for more details see Ludwig and Oanvin, 1971, *Canadian Journal* of Botany. 49 : 1299-1313)

15

BACTERIA

Bacteria are one of the smallest among the microorganisms. Though small in size. They are by no means so, in their significance. With no exception, one can say that human life is being influenced every minute either for good or for bad by these inhabitants of microcosm. Bacteria are highly potent individuals. From the point of view of human welfare, They are of greatest signifiance. In the form of such deadely diseases like cholera, influenza, tetanus, leprosy, tuberculosis, typhoid etc., to name only a few, they are man's worst enemies. They are man's best friends also, in various industries (ex. Preparation ofvarous types of alcohols and acids, curing of coffee, tea and tobacco) and in various other aspects. When such interests are involved, no doubt a separate branch of science 'Bacteriology' is set apart for the study of these.

Historical account:

Credit of discovering Bacteria goes to Antony Von Leeuwenhoek (1676), a Dutch scientist. He was interested in grinding lenses and constructing crude microscopes, with which he used to observe drops of pond water, cells of retina, muscle fibres, nerve cells etc. One day, while he was observing the outer coat of pepper grain with the idea of finding out the reason for the biting taste, he found certain tiny organisms moving to and fro in the fiedl of the microscope. As these were very different from the ones observed before, he made a study of them and called them animalcules, which were later named 'Bacteria'. The next name to be remembered in the history of Bacteria is of Louis Pasteur. Very often and rightly so he is referred as the "father of Bacteriology". He was the one, who found out that Bacteria bring about fermentation. He conducted extensive studies on Bacteria.

Robert Koch, a German Bacteriologist discovered that tuberculosis is caused by Bacteria.

T.J. Burill (1879), an American scientist discovered that Bacteri are responsible for many plant diseases also.

Distribution

With one word perhaps one can explain the distribution of Bacteria ie., 'omnipresent'. There is virtually no place on earth which is free of bacteria. They are present in air, soil, milk, water, in temperate regions, tropical regions. Arctic regions (in snow), in hot springs, in plants, animals not excluding human beings. Bacteria can withstand extreme temperature ranges. For ex. Some can withstand 190°C cooling and heathing upto 78°C.

Morphology

Bacteria are unicellular organisms. On an average, a cell ranges from 0.1 micron to 1.0 micron in size. Exceptionally some forms may be as large as 15 microns (lmicron=l/10,000ofmm).

Some interesting calculations have been made to impress as to how small the bacteria are. Just a drop of water may contain several millions of bacteria. If a Coccus measuring I micron across is enlarged 500 times it would be 0.2" and an average human being enlarged on the same scale would be 2750 feet tall and more than a thousand feet broad at his shoulders. Nearly 500,000,000,000 bacteria would weigh just one gram.

Based on the cell shapes, four morphological forms have been identified in bacteria viz.

1. Bacillus- the cells are rod shaped and elongated. These are the second largest among bacteria eg. Mycobacterium.

2. *Coccus*- the cells are spherical. These are the smallest among bacteria. Eg. Streptococcus, staphylococcus etc.

3. *Spirillum*- the cell will be in the form of a loose spiral. These forms are the largest. Eg. Rhodospirillum.

4. *Comma*- the cell will be in the form of punctuation mark, comma eg. Vibrio. Coccus form exists in various ways. When existing as single isolated cell, it is called Micrococcus, in pairs- *Diplococcus*, in chains- *Streptococcus*, in sheets Staphyloccus etc.

Many bacteria exhibit several shapes or forms. This is known as "pleomorphism" (Eg. *Azatobacter; hizobium* etc).

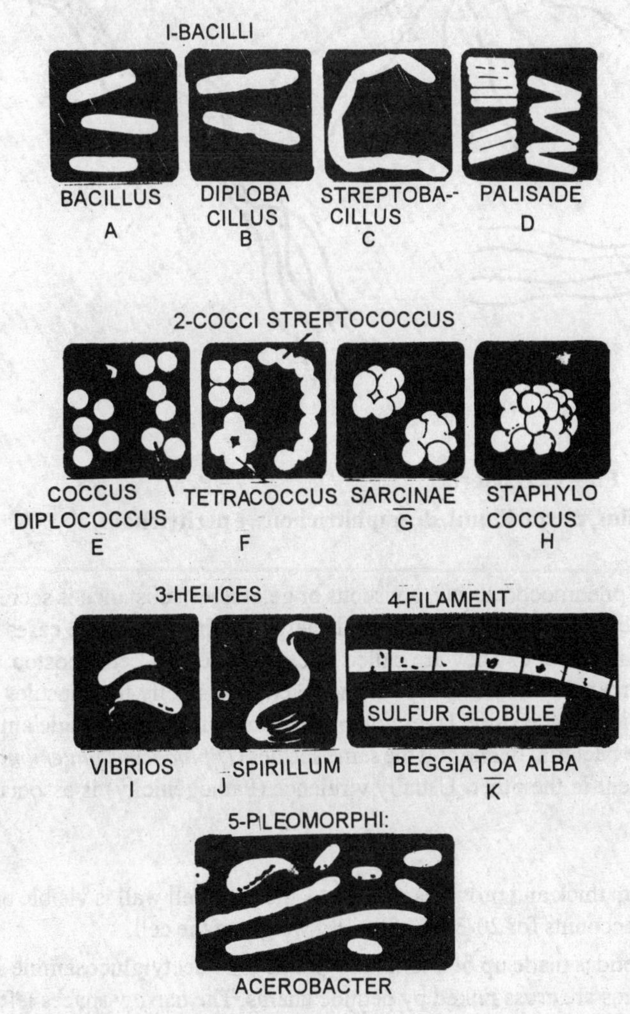

Fig 15.1 Bacteria
Types of morphological forms

Locomotion

Except the coccus, all the others forms of bacteria may posses organs of locomotion ie. Flagella. The flagellum is a long thread like structure whose length usually exceeds that of the cell. It arises from a basal granule. On the basis of the number and position of flagella, bacterial forms are classified into

1. *Atrichous*- no flagella eg. *Cocci*

2. *Monotrichous*- single flagellum present at one end eg. *Vibrio*

3. *Lophotrichous*- many flagella present at one pole eg. *Spirillum undula*

4. *Amphitrichous*- two tufts of flagella attached at either poles of the cell eg. *Spirilla*.

5. Peritrichous- flagella are present all round eg. *Salmonella typhosa*.

Bacteria move with varied speeds. Some can move about 2,000 times their size in an hour. Hay bacillus can travel at a speed of 200 microns per second (Knayasi, 1938).

Ultra structure ofa bacterialcel:

Electron microscopic studies have revealed the ultra structure of the bacterial cell. The outer envelop is the cell wall internal to which is the cell membrane. Sometimes external to the cell wall there will be aloose slimy layer or capsule. Some bacteria are flagellate; others have small pili or fimbriae. The bacterial cytoplasm includes ribosomes, mesosomes,fat globules, vacuoles, inclusion bodies and nuclear material. We shall study each one of them in detail.

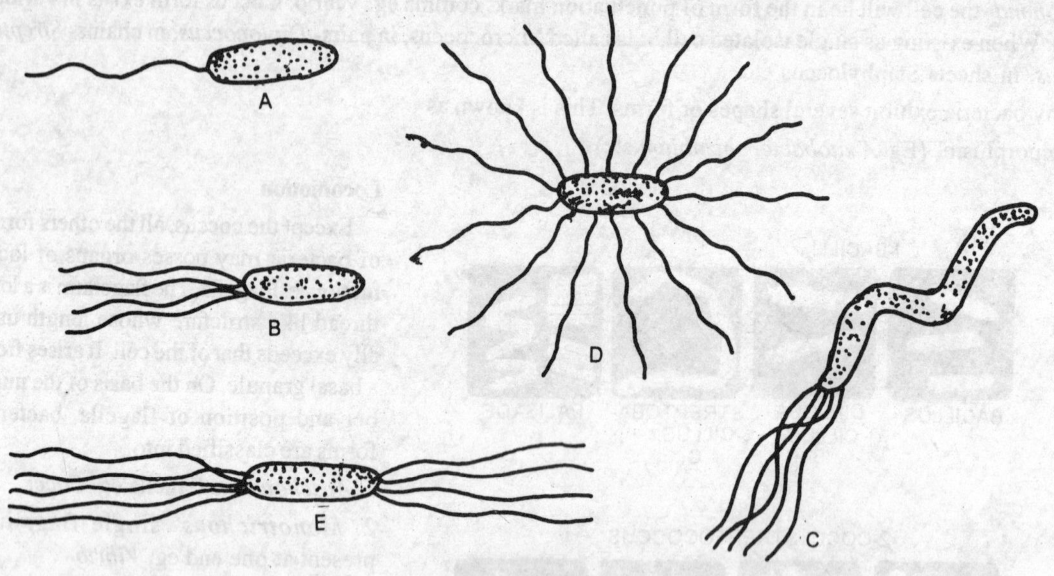

Fig15.2 Bacteria

a, Monotrichous ,b-c. lohotrichous (b= basillus, c= spirillum), d. amphitrichous, e peritrichous

Capsule : In many bacteria like diplococcus, pneumococcus etc. a viscous or gelatinous substance is secreted surrounding the cell wall. This is called the capsule when it is thick and sharply defined. In some cases the capsules are too thin to be seen under light microscope. They are called microcapsules. In Leuconostoc, the viscous substance is very loose and undermarcated and is known as slime layer. Chemically the capsules are made up of proteins, polysaccharides and lipids. Some slime layer polypeptidesare built with a single amino acid. The capsules may not be present in all the bacteria. In one and the same species (*Diplococcus pnuemoniae*) capsule may be present in one strain and absent in the other. Usually virulence (Pathogenicity) is associated with the capsule.

Cell wall: the cell wall is thin about 10-25 mm thick and provides rigidity to the cell. Cell wall is visible only under the electron microscope. The cell wall accounts for 20-30% of the dry weight of the cell.

Chemically, the cell wall consists ofmucopeptides made up of alternating chains of N-acetyiglucosamine and N-acetyl muramic acid molecules. These chains are cross linked by peptide chains. The narrow spaces left by these chains are filled by various other chemical components which vary with the species.

The cell wall composition differs between gram positive and gram negative bacteria. The cell walls of gram positive bacteria contain up to 95% peptidoglycan and about 10% teichoic acids. In gram negative bacteria, the cell wall composition is more complex. It is made up of several layers. Next to the cell membrane is The peptidoglycan layer. Next to this extreroallyisthe periplasmic region consisting of a number of enzymatic proteins.

The outer membranous portion is joined to the peptidoglycan region by many links of lipoproteins. The outer membrane regulates the entry of molecules into the periplasmic space. The following are some of the differences between gram positive and gram negative bacteria in their wall architecture.

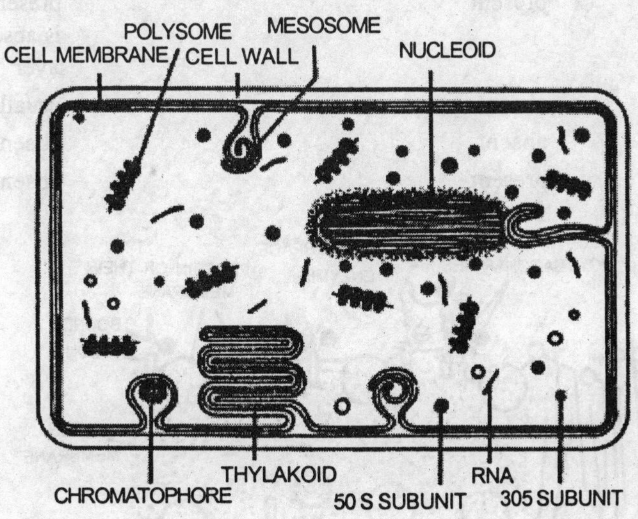

Fig 15. 3 Baacteria

Detailed cell structure (Electron micrographic view)

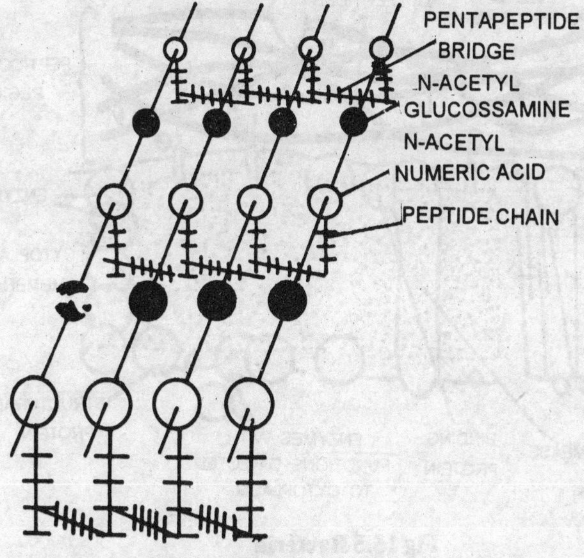

Fig 15.4 Bacteria

Chemical structure of bacterial cell

Feature	Gram positive	Gram negative
1. Thickness	thick	thin
2. Amino acids	few	several
3. Sulphur containing amino acids	absent	present
4. Teichoic acid	present	absent
5. Peptidoglycan	present	present in outer layer is absent in the inner layer
6. Lipoproteins	absent	usually present
7. Lipopolysaccharide	absent	absent
8. Polysaccharide	present	absent

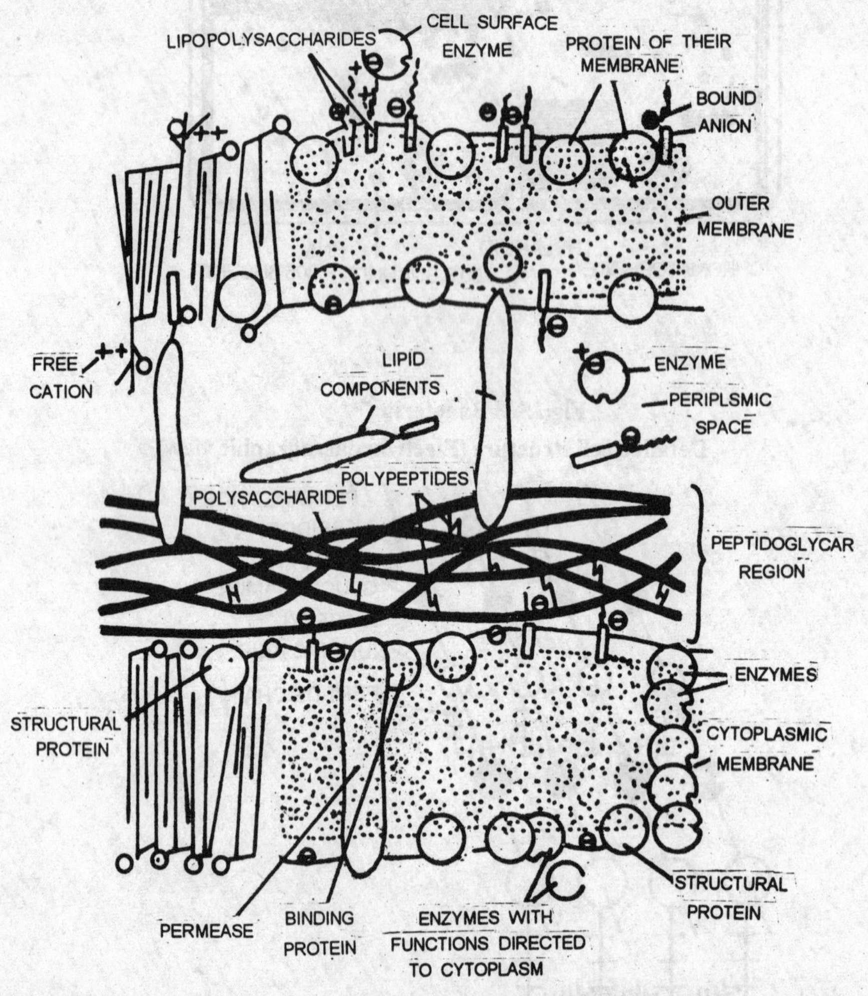

Fig 15.5 Bacteria

Diagramatic represents of wall of Gram -negative bacteria (adopted from Costerton et al 1974)

Cell membrane: Internal to the cll wall is the cell membrane or the plasma membrane. It forms the outer boundary of the cytoplasm and is selectively permeable and it thus regulates the entry and exit of molecules into cytoplasm. The membrane is chemically made up oflipoproteins (70:30 protein and lipid) and practically no carbohydrates. Electron microscopic studies (Costerton *et al* 1974) have shown that the membrane is a three layered one with a unit membrane

Structure. There are two electron dense layers encompassing a central electron transparent layer. The outer layers are about 3.5nm thick and the middle electron transparent layer is about 5nm thick. Lipids found in the membrane are phospholipids such as phosphatidyl ethanol amine. The bacterial cell membrane performs the following functions.

I .Functions as an osmotic barrier 2. Contains enzyme systems involved in biosynthesis of membrane and cell wall components.
3. Contains components of energy generation systems.

Mesosomes:

In some bacteria, particularly in gram positive bacteria, the cell membrane forms vesicles or packet like infoldings into cytoplasm at several regions. These are called mesomsomes. Mesosomes are the reservoirs of respiratory enzymes and are supposed to be analogous to mitochondria ofeukaryotic cells. The presence of more number ofmesosomes particularly during the log phase of the growth when respiratory activity is high supports the involvement ofmesosomes in respiration. In photosynthetic bacteria, mesosome number is related to pigment concentration and photosynthetic activity. In sporulating bacteria, mesosome formation is a prequisite for spore development. According to peberdy (1980), mesosomes are also invovied in cell wall synthesis and are necessary for cross wall formation during cell division.

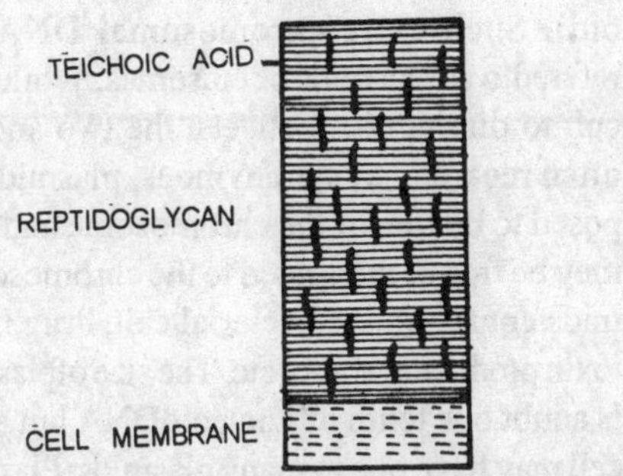

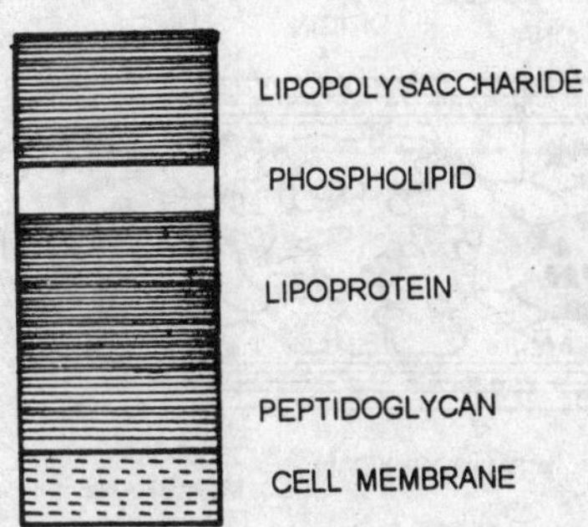

Fig 15.6 Bacteria

A comparison of cell walls of Gram -positive (on left) and Gram -negative (on right) bacteria.

Cytoplasm: Cytoplasm is a viscous substance like in eukaryotes but differs from them in not showing streaming movments. Cytoplasm includes ribosomes, nuclear material, proteins and other water soluble material and reserve food material. Organelles like endoplasmic reticulum, mitochondria, golgi bodies etc are absent. Plasmids (extra chromosomal DNA material) and episomes are also found in cytoplasm.

Ribosomes

They are dispersed throughout the cytoplasm and as in eukaryotes are invoviedin protein sythesis. The number ofribosomes per cell varies and may reach upto 15,000 per cell. Higher number ofribosomes are found during increased activity of protein synthesis.

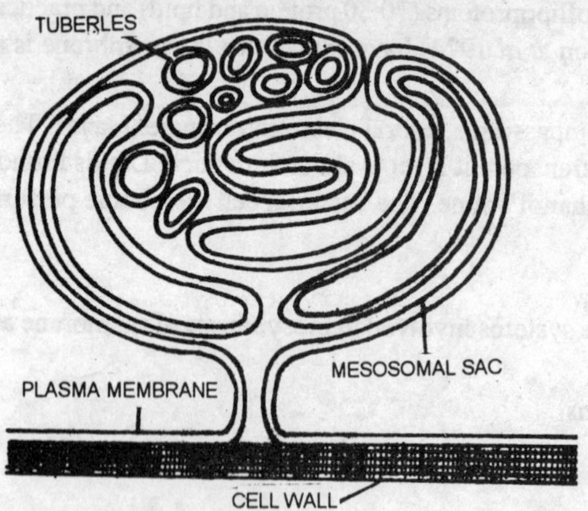

Fig 15.7 Bacteria – Diagram of a mesosome

Each ribosome is a ribonucleo protein particle (60-40 RNA: protein) of about 100-200 A° in diameter. Bacterial ribosomes are of the 70s type as noticed by their sedimentation properties. Each ribosomes is made up of two subunits 50s and 30s. Each 50s sub unit consists of one molecule of 23s r-RNA, one molecule of 5s r-RNA and 32 different kinds of proteins while the 30s sub unit has one molecule of 16s r-RNA and 21 different kinds of proteins. During protein synthesis many bacterial ribosomes are held together by a mRNA chain and are called Polyribosomes or Polysomes.

Nuclear material: Nuclear material is dispersed in the centre. It is called a nucleoid as it has no membrane or nucleolus. Electron micrographs suggest that the nucleoid consists of closely packed fibrillar DNA(Kavenoff and Bowen 1976). Studies indicate that it is a single molecule of double stranded DNA. Some recent works indicate RNA may also be associated with DNA. According to Kieppe, et al (1979), the DNA in Escherichia coli is in a supercoiled state in the centre while at the periphery it is loosely spiralled. In E.coli\he DNA has an average length of lOOOjJLm and contains 5x103 kilobase pairs with amolecular weight of 25x109 daltons. The bacterial DNA molecule is often refered to as the bacterial chromosome.

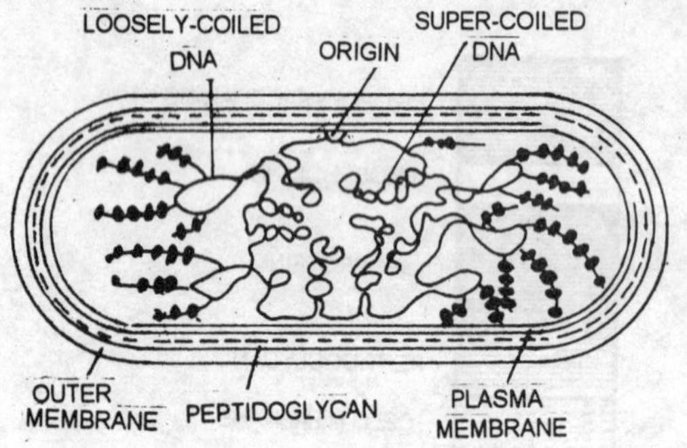

FIg 15.8 Bacteria

A diagramatic model of the organization of the Escherichia coli

Plasmids and episomes: In many of the bacteria, DNA is present outside the chromosome (nucleoid). Such extra chromosomal DNA is often referedto as plasmids or episomes. While it is difficult to dintinguish between the two and they are often regarded as synonymous, plasmids are supposed to be free of the chromosome and episomes may be free or integrated to the chromosome. Plasmids confer on the bacterial cell, drug resistance, toxin producing ability etc. The size ofplasmid DNA is aoubt one tenth of nucleoidDNA but a bacterial cell may have one to many plasmids. Plasmid DNA is also double stranded and circular and it is also not having any membranous envelop.(see later for details)

Cellular reserve materials : Avariety of reserve materials are found in the bacterial cells. Of these starch, glycogen and poly and hydroxy butyric acids are important. In some cases volutin granules are also found. Sulphur bacteria accumulate sulphur transiently during the process ofH_2S oxidation. While bacteria do not have any membrane bound organelles, gas vesicles, *chlorobium* vesicles and carboxysomes bound by non unit membrane have been described in some photosynthetic bacteria.

In the photosynthetic green bacterium *Chlorobium*, the photosynthetic apparatus has a distinct intracellular location and is bound by a series of cigar shaped vesicles arranged in a cortical layer that immediately underline the plasma membrane. These structures are about 50nm wide and 100-150nm long and are enclosed in a single layered 3-5nm thick membrane and contain photosynthetic pigments.

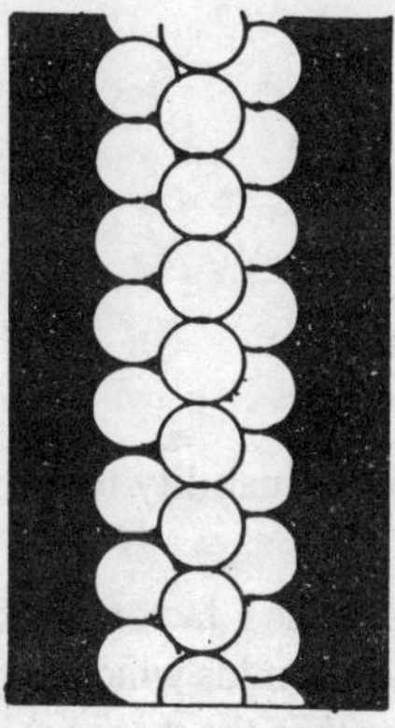

Fig 15.9 Bacteria
A model showing three helically wound strands in a bacterial flagellum

In a number of photosynthetic, chemolithotropic and purple bacteria anumber of polyhderal structures (50-500nm wide) are found surrounded by a single layered membrane of 3.5nm thickness. Theses structures called carboxysomes contain key enzymes involved in Co, fixation during carbon assimilation.

Flagella: These are long thread like appendages attached to the cells found in bacilli and spirillum. They bring about locomotion in moist medium. The flagellum is thin and about 0.02*um* in diameter hence invisible under light microscope. Each flagellum is about 3-20llmlong and terminates in a squarish tip.

Electron microscopic and X-ray diffraction studies have shown that the flagellum is made up of three parts viz., filament, hook and basal body. Filament is made up ofprotein fibres and about 13-14nm in diameter. It is outside the cell and connects to the hook at the cell surface* Hook is bent and is thicker than the filament (17nm) and it penetrates the cell wall. The basal body is a spindle shaped structure that joins the hook to the cell membrane. A short collar like structure attaches the basal body to the hook. In gram negative bacteria, the basal body has two pairs of rings. The outer pair (L and P rings) is fixed to the cell wall while the inner pair (S and M rings) is attached to the cell membrane. In gram positive bacteria the basal body is much simpler. It has only one ring

Funbriae(pili):

These are short, very fine, hair like processes found in some gram negative bacilli. Also called pili they are about 0.5lLlm in length and less than IOnm thick. They originate from the plasma membrane and their fucntion is to adhere to the cells particularly during conjugation. Some pili are longer than the rest in some cells. fSuch cells are 'male' and the pili are called sex pili.

A BRIEF ACCOUNT OF PLASMIDS, TRANSPOSONS AND DRUG RESISTANCE

Plasmids

The bacterial cell has an additional fragment ofnucleci acids in addition to its normal chromosome. This additional genetic material was given the name episome (Jacob, ShcaefferandWollmann, 1960). Episome can exist either independently or it can integrate itself with the bacterial chromosome. Plasmids however are always independent and never integrate with the bacterial chromosome. Plasmids control a number of traits in bacteria such as - toxin production, antibiotic resistance, pilus production etc. Some people regard that episomes and plasmids are synonymus. The word plasmid was first used by J.Lederberg (1952). However, the term is restricted only to bacterial cells where circular DNA molecules exist independent of the main genome.

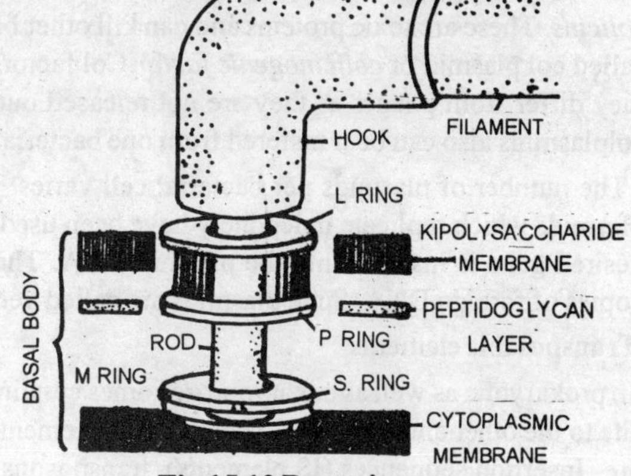

Fig 15.10 Bacteria

Interprective sketch to show the attachement of rings tocell wall and cell membrane of a Gram negative bacterial cell

sPlasmids are circular double stranded molecules of DNA and exist independently of chromosomal DNA in bacterial cells. They undergo replication during cell division and are carried to both the daughter cells. The plasmid has a right handed superhelical coil during the resting stage. There, is one super helical turn for every 400-600 base pairs. The twisted conformation is known as **covalently closed circular DNA (CCD).** Cleavage results in the conversion from CCCD to open circular form.

The following are some of the properties of the plasmids.

1. They are made up of DNA (double stranded).
2. They are capable of independent replication
3. Some plasmids control bacterial conjugation
4. Capable of reversible insertion into bacterial chromosome (only some plasmids)
5. They have the abilitytogettransferedfromonebacterialcell to the other.

Plasmids are refered to as *conjugative* and *non conjugafive* based on their ability/inability to promote their transfer during conjugation.

Classification ofplasmids:Basically plasmids are classified into three Categories1. The F factor or sex factor, 2. TheR plasmid or resistance transfer factor (RTF) and 3. Col factor. There are some plasmids which may share the properties of more than one type of plasmid. Some bacteriologists recognise more than the three types based on some other properties.

F factor. The F factor plasmid confers the ability to the bacterial chromosome to get transfered to another cell (recpient). Hence the donor is said to be F+. F factor integrates with the bacterial genome facilitating its transfer. Because of this integration, it is often refered to as an episome.

R. Plasmid. These plasmids, when present confer on the bacterial cell resistance to several antibiotics such as streptomycin, tetracycline, chloramphenicol, sulphonamid etc. R plasmid mediated antibiotic resistance was first discovered in Japan in 1956 when *Shigella* became resistant to several antibiotics at one step. This was due to the transfer of R plasmid from one strain to the other
resulting in not only antibiotic resistance but also the ability of a donor (it can donate R factor to other sensitive strains making them resistant).

R plasmid is a small extrachromosomal DNA ring. Sometimes the Rplasmid has two parts - a. Part having drug resistant genes and b. Part having plasmid transfer genes.

The col factor (plasmid). Certain bacteria like *Escherichia coli, Shigella, Salmonella* etc., produce toxins called *colicins.* These are toxic proteins and can kill other bacteria. The ability to produce *colicin* rests in the plasmid called col plasmid or *colicinogenic factor.*Col factors resemble phages since they cause death of bacteria, but they differ from phages as they are not released out during the lysis of the bacterial cell. Like the F factors colplasmids also can be transfered from one bacterial cell to the other.

The number of plasmids per bacterial cell varies. It may be *single copy* or *multiple copies*. Multiple copy plasmids which replicate indepdnely have been used in gene cloning experiments. A plasmid is cleaved and a desired gene is inserted into the plasmid DNA. The plasmid DNA during its replication produces multiple copies of foreign DNA. Such plasmids are called vectors.

Transposable elements

In prokaryotic as well as eukaryotic geriomes certain sequences ofnucleotides are capable of moving from one site to the other and are called Transposable elements. There are three kinds of *Transposable elements*. These are- Insertion sequences (IS elements), transposons (TN elements) and retroelements. Only transposons are described here.

Transposons: Many plasmid genes confering drug resistance are capable of dissociating themselves from the plasmid and transfered to another location ie,, on to a different DNA molecule. From there it can shift to another location. In otherwords a gene sequence can hop from a plasmid to a chromosome to another plasroid and also to a phage genome. Such movable particles of DNA are called transposons. Transposons are designated as Tnl, Tn2, Tn3 etc.

Transposons in prokaryotes. These were first discovered by Hedges and Jacob (1974) for the sequence ofampicillin resistant nucleotides which get transfered from one plasmid to the other thus transfering drug resistance- It was demonstrated that the two ends of the two strands of transposons had complementary nucleotides, but in reverse order. For example if one end had GTCTGGG, the other end had CCCAGAC.

Transposons in eukaryotes. Transposons are also found in eurkaryotes such as *Drosophila*, maize etc, Barbara Mcclintock hasrecevied Nobel prize for her work on transposons which she called *Jumpinggenes*.

Mechanism of transposition. Berg (1977) has proposed a mechanism for the transposition of one of the transposons (Tn5). The Tn5 element determines resistance to the antibiotic Kanamycin and can betransposed from the lambda phage to the Esherichia coli chromosome and from one chromosomalsite to the other.

Transposition consists of three stages - cleavage, realignment of ends andligation. An enzyme complex transposase recognises and binds to speci sequences at the ends ofTn5 element and to another DNA molecule (recipici The enzyme cuts the transposon at either ends and also nicks the recipient DIS After insertion the cut ends are ligated in both the termini.

Uses of Transposons. They can beused as genetic markers, because they change the pattern of cuts by restriction endosnucleases. Strains with different types ol transposons can be easily identified because of the specific drug resistance they carry. Insertion of transposons may be used to induce mutations as has beer shown inmaize, where the transposon acts as a mutator gene inducing the mutable genes. Some recent experiments have shown that transposons can also be used as transformation vectors .

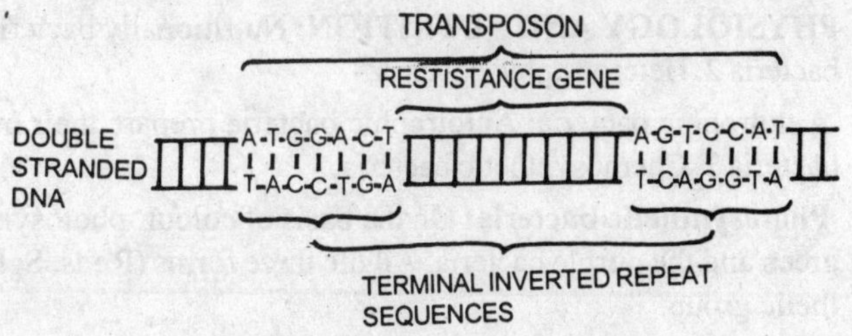

FIg 15.11 Bacteria

Diagramatic representation of a transposon

Durg resistance is a major problem in thechemotherapeutic control ol pathogenic bacteria.

It has already been pointed out that drug resistance is confered on those bacteria which posses **R** plasmids. These are found in both gram positive and gram negative bacteria **R** plasmids consist of two parts-transfer factor (T) and durg resistance factor. T factor is responsible for the independent replication and transfer of the **R** factor. Plasmids carrying **T** factor are called *conjugative R plasmids* while those R plasmids without a T factor are called *non conjugative R plasmids.*

Drug resistance gene sequence can be transfered from plasmids to the bacterial genome. Specific transposons carrying the drug resistant gene transfer them tc differentDNAmolecules.

Mechanism of drug resistance

Davies and Smith (1978) have classified, the various modes of drug resistance mechanism as follows

1. Alteration of target site. In this, the binding of the drug to the target site (in the bacteria) is reduced or removed by the alteration of site, so that the drug can no loger be effective. A specific example is that of erythromycing resistance in *Streptococci* and *Staphylococci.* A plasmid regulated enzyme, methylase causes methylation of 23s r RNA (in the bacteria). This prevents the binding of drug and confers resistance.

Drug	Target	Plasmid enzyme
Erythromycin	50's Ribosomal	
Linomycin	sub unit (235rRNA) methylated; proteins synthesis inhibited	Methylase

2. Blocking of antibiotic transport: Plasmids prevent the entry of antibiotics which are structural analogues of bacterial substrates. Tetracycline resistance is partly due to bloackade of drug transportation.

3 .Enzymatic in activation. Resistance to penicillins and cephalosporms is due to plasmid mediated inactivation chlormaphenicol resistance also is due to acetycation induced by acetyl transferases which again are plasmid regulated.

4. Metabolic change. Change of metabolic pathway that is inhibited by the antibiotic, but producing the same metablite is another method of drug resistance. Resistance to sulphonamides and trimethoprim belongs to this category. Other methods of drug resistance include - a. Saturating the drug by increasing the concentration ofinhitibed enzyme.

b. Production of inhibitor antagonistic metabolite and c. Decreasing the dependence of bacterial cell on the metabolic pathway inhibited by the drug.

PHYSIOLOGY AND NUTRITION: Nutritionally bacteria may be classified into two types. 1. Autotrophic bacteria 2. Heterotrophic bacteria.

Autotrophic bacteria: Autotrophic bacteria prepare their own food. They are of two types I. Photosynthetic bacteria 2. Chemosynthetic bacteria.

Photosynthetic bacteria: On the basis of colour, photosynthetic bacteria fall into two principal groups. The green and the purple bacteria. All the three forms (Rods, Spheres and Spirilla) are found among the photosynthetic group.

The photosynthetic bacteria (like green plants) contain chlorophyll which is chemically very much similar to the chlorophyll of higher plants.

The chlorophyll found in the purple bacteria (Bacterial chlorophyll) differs chemically from green plant chlorophyll in several ways. As a result of this, the pigment will be pale blue or gray. The colour is caused by the presence of various carotenoid pigments.

Chemosynthetic bacteria: These are pigmentless bacteria (but autotrophic) that obtain energy by the oxidation of inorganic compounds. After oxidation, the inoiganic element will be left as residue in the cytoplasm. Depending on the inorganic compound used in oxidation, bacteria may be classified into sulphur bacteria, iron bacteria, hydrogen bacteria, carbon bacteria etc.

One of the sulphur bacteria (Thiobacillus thiooxidans) oxidies elemental sulphur tosulphuricacid.

$S+H_2O+1/2\ O_2 - H_2SO_4$energy

Another group of sulphur bacteria (Beggiatoa) oxidises hydrogen sulphide to elemental sulphur

$H_2S+1/2\ O_2 - S+H_2O$ The oxidation of Iron is carried by Ferrobacilus

$Fe_2\ F^{4}++e^{-}$

the ferric iron is deposited as insoluble Ferric hydroxide.

Heterotrophic bacteria. These are pigmentless bacteria that obtain food either by parasitic mechanism or by saprophytic mechanism.

Parasitic bacteria. They obtain their nutritional requirements by parasitising other organisms both plants and animals. Parasitic bacteria may be either pathogenic or non pathogenic. The pathogenic bacteria secrete toxins which bring about the disease. Not pathogenic bacteria which live in human beings are in fact helpful. For ex. The intestinal bacteria synthesize B vitamin which could be used by human beings.

Saprophytic bacteria: Saprophytic bacteria obtain food anaerobically oxidising organic matter. Depending on the nutritional source they may be classified into I. Fermenting bacteria 2. Putrefying bacteria, Fermenting bacteria. These bacteria use carbohydrates as the respiratory substrate. They anaerobically respire and incompletely oxidise the carbohydrates resulting in the formation,of alcohols. The following is the reaction.

$C_6H_{12}O_6$ - $2C_2H_5OH+2CO_2$+energy

ethyl alcohol

Putrefying bacteria. These make use of protein or nitrogenous matter as the respiratory substrate resulting in the liberation of ammonia, marsh gas (methane) etc. The evolution of these gases could be commercially exploited.

REPRODUCTION:

Bacteria reproduce asexually as well as sexually. Asexual reproduction takes place by the following methods 1. Fission (cell division) 2. Budding 3. Endospore formation.

Cell division. generally bacterial cells do not divide mitotically as is true for all prokaryotic cells. There are tliree aspects of cell division called binary fission. These are. 1. DNA duplication 2. DNA partitioning 3. Cross wall formation. The cell division is completed by the doulbing of all the cell components and their precise distribution and partitioning between the two daughter cells. Unlike in eukaryotic cells there is no spindle formation and no resolution ofchromatin into chromosomes.

Electron microscopic studies have revealed the following details in bacterial cell division.

DNA duplication. The two strands separate and each strand then replicates a new strand.

DNA partitioning. This involves the distribution of DNA between the twodaughter cells. As the DNA is free and not condensed into chromosomes it is a complicated task to assure equal distribution of genetic material between the two daughter cells. It is believed that (Kleppe et al, 1979) the DNA molecule attaches itself at some point to the plasma membrane and after duplication, one strand swings towards one half of the cell keeping itself in the same attached position. Both the havies of cell receive one double stranded DNA each. The attachment of DNA to the plasma membrane is to assure proper distribution without entanglement.

Cross wall formation. After the DNA has been properly distributed between two halves of the cell, cross wall formation begins. The cell wall projects inwardly mid way between the two nuclear materials that have now become two. Cell wall materials are deposited between the membranes and the cell gets divided into two. It is beleived that mesosome plays an important role in the synthesis of the wall material during cross wall formation. Hydrolase enzyme is known to bring about the separation of two daugthercellsin *Bacillus subtilis and Staphylococcus facalis*(Pebrdy, 1980).

Budding. In this process, a protrusion first develops, into which cytoplasm migrates. The nucleus splits into two and one bitenters the protrusion. This is later cut off from the cell by a cross wall. This is the bud. At maturity, this separates from the parental cell and develops into a new individual.

Endospore. Many of the bacteria like *Bacillus, Clostridium* etc., form endospores at certain stages in their life ccle. Strictly endospore formation can not be called reproduction because only one spore is formed per cell. Endospores are highly resistant to environemntal conditions and may be regarded as an attempt by bacteria to withstand the adverse environmental conditions. Endospores ar so highly resistant to desiccation, chemicals, temperature and radiation that they are vialable even for centuries. Some endospores even withstand boiling temperature also for a long time.

As the spores are formed inside the parental bacterial cell, they (spores) are called endospores. On germination each endospore develops into a single bacterial cell. The spore wall assures the endospore high resistance due to the presence of compounds like diplolinic acid, calcium and a very low water content.

Shape aod position of endospore. The spores areof various shapes and occupy various positions in the cell. They may be spherical or oval in shape and in position may be terminal, sub terminal or central. They may be bulging giving a swollen appearance to the cell or non bulging conforming to the outer parameters of the cell.

Ultra structure of endospore. The outermost layer of the endospore is the exosporium which is thick and wrinkled. Chemically it is made up oflipids and proteins with a low content ofmethionine and cysteine. Internal to the exosporium are two spore coats- the outer and the inner. The spore coats may be smooth or show various types of ornamentations like ridges, grooves etc. Chemically the spore coats are made up of lipids, proteins and glycopeptides

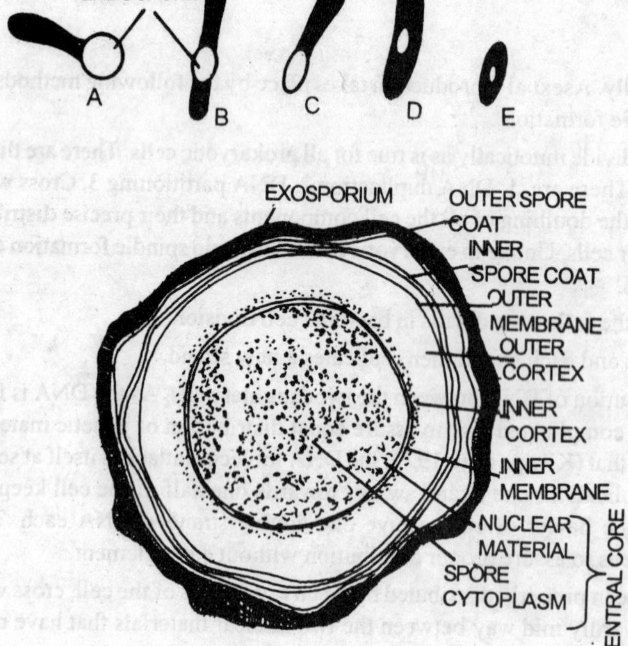

Fig 15.12 Bacterria
A -E shape and position of endospores in bacteria F. electron micrograph of T.S of an endospore Bacillus cerus

The protein of inner spore coat has a high quantity of sulphur containing amir acids. The spore coats surround a space called cortex which can be divided inl outer cortex and inner cortex. Chemically cortex consists of dipicolinic ach peptidoglycan and calcium ions. The cortex is internally lined by the spore wa internal to which lies the cell membrane. Chemically the membrane consists (peptidoglycans but no teichoic acid. The membrane surrounds the central coi of the endospore. The core consists of nuclear material and cytoplasm. Th region has t)NA, RNA and proteins.

Sexual reproduction

Sexual reproduction as shown by genetic recombination has been demonstrated in bacteria. These are three types. Genetic Recombination,Genetictration and Transduction.

Genetic recombination- Lederberg and Tatum (1944) demonstrated this in experimenting with the biochemical mutants of *Escherichia coli.*

These bacteria require the following four nutrients 1. Biotin 2. Methionine 3. Leucine 4. Threonine. But they can be grown on a minimal medium (a medium containing only agarand carbohydrates) because, by using carbohydrates they can synthesize all the above four nutrients. They are called Taratrophs' and can begenotypically designated as follows:

$B^+M^+ L^+T^+$

In this, the letter stands for the nutrient (B=biotin etc) and the sign + stands for the capacity to synthesize that particular nutrient.

By subculturing in the laboratory the paratrophs undergo mutation and lose the capacity to synthesize one or more nutrients. The mutants are cal\edAuxfftrvphs. By careful isolation, Lederberg and Tatum obtained two complementary polyauxotorphic varieteis ofE.coli. They can be designated as follows:

1. $B^+M^+L^+T^+$

2.$B^-M^-L^+T^+$

(The sign-stands for the loss of the synthesizing capacity. For ex:L'means leucine cannot be synthesized and hence it has to be added to the culture.). It is obvious that both strains I aod 2 cannot be cultured on a minimal medium. Stran I cannot be cultured beccause it cannot synthesize leucine and theronine. Strain 2 be cultured because it cannot synthesize biotin and methionine. But when both of them are kept on a minimal medium curiously they survive (normal expectation is, they should not live). On nutritional analysis, it was revealed that they were Paratrophs. The formation of paratrophs from two polyauxotrophic varieties can be explained only on the possibility of sexual recombination between strain I and 2.

$B^+M^+L^+T^+$ $B^-M^-L^-T^-$

1 2

Because of recombination of I and 2 the following combinations may be obtained.

$B^-M^-L^-T^-$ and $B^+M^+L^+T^+$

(a) (b)

Of the above (a) will not survive because it has lost the capacity to synthesize all the nutrients but (b) will survive because it has become a paratroph.

Generic transformation. Every individual born out of Sexual reproduction will have the mixture of traits of both the parents. In genetic terms, this means that there is genetic recombination in the offsping. Genetic recombination or transformation which is an index of genetic change has also been reported in bacteria.

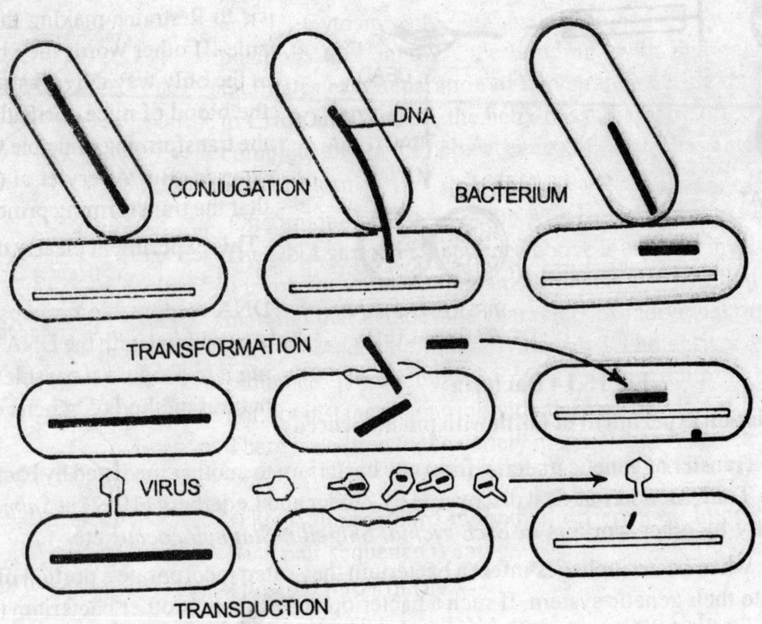

Fig 15.13 Bacteria

Genetic recombinatiom in bacteria

Genetic transformation ie., change of one genetic type to other was first reported by British bacteriologist F.Griffith (1928) while working with pneumococci (Diplococcus pneumoniae).

There are two strains in *Diplococcus penumoniae.* One is capsulated and the other one is non capsulated ones form glistening smooth (S) colonies on the culture medium while the non capsuleatedones form dull, rought(R) colonies. Further, the capsulated (S) strains are virulent while the non capsulated (R) strains are non virulent ie., do not produce toxins. The ability of producing or not producing a capsule is a genetic character.

Griffith injected experimental mice with R strains of pneumococci and found that they survived but whenS strains are injected the, mice died. When heat killed (killing the bacteria by applying heat when DNA gets denatured and will not produce toxin) 'S' strains were injected the mice survived. Griffith next injected both R strain and heat killed 'S' strain to the same experimental mice and they (mice) died. Blood analysis of the dead mice (which were infected withR strain and heat killed S strain) showed the presence of live* S strains

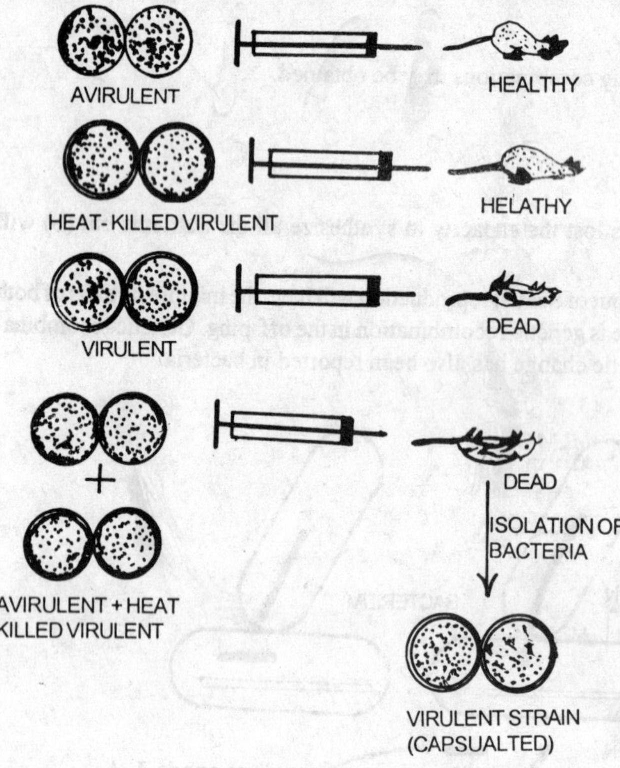

Fig 15.14 Bacteria

The transformation experiment of Gifth with pneumococcus

There are three possibilities how the heat killed 'S' strains can become live again.

I • Heat killed S strain might have become alive in the body of mice- this is an impossibility.

2- A mutation might have occured in the R strain converting them to S strain. This is also not possible because mutations wiB not occur frequently, but everytime the experiment showed the presence of capsulated S strains in the blood of the dead mice.

3. Some factor in the heat killed strain has transferred the genetic formation for capsule synthesis to R strains making them produce the capsule-.If other words they become S strain. This is the only way a live S strain could be found in the blood of mice. Griffith (1928) thought that the transforming principle was bacterial polysaccharide. But Avery et al (1944) demonstrated that the transforming principle was DNA.

This experiment clearly demonstrates that a piece of DNA responsible for capsule syn thesis recombined with the DNA of the R strain inducing it to produce a capsule. This can be called a natural method of genetic engineering.

Transduction. Transfer of genetic material from one bacterium to another mediated by bacteriophages is known as transduction. Transduction was first discovered by Zinder and Lederberg (1952) in *Salmonella typhinmurium* and subsequently by other workers in *Escherichia, Shigella, Staphylococcus* etc.

In transduction, when bacteriophages infect a bacterium they often incorporate a portion of the genetic material of bacterium into their genetic system. If such a bacteriophage infects another bacterium (of the same species) the bacterial genes (of the first infection) may get incorporated into these bacteria. These bacteria now exhibit some of the traits of the first strain. This shows that bacteriophates have picked up some genes from the first bacterium and transfered it on to the second bacterium.

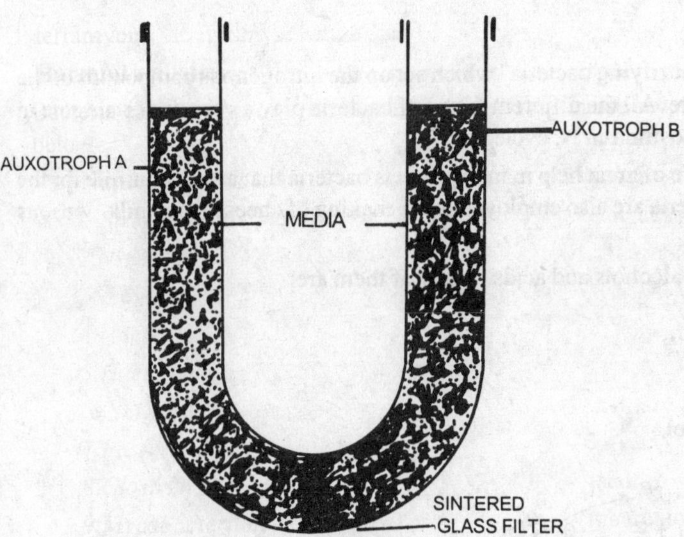

Fig 15.15 Bacteria

The transduction (U tube) experiment of Zinder and Lederberg

In order to prove bacteriophage mediated genetic transfer. Zinder and Lederberg performed a 'U' tube experiment. The experimental set upconsists of a 'U' tube where the two arms are separated at the base by means of a sintered glass filter (Bacterial filter). Zinder and Lederberg (1952) grew two different strains of bacteria *Salmonella typhimurium* in the two arms. One strain was infected with bacteriophates. A little later when the bacteria in both the arms of the 'U' tube were analysed it was noticed that the strins in both the arms had become recombinants. Since the bacterial filter does not allow the migration of bacteria, the only way for transfer of genetic material was through bacteriophages since they can move across bacterial filters.

ECONOMIC IMPORTANCE

From the point of view of human welfare, bacteria are doubly important in that they are friends as well as foes of mankind. Hence a discussion can be conveniently classified into beneficial activites and harmful activities.

Beneficial activities:

7. Disposal of organic matter. Sparophytic bacteria are the most efficient natural scavengers that always maintain the surface of the earth clean. They prevent the accumulation of dead bodies by efficiently decomposing them and returning them to their element stage.

2. Role of bacteria in agriculture. Bacteria help in many ways in enriching the soil nitrogenous substances. While no higher plant can absorb gaseous nitrogen and use it as a source of its nitrogen requirement bacteria can do so. Such bacteria are called "Nitrogen fixing bacteria".

A. Nitrogen fixing bacteria. These are bacteria which absorb gaseous N2 from atmosphere and use if for all their nitrogen requirements. These are of two types:

1. Autotrophic N, fixing bacteria. These live in the soil (Clostridium and Azatobacter) and absorb Nz from air. When these bacteria die, nitrogenous compounds in their body are released to the soil to be used by other organisms.

2. Symbiotic Nt fixing bacteria. (EX. Rhizobium radicicola). These live in the root nodules of leguminous plants in symbiotic association. Originally, these bacteria live in soil. Later they gain entrance into the root of leguminous plants through the root haris and ultimately come to the root cortex where the bacteria multiply in large numbers. Due to this, hypertrophy is formed in the roots resulting in the formation of nodules. The association is symbiotic in that leguminous plants provide food and shelter for bacteria. Inturn bacteria provide the plant with N,. It is for this reason that leguminous plants are rich inproteins. It is for this reason again that leguminous plants are employed in crop rotation (crop rotation is an agricultural practice, in which a legume and a cereal are alternatively grown in the tied. If the same crop is grown year by year, the yield becomes less as the soil becomes deficient in particular nutrients).

B. Nitrifying bacteria. These do not add any extra nitrogen to the soil but convert one of nitrogen compound to the other. For ex. Nitrosomons and Nitrococcus can convert ammonia to nitrate and Nitrobacter convertes nitrate to nitirite. In addition to these, there are ammonifying bacteria which decompose dead bodies and release ammonia.

There are other types of bacteria called "denitrifying bacteria" which act on the nitrogenous compounds of the soil and release gaseous N2 to the atmosphere. All the different forms of bacteria play a very important role in the N, balance of nature by occupying key position in N, cycle.

3. Role of bacteria in industries. Bacteria are of great help in industry. It is bacteria that are responsible for the souring of milk and formation of curds. Bacteria are also employed in the making of cheese from milk. Various types of

bacteria are employed in the manufacture of alcohols and acids. A few of them are:

a. *Lactobacillus acidus* - lactic acid

b. *Acetobacter aceti* - acetic acid

c. *Propionobacterium* - propionic alcohol

d. *Clostridium acetobutylicum* - butyl alcohol .

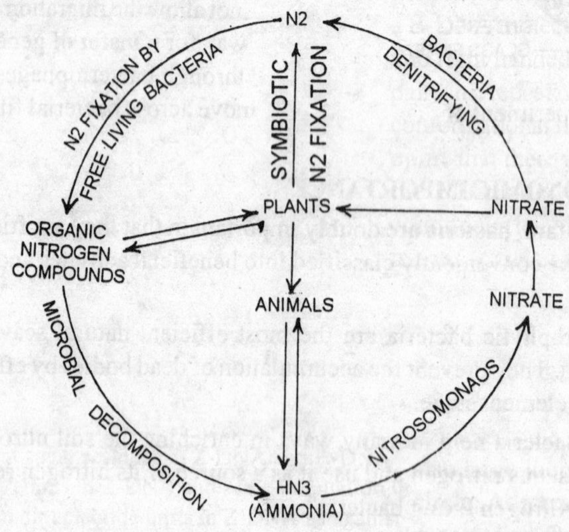

Fig 15.16 Bacteria

The nitrogen cycle

Bacteria are also of great help in curing of coffee, tea and tobacco and in the ning of leather and in the retting of fibres.

The familiar aroma of coffee tea and tobacco will not be there in the raw products. In curing works bacteria eat away the unwanted compounds, producing characteristic flavour. Specific varieties of bacteria are maintained for this purpose and the secret is guarded with utmost care.

In tanneries also, the raw skin is immersed in water containing bacteria. Bacteria eat away the fatty compounds and make the skin soft.

In retting of fibres also, the plant parts with fibres are immersed in water containing bacteria, whereupon they eat away the soft tissues and separate the fibres.

4. Medicinal uses. Bacteria belonging to the order Actinomycetales yield valuable antibiotics.

Streptomycin - Streptomyces griseus

chloromycetin - Streptomyces venezulae

terramycin - Streptomyces rimosus

Harmful activities:

Bacteria can play havoc with human welfare. They cause several deadly diseases. A few examples are given below :

Name of the bacteria	Disease
I .*Streptococcus pygoens*	- pyorrhea (dental decay)
2.*Nisseria gonorrhea*	- gonorrhea
3 .*Salmonella typhosa*	- typhoid
4.*Haemophilus influenZae*	- influenza
5.*Haemophilus pertusis*	- whooping cough
6.*Shigella dysenteriae*	- bacillary dysentry
7.*Pasturella pestis*	- plague
8.*Clostridicrm tetani*	- tetanus
9.*Mycobacterium tuberculosis*	- tuberculosis
9.*Mycobacterium leprae*	-leprosy
11.*Cor~nebacten'um dyphthenae*	- dyphtheria
12.*Vibrio cholerare*	- cholera
13.*Treponemapallidum*	-syphilis
14.*Leptospird icetrohhaemorhageae*	- meningitis
15.*Clostridium botulinum*	- food posionning of canned foods

Bacteria cause disease (to cattle) known as the Anthrax disease (*Bacillus anthracis*).

Denitrifying bacteria bring about loss of fertility of soil by releasing N, to the atmosphere.

Bacterial plant disease:

Bacteria have also not spared the plants. Bacterial plant diseases can be classified as follows.

1. Parenchyma Diseases: These are caused by bacteria which eat away the soft parenchyma tissue of plants. For eg.

A. Angular leaf-spot disease of cotton caused by Xanthomonas malvacearum b. Citrus canker caused by Pseudomonas citri,

2. Vascular disease or bacteriotracheoses. These are caused by bacteria accumulating in the vascular tissues so that the lumen is clogged and the aerial parts of the plant will die due to lack of nutrient supply for eg:Bacterial wilt of Tobacco caused by Xanthomonas solanacearum.

3. Hypertrophies. Due to the infection of bacteria hypertrophies or galls are formed on the plants. For eg. Agrobacterium tumefaciens causes crown gall disease of apples.

BACTERIAL CLASSIFICATION:

Living organisms are classified into various groups based on resemblances and differences. Taxonomic hierarchies are erected starting with kingdom and ending with species. A species is said to be the lowest stable unit in taxonomic classification. With microorganisms however a species can not have the same rigidity and meaning as in higher organisms. Species are not tight genetic units because very frequently a number of mutations occur in microorganisms and all cannot become new species. Further, extrachromosomal genetic elements initiate chromosonal rearrangements. With all these difficulties however species are still identified and groups are erected. It should be understood however that bacterial groups or orders, families etc are not phylogenetic in the senese they represent discontinuous groups. As genetic changes are quite frequent, phenotypic similarly in many unrelated traits in the basis for categorizing a bacterial species. Bacterial classification is varied in that it is based on several character sources. A few of the important approaches to bacterial classification are described below.

Unit of classification in bacteria:

Like in all other living beings bacteria are also named according to the binomial nomenclature. The first name is the generic name and the second name is the specific name. The following are the units of Classification.

Division

class

order

family

tribe

genus

section

series

species

Three terms are commonly employed in microbiological studies with reference to microorganisms-strain, clone and type species. A strain is a culture of a given species while a clone is a strain derived from a single cell. A type species is a standard one that has been studied in detail and is used for comparison with the unknown. The name of type species usually denotes some special traits of the group.

Artificial classification. This represents the earliest attempt towards classifying living beings* Living beings were classified into plants and animals microorganisms were included under both paint and animal kingdom.

Phylogenetic classification. This is a post Darwinian concept in which organisms are classified according to their phylogenetic relationship. Haeckel, a German biologist placed all the microorganisms under a separate group Protista

as different from plants and animals. R.H. whittaker (1969) developed a phylogenetic five kingdom classification of living beings. Monera, Protista, Fungi, Plantae and Animalia are the five kingdoms identified by Whittaker (1969). Microorganisms are included in the first three kingdoms while all prokaryotes are included under Monera.

Morphological classification. Based on the cell shape, size etc bacteria. Are divided into

a. Higher bacteria - filamentous forms

b. Lower bacteria- unicellular forms

1. Bacillus - rod shaped

2. Spirillum - spiral form

3. Cocci - spherical 4.

4. Comma - comma shaped

5.Cocci can be of the following types

a. *Diplococcus* - cells in groups of two

b. *Tetracoccus* - cells in groups of four

c. *Streptococcus* - cells in chain

d. *Staphylococcus* - cells in sheets

Bacteria may also bse calssified based on various criteria such as staining reaction, nutrition, base composition, nucleic acid hybridization, biostatistical principles etc.

Bergey's manual of bacteriology is the most authentic work on the classification and nomenclature of bacteria.

16

VIRUSES

The inhabitants of microcosm have a profound influence on human life. They are both friends as well as foes of man. While many have a prominent role to play in human welfare, others have caused immense harm to human life. Among the most important of microbial enemies of human beings, viruses occupy a prominent place. Simple and tiny enough to pass through the minutest bacterial filters, yet potent enough to change the destiny of human life, viruses are today the centre of attention in causing one of the deadliest of human diseases. What are these viruses? We will try to find answers to some of these in this chapter.

Discovery of Viruses

Adolf Meyer (1886), a dutch Agricultural Chemist, observed mottling disease in the leaves of tobacco plant and named it *Mosaikkranket* (Mosaic). He was able to demonstrate that the mosaic disease could spread from plant to plant if the juice of the infected leaves is applied to healthy leaves. He also showed that the juice even after passing through double filter paper retained the infectivity. He further demonstrated that heating the juice to 80°C would render it ineffective. He, therefore, concluded that the bacteria are responsible for the disease.

A further step in finding out the causative agent of tobacco mosaic disease was taken when Iwanowsky (1892), a Russian scientist, showed that the juice even after passing through the fine bacterial filter (chamberland filter candles), retained the capacity of infection. This means that bacteria are not responsible for the disease. But when the filtered sap was put on the cultured medium no living organism grew. The filterable agent was called a Virus. The term virus was, however first introduced by Louis Pasteur to indicate the cause of canine rabies. But it was a general term used for various infectious agents.

Beijerinck (1898), a Dutch microbiologist, showed that the infective agent could diffuse into agar gel, like a liquid and he called it *Ucontagium vivumfluidum* (contagious fluid).

Loeffler and Frosch (1898), showed that infective agent of foot and mouth disease in cattle (like tobacco mosaic agent), passes through bacterial filter, and neither can it be seen under the microscope not can it be cultured on the culture media.

Viruses attacking bacteria were first discovered by a British scientist, Twort, and later in more detail independently by a French scientist 'd' Herelle, who named them bacteriophage. Subsequent studies showed the existence of hundreds of other viruses in plants, animals and human beings.

Schelsinger (1933), was the first to study the chemical composition of a virus. He showed that the bacteriophage consisted of only nucleic acids and proteins. A few years later, Stanley (1935), isolated tobacco mosaic virus in paracrystalline form. Later, Bawden and Pirie, (1938), identified the chemical nature of TMV and other viruses.

Gierr and Schramm (1956), showed that the nucleic acid in viruses is the actual infective agent. Viroids and satellite viruses were discovered by Kassanis (1966), and Dicner and Raymer (1967), respectively.

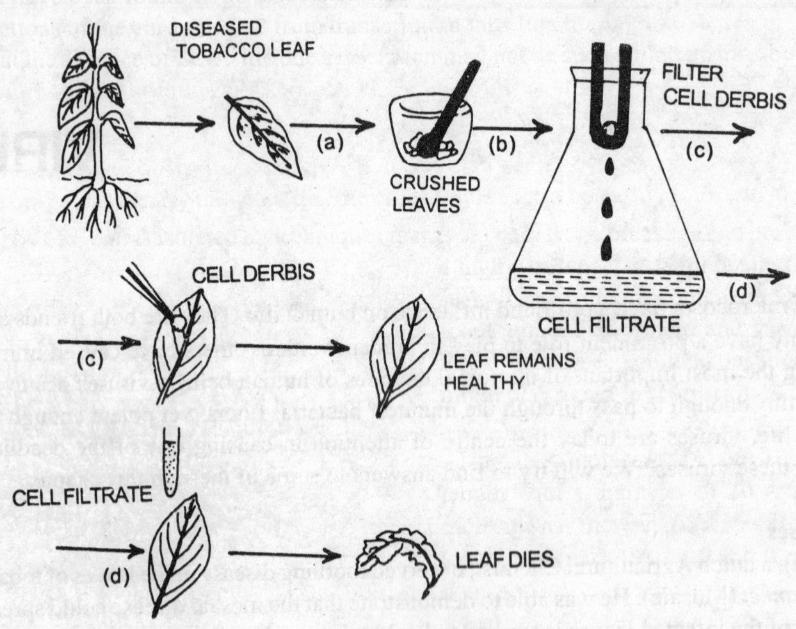

Fig. 16.1 Viruses

(a) Iwanowski's discovery of the filterable virus. Infected tobacco leaves were crushed and filtered through Chamberland's bacterial filter (b). The cell debris applied to healthy leaves without effect (c). When the clear filtreate was placed on the leaves (d). The leaves shriveled and died. This experiment showed the presence of a "filterable virus" that was smaller than any known bacteria

Some important milestones in the history of the study of viruses (Virology) arc as follows:-

Year	Event	Name of the Scientist
1880	Discovery of rabies	Louis Pasteur
1886	Juice of tobacco mosaic is infectious	Adolf Mayer
1892	Infectious agent of Tobacco mosaic can pass through bacterial filter	Iwanowsky
1896	Concept of Contagium vivum fluidum proposed	Beijernick
1898	Foot and mouth disease in animals	F. Loeffler and A. Frosch
1915	First viral infection of bacterial discovered	F.W. Twort
1917	The term 'bacteriophage' coined for the first time	d' Herelle
1935	Virus isolated in crystalline form for the first time	W.M. Stanley
1938	Chemical nature of TMV discovered	F.C. Bawden and N.W. Pirie.

Year	Event	Name of the Scientist
1939	Viral mutations discovered	M. Delbruck
1946	Electron microscope study of plant virus	W.M. Stanley
1949	Cultivation of Polio virus in tissue culture medium	J. Enders
1951	Discovery of cyanophages	R.S. Shafferman and M.E. Morris
1957	First successful polio vaccination	J. Lindermann
1959	Ultrastructure of Ts bacteriophage	S.G. Brenner
1967	Discovery of Viroids	Dinner and Raymer
1973	Discovery of Viral cancer in primates	D. Schidolvski
1976	DNA Plant Viruses study	R.J. Sheperd
1979	Replication in vaccinia Virus	L.A. Gaurina and J.R. Kates

Are Viruses living non living?

Viruses being the simplest and most primitive, it is difficult to assign to them any rank in the living kingdom. It is also not easy to define the features of viruses within the accepted frame work of living beings. Hence, it is better to regard them as an intermediate stage between living and non living. Some of the characters of living as well as non living, exhibited by viruses are as follows:-

Characters of living beings

1. Viruses have genetic material (DNA or RNA).
2. They mutate.
3. They can grow.
4. They can be transmitted form one host to another.
5. They are capable of multiplication within a host.
6. They react to heat, radiation and chemicals.
7. They show irritability.
8. They bring about enzymatic changes in *vitro*.
9. They are able to infect and cause disease to living beings.
10. The DNA and proteins of viruses are similar in composition and structure to those of higher organisms.

Characters of non living

1. They can be crystallized like an ordinary chemical and stored in a bottle or test tube indefinitely.
2. Outside the host, viruses are inert.
3. There is no cell wall, membrane or cytoplasm.
4. There are no cell organelles, and there is no metabolism.
5. Sedimentation of viruses is according to their molecular weight like non living beings.
6. They do not have functional autonomy i.e., they are not capable of any function unless they obtain metabolic products from others.
7. Energy producing enzyme system is absent.

Some unique characters of viruses

The following characters are unique to viruses and are not seen either in living or non living.

1. Presence of only D.N.A. or R.N.A.
2. Capacity to reproduce from the sole nucleic acid.
3. They do not show cell division.

4. They use the metabolic machinery of the host cell to replicate.

In view of the above, as many virologists believe, it is better to regard that "viruses are chemicals in a test tube, but living beings inside the host"

Properties of Viruses

1. Viruses are called acellular as they do no have cellular organization like other microorganisms.
2. They are ultra microscopic (invisible under ordinary microscope).
3. The genetic material in viruses is either DNA or RNA but never both.
4. Viruses are obligate parasites. They can not be cultured on inanimate media.
5. Viruses can be cultured on cell media (bacteria, cells of chick embryo etc.,).
6. Viruses are filterable; they can pass through bacterial filters. They can, however, be filtered on molecular filters.
7. Even though viruses are made up of nucleoproteins, they lack the enzyme necessary for their synthesis. For this, they (Viruses) depend on the host enzymes.
8. Viruses do not show cell division.
9. They can be crystallized.
10. Viruses can be transmitted from one host to another, either directly or through vectors.
11. They can be transmitted from one host to another either directly or through fectors.
12. The capside (outer coat) of viruses is mostly made up of proteins except in some animal viruses, where polysaccharides are also present.

Classification of Viruses

A variety of criteria are used in classifying the viruses. Earlier, viruses were classified into several categories based on their hosts.

1. Plant Viruses – Infect plants
2. Animal Viruses – Infect animals
3. Bacteriophages – Infect bacteria
4. Cyanophages – Infect cyanobacteria (blue green algae).
5. Mycoviruses – Infect fungi
6. Mycoplasma viruses – Infect mycoplasma
7. Phycoviruses – Infect algae.

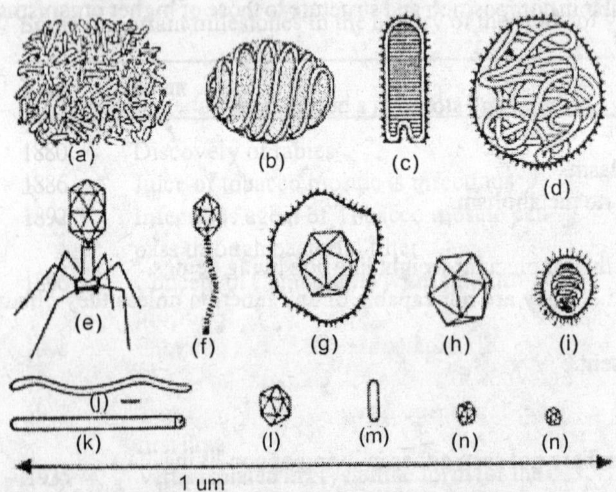

Fig 16.2 Viruses

Diagram of virus morphology and size range (a). Poxvirus (vaccinia) (b). Poxvirus (orf). (c). Rhabdo virus, (d). Paramyxovirus(Mumpsvirs), (e) T-even phage, (f). Flexous taialed phase

Modern system of classification of viruses proposed by Lowaff (1966), and accepted by the provisional committee on nomenclature of viruses (PCNV), is based on the following criteria.

(i) Type of Nucleic acid – DNA *or* RNA.

(ii) Symmetry of the capsule – Helical, cubical and complex.

(iii) Presence or absence of envelope around the capsule.

(iv) Diameter of the helical capsule.
(v) Number of capsomeres (Units) in cubical viruses.

The following is an outline of the classification of viruses as proposed by Lowaff, Home and Toumier (1966).

This system is known as the LHT (Lowaff, Home and Toumier) system based on the name of the scientists who proposed the classification.
Phylum ***VIRA*** (includes all viruses) divided into two sub Phyla.

Sub phylum *Deoxyvira* (DNA viruses)
Sub phylum *Ribovira* (RNA viruses)
Sub phylum *Deoxyvira is* divided into three Classes–

Class 1. Deoxyhelica : Includes viruses with helical symmetry. It has only one order, Chitovirales (enveloped), and one family Poxviridae.

Class 2 *Deoxycubica* : Includes viruses with cubical symmetry. It has two orders *Haplovirales* (no envelope) and *Peplovirales* (enveloped). Haplovirales as five families – *Microviridae. Parvoviridae. Paploviridae, Adenoviridae* and *Iridoviridae.* Order peplovirales has only one family, *Herpesviridae.*

Class 3 *Deoxy binala* : Viruses have complex (Binal) symmetry. There is only one order *Urovirales* including a single family *Phagoviridae* (bacteriophages).

Sub phylum Ribovira is divided into two Classes Viz., *Ribohelica* and *Ribocubica.* Ribohelica includes viruses with helical symmetry, while ribocubica includes viruses with cuboid symmetry.

The class Ribohelica has two orders -- Rhabdovirales andSagovirales. Rhabdovirales do not have an envelope and is further divided into two sub orders -- Rigidovirales (rigid viruses) with three families (Dolichoviridae, Protoviridae and pachyviridae), and Flexivirales (Flexible viruses), with three families (Leptoviridae, Mesoviridae and Adroviridae).

The order Sagovirales has enveloped viruses, and is divided into three families viz., Myxoviridae, Paramyxoviridae and Stomatoviridae.

The class Ribocubica has two orders Gymnovirales (viruses without envelop) with two families (Napoviridae and Reoviridae), and Togavirales with a single family Arboviridae.

PHYLUM VIRA

Subphylum Deoxy vira
- Class : Deoxyhelica
  - Order : Chetovirales
- Class : Deoxycubica
  - Order : Haplovirales
    Peplovirales
- Class : Deoxybinala
  - Order : Urovirales

Subphylum Ribovira
- Class : Ribohelica
  - Order : Rhabdovirales
    Sagovirales
- Class : Ribocubica

Order : Gymnovirales
Tugavirales

In another system of classification proposed by Casjens and King (1975), viruses are classified based on – type of nucleic acid, presence or absence of an envelop and site of assembly of virus particles in the host (whether in nucleus or cytoplasm). They have classified viruses into four divisions ssRNA viruses, dsDNA viruses and ssDNA viruses (For more details see Casjens and King J., 1975 - Annual Review of Biochemistry 44:555 - 611).

Structure and chemical composition

Viruses are not cellular, hence they do not have any cell wall, membrane and cytoplasm. They are extremely small and smaller than bacteria. The technical name for a virus particle is Virion. Virions vary in size. The smallest virus (causing foot and mouth disease) has a diameter of IOnm. Some of the large viruses may reach the size of small bacterium. For e.g., Pox virus is about 250nm. Lympho granuloma virus is 300 - 400nm in size.

Viruses vary in shape also. They may be rod shaped, bullet shaped, oval, irregular or pleomorphic. Based on shape, vinises may be categorized into two groups – polyhedral forms (Adeno virus), and helical forms (Tobacco mosaic virus).

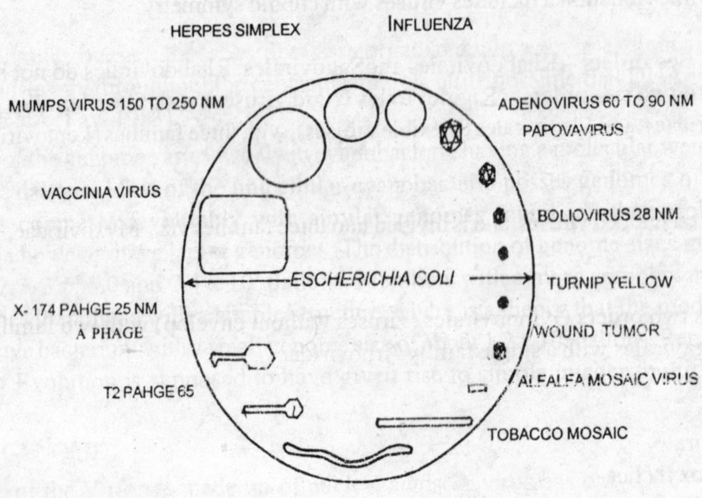

Fig. 16.3 Viruses

Diagrammatic comparison of size of viruses and related individuals. The largest circle, represents the diameter of E.coli, about 1.0μm. The other organisms are drawn approximately to the same scale

Structurally each virion is very simple. It consists of a nucleic acid core surrounded by a protein coat (capsid), to form the nucleocapsid. The capsid is made up of many independent units called Capsomeres. The nucleocapsid may be naked or as in some cases surrounded by a loose membranous envelop. The capsid offers protection to the nucleic acids against the action of nucleases.

Shape of capsid The Capsid mainly exhibits three kinds of symmetry or shape. These are– polyhedral, helical and complex (binal). Polyhedral and helical viruses may or may not have envelopes.

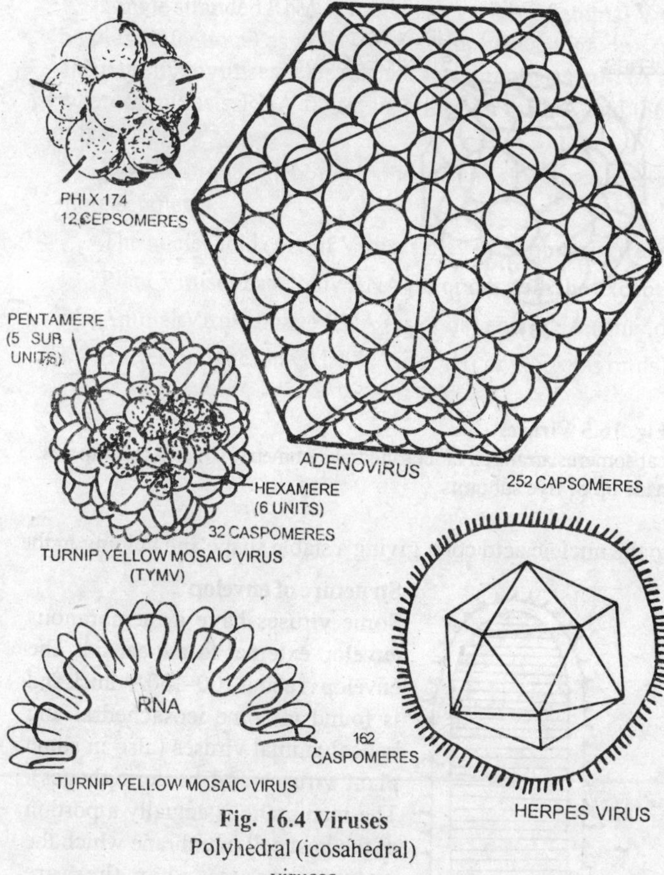

Fig. 16.4 Viruses
Polyhedral (icosahedral) viruses

Polyhedral capsids are also called icosahedral and have tetrahedral, octa hedral or Polyhedral (more than eight) corners or vertices and 20 facets or sides. The shape of each facet is like an equilateral triangle. As has already been mentioned, each capsid is made up of a number of capsomeres and each capsomeres has a number of monomeres which form polygonal rings with a central space of up to 40A.

The capsomeres are of two types – pentameres and hexameres. The pentamere (pentagonal capsomere), is made up of five monomeres while hexamere has six monomeres. The monomeres are held together by bonds (e.g. Turnip yellow virus).

The helical capsids have monomere arranged in the form of a helix around a single rotational axis. The monomeres curve into a helix because they are thicker at one end than the other e.g. TMV, Influenza virus etc.

The complex capsids are divided into two categories. Those without identifiable capsids and those with capsids to which additional structures are attached. In vaccinia virus, there is no identifiable capsid.

Bacteriophages of T even series have capsids with additional attachments. Usually the capsid has two regions – head and tail. The head capsid is a icosahedron and may be made up of 2000 identical subunits. The tail is connected to the head through a collar. At the free end, the tail has a base plate with 6 corners. At each corner is a spike. In addition, from each corner of the hexagonal plate a long tail fibre is given out.

Chemically the capsid is made up of proteins. Each capsomere in TMV is made up of 158 amino acids.

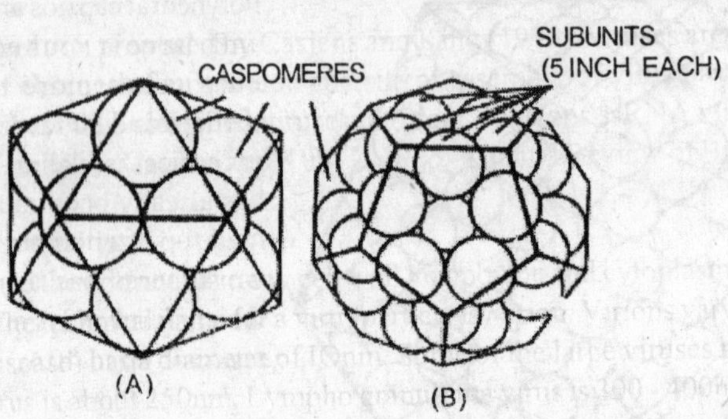

Fig. 16.5 Viruses

A cubic virus fX174 (a). A particle consists of 12 capsomeres arranged in icosahedral symmetry, (b). Six capsomeres, each made up of five subunits

Capsid helps the virus – in offering protection to the nucleic acid core, giving a stable shape and helping in the initial attachment of the virus to the host cell.

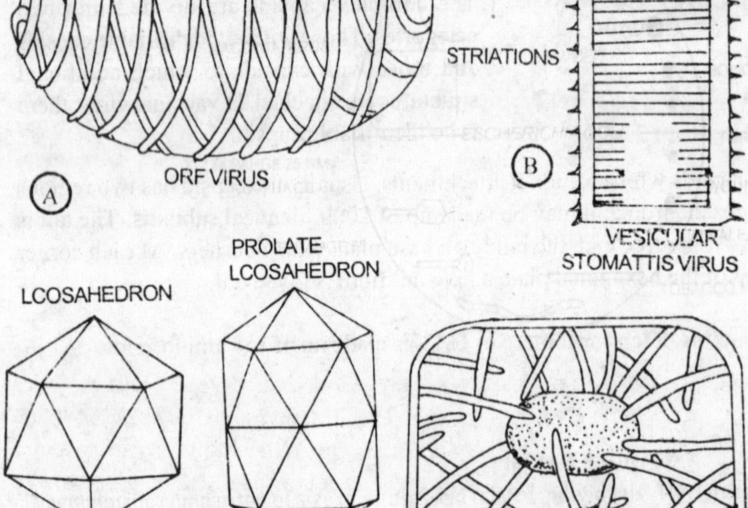

Fig.16.6 Viruses

Complex viruses

Structure of envelop

Some viruses have a membranous envelop external to the capsid. The envelop is about 100–150A^0 thick and is found in some icosachedral and helical animal viruses (also in some plant viruses and bacteriophages). The membrane is actually a portion of the host cell membrane which the viruses acquire when they are released from the host cell after multiplication. The memberane has the typical bilayer structure with phospholipids and embedded proteins. In addition to the above, membranes may also have a significant amount of carbohydrates like galactose, mannose, glucosamine and galactosamine etc.

Nucleic acids

The core of the virion is made up nucleic acids. A virus has only DNA or RNA, never both together. Within these parameters, however, considerable diversity exists. The nucleic acids may be double stranded, single stranded, linear or circular. Some have plus polarity; other may have minus polarity. Four types of nucleic acids are found in viruses with reference to the number of strands. These are:

(i) Single stranded DNA'(ssDNA) E.g. Colliphage virus
(ii) Double stranded DNA'(dsDNA) E.g. Herpes virus
(iii) Single stranded RNA-(ssRNA) E.g. TMV
(iv) Double stranded RNA'(dsRNA) E.g. Reovirus

Single stranded DNA is found in ϕX 174 virus. It was first discovered by Sinsheimer et al, in 1950's. The ssDNA may be linear (paroviruses), or circular (ϕX174 virus). The ssDNA becomes double stranded during replication and at that time it is known as the replicative form.

Double stranded DNA is found in a number of animal viruses and bacteriophages. It has a variety of forms – linear (bacteriophages), cross linked (vaccinia virus) or closed circular duplex as in papova viruses.

Single stranded RNA is found in a variety of animal viruses and icosahedral plant viruses. The strand may be plus (infectious), as in RNA bacteriophages, togavinises etc., or minus (non infectious), as in rhabdoviruses and paramyxoviruses. Plus ssRNA acts directly as mRNA and translates proteins on the ribosomes of bacteria, whereas minus ssRNA first transcribes an mRNA and then it (mRNA) translates proteins (hence called non infectious).

Double stranded RNA is found in animal viruses like reovirus, blue tongue virus etc. The ssRNA has 10 or more segments.

The nucleic acid core of viruses may be summarised as follows

(1) Plant viruses have only RNA (ss or ds) with the exception of cauliflower mosaic viruses (DNA virus).
(2) Animal viruses have RNA (ss or ds) and DNA, (only ds no ss).
(3) Bacteriophages have DNA (ss or ds). However, most phages are DNA viruses .

Replication of viruses

Like any other living being, viruses also multiply but with a difference. They cannot multiply on their own. They do not reproduce by division independently because the viral nucleic acid cannot duplicate by itself. There are no raw materials to produce fresh nucleotides, neither is there an enzymatic machinery to assemble them together into a duplicate molecule.

Viruses multiply only when they are in a living, cell. The DNA or RNA of the virus takes control of the host cell metabolic machinery and new viral particles are produced utilizing the raw materials from the host cell.

Multiplication of tobacco mosaic virus (TMV), takes place inside the leaf cells of the host. The virus reaches the inside of the cell either through a wound or abrasion. The protein coat of the virus is dissolved by host enzymes and the synthesis of mRNA, proteins and genome as well as the maturation of the viral particles occur in the nucleus of the host cell and not in cytoplasm. The cells do not breakdown (no lysis) and the virus particles move into other cells of the plant through plasmodesmata. Viral particles are transmitted through sop from injured tisssues.

Replication of viruses is studied in great detail in bacteriophages. Two kinds of replication cycles occur in phages. These are the *virulent* or *lytic cycle,* and the *temperate* or *lysogenic cycle.* In the lytic cycle, the viral particles multiply inside the bacterial cell. After the formation of new viral particles the host cell breaks down (lysis), and the viral particles are released. In the lysogenic cycle, the viral genome (after entry) gets integrated with the bacterial genome and is called prophage. The prophage, multiplies along with the bacterial genome (no independent multiplication) without causing any damage to the host cell. At some stage, however, depending on environmental conditions, the prophage may separate from the bacterial genome and start a lytic cycle.

As the lytic bacteriophages (which efect E. coil) have been studied in great detail with reference to viral multiplication. The following acccunt is mainly based on them. However, in other viruses also multiplication is same as in T even phages.

The main stages involved in the replication of T even phages are–

(i) Attachment of phage particle to the host.
(ii) Adsorption of virus particle.
(iii) Separation of nucleic acid from coat.
(iv) Penetration into the host.
(v) Effect of phage attachment on the host cell metabolism.
(vi) Replication of viral of genome (nucleic acid).
(vii) Synthesis of viral protein (capsid).
(viii) Assembly of new viral particles and
(ix) Release of viral particles from the host cell.

Attachment (of phage) to the host cell

Viral particles come into contact with host cell surface. The tail plate of the virus attaches to the surface of the host cell, and anchors itself firmly with the help of the tail fibres. The attachment of tail plate (of the virus) to the host cell, is a highly specific chemoreceptor regulated process. Certain protein receptors present in the tail capsid of the virus can recognise receptor sites on the host cell surface. It has been noticed that in *E.idwrichia coU,* the outer lipoprotein layer in the peptidoglycan covering has many receptors for pliage attachment.

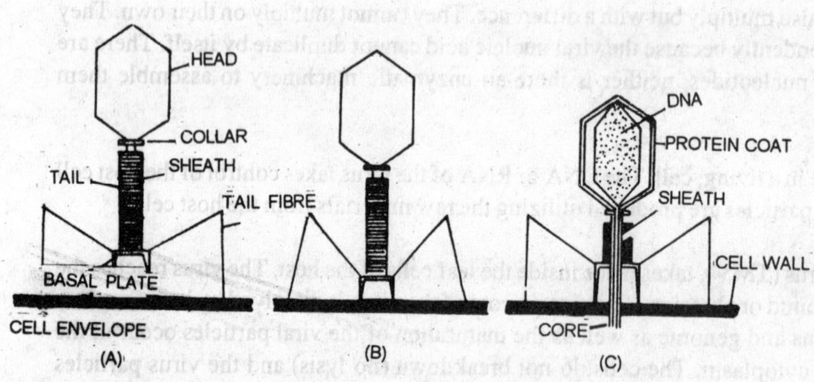

Fig.16.7 Viruses

Mechanism of attachment and introduction of DNA intohost cell in a T-even bacteriophage. (a). The tail fibers get fixed to the surface of host cell. (b). The base plate is brought into contact with the cell surface (c). The sheath contracts driving the tubular core through the cell wall of bacteria and the DNA from the head enters the host cell through the core

Absorption of pliage

Initially the attachment of the phage to the host cell is reversible, but later it becomes irreversible (virus can not be removed). Some cations are known to play a role in the absorption of the Phage. The receptors of the host cell surface are complex polysaccharides which can bind the phage on antigenic specificity. The absorbed phage has its head perpendicular to the cell surface. If the host cells ar capsulated, and enzyme at the tail top of the phage hydrolyses the polysaccharides of the bacterial capsule boring a tunnel through which cell the wall can be reached. The receptor in phage is a mannose transport protein. The receptor in bacteria some times may be present in pili (as in males), in which case some coliphages are male specific i.e., they attach to the pili sex only.

Separation of nucleic acid from coat

Hershey and Chase (1952), have shown that phage DNA carries the genetic information into the cell. Injection of DNA into the cell does not require any energy from the host. After the DNA is injected, the coat remains outside. The naked DNA temporarily loses the infectivity until the production of new viral particles. This temporary phase is called *eclipse.*

Penetration into the host

The nucleic acid together with some internal proteins is released from the capsid after the virus is irreversibly attached to the host cell. It penetrates into the cell at the sites were the inner and outer membrances are in contact with each and remains associated with the membrane physically. The phage DNA is protected against the membrane nucleases of the host by its asociated proteins and by DNA modifications. A highly specialilzed mechanism of injection of DNA is seen in the T–even phages. Electron microscopic studies have revealed that after the tail plate is fixed to the cell, the tail sheath protein becomes contractile and pulls the collar and the phage head towards the basal plate and pushes the tube through the cell wall. When the tube reaches the plasma membrane, DNA is ejected. The energy necessary for contracion is derived from ATP present on the phage tail. Lysozyme present in the tail may also help DNA penetration by the opening in the bacterial cell wall.

(Experimentally purified phage DNA has been used (after pretreatment with Ca^{2+}) to bring about infection of the bacterial cell. This process is known as transfection.)

After the infection of DNA the empty head and tail of the phage remain outside the bacterial cell.

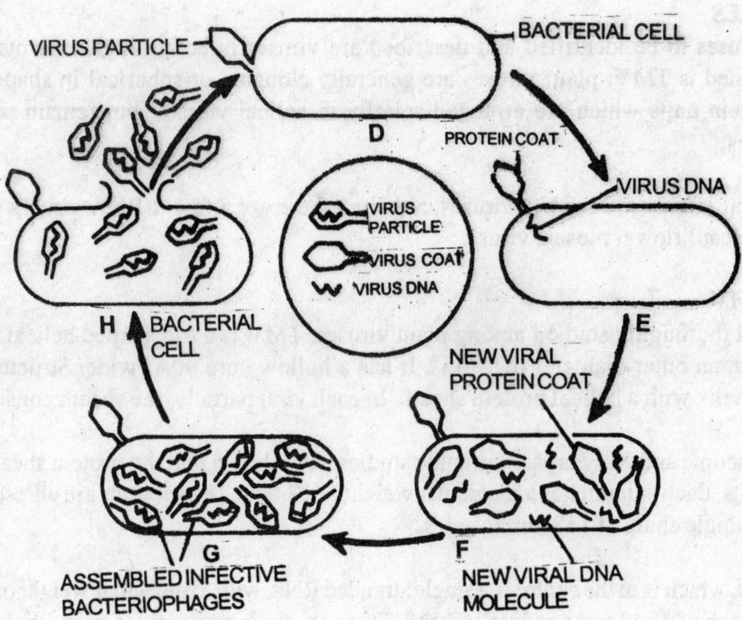

Fig. 16.8 Viruses

Attachment, adsorption and penetration (A-C) of a bacterial host cell by a T-phage; D-H stages of the cycle of infection of bacteria cells by bacteriophages

Effect of phage attachment on host metabolism

The attachment of T. phage by itself (i.e. without expression of viral genes) causes disorganization of the cell membrane. This brings in various metabolic changes including stoppage of protein synthesis of the host cell. These changes sometimes lead to cell lysis even without viral multiplication.

Replication of viral nucleic acid

Viral nucleic acid molecules produce many replicas using the enzyme machinery of the host cell. Host cell nucleus is the site for the multiplication of viral DNA.

Protein Synthesis of virus

In order to produce the'capsid', fresh viral proteins are to be produced. First, some enzymes are synthesized. These produce proteins peculiar to phage called'early proteins' and later the late proteins' are produced which develop into subunits of head and tail. At this stage, bacterial metabolism including DNA, RNA and protein synthesis comes to a halt.

Assembly of new viral particles

Phage DNA and capsid proteins are synthesized separately in the bacterial cell. The phage DNA is then condensed and pushed into the head, and tail is added. The process of assembling the parts of a virus into a new virion is called 'maturation'.

Release of (new) viral particles

When the phage policies are being produced in the cell, the wall of the bacteria gets weakened considerably. Further, viral enzymes (produced in the bactrial cell) act on the cell wall resulting in its bursting open. As a result, phage particles are released outside and they infect other cells repealing the cycle.

PLANT VIRUSES

The earliest viruses to be identified and described are viruses infecting higher plants and one of the most thoroughly studied is TMV. plant viruses are generally elongate or spherical in shape. Their surface has a number of protein units which are arranged spirally in helical viruses, but remain packed on the sides in spherical viruses.

Most of the plant viruses are ssRNA viruses, although there are a few dsRNA viruses (rice dwarf virus) and dsDNA viruses (cauliflower mosaic virus).

Structure of TMV

This is the most thoroughly studied among plant viruses. TMV is a rod shaped helical virus. Each particle is 300A^0 long with an outer diameter of 180A^0. It has a hollow core 40A^0 wide. Structurally there is a single stranded RNA helix with a helical protein sheath. In each viral particle, the sheath consists of 129 spirals.

Electron microscopic and Xray crysallographic studies have shown that the protein sheath is made up of 2,130 identical subunits. Each subunit has a molecular weight of 17,000. The subunits are ellipsodidal in shape and are composed of a single chain of 158 amino acids.

The nucleic acid, which is in the centre is a single stranded RNA wilh a molecular weight of 2.4 million. The helix of RNA has a pitch of 23A^0 and a radius of 40A^0. There arc three nucleotides per particle of TMV.

SOME MAJOR GROUPS OF PLANT VIRUSES

Virus group	Nature of N.A.	Representative viruses	Propertiesof various
Tobacco mosaic	RNA	TMV; tomato aucuba Mosaic; Holmes rib grass; cucumber viruses 3 and 4	Rigid rods, 3000^{0}Along
Potato X	RNA	Strains of H, G, L etc, potato mottle; potato ring spot	Flexible rods,about 5000A^0
Potato Y	RNA	Leaf drop streak; rugose mosaic; potato virus C	Flexible rods, about 7000A^0 long
Tobacco rattle	RNA		Rigid rods, 1750A^0 long
Tobacco etch	RNA		Flexible rods
Soyabean mosaic	RNA		Flexible rods
Tobacco necrosis	RNA		- do -, 170 A^0
Tobacco bushy stunt			- do -, 280 A^0
Southern bean mosaic	RNA		- do -
Turnip yellow mosaic	RNA		- do -
Potato yellow dwarf	RNA		Enveloped, bacilliform 750 X 3800A^0
Cauliflower mosaic	DNA		diameter

Viral diseases of plants

Viral plant diseases may be systemic, or local. In systemic infections, the disease spreads all over the plant through the vasculture, while in local, the disease is restricted to certain areas. The following are some of the important symptoms of viral plant diseases.

1. **Mosaic :** Mixed light yellow and green pattern appear on the leaves making a mosaic pattern. This is a systemic infection, e.g., TMV, mosaic of cucurbits, mosaic of potato etc.

2. **Yellow :** Uniform chlorosis all over the leaf. It is also called chlorosis disease,e.g., Rice yellow.

3. **Enantions :** Small hair like outgrowths appear on the leaves called enantions. These outgrowths often accompany mosaic symptoms , e.g.. Bean enaiition mosaic.

4. **Vein clearing, vein banding and vein thickening :** Veins and veinlets become light yellow and transparent while the lamina remains normal. It is *vice versa* in vein banding–the whole lamina turns yellow, e.g., vein clearing in Lady's finger.

5. **Leaf curling or leaf rolling.** The infected leaves curl or roll back, e.g., leaf curl of papaya, tomato, etc.

6. **Little leaf :** In this, the affected leaves become small and crowded giving a rosette appearance, e.g., Little leaf of brinjal.

7. **Stunting :** The infected plants have overretardation of growth, e.g.. Bunchy lop of banana.

8. **Breaking and greening of plants;** Petals of flowers change colour, become small and green and appear variegated. Some times it may enhance the beauty of the flowers, e.g., Broken tulip flowers.

9 TiimiHirs : Some viral infections produce tumours on leaves or roots.

10. Witche's broom : Leaves get reduced and so also the internodes, resulting in crowded clusters looking like a broom.

Transmission of plant viruses

Viruses are transmitted from plant to plant in a number of ways. The chief methods of transmission of plant viruses are:

(1) Transmission by vegetative propagation

It is one of the chief methods of transmission, whenever plants are propagated vegetatively by means of cuttings, tubers, bulbs, rhizomes etc., viruses present in the other plant are easily carried over to the propagated plant. This method of transmission is chiefly seen in Potato, Strawberry, Raspberries etc.

Vectors and the Plant Viruses they Carry

Vector	Virus transmitted
Nematode	
Xiphinema index	Grape 'fan-leaf' virus
Fungi	
Olipidium brassicae	Tobacco mosaic virus (TMV) Tobacco stunt virust (TSV) Lettuce big-vein disease
virus	
Synchytrium endobioticum	Potato virus X (PVX)
Insects	
Myzus persicae (aphid)	Potato leaf-roll virus.
Nephotittix apicalis	Rice dwarf disease virus
Circudulina sp (leaf-hooper)	Maize streak virus.
Thysanoptera (thirp)	Wheat spotted - wilt virus.
Bemisa tabaci (white flies)	Tomato leaf-curl virus
infecting tobacco	

(2) Seed transmission

Transmission of viruses through seeds is not very common. It has been reported in very few cases such as: legumes, tomatoes, bean-mosaic, cucumber mosaic etc. They seem to come from ovules of infected plants. The viruses however, do not enter the embryo.

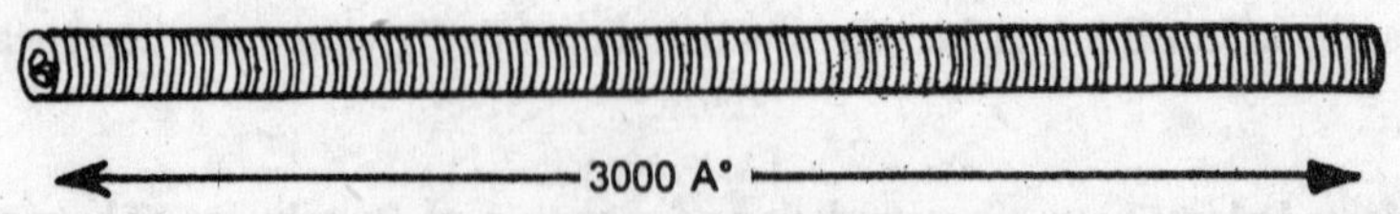

Fig. 16.9 Viruses

A rod of Tobacco mosaic virus, showing arrangement of spirals forming the protein sheath

(3) Soil transmission

Soil transmission has been observed in few cases such as tobacco mosaic virus, potato mosaic virus, wheat mosaic virus, oat mosaic virus etc. These viruses live in the soil along with the plant debris after the harvest of the crop when sown in the same field.

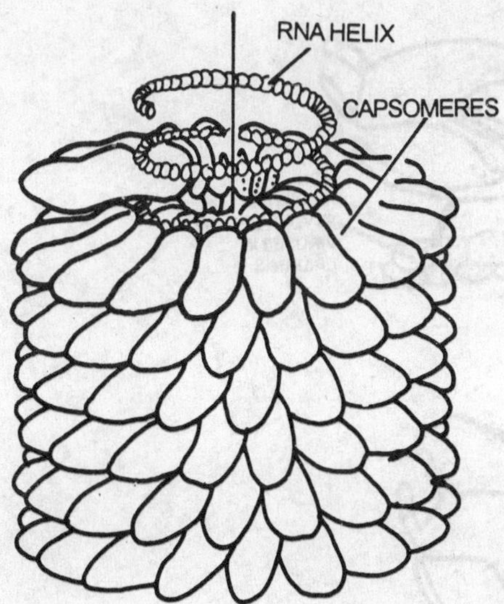

Fig. 16.10 Viruses

A helical virus, TMV. Proteinaceous capsomeres arranged in helical manner around hollow core of RNA The hilical RNA(a) Adeno virus (b) Herpes virus

(4) Transmission by contact

This is brought about in nature, by contact of one plant with the other. This is usually brought about by strong wind, which makes the infected part to come in contact with a healthy plant.

(5) Tobacco mosaic virus can remain infective in dry tobacco leaves as long as 25-30 years. They can be easily transmitted by the fingers of the smokers and also by the unburnt pieces of cigarettes, bi8dies, cigars etc.

(6) Transmission hy Insects

Undoubtedly, the most common method of transmission of viruses is by the activity of insect, which are termed vectors. Some of the insects which play an important role in the transmission of viruses are Scale Insects, *White* flies, Leafhoppers, Aphids etc. Aphids transmit more plant viruses than any other insects. Insects with their well developed mouth part carry viruses on their styles (Style borne virus), or may accumulate them internally, and after passing through the tissues, they are introduced again through the mouth part (circulative virus). Some of these viruses may multiply forming the propagative viruses. In some cases, the vector fails to infect a healthy plant soon after it is fed upon a diseased plant, but it can infect the healthy only after some time. This period of development of infectivity for the virus within the vector is called the *Incubation* period. This period varies with different viruses from hours to days.

There seems to be some relationship between the plant viruses and the insect vectors which transmit them. For eg:-Stunt virus of paddy is transmitted only by a Icaft hoper *Nepholetlix apicalii*.. Another type of leafhopper namely, *Circulifer tenellus* transmits only the curly top virus of beet root. The exact nature of the relationship between the virus and the vector is still unknown.

(6) Fungal transmission

Certain fungi also take in the transmission of viruses. For eg:- the root infecting fungus *Olipidiwn hrassicae* ,transmits at least 2 plants viruses namely, (1) Tobacco necrosis virus and (2) Tobacco stunt virus.

Control

1. Field sanitation measures
2. Growing disease resistant varieties.
3. Diseased plants should be uprooted and burnt to avoid further infection.
4. Insect vectors should be kept in check by spraying.

ANIMAL VIRUSES (including human)

Animal viruses bring about a number of diseases in animals and also in human beings. Animal viruses have either DNA or RNA as the genetic material. DNA is mostly double stranded while RNA is either single standed or double standed. Given below is an account of some animal viruses.

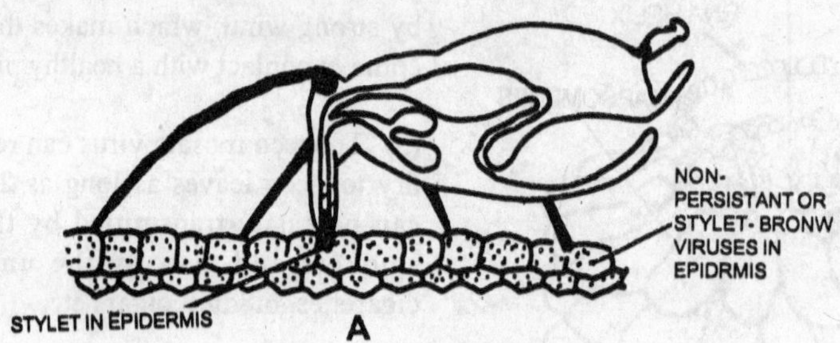

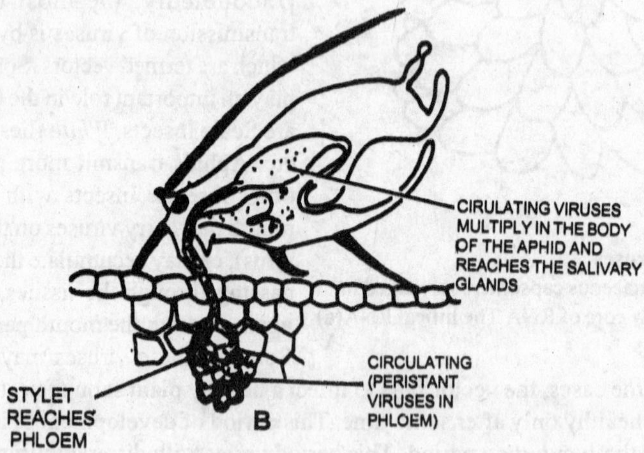

Fig. 16.11 Viruses

Diagram showing acquisition of viruses by aphids, A. Stylet borne virus, B. Circulating virus

Foot and mouth disease virus

This is a pathogen of cattle. .The virus is a single standed RNA particle having about 7,500 bases with a coding capacity of 2,500 amino acids. *The* capsid is icosahedral.

Rabies Virus

The pathogen is called a rhabdovirus and is bullet shaped. It is found in plants, invertebrates and vertebrates. In dogs it causes rabies, the virus is enveloped which is made up of two layers of lipids. The hereditary material is ss RNA. (for more details see under human viruses) Another example of habdovirus is the vesicular stomatitis virus which is a mild pathogen of cattle.

Human pathogenic viruses

These are artificially classified into the following four groups based on the organs in which disease symptoms are produced.

1. **Pneutropic Viral diseases :** Diseases involve respiratory tracts, e.g. Influenza, Common cold etc.
2. **Dermotrophic viral diseases :** Diseases involve mainly skin e.g. Measles, Chicken pox. Small pox, mumps etc.,
3. **Viscerotropic viral diseases :** Diseases involve blood and visceral organs, e.g. Yellow fever, dengue, hepatitis etc.,
4. **Neurotropic Viral diseases :** Diseases involving the central nervous system, e.g. Rabies, polio. Encephalitis etc. We shall study some of these briefly.

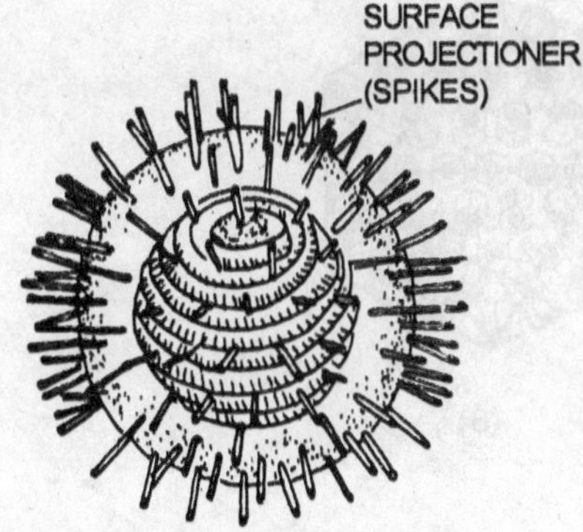

Fig. 16.12 Viruses
A typical myxovirus
(influenza virus)

Influenza Virus : The pathogen is a helical virus containing ssRNA and it is enveloped. The envelope contains a series of projections or spikes. The virus causes disease of the upper respiratory tract.

There are three types of influenza virus–Types A,B and C. Type A and B causes epidemics, while type C occurs sporadically. Each type is characterized by its antigenic variation in which changes occur in the protein of the capsid. As a result, innumerable variants are produced which are unrecognizable by the antibodies produced during a previous influenza infection. The nomenclature of influenza virus is therefore based on changing antigenic pattern.

Viral infection causes sudden chills, fatigue, headache and pain. Fever of upto to 103° to 104°F is very common, influenza A is treated with amantadine, a synthetic drug which is known to prevent viral penetration into the host cells.

Adenoviruses : These are a group of about 35 icoscahedral particles having ds DNA. The viral particle is about 60-90nm indiameter having a dense central core and the capsid composed of 252 capsomeres (240 hexons and 12 pentons).

The viruses owe their name to the adenoid cells from which they were first isolated. They multiply in the nuclei of the host cells.

Adenoviruses cause diseases in respiratory tracts, eyes and meninges. Type 8 adenovirus is the main agent of Kerato conjunctivitis inflammation of the eye.

Rhinoviruses : These represent a collection of about 100 icosahedral RNA viruses (Rhinonose) ,are involved in respiratory infections causing headache, sneezings blocked nostrils etc. Common cold is caused by about 15-29 types of rhinoviruses.

Chicken pox virus : The pathogen is a icosahedral virus having ds DNA as the hereditary material. The capsid is made up of 162 capsomeres. Chicken pox virus belongs to herpse virus group. In adults the same virus causes herpes (Virus - herpes Zoster), which is a painful disease where the spinal cord is affected with red patches on the skin.

Measles Virus : This is a common pathogen among children, causing a communicable respiratory disease, the symptoms of which develop on the skin. The pathogen is called *Rubela* (meaning red), owing to the appearance of red rashes on the skin.

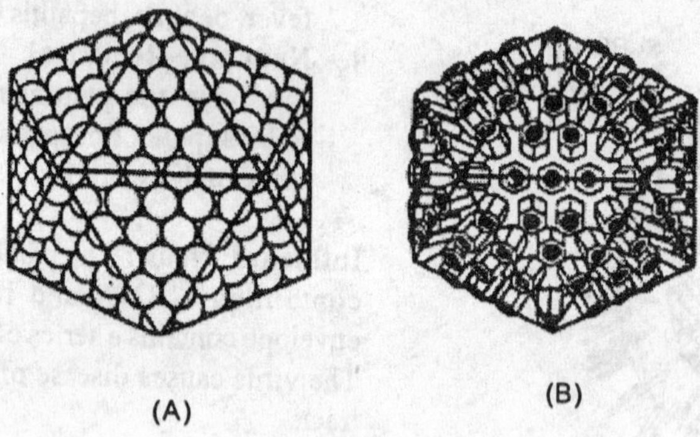

Fig. 16.13 Viruses
(a) Adeno virus (b) Herpes virus

Rubela is a helical ssRNA virus with an envelop in addition to the capsid. An effective vaccine is available against the disease.

Mumps virus : This is disease in children mainly affecting the parotid glands (hence the technical *name-epidemic parotitis* for the disease) There is a swelling visible on either side of the cheek. The virus has ss RNA and is helical having an envelop in addition lo 'capsid'. *This is similar to measles virus except that there are no nenraminidase spikes on the envelop.* A vaccine is available for the disease.

Small pox virus : Commonly called variola virus it has been eradicated the world over. The virus causing the dreadful disease has a complex archiecture, brick shaped and is about 270nm in size. It has no envelop but the capsid is surrounded by a series of fibres.

Fig. 16.14 Viruses
Poxvirus a brick-shaped virus

Early symptoms of the disease are chill, fever, and fatigue. Pink red spots (mancles) appear after fall in temperature on the scalp, forehead and all parts of the body. The spots turn into fluid filled vesicles and then into pustules. By preventive vaccination the disease has been completely contained. The last case of small pox reported (as per WHO) was in October 1977. On October 1979, an announcement was made that small pox has been eliminated from the earth.

Rabies virus : The rabies virus not only causes disease to dogs, but to human beings also with a high rate of mortality if not treated properly. Besides dogs, rabies virus also occurs in cats, horses, rats etc. The virus has RNA as the genetic material and is enveloped. It is rounded at one end and flattened at the other to give the appearance of a bullet. The virus gains entry into the human tissue through wound or abrasion caused from an infected animal. Though the disease is mainly transmitted through saliva, urine, lymph, or even milk may be equally infective. The virus after gaining entry spreads rapidly through the neural pathway. Disease symptoms include, aggressive behaviour, difficulty in swallowing, excessive salivation, dislike for water (hydrophobia) etc.

Poliovirus

One of the smallest among viruses, the polio virus (diameter 27-30nm) has an icosahedral capsid with a ssRNA as genetic material. The multiplication of the virus is phenomenal; within a six hour cycle it produces as many as 10,000 new particles.

The virus after gaining entry through contaminated food or water multiplies in the tonsils and the lymphoid tissues. After entering the blood stream, the virus localizes in the grey matter of the spinal cord. (Greek polis = grey). Three types of polio are known to exist–Brunhidle strain-epidemics and paralysis; lansing strain occurs sporadically but causes greatest number of paralytic cases and Icon strain occurs occassionally confining itself to intestine rarely causing paralysis.

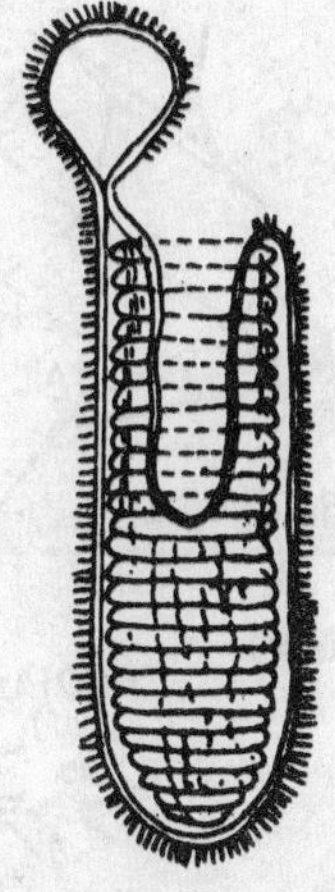

Fig. 16.15 Viruses
Rhabdovirus, a bullet-shaped virus

AIDS

AIDS or Acquired Immunity Deficiency Syndrome is caused by a virus belonging to Retrovirus group. It has been given the name HIV (Human immunodeficiency virus). In this disease, the viruses attack the lymphocytes (T4), and destroy them. Lymphocytes are necessary to provide immunity against any disease. Whenever our body gets any kind of infection lymphocytes are produced in large numbers and they kill the infecting microbes. With the destruction of the lymphocytes, the entire defence mechanism (immunity) is disturbed and the body will not be able to combat even ordinary infections. While AIDS does not kill the individual directly, it renders the body weak and susceptible to all kinds of infections.

The first few AIDS cases were reported in 1981 in Atlanta, USA, since then it has spread to all countries including India.

Transmission of AIDS : The following are the methods of viral transmission, (a) through sexual intercourse with infected persons (Homosexuals are more succeptible), (b) through contaminated injecting needles, (c) through blood transfusion and (d) infants of infected mothers also get the disease through maternal blood.

Symptoms : All HIV infected persons need not show visible symptoms. The virus after gaining entry may remain dormant for a period ranging between 9-300 months. In blood transfusion cases, this may be between 4-14 months. Persons with AIDS infection can be classified into 3 categories.

1. **Healthy Carriers :** This is the first stage; there are no symptoms. But they are capable of infecting others.

They may develop full blown symptoms or may remain healthy for a long period.

2. Prodromal phase : This is called AIDS related complex (ARC). This group has a milder from of the disease with fever, diarrhoea, weight loss and enlarged lymph glands. They exhibit depression of immune system.

3. The End Stage : Here AIDS is fully developed and the patient dies of incontrollable infections of lung, exertion, chest pain, fever, loss of weight, severe headache, loss of memory, convulsions, and development of sidii malignancy (Kaposis sarcoma).

Identifications of AIDS : Serological test of the blood by ELISA or western blot will indicate the presence of antibodies against HIV in infected patients.

Cure : Till now no cure has been found for AIDS. Drugs Azidothymidine, Suramin etc. may give, a temporary relief. Attempts are being made to develop a vaccine against the virus. However, chances of contacting AIDS.may be avoided by taking the following precautions.

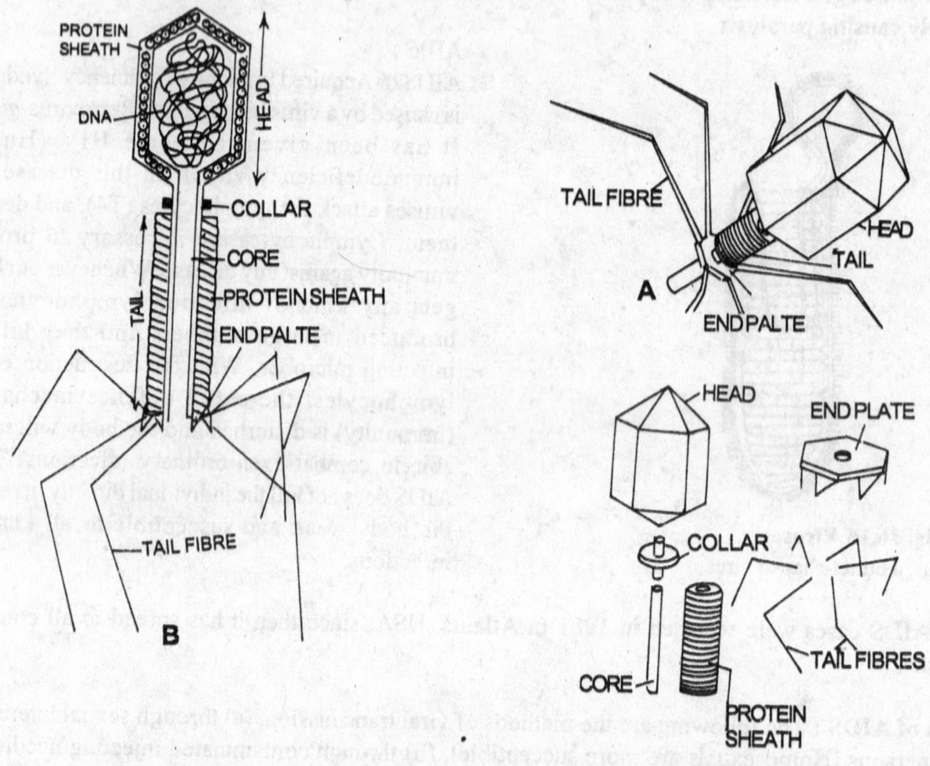

Fig. 16.16 Viruses.
Morphology of T-even bacteriophage virus
A. External morphology; **B.** Diagrammatic representation of L.S. of a bacteriophage; **C.** Various components of a bacteriophage

(a) Sterilizing the needles properly before injection or use of disposable syringes.

(b) Taken utmost care during blood transfusion; only HIV screened blood should be used. This is a moral step. Avoid sexual contact with several partners. Sexual contact to be had with only one known partner. Homosexuality should be strictly avoided.

BACTERIOPHAGES (BACTERIAL VIRUSES)

Viruses that infect bacteria are called bacteriophages. The word phage means eating, hence bacteriophages are bacteria eaters. Bacteriophages may have ssDNA, dsDNA, ssRNA or dsRNA as the genetic material. The word bacteriophage was first coined by d'Herelle (1917).

ssDNA phages are of two morphological types–icosahedral and helical. The best known example for icosahedral in <j>X174. The helical phages are filamentous and are divided into three group– (i) Ff group: Attach to the F type sex pili in bacteria, (ii) If group: Attach to I type sex pili specified by the resistence (for drug) factor and (iii) Ike group which is physically similar to Ff group but serologically distinct.

Among the dsDNA phages the important ones are-T even phage of *Escherichia coli,* N_1 phage of cyanobacteria etc.

ssRNA phages include over thirty viruses belonging to three or four serological groups. 06 is an example of dsRNA phage.

Among the phages, T-even phage has been studied in detail and we shall describe it is an example of a typical bacteriophage. T-even phage is a coliphage (attacking *E. coli),* and includes T, T, and Tg phages.

Morphologically, the T even phage has a hexagonal head and a cylindrical tail. The head is made up of a protein capsid which has icosahedral symmetry enclosing a tightly packed molecule of dsDNA. The tail consists of a hollow core surrounded by a contractile sheath. At the free end of the tail is present an end plate with a number of tail fibres. The end plate also has a number of spikes. The tail connects to the head with help of a disc or collar. The size of the heades 95nm X 65nm, while the tail is 115nm long with a diameter of 15nm.

The genetic material is a DNA with a molecular weight of 120 x 10* daltons. Some virologists believe that the DNA has hydroxymethyl cytosine instead of the normal cytosine.

(Life cycle and multiplication of bacteriophages is discussed under virus multiplication.)

CYANOPHAGES

Viruses that attack cyanobacteria(blue green algae) are called cyanophages. The first suggestion that viruses may be the cause of sudden inexplicable disapperance of blue green algal blooms was first made by Krauss (1961). Earlier (1960), Lewin reported a phage attacking *Spirochaeta rosea* (considered to be a blue green alga by some phycologists).

The first cyanophage was isolated in 1963, by Saffermann and Morris . Since then there have been several reports of cyanophages including one from India (Singh and Singh 1967).

NOMENCLATURE

Cyanophages are named after the hosts they attack. If they attack more than one host, the names of the hosts are abbreviated. For example, cyanophage attacking *Lyngbya, Phormidium and Plectonema* is named LPP-I taking the first letters of the three hosts.C-l cyanophage attacks *Cylindrospermum. So* far about 14 genera of cyanobacteria are known to harbour viruses-These (hosts) are *Anacystis, Chroococcus, Microcystis, Nostoc, Lyngbya, Phormidiutn, Cylindrospermum etc.*

Structure

Cyanophages may be icosahedral or helical. The capsid protein has a molecular weight of about 4X *W* daltons.

The LPP-virus has a distinct head and a short tail similar to a bacteriophages. The capsid head is about 600nm and the tail 200nm long. The capsid is covered by an envelop. The genetic material is DNA. It has a *C/G* ratio of 55-57%.

The infection and multiplication of cyanophages is almost similar to bacteriophages. The viral DNA multiplies in the host cell in between photosynthetic lamellae, and in the region where host DNA lies (nucleoplasm). It has been reported that LPP- 1 phage breaks down half of the host DNA, and incorporates it into its own DNA.

Viruses attack the host resulting in the rates of photosynthesis and respiration coming down. Lysis of the cells of the host is restricted to vegetative cells while heterocysts and spores are affected very little.

Isolation

Samples of sewage water are added to algal culture which is incubated for about a week and centrifuged. The supernatant is treated with chloroform and screened for phage activity by allowing it to grow on a test blue green alga.

Uses

Cyanophages can be used to destroy blue green algal blooms. Goryush and Chaplinskaya (1968), have reported the destruction of blooms by using cyanophages. Daft and Stewart (1971), have also demonstrated a similar phenomenon.

17

CHEMICAL BASIS OF HEREDITY

The cells have diverse chemical components of which most were known to scientists by the first quarter of the nineteeth century. These are classified into three major groups namely-carbohydrates, proteins and lipids. Of these three-proteins are most complex and diverse and many biologists through that they may constitute the basis of biological diversity.

Eventhough the other group of macromolecular compounds - Nuclei acids were discovered in 1869 by Sweedish biochemist Fredrich Miescher, it was only in 1952, their (nuclei acids) role in inheritance was proved.

Meischer collected pus filled hospital bandages and from them he isolated and purified nuclei. On chemical fractionation he found that a cellular chemical content had a very high component of phosphorous compound than what was known before. As the new compound was obrtained from nucleus, Miescher named it *nuclein.* Subsequently nuclein was isolated from many sources and due to its acidic properties, it was named nucleic acid. Upto 1952 however, interest in nuclei acids was limited to only its physiochemical studies. When its role in inheritance was proved nucleic acid became the focus of intensive research. Comparative studies of nuclei acids and proteins revealed that while proteins are of diverse types, there are only two types of nuclei acids. It was hard to believe that nuclei acids which apparently had no diversity could be the basis of heredity and not the proteins.

Presently however, there is not even an iota of doubt as to the fact that nucleic acids alone constitute the hereditary factor. Three major experiments helped provide an evidence for this.

Bacterial Transformation

Fred Griffith and English bacteriologist worked with *Diplococcus pneumococci* and found out that there are two strains in this - *Virulent and Avirulent.* When injected to experimental mice the virulent strains cause the death of the mice while avirulent strains do not casue any harm. Griffith showed that heat killed virulent strains injected into the mice along with avirulent strains cause the death of the mice and from the bodyof the mice live virulent strains can be obtained. Dawson and Sia (1931) showed that the heat killed virulent strains can transform avirulent strains on a nutrient broth. Alloway (1933) showed that even a lysed cell of virulent strain retains the ability of transformation thereby indicating that some component of the cell rather than the entire cell was involved in transformation.

Nature of the transforming agent

Three scientists - O.T. Avery, Colin Macleod and Maclyn McCarty (1944) went into serious search of the transforming agent from the lysed cells of bacteria. Even after digestion of polysaccharide coat, the cell retained the ability of transformation. This shows that the capsule has no role in transformation. Similarly protein degradation also had no adverse effect on transformation. With the two important components out of the way, it thus became clear that nucleic acids perhaps might play a role in transformation. It was shown that the DNA fraction of nucleic acids was responsible for transformation. It was also shown that if the cells are treated with DNAse (a DNA degrading enzyme), the transforming ability will be lost. Enzymes which digest RNA do not disturb transformation.

Bacteriophages and chemical basis of heredity

Bacteriophages are viruses that attack a bacterial cell and multiplty inside the (bacterial) cell. The phage attaches itself to the bacterial cell and ingects its components into the cell where replica of phages are produced. Hershey and Chase (1952) conducted an experiment to find out which component of the bacteriophage carries the genetic information necessary for the replication of viruses. They cultured the host bacterium *Escherichia coli* in a medium containing radio isotopes of Sulphur (S^{35}) and phosphorus (P^{32}). When T_2 phages attacked these cells subsequently, the labelled compounds get incorporated in phage DNA which always has radioactive phosphorus but not any sulphur. In the next set of experiment only the protein coat of T_2 phage was labelled with S^{35} and the DNA core of another was labelled with P^{32}. Hershey and Chase separated the radioactive phages by centrifugation and allowed them to infect a population of n‹ n radioactive bacteria. A few minutes after initial infection they separated the phages from bacteria by agitating the medium. After centrifugation, the host and the parasite were separated by centrifugation and their radioactivity measured. About 95% of P^{32} was found in bacteria while about 95% S^{35} was seen in phages. This experiment conclusively proved that during infection only DNA is injected into the medium. This is the reason why the non radioactive bacteria showed labelled P^{32} (as they obtained it from labelled phage DNA). For this work Hershey was awarded Nobel prize in 1969.

Thus the experiments of Hersehy and Chase. Avery, Mcleod and McCarty. Griffith etc., clearly proved that DNA constitutes the chemical basis of Heredity.

TMV and Chemical Basis of Heredity

The discovery that DNA is the hereditary material posed some problems to explain the inheritance in instances such as TMV (Tobacco mosaic virus) where there is no DNA (TMV is a RNA virus). But the absence of DNA has not prevented TMV from transmitting its characters from one generation to the other. TMV is made up of a RNA core and a helical protein coat. The protein RNA component of TMV can be easily separated by shaking the viral particles in a mixture of water and phenol. Frankek Conrad and Wichams (1955) separated the protein and RNA from TMV and reassembled the particles from the individual components thus indicating the purification process (of the components) does not affect their function. Subsequently Gierr and Shramm (1956) showed that on contact with the host only the RNA can bring about the disease and not the protein coat. In TMV therefore it is RNA which is the hereditary material. It can therefore be concluded that nucleic acids in general and DNA in particular is the hereditary material in most of the organisms and in the absence of DNA (as in some viruses) RNA can constitute the genetic material.

Nucleic Acids

These are the most important constituents of the nucleus playing a vital role in the inheritence of characters from one generation to the other. There are two types of nucleic acids - Deoxyribose nucleic acid (DNA) and Ribose nucleic acid (RNA). While both constitute the genetic material in different organisms, it is predominantly the DNA which is the genetic material in most of the organisms. The following are some of the important milestones in the discovery of Nucleic acids.

1. Meischer (1869) isolated nuclei acids and called them nuclein.
2. Hertwig (1884) suggested a heriditary role to the nuclein
3. Kossel (1884) recognised purines and pyrimidines in the nucleic acids.
4. Feulgen (1912) devised basic fushsin dye positive only to DNA.
5. Astbury and Bell (1938) showed that the bases in DNA molecule are arranged one above the other.
6. Avery, Macleod and Macarty (1944) presented evidence for DNA to be the genetic material.
7. Chargaff (1950) showed that DNA contains equal amount of purines and pyrimidines.
8. Wilkins *et al* (1950) identified the molecular distance between the bases.
9. Furberg (1952); DNA is a coil of single chain.
10. Pauling and Corey (1952); DNA is a helix of three chains.

11. Watson and Crick (1953) demonstrated the double helix model for DNA molecule.
12. Meselson and Stahl (1958) demonstrated the semi conservative replication in DNA.
13. Jacob and Monod (1961) postulated the presence of RNA in protein synthesis.

DEOXY RIBOSE NUCLEIC ACID (DNA)

DNA is the main genetic material in almost all organisms except for certain viruses. In prokaryotes like *Escherichia coli,* the genetic material is a single giant molecule of *DNA* is mainly concentratedin the nucleus, it has also been reported in cytoplasmic organelles like chloroplast, mitochondria etc.

Chemical Composition of DNA

DNA extract from many living organisms has revealed it to be a complex macromolecule. It is a long chain polymer cmposed of monomers (units) called nucleotides. Each nucleotide in turn is composed of three sub units. These are - (a) nitrogen base, (b) sugar and (c) phosphate.

Nitrogen bases

The nitrogen bases are of two kinds viz., pyrimidines and purines. Both pyrimidines and purines are heterocyclic compounds, with a basic six membered ring in which one or the other carbon position is replaced by nitrogen.

The pyrimidines have a basic six membered ring of benzene in which, position 1' and 3' are replaced by nitrogen. There are two kinds of pyrimidines - *Cytosine* and *Thymine.* Structurally cytosine has a single six membered ring with a hydroxyl group at the 2' position and an amino group at the 6' position *(2 hydroxy, 6 amino pyrimidine).* Thymine also has a single six membered ring with hydroxyl groups at the 2' and 6' position and a methyl group attached to the 5' position *(2, 6 hydroxy 5 methyl pyrimidine).*

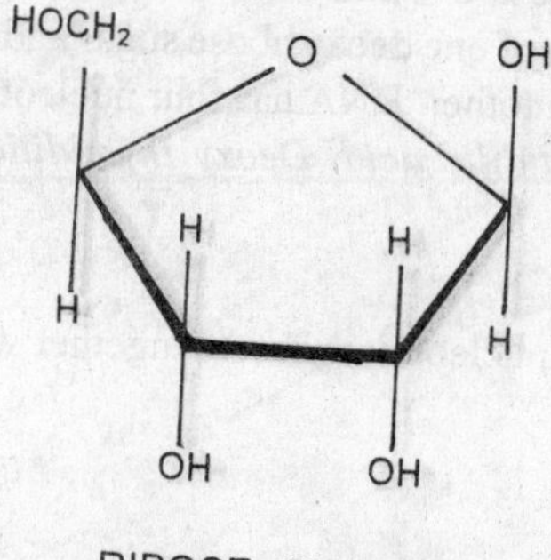

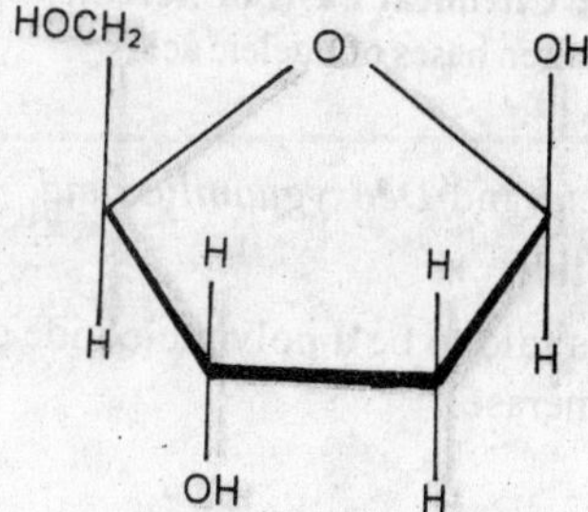

Fig. 17.1 Chemical basis of heredity

Pentose sugars of nucleic acids (left, robose; right dexyribose)

The purines have a double ringed structure. Attached to the six membered benzene ring is a 5 membered imidazole ring. The imidazole ring is joined to the hexagonal ring at 4' and 5' positions. The benzene ring has nitrogen in the 1' and 3' position (replacing Carbon).

There are two types of purines viz. *Adenine* and *suanine.* Adenine has an amino group at the 6' position (*6 aminopurine),* while guanine has an aminogroup at the 2' position and a hydroxyl group at the 6' position (*2 amino, 6 hydroxy purine).*

Sugar

The sugar molecule in DNA is basically a ribose sugar. The ribose sugar is a *Pentose Sugar.* It is a 5 membered ring. The ring consists of four carbon atoms while the 5' C atom is outside thering. Positions 1', 3' and 5' in the ring have OH groups while the 2' carbon position has only H. There is no O; hence the sugar is said to be Deoxyribose (no oxygen in the seconc carbon position).

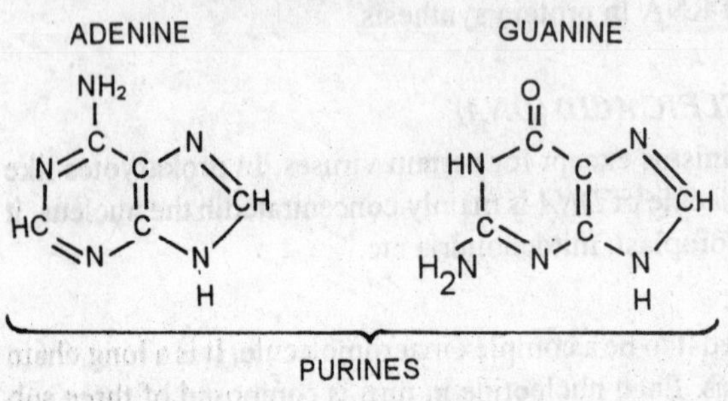

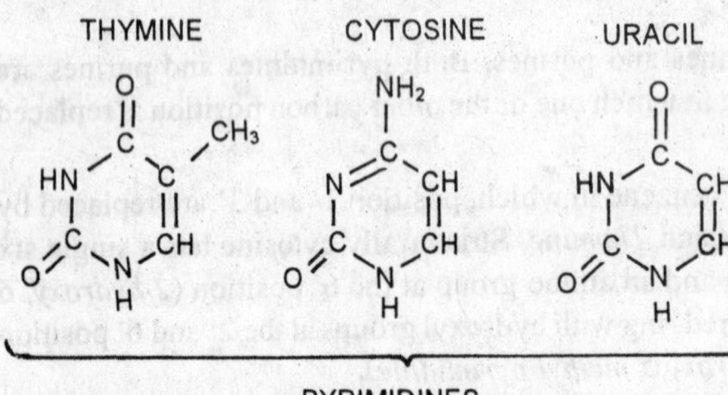

Fig.17.2 Chemical basis of heredity
Nitrogen bases of nucleic acids

Phosphate

It has one central phosphorus with four oxygen atoms.

Nucleosides

A nitrogen base coupled with sugars forms a nucleoside. DNA has four nucleosides -

(a) Deoxycytidine
(b) Deoxythymidine
(c) Deoxyadenosine and
(d) Deoxyguanosine

The sugar molecule is attached to the nitrogen base to the 3' position in the benzene ring in the case of pyrimidines and to the 9' position of the imidazole ring in the case of purines.

Nucleotides

A nucleotide is derived from the nucleoside by the addition of phosphoric acid molecule to the latter. The phosphate group is usually linked to the two sugar molecules above and below. The linkage is with the 3'C atom of one deoxyribose sugar and with the 5' C atom of another. DNA has four nucleotides - *Deoxycytidilic acid, Deoxy thymidilic acid, Deoxy adenylic acid* and *Deoxyguanilic acid.*

Polynucleotide

DNA is said to be a polynucleotide chain with a number of nucleotides joining together with the help of DNA polymerase.

Molecular model of DNA structure

A molecular model for DNA structure first proposed by J.D.Watson and F.H.C Crick (1953) has been universally accepted. They were awarded Nobel prize for their discovery. The work of Crick and Watson has been supported by X-ray diffraction picture obtained by Wilkins and his associates.

According to this model, a DNA molecule is composed of two chains that are spirally coiled and form a double helix, much like a spiral stairway. The distance between the two chains is uniform and it is maintained by base pairing. The back bone of the chain is made up of alternating sugar and phosphate molecules. At specific intervals, the chains are joined by 'steps'. Each step is made up of two nitrogen bases. Each nitrogen base is attached to the sugar molecule of the opposite chain. Between themselves, the nitrogen bases of a set are joined by hydrogen bonding. The pairing of the nitrogen bases is always specific. It is made up of a purine and a pyrimidine i.e., a double ring and a single ring. Two purines would occupy too much space and can not be accomodated between two chains; similarly two pyrimidines would form a short step and leave a gap between two chains. Hence the pairing is always between a purine and pyrimidine. Because of this pairing,in a DNA molecule, the total purine content is always equivalent to total pyrimidine content.

Fig. 17.3 Chemical basis of heredity
F.H.C. Crick (left) and J.D. Watson (right)

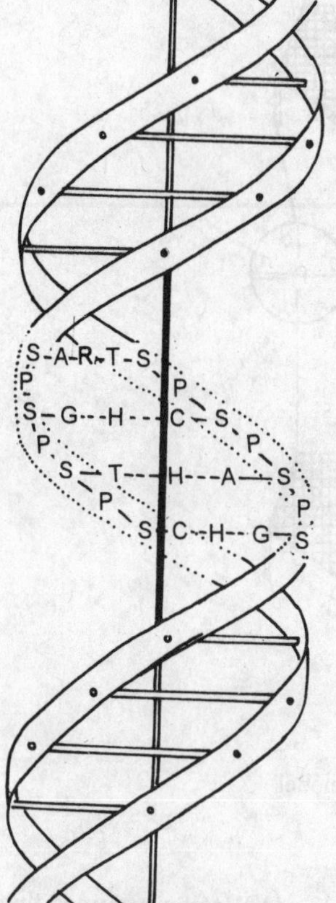

Fig. 17.4 Chemical basis of heredity
The double helix model of DNA

The purine, pyrimidine base pairing is further specific. It is always Adenine with Thymine and Guanine with Cytosine. Adenine and Thymine are joined by two hydrogen bonds through atoms attached to position 6' and 1'. Cytosine and Guanine are joined by three hydrogen bonds at positions 6' 1' and 2', the bonding with hydrogen is weak; but this facilitates easy separation of two strands during replication. According to Crick and Watson, the helix has a diametre of 20 A^0 and there is one complete turn to the helix at every 34 A^0 and has a stack of 10 nucleotides at every turn.

One unique character of DNA is its unity and yet infinitismal variety. The DNA structure mentioned above is uniform in all organisms, whether it is man or virus. At the same time it is also true that DNA is the genetic material that decides the characters of individual organisms. Then how does DNA decide the individuality? The answer lies in the base pair sequence. The combination of A/T, G/C is uniform, but their sequence is not; for instance in one individual it may be AT, G/C, GC, AT, GC AT etc. There is no limit to the variety in which base pair sequences are arranged.

This is the secret of individual differences. In no two individuals, the base pair sequence is same.

Unusual bases in DNA

The normal bases in DNA are A, T, G and C. In some viruses like PBS1 and PBS2 uracil replaces thymine. In some bacteriophages C is replaced by 5 hydroxy methyl cytosine.

Single strand DNA Although DNA in most organisms is double stranded, in some phage viruses (ϕ x 174), single stranded DNA is found. The following table summaries the differences between double stranded DNA and single stranded DNA.

Character	Double Strand DNA	Single Strand DNA
1. U.V. Absorption	Constant from 10-80^{0}C and then rises rapidly	Steady increase from 20^0 to 90^{0}C
2. Action of formaldehyde	Resistant	Not Resistant
3. Ratio of	equal, A=T and 1.33:0.98:0.75	A:T:G:C is
4. Shape	Linear	Circular
5. Respiration	Two strands separate and replication begins	Becomes double (replicative form) temporarily.

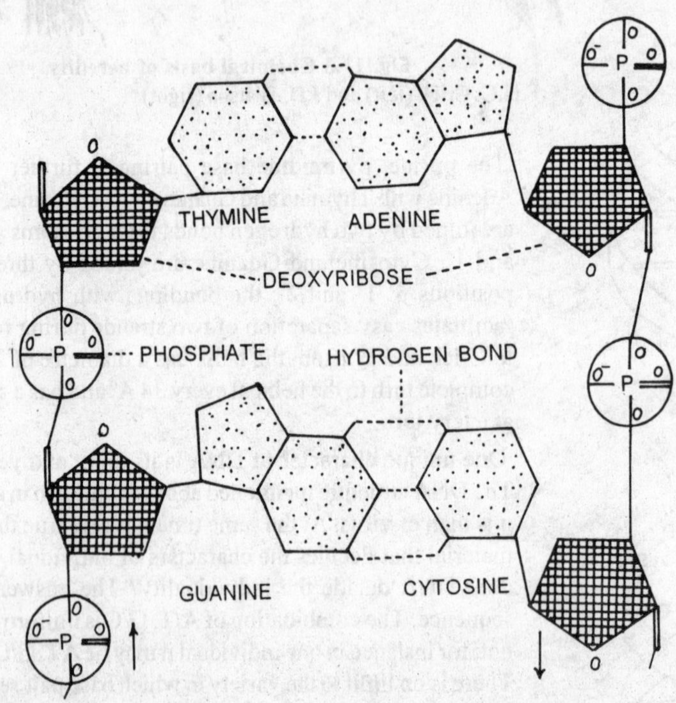

Fig.17.5 Chemical basis of heredity
Arrangement of nucleotides in a segment of the DNA model

Alternative forms of DNA double helices

The structure of DNA molecule described above as visualized by crick and Watson assumes that the helical coiling of the molecule is right handed and this form of DNA has been called the B-form. Subsequent researches however have indicated that other types of coiling are also possible for the DNA helex. These alternative forms of DNA differ from one another inthe following respects.

1. Number of residues per turn (<u>n</u>) and
2. Spacing of residues along the helix (<u>h</u>)

The existence of DNA as double helix has also been confirmed by experiments used for measuring the number of base pairs per turn of the helix and this works out to be 10.4 instead of 10 in the B-DNA. This leads to the modification of angle of rotation from 36 to 34.6. Thus it is possible that there could be structural variants of DNA with varying values for <u>n</u> and <u>h</u>. Three structural variants of DNA A,B and C are known since a long time and transitions between them are also possible. The following table gives a brief summary of the characters of these forms of DNA.

Helix type	Condition	Basepair	Rotation per bryophytes
A	75%	11	+32.7 (right handed)
B	92%	10	+36.0 (right handed)
C	66%	9.33	+38.6 (right handed)
Z	very high salt	12	-30.0 (left handed)

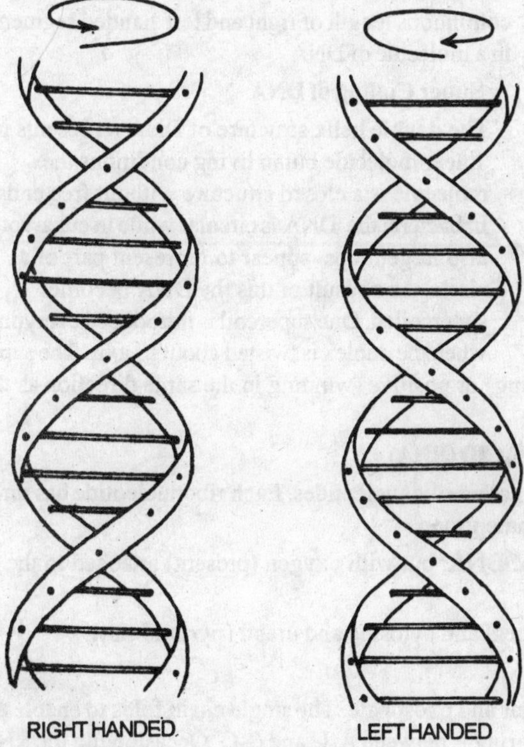

Fig. 17.6 Chemical basis of heredity
Two helices, one showing right handed coil as found in B-DNA and the other showing left handed sense as found in Z-DNA

The forms of DNA are assumed to be found in all molecules depending on the surrounding conditions. In addition to the above D and E forms of DNA are also found very rarely with only 8 and 7.5 base pair per turn. These are found only in some molecules of DNA lacking in Guanine. In contrast A,B and C forms are found in all molecules irrespective of DNA sequence.

Z-DNA

This type of DNA was first synthesized artifically but subsequently has been reported from living cells also. As this molecule follows a zig-zag course it is called Z-DNA. The resemblances and differences between B-DNA and Z-DNA are as follows -

Resemblances between B-form and Z-form

1. Both are double helical
2. The two strands are antiparallel in both forms
3. Both forms exhibit A=T and G=C pairing

Differences between Z-DNA and B-DNA

1. Z-DNA is left handed while B-DNA is right handed
2. The phosphate back bone is regular in B-DNA while it is zig in Z-DNA.
3. In Z-DNA sugar residues have alternating orientations as a result of which repeating units are dinucleotide whereas in B-DNA they are monocleotides.

4. Z-DNA has 12 bp for every twist of 360⁰ while B-DNA has only 10bp.
5. As more base pairs are accommodated in one helix in Z-DNA the angle of twist is 60⁰ as against 36⁰ in B-DNA.
6. One complete helix is 45A⁰ in Z-DNA while it is 34A⁰ in B-DNA
7. The diameter of the helix is 18A⁰ in Z-DNA while it is 20A⁰ n B-DNA

The role of Z-DNA is not certain, even though it is thought to play a role in regulation of gene activity, recombination or rearrangement of DNA.

RL Model

A new model for the molecular structure of DNA has been proposed by Sasisekharan and associates (1976-89) working at the Indian Institute of Science Bangalore. According to them the right handed (R) helix of B-DNA cannot easily explain the unwinding of the coil due to topological differences. Hence they suggested a combination of right handed (R) and left handed (L) helices which they called the RL model. In this model, a B-DNA molecule will have alternating left handed and right handed segments of approximately 5 base pairs in a repeat of 10 base pairs. This model provides conformational flexibility. Many molecular biologists opine that there is strong experimental evidence for the RL model although one is not sure of the continuous length of right and left handed segments in a molecule of DNA.

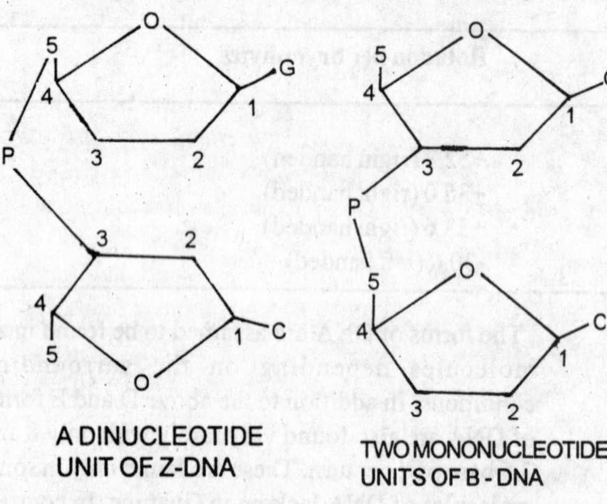

Fig. 17.7 Chemical basis of heredity
Orientation of adjacent sugar residues in Z-DNA and B-DNA showing opposite orientations in Z-DNA and same orientation in B-DNA. This results in dinucleotide units in Z-DNA as against mononucleotide units in B-DNA

Super Coiling of DNA

The double helix structure of DNA represents in linear molecule but in living conditions the molecule is a closed structure without free ends. In bacteria the DNA is circular while in eukaryotes also large loops appear to represent part of a circle. As a result of this the DNA becomes supercoiled. One supercoil is introduced everytime when the duplex is twisted about its axis. The super coil may be negative (winding opposite to the helix coiling) or positive (winding in the same direction as the helix coiling)

RIBONUCLEIC ACID (RNA)

The RNA is also a polymer of nucleotides. These are called ribonucleotides. Each ribonucleotide has three parts viz., a pentose sugar, a nitrogen base and a phosphate group.

The pentose sugar is a ribose sugar much like that of DNA, but with oxygen (present) attached to the 2' carbon position of the ring.

The nitrogen bases are - Adenine and guanine (purines) and cytosine and uracil (pyrimidines).

Molecular structure of RNA

Here also the back bone is made up of alternating sugar and phosphate. The single chain folds to enable the base pairs to meet here and there to form the steps. The pairing is between A-U and C-G. Occasionally the RNA may be double stranded but does not form a helix like DNA.

Single stranded RNA is the genetic material in plant viruses, poliomyletis viruses etc. Double stranded (non helical) RNA occurs as the genetic material in reoviruses.

Types of RNA

The following are the types of RNA found

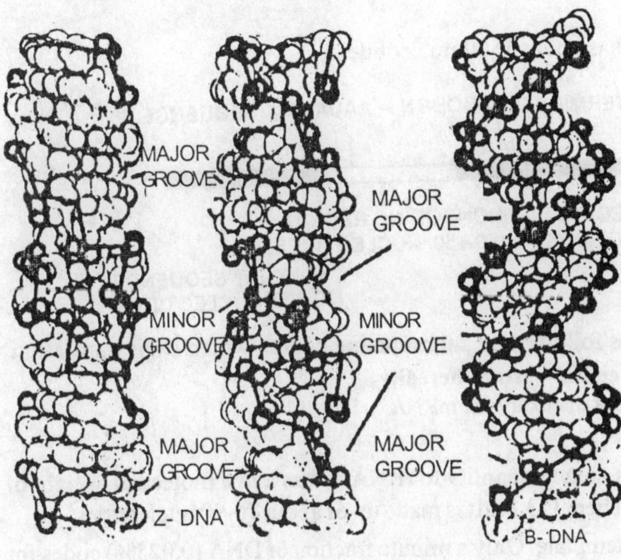

Fig. 17.8 Chemical basis of heredity

Side views of A-DNA. Two views of Z-DNA are 300 apart. Irregularity of backbone in Z-DNA is shown by heavy lines showing the path of phosphate residues, which is quite regular and smooth in B-DNA

(a) Genetic RNA
(b) Ribosomal RNA
(c) Messenger RNA
(d) Soluble or Transfer RNA.

Genetic RNA

In certain viruses, RNA is the only genetic material. DNA being absent. This RNA is self replicating; it is called RNA dependent RNA synthesis. The viral RNA directly functions as a messenger and produces RNA polymerase enzyme as well as proteins.

Ribosomal RNA (rRNA)

This is found in the ribosomes. It constitutes about 80% of total RNA of the cell. The base pairs ofrRNA are complementary to that part of the nuclear DNA from where they are produced rRNA consists of a single strand folding upon itself at some regions. The folded regions have paired bases, while the other regions have unpaired bases. Due to this in rRNA there is no purine pyrimidine equality.

There are three types of rRNA bases on sedimentation and molecular weight. These are (a) rRNA with molecules weight over a million, e.g. 21s-29s rRNA, (b) rRNA with molecular weight less than a million, e.g. 12s-18s rRNA, (c) rRNA with very low molecular weight (40,000) e.g. 5srRNA.

Messenger RNA (mRNA)

mRNA carries the message from DNA to ribosomes during protein synthesis. It is about 3-5% of total RNA content of the cell.

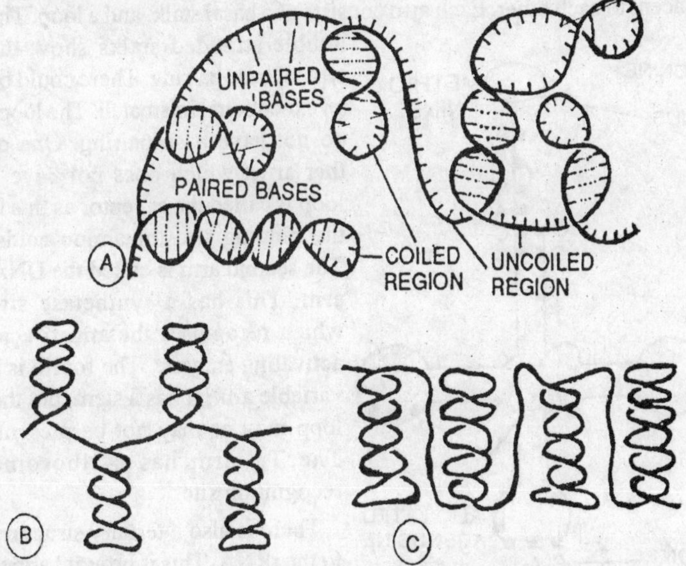

Fig.17.9 Chemical basis of heredity

Secondary structure of ribosomal RNA (diagrammatic)

A. Extended strand, **B.** Negative interaction showing repulsion, C. Positive interaction showing association

mRNA has a molecular wieght of about 5,00,000 with a edimentation cefficient of 8s. This however is highly variable mRNA is always single stranded and there is no base pairing, even though random coiling may be observed. The mRNA molecule is produced from the coding strand and its base sequence is complementary to the coding strand. The following are some of the general features of mRNA.

1. *Cap* : This is a blocked methylated structure found at the 5' end of RNA.

2. *Non coding region* : This follows the cap and consists of 10-100 nucleotides. There are no codes for proteins in this region.

3. *Initiating codon* : This region

begins the coding

4. *Coding region :* Codes for proteins and has about 1500 nucleotides.

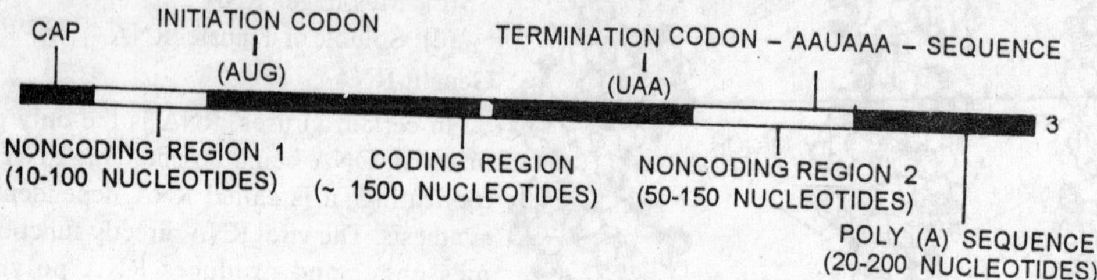

5. *Termination codon :* The coding region is followed by a terminating codon, which stops translation.

Fig. 17.10 Chemical basis of heredity
General features of eukaryotic mRNA

Transfer RNA (rRNA or sRNA)

The transfer RNA also called soluble RNA, is next in quantity to rRNA. It has a low molecular weight of about 25,000-30,000 and the sedimentation co-efficient is 3.8s. It is made up of about 70-90 nucleotides.

tRNA is synthesised in the nucleus on a DNA template. Only a minute fraction of DNA (0.025%) codes for tRNA. The tRNA molecule consists of a single strand looped around itself with the 3' end always terminating in a C-C-A sequence. Some bases are paired.

The function of tRNA is to carry aminoacids to mRNA during protein synthesis. Each aminoacid couples with a specific tRNA. Since the protein molecule consists of about 20 types of amino acids, there must at least be 20 types of tRNAs. But there could be many more, sinced there are two tRNAs for every amino acid in several cases.

Secondary structure of RNA

This refers to the arrangement of the chain of the molecule to enable it to function in a proper way. Many models have been proposed for the secondary structure of tRNA. Of these, clover leaf model proposed by Holley (1966) is most widely accepted.

According to the clover leaf model, the single chain of tRNA is folded on itself to form five arms. As a result of this folding, the 3' end and 5' end lie adjacent to each other. Each arm consists of a basal stalk, and a loop. The double stranded stalks show the typical base pairing. There could be an exceptional mismatch. The loops do not have base pairing. One of ther arms which does not have a loop is called the acceptor as this is the binding site for amino acids. The second arm is called the DNA arm. This has a synthetase site which recognises the amino acid activating enzyme. The fourth is a variable arm; it has a stem, but the loop may or may not be present. The Tψ arm has a ribosome recognition site.

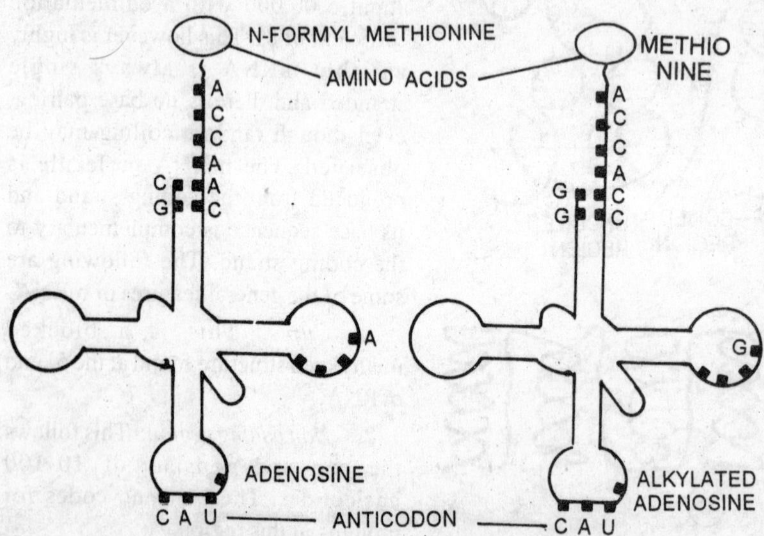

Fig. 17.11 Chemical basis of heredity
econdary structure of tRNA (Clover leaf model)

There is also a tertiary structure to the rRNA. This is brought about by H bonding between bases, and between bases and sugar phosphate backbone.

18

GENOME IN PROKARYOTES

The hereditary material in all organisms is remarkably similar. Having been evolved in the remote past, life has been able to maintain itself because of the ability of the hereditary material to maintain a basic uniformity in structure and function. The hereditary material in all living beings (whether it is a border line entity like a Virus or an evolved human being) is Nucleic acids without exception, even though there may be some stray instances of proteins (as prions) believed to carry out inheritance. The structure of the nucleic acids is unique; it has a basic uniformity, yet sufficient plasticity to allow for its manifestation from most primitive to the advanced organism.

In eukaryotic organisms, the genetic material is organised into large entities called chromosomes which allow for proper replication and distribution of genetic material during cell division and reproduction.

In prokaryotic organisms also,the nucleic acids constitute the chromosome, but the chromosome of prokaryotes is a naked strand of DNA (not associated with proteins) lying free in cytoplasm. The entire genetic system of each species is encoded in a single chromosome. Even in Viruses, the genome is either DNA or RNA.

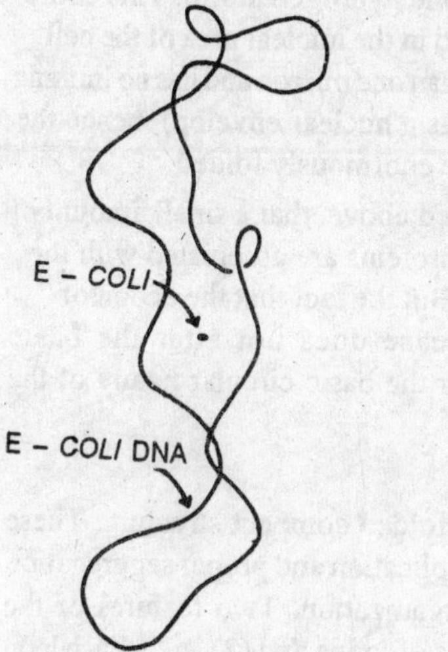

Fig. 18.1 The genome in Prokaryotes Schematic diagram showing the relative sizes of E.coli and its DNA molecule

The prokaryotic genome is refered to as the chromosome, centrosome, nucleoid, chromatin body or the gneophore. Some traditional cytogeneticists object to the word chromosome and prefer to use the term genophore or nucleoid. However, as Swanson (1981), points out "this is more a matter of semantics than of cytogenetics, since the prokaryotic chromosome not only has the same fundamental structure based on nucleic acid sequences, but it also carries out, the same three basic functions of its eukaryotic counter part. That is, it is the organelle which stores, replicates and transmits the coded information of biological inheritance; which is so regulated in its action that its primary products appear at appropriate times and in appropriate amounts during the life cycle of the organism; and which is capable of recombining its genetic content with other similar individuals in such a manner that the seggregating progeny of a hybrid union can have different genetic properties". Swanson (1981) therefore opines that the term "chromosome" can be applied to both eukaryotic and prokaryotic species with distnctions in structure being made at appropriate times.

The chromosome of *Escherichia coli* and *Salmonella typhimurium* are probably the most throughly mapped genomes of any cellular organism. Hundreds of identifiable genes are located in a single circular linkage group. The entire genomeof a bacterium is made up of a single chromosome. Occasionally a single bacterial cell may have more than one chromosome at a given time. This should not be taken as the cell having two chromosomes. This represents a stage wherein the chromosome has replicated and formed two units but the cell is yet to divide.

The chromosome of *E.coli* is a single, large double helix which is circular in outline. The DNA molecule is supercoiled. It (DNA) has no association of histones as in eykaryotic chromosomes. However, small amounts of basic proteins have been found to be associated with the isolated DNA. It is quite likely that these might prevent some sections of the chromosome from transcription thus functioning like repressor proteins. Thus it can be argued that the absence of DNA-histone association may not be such a fundamental point of difference between prokaryotic and eukaryotic DNA.

In *E.coli* the total length of the circular DNA is 1300μm, whereas cylindrical bacteria have a maximum diameter of 1μm only. This clearly indicates that the chromosome must be highly folded in the cell.

When the DNA of *E.coli* is isolated by techniques that avoid both DNA breakage and protein denaturation, a highly compact structure called *nucleoid* can be seen. This structure contains a DNA molecule, a fixed amount of protein and variable amounts of RNA. As has been pointed out already, the proteins might play a part in repression of transcription.

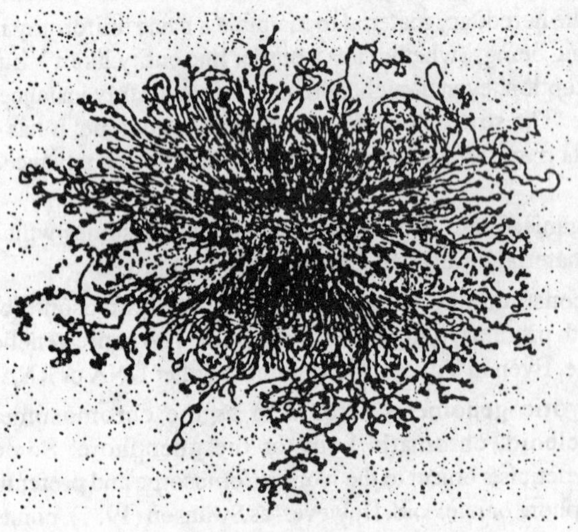

Fig. 18.2 The genome in Prokaryotes
An electron micrograph of an E.coli chromosomeshowingthemultiple loops merging from a central region

The DNA molecule is about 1000 to 1100μm in length with a diameter of 20 Å to 40 Å. Such a molecule has more than 3.235×10^6 nucleotide pairs (each pairs is 3.4 Å long and there are 2941 base pairs per micrometer). The molecule has a molecular weight of 2.5×10^9 daltons (one dalton is equivalent to the weight of one hydrogen atom). This entire molecule is limited in the nuclear area of the cell which is no more than one micron and has no limiting boundary (such as a nuclear envelop), hence the molecule must be enormously folded.

It has been pointed above, that a small amount of RNA and basic proteins are associated with the DNA molecule. But the fact that the action of RNAse or protease does not alter the basic structure of DNA, indicates that neither RNA nor proteins are necessary for the basic circular nature of the chromosome.

Detailed Structure of the Chromosome

The chromosome (DNA) of *E.coli* on isolation appears to be a highly folded compact structure. These folds appear to be highly regular and not random. If they were to be random, replication and proper seggregation would not be possible. The chromosome obviously has a high level of organization. Two features of the chromosome are noteworthy. These are - (1) The DNA is arranged in a series of loops and (2) Each loop is supercoiled. A dense region containing membranous material is seen in the central part. Some people think it to be an artefact resulting out of the isolation procedure. But it is quite likely that it represents fragments (mesosomes) of the plasma membrane to which the chromosome is attached in the intact cell. It is possible that enzymes required for the replication are located at this point of contact (with the membrane). Mesosomes do play a role in the separation of the replicated molecules of DNA during cell division.

The DNA has the four usual bases - adenine, guanine, thymine and cytosine. In addition, the genome of bacteria may also have some amounts of methylated bases such as *6-methylaminopurine* and *5-methylcytosine.*

Introduction of a single strand break into the supercoil by a DNAse enzyme results in an abrupt change from supercoiled to non supercoiled condition. This is because of free rotation about the opposing sugar phosphate bond. If the nucleoid however is nicked (cut), it does not become an open circle. But with each nick it passes through 40-50 intermediate stages in which the supercoiled loops turn into open loops one by one. This is a clear pointer to the fact that a single strand break does not cause free rotation of the entire molecule. The DNA is assumed to be fixed to the proteins in such a way that does not allow overlapping of the individual loops.

The enzyme (DNA gyrase) that brings about DNA replication is also responsible for supercoiling. This can be proved by the addition of *coumermycin* an inhibitor of DNA gyrase, to the culture medium containing *E.coli* cells. The chromosomes of *E.coli* quickly losesupercoiling (on additionof coumerycin)

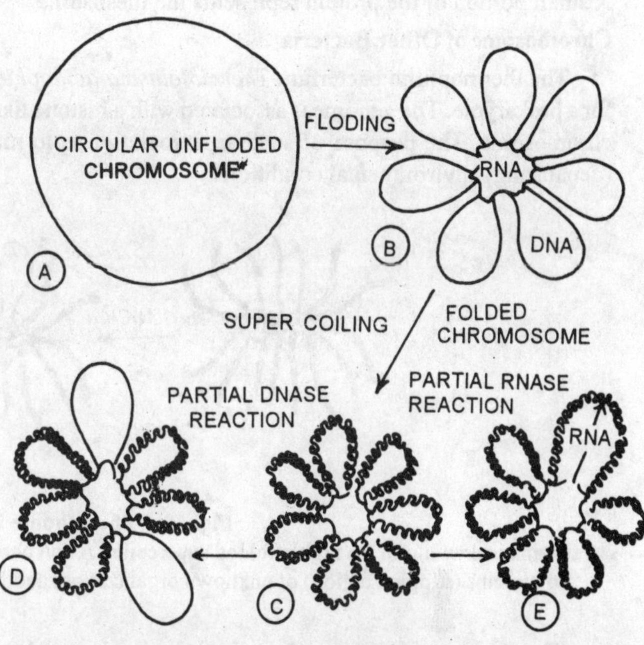

Fig. 18.3 The genome in Prokaryotes
Model of bacterial chromosome

. In addition to DNA gyrase another enzyme *topoisomerase* I has the ability of unwinding the supercoils. Hence it is believed that the two enzymes -*DNA gyrase* and *topoisomerase* I act in a coordinated fashion and regulate the degree of coiling. *In vitro* experiments have shown that purified topoisomerase I enzyme relases supercoiling to a non supercoiled covalent circle. Mutants of *E.coli* lacking *topiosomerase* I have been seen to posses increased supercoiling to the extent of 32 percent more than the normal.

Some interesting observations have been made on these mutants lacking in topoisomerase I. In *in virto* conditions, the mutants undergo a further secondary mutation in or near the gyrase gene which seems to reduce the gyrase activity and return the supercoiling to near normal state.

A. Circular unfolded chromosome
B. Folded chromosome
C. Folded and supercoiled chromosome
D. Supercoiling relaxed into loops by DNAse nicks
E. RNAse cuts one RNA molecule

Other Associated Structures of Chromosomes

The isolated genome of *E.coli* has about 30 percent by weight RNA and 10 percent protein. The RNA is largely a newly transcribed mRNA, tRNA or ribosomal RNA which is yet to be released into the cell machinery. A small amount of additional RNA in this is responsible for the folded character of the genome. RNAse activity can partially unfold the genome.

Most of the proteins have been identified as RNA polymerase engaged in transcription. This is quite understandable as a metabolically active bacterial cell would be virtually transcribing throughout its life cycle.

A small portion of the protein represents the mesosome.

Chromosome of Other Bacteria

The thermophilic bacterium *Thermolplasma acidophilum* possesses a chromosome that is quite unusual for a prokaryote. The genome is associated with a histone like protein which can complex with about 20% of the chromosome. The purpose of such an association is to maintain the stability of DNA during unfavourable (denaturing) environmental conditions.

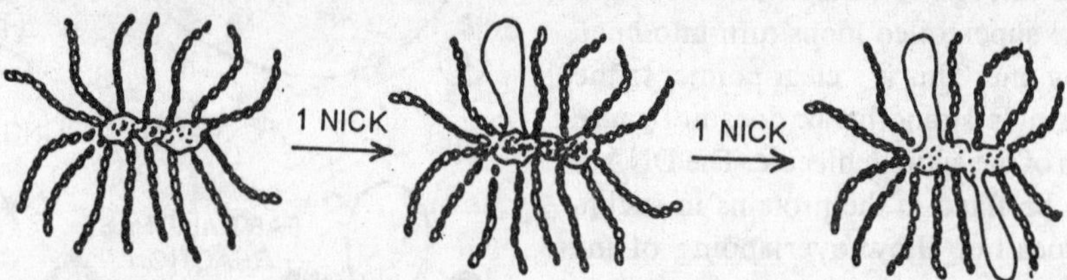

Fig. 18.4 The genome in Prokaryotes
A schematic drawing of the highly folded supercoiled *E.coli* chromosome, showing only 15 of the 40-50 loops attached to proteins (stippled region) of unknown organization, andthe opening of a loop by a single-strand break (nick)

The chromosome of *Bacillus subtilis* is circular and has approximately the same dimensions and molecular weight like that of *E.coli.* There are suggestions that the genome of *B.subtils* replicates as a linear molecule rather than as a circular molecule. *Streptomyces coelicolor* and *Salmonella typhimurium* are also known to possess circular genome with the latter acomodating 323 identified genes. Species of *M.mycoplasma* have a small circular chromosome of about 265μ in length. *Haemophilus infleuzae* the whooping cough causing bacterium, has a chromosome as long as 865μm which often appears linear with two free ends(?). Whether these ends really represent the points of a linear molecule or represent the ruptures of a circle during extraction it is difficult to say.

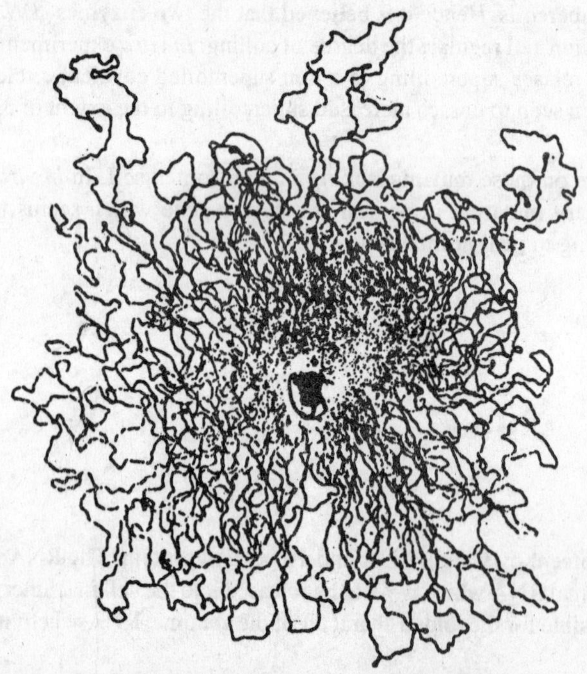

Fig. 18.5 The genome in Prokaryotes
An electron micrograph of the chromosome of the bacterium *Hemophilus influenzae.*

PLASMIDS

The bacterial cell has an additional fragment of nucleic acids in addition to its normal chromosome. This additional genetic material was given the name episome (Jacob, Sccaeffer and Woolmann, 1960). Episome can exist either independently or it can integrate itself with the bacterial chromosome. Plasmids however are always independent and never integrate with the bacterial chromosome. Plasmids control a number of traits in bacteria such as -toxin production, antibiotic resistance, pilus production etc. Some people regard that episomes and plasmids are synonymous.

The word plasmid was first used by J.Lederberg (1952). However, the term is restricted only to bacterial cells where circular DNA molecules exist independent of the main genome.

Plasmids are circular double stranded molecules of DNA and exist independently of chromosomal DNA in bacterial cells. They undergo replication during cell division and are carried to both the daughter cells. The plasmid has a right handed superhelical coil during the resting stage. There, is one super helical turn for every 400-600 base pairs. The twisted conformation is known as **covalently closed DNA (CCD).** Cleavage results in the conversion from CCD to open circular form.

The following are some of the properties of the plasmids.

1. They are made up of DNA (double stranded).
2. They are capable of independent replication
3. Some plasmids control bacterial conjugation
4. Capable of reversible insertion into bacterial chromosome (only some plasmids)
5. They have the ability to get transfered from one bacterial cell to the other.

Plasmids are refered to as *conjugative* and *non conjugative* based on the ability/inability to promote their transfer during conjugation.

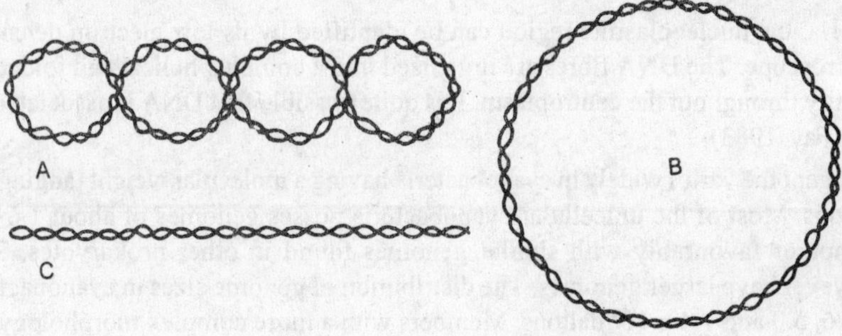

Fig. 18.6 The genome in Prokaryotes
Plasmid DNA, **A.** Supercoiled form, **B.** Open nicked circle, **C.** Linear duplex

Classification of Plasmids

Basically plasmids are classified into three categories -

1. The F factor or sex factor
2. The R Plasmid or resistance transfer factor (RTF) and
3. *Col* factor. There are some plasmids which may share the properties of more than one type of plasmid. Some bacteriologists recognise more than the three types based on some other properties.

F factor

The F factor plasmid confers the ability to the bacterial chromosome to get transfered to another cell (recepient). Hence the donor is said to be F^+. F factor integrates with the bacterial genome facilitating its transfer. Because of this integration, it is often refered to as an episome.

R Plasmid

These plasmids, when present confer on the bacterial cell resistance to several antibiotics such as streptomycin, tetracycline, chloramphenicol, sulphonamide etc. R plasmid mediated antibiotic resistance was first discovered in Japan in 1956 when *Shigella* became resistant to several antibiotics at one step. This was due to the transfer of R plasmid from one strain to the other resulting in not only antibiotic resistance but also the ability of a donor (it can donate R factor to other sensitive strains making them resistant).

R plasmid is a small extrachromosomal DNA ring. Sometimes the R plasmid has two parts - (a) Part havin drug resistant genes and (b) Part having plasmid transfer genes.

The col factor (Plasmid)

Certain bacteria like *Escherichia coli, Shigella, Salmonella* etc., produce toxins called *colicins*. These are toxic proteins and can kill other bacteria. The ability to produce *colicin* rests in the plasmid called col plasmid or *colicinogenic factor*. Col factors resemble phages since they cause death of bacteria, but they differ from phages as they are not released out during the lysis of the bacterial cell. Like the F factors col plasmids also can be transfered from one bacterial cell to the other.

The number of plasmids per bacterial cell varies. It may be *a single copy* or *multiple copies*. Multiple copy plasmids which replicate independently have been used in gene cloning experiments. A plasmid is cleaved and a desired gene is inserted into the plasmid DNA. The plasmid DNA during its replication produces multiple copies of foreign DNA. Such plasmids are called vectors.

GENOME OF CYANOBACTERIA

In cyanobacteria, the nucleoplasmic region can be identified by its low electron density when observed under electron microscope. The DNA fibres are organized into a complex helical and folded structure and are distributed uniformly throughout the centroplasm. It is quite possible that DNA is associated with histone like proteins and RNA (Fay, 1983).

The size of the genome varies widely in cyanobacteria having a molecular weight ranging between 1.6×10^9 and 8.6×10^9 daltons. Most of the unicellular cyanobacteria posses genomes of about 1.6×10^9 to 2.7×10^9 daltons. This compares favourably with similar genomes found in other prokaryotes. Some filamentous cyanobacteria however have larger genomes. The distribution of genome sizes in cyanobacteria falls into four categories - 2.2, 3.6, 5.0 and 7.4×10^9 daltons. Members with a more complex morphology obviously posses larger genomes than those with a simple structure. There is a theory that the modern day prokaryotes evolved from a primitive bacterium with a small genome size of about 1.2×10^9 daltons. Replication of this genome during the course of Evolution is supposed to have given rise to simple integer multiples seen in modern bacterial genome.

THE VIRAL GENOME

The core of the Virion is made up of nucleic acids. A virus has only DNA or RNA never both together. Within these parameters however considerable diversity exists. The nucleic acids may be double stranded, single stranded, linear or circular. Some have plus polarity; others may have minus polarity. Four types of nucleic acids are found in viruses with reference to the number of strands. These are

1. Single stranded DNA — (ss DNA) e.g., Colliphage virus
2. Double stranded DNA (ds DNA) e.g., Herpes virus
3. Single stranded RNA (ssRNA) e.g., TMV
4. Double stranded RNA (dsRNA) e.g., Reo Virus

Single stranded DNA is found in ϕX174 Virus. It was first discovered by Sinscheimer *et al* in 1950's. ssDNA may be linear (Paroviruses) or circular (ϕX174 Virus). The ssDNA becomes double stranded during replication and at that time it is known as the replicative form.

Double stranded DNA is found in a number of animal Viruses and bacteriophages. It has a variety of forms - linear (bacteriophages), cross linked (vaccinia Virus) or closed circular duplex as in papova Viruses.

Single stranded RNA is found in a variety of animal Viruses and icosahedral plant Viruses. The strand may be plus (infectious) as in RNA bacteriophages, togaviruses etc or minus (non infectious) as in rhabdoviruses and paramyxoviruses. Plus ssRNA acts directly as mRNA and translates proteins on the ribosomes of bacteria, whereas minus ssRNA first transcribes an mRNA and then it (mRNA) translates proteins (hence called non infectious).

Double stranded RNA is found in animal Viruses like reovirus, blue tongue Virus etc. The dsRNA has 10 or more segments.

The nuclei acid core of Viruses may be summarised as follows -

1. Plant Viruses have only RNA (ss or ds) with the exception of cauliflower mosaic Viruses (DNA virus)
2. Animal Viruses have RNA (ss or ds) and DNA (only ds no ss)
3. Bacteriophages have DNA (ss or ds) or RNA (ss or ds). However most phages or DNA Viruses.

GENES

The possibility of the existence of discrete hereditary particles came to be known due to the path breaking work of John Gregor Mendel. It was he who provided some order into the concept of heredity which was otherwise an assemblage of weird theories proposed by many biologists others. Mendel, by his simple but ingenious experiments proved that the heredity is controlled by discrete particles called 'factors' which seggregate, recombine and seggregate again producing a variety of characters. Thus Mendel provided the proof of existence of a physical basis for heredity. The factors of Mendel later came to be known as genes. The concept of gene has undergone a sea change eversince Mendel introduced it, but its essentials remain the same.

The term *gene* was coined by Johannsen in the year 1909 for Mendelian factors. He however knew very little regarding the nature and structure of the gene.

Gene concept

The gene concept was first introduced by Sutton and works of Morghan, Bridges, Boveri etc proved this hypothesis. The essential features of gene concept are summarised below -

1. Genes decide the physical as well as physiological characters. These are transmitted from parents tooffspring, thus serving as a genetic link between one generation and the other.

2. Genes are located on the chromosomes. Many genes are located on a single chromosome. This is because, there are a large number of traits and consequently large number of genes, while the number of chromosomes are far less.

3. Each gene occupies a specific position or *locus* on the chromosome. Any alteration in the position and number of genes on the chromosome will bring in large scale genetic changes. (Eg., bar locus in *Drosophila).*
4. Genes are arranged in a linear order on the chromosome much in the same way as beads on a string.

5. A gene may exist in only one state or may exist in or many states. The alternative state of existence of the gene is known as *allele.* Most of the genes are *diallelic* i.e., they exist in two states - *dominant* and *recessive;* of these one is the original and the other is mutant. In some genes, due to series of mutations in the same locus instead of two alleles, there will be a number of alleles. This is known as *multiple alleles* (eg., blood groups in man). They represent a series of less and less dominant genes starting with the wild one.

   In all cases however, even in dialletic series a gene need not be completely dominant or recessive.

6. Genes undergo change, resulting in their altered structure and function. This is known as mutation.
7. Genes duplicate themselves precisely andproduce copies. This is known as replication
8. Genes control the production of proteins (enzymes) and thereby regulate metabolic activity.

9. Chemically genes are made up of segments of DNA. This has been proved by the genetic transformation experiments of Avery Macleod and McCarty (1944). In some plant viruses however genes are composed of only RNA (eg., TMV).

Modern Definitions of Gene

Benzer clearly established the relation between various genetic phenomena and the DNA molecule. He coined different terms to denote the various aspects of gene structure and function. These are - *Recon, Muton* and *Cistron.*

Recon

Genetic research has indicated that on the DNA molecule, short segments exist which can be separated from the rest during crossing over. These segments are refered to as genes. In other words, gene represents a unit of recombination of the DNA molecule. Within a gene crossing over or recombination is not possible. But studies have indicated that within a gene also, recombination is possible. Hence a sub unit of gene which is capable of recombination is called a *recon.* Recombination studies on microbes have indicated that structurally, recon consists of one or two pairs of nucleotides.
Crossing over within a gene was demonstrated by Benzer in T4 bacteriophages.

Cistron

Garrod (1908), thepioneer who established the relationship between gene and its biochemical function argued that the gene directly controls enzyme synthesis. This was later elaborated by Beadle and tatum (1941) as '*one gene one enzyme hypothesis*'. The gene was regarded as a single unit to code for a single enzyme protein.

According to Benzee, the cistron represents a segment of the DNA molecule and consists of a linear sequence of nucleotides to control the production of a protein. In most cases thecistron is equivalent to the gene structurally. In *Escherichia coli,* eachcistron has 1500 base pairs. Some cistrons may be as long as 30,000 base pairs. As the genetic code is a triplet, each gene has 3 times the base pairs as compared to thenumber of aminoacids in the polypeptide chain (it codes).

Researches have however indicated that not all genes code for proteins, the hypothesis one gene one enzyme is no longer tenable.

Complon

This is a term used to replace cistron. A complon is said to be a unit of complimentation. Complimentation is a phenomenon, where in the active groups of enzyr ›s (when they are made up of more than one polypeptide chain) are complimentary to one another.

A few other aspects of gene structure and function are given below.

Discontinuous genes

Researches carried out in 70s indicated that biological informationmay nor be carried by the gene continuously all along its length. A gene is split into several sections separated by non coding segments of DNA. Such genes are called *discontinuous genes* or *split genes* or *mosaic genes.*

In split genes there are two sections - *introns* and *exons.*

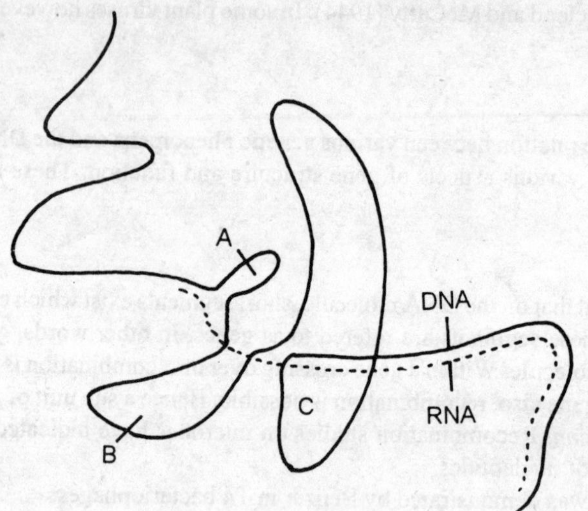

Fig. 19.1Genes
The hexon gene of Adenovirus. The loops of DNA represent regions of intron

Introns are non coding sections of DNA segment. These are present in genes of eukaryotes and their viruses, but absent in viruses, but absent in prokaryots and their viruses. In eukaryotes mitochondrial genome does not have introns. The mammalian α collagen gene has as many as 52 introns.

Gilbert, has proposed a hypothesis for the origin hypothesis for the origin of new genes by the shuffling of exons between different discontinuous genes.

Transcription from a split gene which has several introns would initially produce a precursoe mRNA which will have several non coding regions. Before translation, the introns have to be removed and the exons reatached to produce a final copy of mRNA with continuous codons. This is called *gene splicing.* Examples of split genes include *Hexons gene* of adenovirus, *ovalubumine gene* of chicken, β globin gene of mice and rabbits. In the ovalbumin gene the intron proteins of mRNA loopout and then these loops are excised.

Overalpping genes
According to the one gene one protein hypothesis of Beadle and Tatum (1940), each gene will separately code for one protein and that genes do not overlap. But the work of Barrel *et al* (1976) gave the first evidence of overlapping genes in the bacteriophage ϕ X174. They found that two genes code for a particular protein. Usually in organisms the length of DNA should be more than the length of the polypeptide as three nucleotides code for one aminoacid. In phage ϕ X174, however the length of the polypeptide chain is more than that of DNA. This indicates that the proteins coded by two genes is specified by the same of DNA.

Jumping genes
Generally, the gene loci on the chromosome are fixed. But the work of Nobel laureate Barbara Mclintock has shown that genes keep changing their location on the chromosome. The occurrence of jumping genes has also been discovered in bacteria. Streches of DNA in bacteria can move from one location to another, therby altering the expression. The jumping segments of DNA have been called **insertion elements.**

Barbara Mclintock, working on maize has shown the movement of certain streches of DNA known as controlling elements from one location to another.

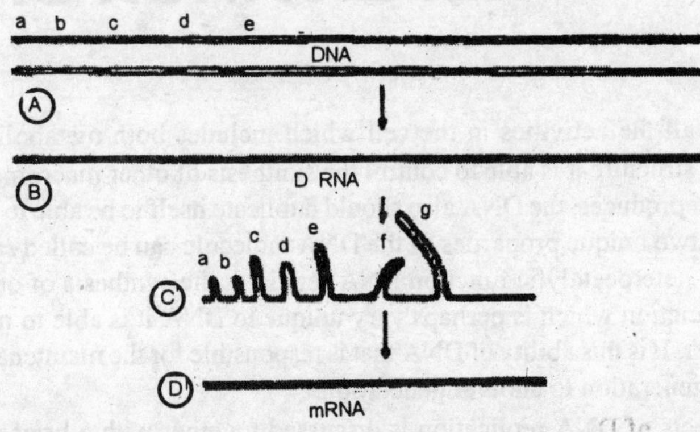

Fig. 19. 2 Genes

Transcription and mRNA in ovalbumin. gene-Differentstages.(In A and B blank representsnon coding regions while black represents coding regions)

Pseudogenes

In multicellular organisms, a wide variety of DNA sequences are found which apparently have no function. Some of these sequences are defective copies of functional genes and are therefore called pseudogenes. Pseudogenes have been reported from mouse, *Drosophila* and human beings. A typical example of human pseudogene is α globin and β globin pseudogene.

20

REPLICATION OF DNA

The DNA molecule regulates all the activities in the cell which includes both metabolism as well as reproduction. Because of its unique structure it is able to control the synthesis of other macromolecules of the cell. At the same time when the cell reproduces, the DNA also should duplicate itself to be able to get distributed equally to the daughter cells. These two unique properties of the DNA molecule can be called *Heterocatalytic* and *Auto catalytic* functions. In the heterocatalytic function DNA regulates the synthesis of other molecules like proteins. In the autocatalytic function which is perhaps very unique to DNA it is able to make copies of itself in a precise and accurate manner. It is this ability of DNA that is responsible for the maintenance of life and inheritance of characters from one generation to another generation.

In this chapter the various aspects of DNA replication is discussed together with a brief account of the enzymes responsible for replication.

Types of Replication

Three different types of replication have been proposed at the molecular level for DNA; these are - Semiconservative, Conservative and Dispersive.

1. Semiconservative

In this model of replication the DNA helix unwinds but there is no rupture of a separated polynucleotide strands. Thus the primary helices of DNA are not completely changed. According to this mechanism the unwinding of the original strand would be followed by the synthesis of complementary strands for each of the two separated original polynucleotide strands. This mechanism was proposed by Crick and Watson the discoverers of the molecular structure of DNA. In this method of replication, after replication of the original molecule, two molecules of DNA are formed. Each one will have one parental strand and one newly synthesised strand. Thus in the semiconservative process, after replication the molecules of the next generation can be regarded as hybrids. The mechanism is called semiconservative because of the two strands of the parental molecule only one strand remains in the molecule whereas the other goes to another molecule.

2. Conservative

In this mechanism the original molecule is completely conserved and thus form a part of the second generation molecule. Both the strands of the parental helix will synthesis an entirely new molecule the new molecule synthesized would have both of its strands produced newly. After one generation of application 50% of the molecules will be parental and 50% newly produced molecules. This is called conservative because the original combination of both the parental strands will remain together and will not be separated.

3. Dispersive

In this method of replication the parental DNA helix breaks up at several points forming many fragments. Each fragment of the helix would replicate and recombine at random to produce two daughter molecules of DNA. This is called dispersive because the original combination of the strands is completely disturbed as they get fragmented into bits.

Evidence for the Semiconservative Mode of Replication

Of the three types of replication mentioned above there are overwhelming evidences to support the semiconservative mode of replication. Taylor *et al* (1957) demonstrated by autoradiography that the two chromatids during prohpase have one strand old and one strand new material. This clearly indicates that the replication is semiconservative. The following three experiments have clearly given experimental evidence for the semiconservative mode of replication.

1. Meselson and Sthal's Experiment

Meselson and Stahl (1958) conducted experiments on *E.coli* and demonstrated that the semiconservative mode of replication is correct. In their experiments they used the following two basic principles.

(i) They cultured *E.coli* on a culture medium containing N^{15} (N^{15} is an isotope while normal nitrogen is N^{14}). The bacteria replicated in this medium for a number of generations and as a result both the strands of the DNA molecule now contain N^{15}. In other words, the DNA of the bacterial populaton was now labelled with heavy nitrogen (N^{15}). The labelled bacteria were then transfered to a culture medium containing N^{14} and are allowed to replicate. After the first generation of replication, the molecules of DNA each one possesed one strand with N^{15} and the other strand with N^{14} (this should be the composition of the molecules if the mode of replication is semiconservative). In these molecules the N^{15} strand represents a parental strand and the lighter N^{14} strand presents a newly synthesised strand. Thus each molecule can be regarded as a hybrid DNA (having both N^{15} and N^{14}).

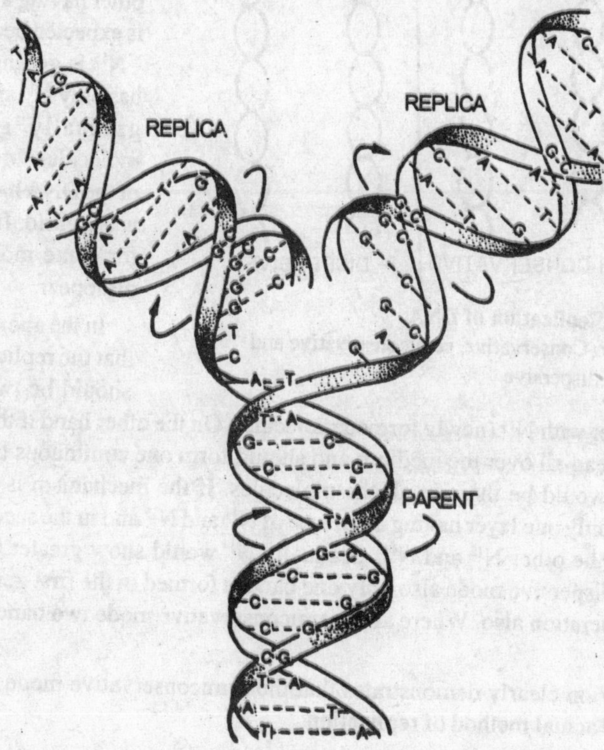

Fig. 20.1 Replication of DNA
Model of replication of DNA

(ii) Meselson and Stahl used the principal of density gradient centrifugation and prepared gradients of Caesium chloride. This density gradient is based on the fact that when heavy salt solutions are subjected to ultra centrifugation a continuous density gradient is set up (heavy would be at the bottom and lighter ones will be at the top with a range in between in the centrifuge tube). When a substance having a density within the range of this gradient is dissolved in salt solution, the substance will find place at its own level of density. Thus with this technique even very slight differences in density can be detected.

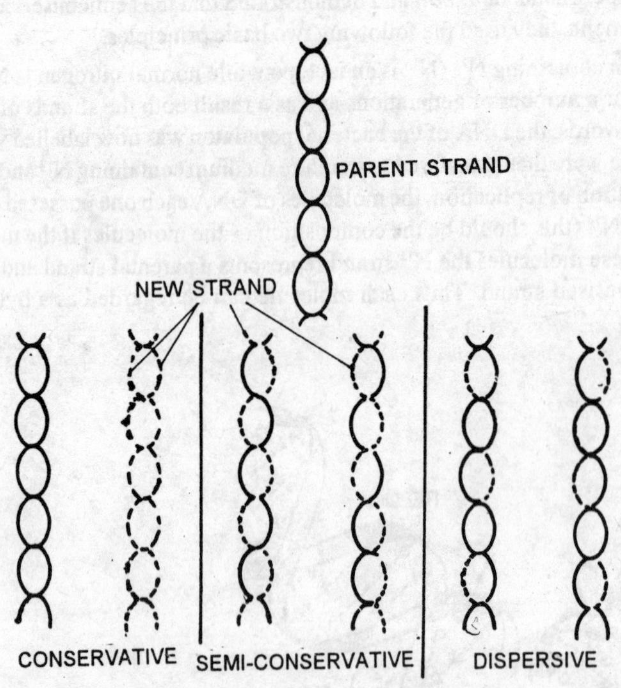

Fig. 20.2 Replication of DNA
Models of DNA replication-Conservative, semiconservative and dispersive

Based on the principles mentioned above Messelson and Stahl subjected the bacterial cultures for density gradient centrifugation. At the first instance N^{15} labelled bacteria were allowed to replicate on a medium containing N^{14}. After the first division the centifugation showed the hybrid DNA had an intermediate density and formed a band inbetween N^{14} and N^{15} indicating clearly that each molecule contains one N^{14} and N^{15}. in the second generation two bands were found one having normal N^{14} and the other having a hybrid (N^{14} and N^{15}). This is expected because the labelled nitrogen - N^{15} is getting reduced as the medium has only N^{14} and the original labelled N^{15} gradually gets decreased in the molecules. In the third generation 3/4ths of the DNA had N^{14} and 1/4th of the DNA was hybrid. If the replication continues for some more generations N^{15} would disappear.

In the above experiment if we assume that the replication is conservative there should be two bands in the centrifuge one with N^{15} and the other with N^{14} (newly formed molecule). On the other hand if the mechanism is dispersive the molecule should spread all over the medium and should form one continuous band as both N^{14} and N^{15} in different combinations would be there in all the molecules. If the mechanism is semiconservative the first generation would show only one layer having a mixture of N^{14} and N^{15} and in the second generation there would two layers one N^{14} and the other N^{15} and N^{14}. Gradually N^{14} would show greater band width and N^{15} would decrease. (Note. In the dispersive mode also only one band is formed in the first generation and it continues to be so for the second generation also. Where as in semiconservative mode two bands are formed in the second generation).

The above explanation clearly demonstrates that the semiconservative mode of replication proposed by Watson and Crick is the actual method of replication.

2. Cairn's autoradiographic experiment

The semiconservative mode of DNA replication has been proved in the chromosomes of *E.coli* by autoradiographic experiments conducted by Cairns.

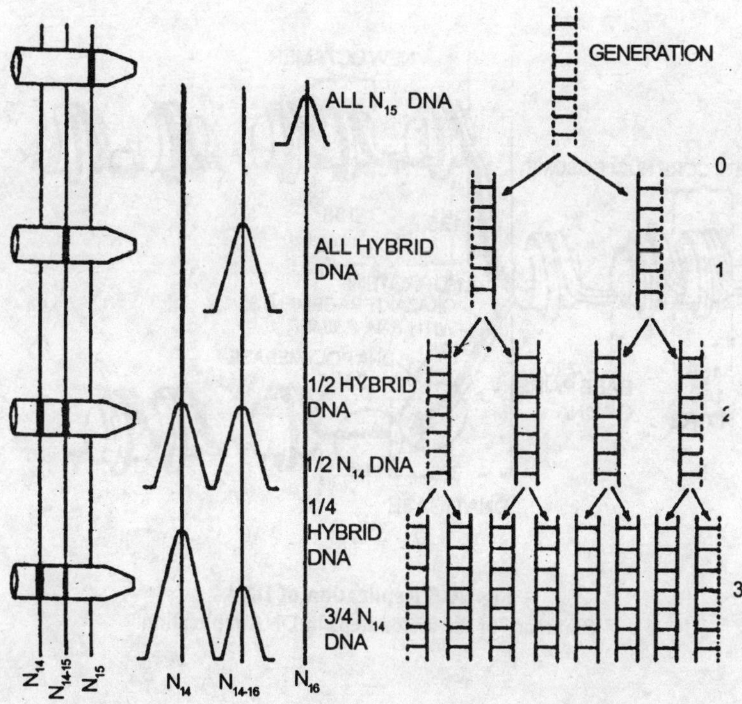

Fig. 20.3 Replication of DNA
Meselson and Stahl experiment in support of the semiconservative model of DNA replication

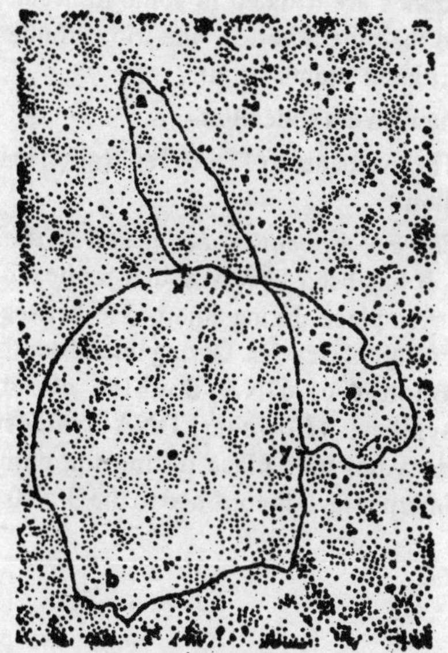

Fig. 20.4 Replication of DNA
Autoradiography of the single circular chromosome of *E.coli* labelled with tritiated thymidine-Cairans experiment

Autoradiography is based on the fact that decaying radioactive elements emit electrons which collide with the silver grains of a photographic plate and turn it black. Thus in a autoradiogram black dots represent positions of decaying radioactive elements. DNA of an organism can be labelled selectively by growing cells in a medium containing radioactive thymidine. Radioactive thymidine is obtained by using H_3- a heavy isotope of hydrogen known as tritium. Cairns used tritiated thymidine as it selectively labels only DNA and not RNA, because thymidine base is found only in DNA and not in RNA. By growing *E.coli* in the culture medium containing tritiated thymidine, radioactivity was incorporated in the daughter DNA molecules. The material (DNA) was extracted and mounted on a photographic emulsion for autoradiography. The labelled DNA exposes the film producing a diffused profile of DNA molecule. The duplicating molecule showed a replication fork. After the replication the radioactivity was found incorporated in only one strands of DNA and in the second generation both the strands become labelled and this will be clearly seen on the photographic emulsion because of the presence of black dots.

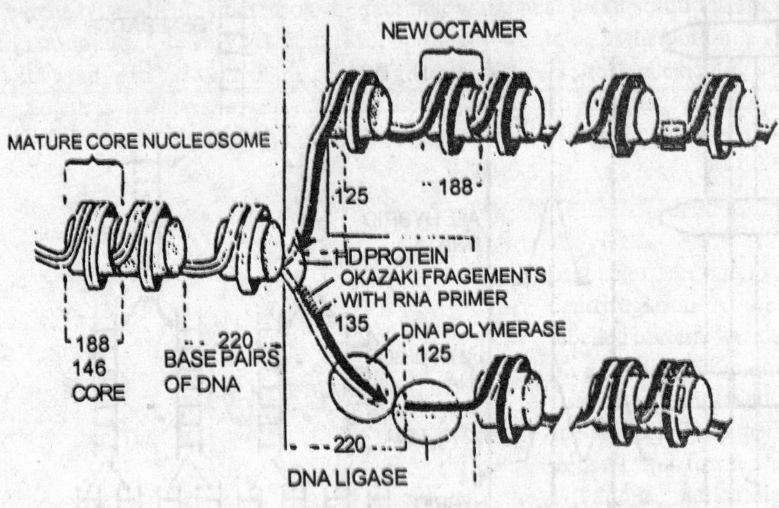

Fig. 20.5 Replication of DNA
Summary of the major steps in DNA replication

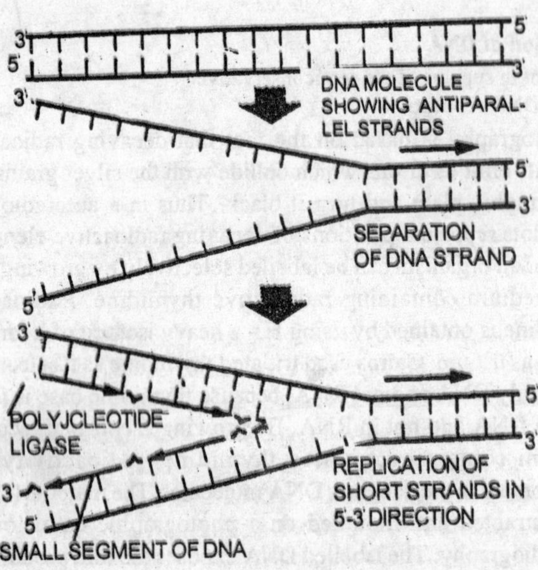

Fig. 20.6 Replication of DNA
Diagrammatic representation of discontinuous synthesis of DNA

This experiment also clearly shows the mode of replication is semiconservative because after the first replication, each molecule will have only one labelled and the other non tritiated strand. In the second generation both the strands are tritiated in some molecules and in the other hybrid - one tritiated other non tritiated strands occur.

Mechanism of replication

The mechanism of DNA replication in is as follows -

1. *Stage of replication during cell division* - It has been observed that replication takes place during interphase between two mitotic cycles. During interphase the amount of DNA doubles due to synthesis. However, DNA synthesis does not take place during the entire interphase. It is confined to the S phase.

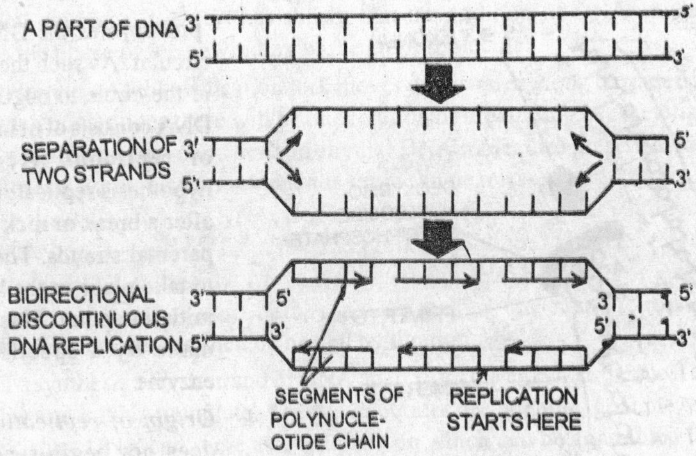

Fig. 20.7 Replication of DNA
Diagram showing bidirectional DNA replication

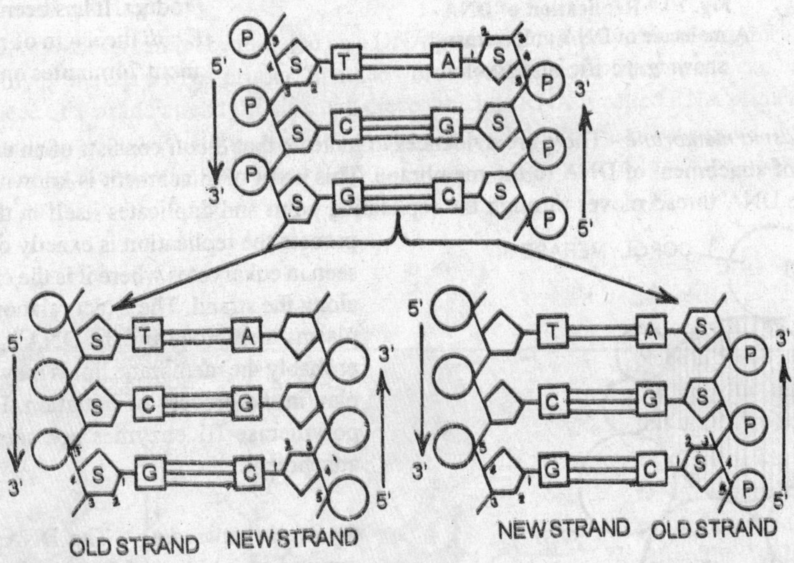

Fig. 20.8 Replication of DNA
Replication of DNA molecule. The two strands of DNA separate and each of them synthesizes its counterpart, thereby forming two similar molecules of DNA

2. *DNA Synthesising Genes* - In *E.coli* the overall process of replication requires the products of the genes *dnaA, dnaB, dnaC-D, dnaE,* and *dnaG.* The gene products of A, C and D ae required for the initiation of fresh replication and the products of E and G are required for the fork movement for replication. The product of G appears to be necessary for initiating the synthesis of DNA fragments. The *E.coli* also possesses a gene for the origin of DNA synthesis which is about 74 minutes from the beginning. There is also a gene for the termination of the replication. There are additional genes to control the synthesis of the three DNA polymerase enzymes. PolA and polB regulate the synthesis of polymerase I and polymerase II enzymes while G and E regulates polymerase III enzyme.
3. *Initiation of replication with a nick* - Unlike in the case of eukaryotes when the molecules are linear in

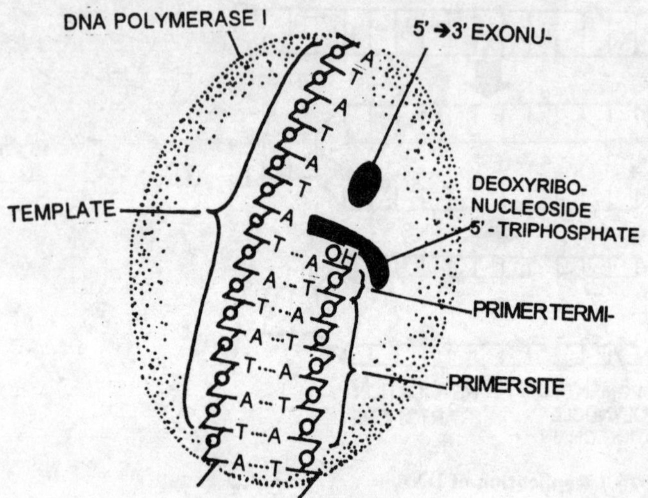

Fig. 20.9 Replication of DNA
A molecule of DNA polymerase-1 showing specific binding sites

prokaryotes the DNA molecules are circular. As such there has to be break in the circle to begin replication. The DNA consists of many replicating units or replicons. According to one hypothesis replication is initiated only after a break or nick in one of the two parental strands. The cut is presumed to take place in the helix at a specific initiation point. This cut or incision is made by a specific endonuclease enzyme.

4. *Origin of replication* - Replication does not begin randomly it always begins with a specific site called *ori* site and is regulated by a gene and its product. It has been discovered that in *E.coli* theorigin of replication point as about 74minutes on the map.

5. *Role of plasma membrane* - There are evidences to indicate that *E.coli* consists of an enzyme complex at the point of attachment of DNA to the membrane. This point of attachment is known as the replicating point. The DNA thread moves through the replicating point and duplicates itself in the process. In this process the replication is exactly opposite of what is seen in eukaryotes where it is the enzyme that moves along the strand. The exact relationship between the plasma membrane and the DNA is not known. Quite probably the membrane lipids may have some role to play in the process of initiation. Polymerase II and polymerase III enzymes are seen at the point of attachment.

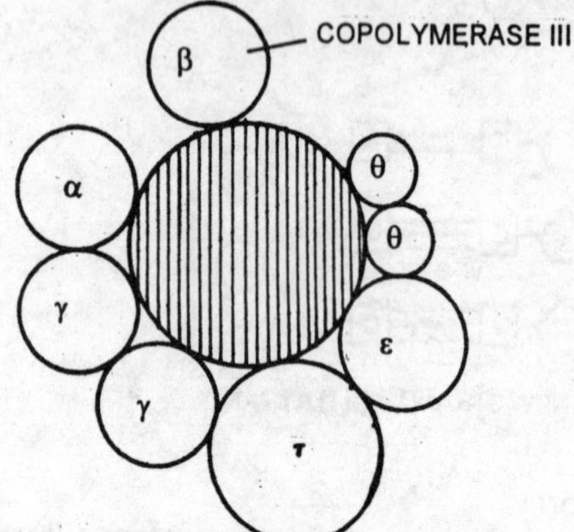

Fig. 20.10 Replication of DNA
DNA polymerase-III holoenzyme, it consists of subunits α, β, γ, ε, θ, δ, and ε

6. *Strand unwinding* - The DNA helix has to first unwind in order to separate the two strands for beginning the replication. A study of the unwinding process in T_2 bacteriophages shows that each DNA molecule has to undergo 20,000 rotations in unwinding during replication. As the molecule itself consists of 2 lakh nucleotide pairs, in these phages replication must go on simultaneously along with unwinding.

The unwinding of the helix is controlled by enzyme proteins commonly referred to as DNA unwinding proteins. These proteins selectively bind to the single strands of DNA and promote unwinding of the coil. The region of the double helix where the proteins bind forms a Y shaped structure called the replication fork. It has been found out that nearly 200 molecules of the proteins are found at each fork.

Chemical analysis of the DNA unwinding protein shows that each molecule is a tetramer having 4 sub units. Experimental evidence in *E.coli* has shown that the addition of DNA unwinding proteins to the medium increases the phase of the unwinding of the DNA molecule.

7. *Role of Swiveling Protein* - As the replication fork moves downwards along the parental strands, unwinding begins. For every ten nucleotides there will be one rotation of unwinding. As the original configuration of the molecules is in the form of a wound helix in circular DNA molecules present in prokaryotes untwisting or unwinding imposes a strain. To some extent this strain can be releived by additional twisting called super twisting in the unreplicated part of the DNA. However this super twisting has a limit and ultimately even the supertwisting should be unwound for the replication fork to continue. The releiving of the strain imposed by unwinding in brought about by proteins called superhelix relaxing proteins. These proteins introduce small cuts in the non replicating region of DNA thereby making one strand to rotate upon the other; thus releiving the strain. The breaks can then be healed by ligation.
8. *DNA Template* - To synthesis a fresh strand of DNA apart from the enzymes the most important requirement are the nucleotides and a DNA template. The 4 nucleotides are adenine, guanine, cytosine and thymine. The previously existing DNA can serve as a template on which can be assumbled fresh nucleotides.

   The template DNA can be either single stranded or double stranded. Single stranded DNA is found naturally in some bacteriophages. Template DNA acts as a mould on which new strand of DNA is synthesised by complementary base pairing.
9. *RNA Primer* - DNA synthesis cannot be initiated on a DNA primer. DNA polymerase enzymes can initiate DNA synthesis only if there is a RNA primer. In other words synthesis of a fresh strand can begin only if there is a small piece of a strand already. This small piece which is RNA is called RNA primer. The RNA primer is a short polynucleotide chain synthesised by the DNA template close to the point of replication. In all probability this is catalysed by a modified RNA polymerase enzyme. It has been noticed that in *E.coli* RNA polymerase is necessary to begin the replication.
10. *Elongation of chain* - Synthesis of new DNA strand takes place by the addition of fresh nucleotides to the 3'-OH group of the last nucleotide in the RNA primer. This synthesis takes place in 5'→3' direction and the enzyme that catalyses this is DNA polymerase III.

    The newly synthesised DNA chain has an RNA primer attached to its 5' end. After the synthesis of the chain, the RNA primer is hydrolysed by the exonuclease activity of DNA polymerase I. The resulting gap left by the RNA primer is filled by DNA nucleotides by the catalytic activity of DNA polymerase I. The freshly formed nucleotide chain is joined to the existing chain by the lygase enzyme.

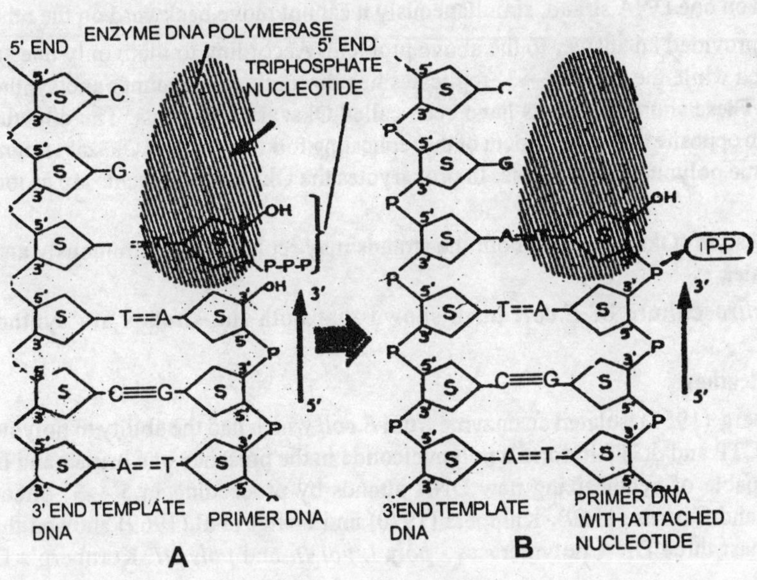

Fig. 20.11 Replication of DNA
Diagram showing action of DNA polymerase in DNA replication

Several opinions have been put forward to account for the presence of an RNA primer to initiate the synthesis of DNA chain. According to one opinion if any errors occur during the initial synthesis it can be eliminated when the RNA primer is removed.

11. *Direction of Replication* - It has been seen that in prokaryotes the replication may proceed in one or both directions from the point of origin. Thus replication may be unidirectional or bidirectional. Bidirectional replication has been seen in *E.coli, Bacillus subtilis, Salmolella, typhimirum, E.coli* phages etc. Unidirectional replication has been reported bacteriophages P2 and 186 and also in the mit DNA of mouse LD cells. Bidirectional replication has an advantage in that the origin is completely duplicated before the complication of replication.

12. *Formation of Replicating Forks* - At the point where the two strands get separated to initiate replication a replicative fork is found. The shape of the replicating fork is in the form of the English alphabet Y. When bidirectional replication takes place, the separated strands between the two forks appear like the bubble or a convex lens under the electronic microscope. The speed of movement of the two forks need not be the same.

    In organisms like *E.coli* where the chromosome is circular there will be only two replicating forks, where as in eukaryotes there could be several thousand replicating forks. These large number of forks is due to the fact that replication begins simultaneously at many points along the length of DNA. Occasionally however even in bacteria when replication is going on at one point another replication may start at another point. This will facilitate in completing the replication in shorter period.

13. *Discontinuous Replication* (Okazaki fragments) - The duplication of the strand of DNA takes place by the movement of replication forks. Replication of the two parental strands will take place at the same time as the fork moves along. However all the DNA polymerase enzymes can enhance the length of the strands only in the 5'→3' direction. The strands of DNA on the other hand polarise in opposite directions. This naturally possess a question as to how can DNA polymerase bring about simultaneous replication. While the enzyme moves forward on one DNA strand, simultaneously it cannot move backward on the other.

Okazaki et al (1968) have provided an answer to the above problem. According to them only one strand 3'→5' is continuously replicated while the other 5'→3' replicates in a discontinuous manner and synthesize only short fragments of DNA. These short fragments have been called Okazaki fragments. The direction of synthesis of Okazaki fragments in opposite to the movement of the replicating fork. Eventually Okazaki fragments are joined together by the enzyme polynucleotide ligase. In prokaryotes the Okazaki fragments are as long as 1,000- 7,000 nucleotides.

According to another suggestion (Okazaki et al) both the strands may replicate discontinuously and the fragments formed may join up later.

Evidence obtained form *in vitro* culture of *E.coli* have shown that both the strands are synthesisd discontinuously.

Enzymes involved in DNA Replication

1. **DNA Polymerases** - Kornberg (1957) isolated an enzyme from *E.coli* which had the ability to polymerise the nucleotides dATP, dTTP, dCTP and dGTP to form a polynucleotide in the presence of a primer and DNA template. This enzyme was capable of synthesizing new DNA strands by proceeding in 5'→3' direction. Subsequent works of DeLucia and Carirns (1969), Knippers (1970) and Gefter et al (1969) showed that in prokaryotic cells there are at least three DNA polymerases - *poly I, pol II,* and *poly III.* Kornberg's DNA polymerase is equivalent to *Pol I.*

Polymerase I - This enzyme does not seem to be necessary as bacterial mutatants lacking this are able to synthesize DNA. *Poly I* seems to be taking part in DNA repair. Structurally poll is a single polypeptide chain with a molecular weight of 1,09,000. Electron microscopic studies of the enzyme have shown that the enzyme has a number of functional sites -

(i) A template site for binding template DNA
(ii) Primer site for binding primer RNA segment
(iii) A primer terminus site for 3' - hydroxyl terminus of primer
(iv) A triphosphate site - locus for incoming deoxy ribonucleotide 5' group
(v) Exonuclease site for 5'→3' exonuclease activity.

Polymerase II - This is a single polypeptide chain with a molecular weight of 90,000, *E.coli* cell has about 40 molecules of *pol III.* It helps in 5'→3' polymerization.

Polymerase III - Discovered by Kornberg and Gefter (1972) it is the most active among all the three enzymes. The enzyme is chiefly responsible for DNA chain elongation. It has a molecular weight of 5,5000. The enzyme also acts as 5'→3' and 3'→5' exonuclease..

DNA ligase. Also known as polynucleotide ligase, this enzyme joins the two fragments by catalysing the synthesis of a phosphodiester bond between a 3'- OH group at the end of one chain and 5'→3' - OH group at the other end of the other chain. The enzyme is necessary to join Okazaki fragments which are produced after discontinuous replication.

The enzyme has been isolated from *E.coli.* It is made up of a single polypeptide chain and has a molecular weight of 70,000. Each cell of *E.coli* is known to posses 2000-4000 molecules of this enzyme. DNA ligase has the following functions.

1. Joins Okazaki fragments during DNA replication
2. Helps in sealing single strand nicks in DNA duplex
3. Joins free ends of DNA (linear molecule) to form a circular chromosome.
4. Joins the segments of DNA during recombination during meiosis.

MODELS FOR DNA REPLICATION (IN PROKARYOTES)

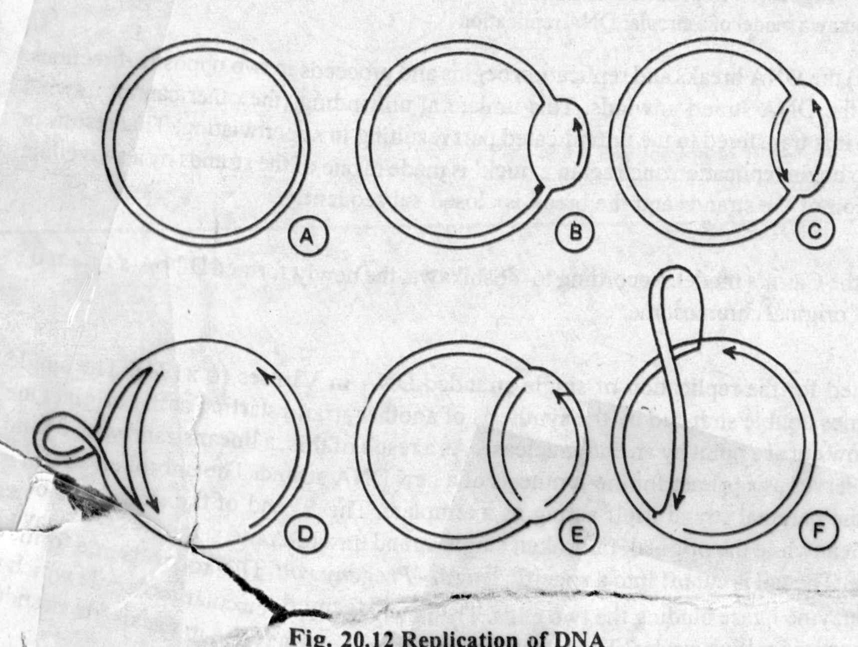

Fig. 20.12 Replication of DNA
Cairn's model for circular DNA replication

Jaboc and Brenner's Model

Studies with phase contrast microscope have shown that in the dividing cell, the sister chromosomes get oriented equidistantly inside the cell. In eukaryotic cell, this separation is achieved by the mitotic spindle. In prokaryotes like bacteris such a mechanism is lacking. Jacob and Brenner have proposed a model to explain the replication and separation of daughter chromosomes. Accordingto this model, the chromosome gets attached to the mesosome at the point of origin of replication. After the completion of replication, a new membrane attachment site is formed adjacent to the old one. The strand which has acted as a template remains a circle, while the newly formed one is broken. One of the free ends of the newly formed chromosome attaches to the new binding site. During replication, the replicating fork and the enzyme system remains stationary while the DNA molecule moves along and gets replicated. Simultaneously a new membrane is synthesized between the two and the two genomes get separated. The newly synthesized DNA becomes circular eventually.

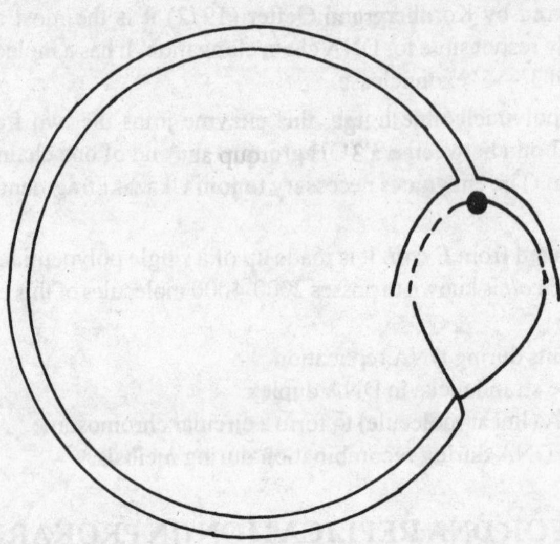

Fig. 20.13 Replication of DNA
Yoshikawa model of a circular DNA replication

Carin's Model

At a specific point (origin) the DNA breaks and replication begins and proceeds in two opposite directions. As the replication proceeds, the DNA strand unwinds. This unilateral unwinding (the other cannot unwind freely) creates a pressure and it is transfered to the unreplicated part resulting in supertwisting. This results in the stoppage of replication. To begin replication once again a 'nick' is made in one of the strands by a swivelling protein. This facilitates rotation of the strands and the break is closed subsequently.

Yoshikawa Model

This is a modification of the Cairn's model. According to Yoshikawa, the newly formed DNA strands covalently join to the ends of original chromosome.

Rolling Circle Model

This is the model accepted for the replication of single stranded DNA in Viruses (ϕ x174). The single stranded circular DNA becomes double stranded by the synthesis of another strand starting at the origin. One strand of this duplex ring is now cut at a point by an endonuclease. As a result of this, a linear strand with 3' and 5' end is created. The 3' end serves as a primer for the synthesis of a new DNA strand. The unbroken strand is used for this purpose; with the parental strand itself acting as a template. The 5' end of the strand [illegible]omes attached to the membrane. Meanwhile the original, unbroken single strand unwinds and separates [illegible] the tail attached to the membrane. The tail is cut off into a specific length - *Progeny rod*. The rod rolls and for[illegible] circular molecule, with the enzyme ligase binding the two ends. The newly formed circular molecule which is double stranded can in turn become rolling circles. The genetic information is preserved in the single stranded template which remains circular always. New DNA molecules may be formed either from the single stranded original molecule or from the double stranded rolling circles. Evidence for the rolling circle model has been obtained from the replication of several viruses (M13, P2 T4, λ etc).

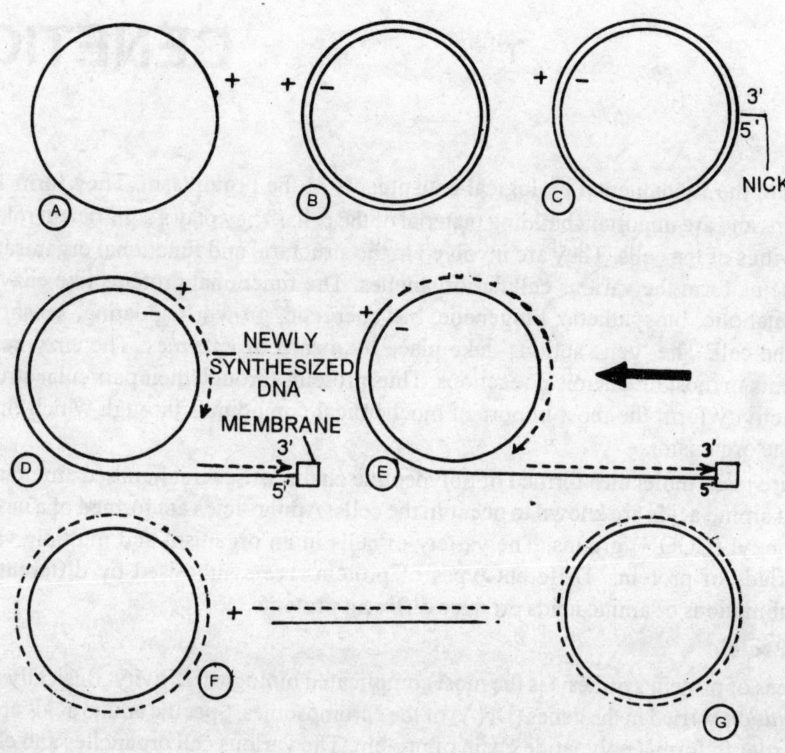

Fig. 20.14 Replication of DNA
Rolling circle model of a circular DNA replication

21

GENETIC CODE

Proteins are the fundamental biological constituents of the protoplasm. They form 15% of the body in living organisms and are important building material of the cells. They play a significiant role in the physical and chemical activities of the cells. They are involved in the structural and functional organisation of the cell. The structural proteins form the various cellular organelles. The functional proteins like enzymes and hormones control the metabolic, biosynthetic, biogenetic, bio-energetic, growth regulating, sensory and reproductive functions of the cell. The 'gene actions' take place by means of enzymes. The enzymes are the important catalysing agents in most biochemical reactions. Thus proteins through their particular structure, composition and enzyme activity form the most important biochemical compounds through which 'hereditary traits' are expressed in the organism.

Proteins are giant molecules formed of polypeptide chains of several hundred amino acids. About 20 and odd number of amino acids are known to occur in the cells. Amino acids are formed of a basic amino (NH_2) and an acidic carboxyl (COOH) groups. The variety of cells in an organism and multiple variety of organisms represent myriads of proteins. Different types of proteins are synthesised by different cells of the body. Different combinations of amino acids produce different proteins.

Synthesis of Protein

The process of protein synthesis is the most complicated biological activity. Basically it is governed by the *genetic information* carried in the genes (DNA) of the chromosomes. Specific amino acids are linked in particular sequence or order to form a polypetide chain of protein. The various cell organelles and chemical compounds involved in the synthetic proces are

1. Deoxyribonucleic acid .... DNA
2. Three type of ribonucleic acids .... RNA
3. Amino acids
4. Enzymes and
5. Ribosomes

The discussion above will clearly reveal that proteins are the representative of the master molecule of the cell DNA to carry out and regulate all the metabolic activities. Proteins are important in another way also. They are important for the existence of DNA itself. A number enzymes (proteins) are necessary for the replication of DNA itself. When proteins are of such fundamental biological importance, no wonder then the cell machinery to synthesize them (proteins) is not only complicated but precisely regulated.

Which molecule(s) present in the cell can regulate the synthesis of proteins? Obviously there cannot be another compound than DNA. The structure and properties of the DNA molecule are eminently suited to exercise such a regulation over the synthesis of proteins.

Having conceded that DNA directly is involved in the regulation of protein synthesis we may now try to answer the two following questions - which aspect of protein structure is controlled by DNA? And how and by what mechanism DNA regulates the assembly of amino acids to synthesize the proteins?

Proteins are extremely complicated biological macromolecules. There are different dimensions for its structural complexity and integrity. These are theprimary, secondary, teritary and quaternary structures (for details - see later in protein synthesis). DNA molecule specifies only the primary structure, while othe ther dimensions of protein structure like folding etc is controlled by proeins themselves.

The primary structure of proteins involves the following aspects - quality, quantity and sequence of amino acids and it is this that is regulated by DNA.

What is the mechanism adopted by DNA to regulate the synthesis of proteins? Obviously proteins cannot be synthesized *in situ ie* within the nucleus, because all the components needed for proteins-amino acids - ribosomes etc are not present in the nucleus. It is not possible also for the DNA itself to migrate to cytoplasm to synthesize the proteins. The only alternative is for DNA to send a messenger or representative which carries the message to specify the quality, quantity and the sequence of amino acids to be asembled to produce the protein. This message is called the genetic code, because it specifies in genetic language the amino acids to be assembled.

The language of genetic code

A code is said to be a cryptic or hidden language carrying a message and upon deciphering or translation, the message is understood. In the genetic code, the message is translated as the type of aminoacid to be assembled. What is the composition of the genetic code that comes out of DNA? To answer this we must first understand the manner in which genetic information exists in the DNA molecule.

It is well known that eventhough the DNA molecule has three components namely sugar, phosphate and nitrogen bases, it is the last one that is unique to each DNA molecule in the sequence of their arrangement. The nitrogen bases vary from one DNA segment to another in arrangement so the information is based upon the sequence of their arrangement. The sequence of nucleotides is then the code in which the DNA sends the message. How is the message sent? The DNA molecule replicates in a particular segment and the replicated segment in the form of RNA now called mRNA carries the genetic message in the form of nucleotide sequence (for details see protein synthesis). The sequence of nitrogen bases have been found to be identical to the linear sequence of amino acids in a peptide chain. In otherwords there is colinearity between mRNA and polypeptide chain (see later in the chapter for details of colinearity).

It is thus clear that the genetic code is in the language of nitrogen bases and the reading of the code is in their sequence of arrangement.

Amino acids involved in protein synthesis

A large number of amino acids - almost upto 150 are found in nature. Of these however quite a number of them remain as free amino acids in the sense they do not join together to form a molecule of protein. It has been found out that only about a little more than 20 amino acids are protein amino acids in that all the known proteins are made up of only these amino acids. Some amino acids however are found in a almost all protein chains while a few others are found only in very few proteins. For example cystine is present only in proteins which have a disulphide linkage. Hydroxylproline is another amino acid found only in few proteins.

The code is triplet (basis of crypto analysis)

A DNA molecule from every source contains only 4 types of nucleotides- adenine, guanine, cytosine and thymine. On the other hand the basic building blocks of proteins namely amino acids are more than 20 types. A basic problem that one has to analyse before finding about a genetic code is how many bases specify a unit of amino acis. Sevral opinions have been put forward in this regard.

Singlet code. According to one of the opinions each nucleotide will code for one amino acid. But this would be insufficient to code for all the 20 amino acids as there are only 4 nucleotides. With 4 nucleotides only 4 amino acids can be coded. What about the rest?

Doublet code. This asumes that each code is formed of two letters i.e., 2 nucleotides. In other words, 2 nucleotides will specify one amino acid. In this type of a code all the 20 amino acids cannot be coded but 16 codes (4 x 4) are possible. But there will be a number of ambiguous determination of amino acids.

Possible singlet, doublet and triplet codes of mRNA

Single Code (4 words)	Double (16 words)				Code Triple Code (64 words)			
					AAA	AAG	AAC	AAU
					AGA	AGG	AGC	AGU
					ACA	ACG	ACC	ACU
					AUA	AUG	AUC	AUU
					GAA	GAG	GAC	GAU
					GGA	GGG	GGC	GGU
					GCA	GCG	GCC	GCU
A	AA	AG	AC	AU	GUA	GUG	GUC	GUU
G	GA	GG	GC	GU	CAA	CAG	CAC	CAU
C	CA	CG	CC	CU	CGA	CGG	CGC	CGU
U	UA	UG	UC	UU	CCA	CCG	CCC	CCU
					CUA	CUG	CUC	CUU
					UAA	UAG	UAC	UAU
					UGA	UGG	UGC	UGU
					UCA	UCG	UCC	UCC
					UUA	UUG	UUC	UUC

Triplet code. This was first suggested by the physicist George Gamow (1954). According to this 3 bases (nucleotides) will code for one amino acid. In other words each code will have 3 nucleotides which will specify the particular amino acid. These triplets of codes are given the name codons. With triplet codes the possibilities would be 64 (4x4x4) codons. This would be more than the number of amino acids (which are only 20). Obviously then each amino acid will have more than one code. Experimental evidence shows that the code is a triplet and about 61 of the 64 codes are used to specify amino acids during protein synthesis.

The maximum possible number of codons in the singlet, doublet and triplet codes

Type of code	Number of bases in codon	Number of codons	Ambiguous/ degenerate
Single code	1	4	Amgiguous
Double code	2	4 x 4 = 16	Amgiguous
Triplet code	3	4 x 4 x 4 = 65	Degenerate (more than required number)

Properties of code

1. The code is non overlapping

The DNA molecule is a very long chain and has series of nucleotides. There is gap between the nucleotides. Naturally a question arises if three nucleotides constitute a codon will there be any gap between two successive codes or do the codes overlap. In otherwords is the code overlapping or non overlapping? If the code is over lapping from where the reading will begin. Because the beginning of the code reading will alter the nucleotide combination and hence will alter which amino acid is specified by the code. When Gamow proposed the code he assumed it to be overlapping. Many people attempted to rationalise the excess number of codes i.e., 64 for 20 amino acids by assuming that the codes are overlapping and the codons can be reduced to about 20. But this involves a lot of problems. If we take for instance the codes for lysine (AGG and AAA) and codes for phenyl alanine (UUU and UUC). The code letters of these two amino acids are mutual exclusive i.e., lysine can not follow phenyl alanine and *vice versa* in any protein. This constitutes a forbidden combination of amino acids and proteins. But on the other hand these amino acids are found in proteins proving that the genetic code is non overlapping. Further the overlapping code assumes that a nucleotide is read twice in different combinations to code for amino acids. If this is true the entire precision of code would be disturbed. Further, any mutational change in an overlapping code would affect more than one code thus disturbing a number of amino acids.

Mutational studies have shown that the code is non overlapping. In TMV (Tobacco Mosaic Virus) the change of one base in the DNA due to mutation would alter the composition of protein by a single amino acid proving that the code is non overlapping. If the code is overlapping change of one base should have disturbed more than one amino acid. Studies on normal and sickle cell haemoglobins have shown that single base mutations result in the replacement of only one amino acid.

2. The code is degenerate

(Synonym codons). As it has been mentioned above there are 64 possible codons considering the code to be a triplet one. Of these 64 codes about 61 code for amino acids. As they are only 20 amino acids which form the constituents of proteins, it is obvious that the number of codons far exceed the number of amino acids. This leads to the conclusion that each amino acid must have more than one code. Of the 20 amino acids only tryptophan and methionine have a single codon each while all the others have more than one codons. *Phenylalanine, tryosine, histidine, glutamine, asparagine, lysine, aspartic acid* and *glutamic acid* have two codons each (see the codon table). *Leucine, argenine* and *serine* have 6 codons each. Thus there is no specificity as to the number of codons that decide an amino acid. The number of codons (whether one or many) specifying the amino acid possibley have been evolved over a long period of time during the process of evolution. The variability in the number of codons perhaps may explain the differential distribution of amino acids in proteins (those with multiple codons occur more frequently).

When there are multiple codons for an amino acid, it must be noted that these codons are not totally different. Synonym codons specifying an amino acid have the first two bases of the triplet being constant. Whereas only the third will vary. For instance proline is coded by the following codons - CCU, CCC, CCA and CCG. It is obvious here in these synonm codons the first two bases are common.

The multiplicity or degeneracy of the code may pose a question as to why nature has provided alternate codons for an amino acid where a single codon would have been more precise and specific. One possible reason for the degenrecy of the codes is that multiple codons protect the organism from mutations which might eliminate an amino acid altogether if there were to be only one code.

Number of codons coding for different amino acids

	Amino acids	Number of codons
1.	Tryptophan, methoionine	1
2.	Phenylalanine, tyrosine, histidine, glutamine, asparagine	2
	Lysine, aspartic acid, glutamic acid, cysteine	2
3.	Isoleucine	3
4.	Valine, proline, threonine, alanine, glycine	4
5.	Leucine, agrinine, serine	6

3. Polarity of the code

It is very essential that a particular gene shoule specify an amino acid always and in all situations unless it undergoes mutational changes. It is necessary therefore that the code must be read between a fixed beginning point and a fixed end point. These points are referred to as the initiation and termination codons respectively.

Another necessity is the reading of the direction of the code because any change in the direction would alter the amino acid. All these features clearly point out that the code must have a polarity i.e., a fixed start, a fixed end point and a predetermined direction. For instance if the codon direction as given below varies the composition of the amino acids would completely vary.

Leucine	Isoleucine	Valine	Serine →
UUG	AUG	GUG	UCG
← Valine	Leucine	Iso Leucine	Alanine

It is obvious from the above that the direction of reading the code would alter the sequence of the amino acids and obviously changes the composition of the proteins. Available molecular biological evidences indicate that the message in mRNA is readin 5' →3' direction. Based on this message the polypeptide chain is synthesised from the direction of amino group to carboxyl group.

4. The code is comaless

One of the most interesting questions with reference to the non overlapping code is, are there any gaps between the codes in the form of punctuation marks so that one code is separate from the other. In genetic terms this means that there could be bases between codes indicating where one code ends and the other begins. For instance codes with intervening punctuation marks can be represented as follows -

UUG	X	AUG	X	GUG	X	UCG
Valine		Leucine		Leucine		Alanine

In the above X can be regarded as a base separating the two codons. If a mutation occurs in any one of the codes, it will not affect other codes and consequently the disturbance in the protein synthesised would be to the extent of only one amino acid. On the other hand if the code is commaless (no punctuation mark-base between codes) mutation in one base would result in a drastic change of the genetic message. Researches of Dr.Khorana and others had clearly indicated that the code is commaless i.e., there is no delimiting point between one codon and other. Khorana and his associates used artifically synthesised long polynucleotide chains which had repeating sequences to produce polypeptide chains. For example the repeating sequence CUCUCU contains the codons for leucine (CUC) and serine (UCU). When this sequence was used for producing the proteins neither amino acid was incorporated with the protein unless the other one is also present. This clearly explains that the triplet code has to be commaless and only then there would be alternate translation of the codes into leucine and serine.

5. Universality of the code

The code was first deciphered and worked out in microorganisms but subsequent researches have clearly shown that these codes are same in their amino acid specification from bacteria to man. Thus, the code is said to be universal. Marshall Caskey and Nirenbrg (1967) have shown by their experiments that diverse organisms such as *E.coli* (bacterium) *Xenopus laevis* (amphibian) and Guinea pig which is a mammal use the same code for their amino acyl tRNAs.

Gene mutations provide additional evidence for the universality of codes. These mutations result in the substitutionof one amino acid for other. Studies have indicated, gene mutations bringing about amino acid substitutions affect the same amino acid in TMB, *E.coli* and also in man. For instance a single point mutation, changes the amino acid in the coat protein of TMV, alpha chain of tryptophan synthetase in *E.coli* and also in the haemoglobin of man.

Another remarkable feature of the code is not only that it universally occurs in all organisms but it has also remained probably without any change ever since it evolved in some bacteria a few billion years ago. Any mutation that occurs would change the composition of mRNA and consequently the amino acid composition in the protein. Changes such as these apparently have proved to be harmful to the organisms and as such there will be a strong selection pressure against the occurrence of such mutations. Hence natural selection has persisted with the codes perhaps ever since they originated.

Ambiguity in the genetic code

There is no evidence to indicate that the genetic code is ambiguous *in vivo*. Ambiguity means that a single codon may code for more than one amino acid. The only exception to the non ambiguity is the AUG codon in prokaryotes -it codes for formyl methionine at the initiation site while at other positions it specifies methionine. It is not clear as to how this distinction is always met precisely but it is believed that the secondary structure of the mRNA at the initiation site may have got something to do with the coding for formyl methionine. In *in vitro* conditions however genetic codes sometimes appear to be ambiguous. For example poly-U codon is known to incorporate some amounts of leucine in addition to phenylalanine. Treatment of ribosomes with streptomycin is known to introduce ambiguity to codons. For example in addition to coding for phenylalanine UUU is also known to code isoleucine, serine and leucine.

Codons and anticodons

During the process of translation the message carried in the mRNA would be converted into the sequence of amino acids which are picked up by tRNA. This is achieved by the process of binding. For this binding to take place, obviously then there should be a specific relationship between the nucleotide of mRNA and those of tRNA. In otherwords the nucleotides of mRNA should be complementary to the nucleotides of tRNA. The nucleotide sequence of mRNA is called the codon and that of tRNA is called anticodon. The reading of mRNA is in the 5'→ 3' direction and the codons are also written in the same direction. For example codon AUG is written as 5' AUG 3'. The corresponding anticodon and tRNA should therefore represented as 5' CAU 3'. In such a configuration the first nucleotide of both codon and anticodon would be the ones at the 5' end and third nucleotides would be at the 3' end.

Very often the anticodon is written in the 3' → 5' direction so that it will be easy to correlate between the bases of codon and anticodon.

Initiation codons

In most of the polypeptide chains the first amino acid which begins the chain is methionine in eukaryotes and N-formlymethionine in prokaryotes. Methionine or N-formylmethionine tRNA complex specifically binds to the initiation sites in mRNA containing the codon AUG which codes for methionine. The AUG code therefore is referred to as the initiation codon. Less frequently however GUG (code for valine) may also serve as the initiation codon in bacterial protein synthesis. In the bacteriophage MS2, GUG is the initiation codon for the protein referred to as A protein. Researches have indicated that the GUG may replace AUG as initiation codon when the latter is lost due to chromosomal abbreation. It has also been found out that chain initiation by GUG is not as efficient as AUG possibly because the former has less affinity for formyl methionyl tRNA.

Termination codons (nonsense codons)

It has been mentioned above that AUG (occasionally GUG) initiates the synthesise of a protein chain. The natural question would be which codon specifies or signals the end of termination of the synthesisof the protein chain. It has been found out that three of the 64 codons do not specify any tRNA and as such they cannot assemble any amino acid. As such these codons whenever they are found in mRNA would signal the end of the chain. They are called termination codons as they terminate the synthesis of the polypeptide chain. They are also called *nonsense codons* as they donot specify any amino acid. The nonsense codons are UAG (*amber*) UAA (*ochre*) and UGA (*opal*). Of the three termination codons UAG was the first termination codon to be discovered. It was named amber after a graduate student in Germany named Bernstein (means amber in German). This graduate student helped in the discovery of the class of mutations. The other two termination codons were obviously given the name of the colours only to maintain uniformity.

Termination codons found in mRNA will not specify any amino acid and as such stop the process of translation and bring about the release of the peptide chain from ribosomes. No tRNA molecules have anticodons complementary to the termination codons found in mRNA. Termination codons seem to have some affinity to proteins called release factors which bring about the release of the protein chain. In prokaryotes there are three release factors RF-1, RF-2 and RF-3. RF-1 recognises the amber and ochre codons while RF-2 recognises ochre and opal codons, RF-3 stimulates RF-1 and RF-2 ultimately releasing the newly synthesised peptide chain. In eukaryotes there is a single factor (RF protein) which is capable of recognising all the three termination codons.

Evidence for genetic code (deciphiring the code)

Earlier to the pathbreaking experiments of Nirenberg, Matthai and H.G. Khorama, work on the genetic code was largely speculative and restricted to development to theoritical concepts. The great physict George Gamow who proposed the triplet code gave the comprehensive theoritical concept for various aspects of genetic code. The main features of his proposals are -

1. The code is a triplet
2. Translation is a direct procedure between codon and amino acid
3. Translation of codes is of overlapping type
4. Code is degenerate
5. Nuclei acid and the peptide chain are colinear
6. Code is universal

The works of Brenner (1957) showed that the triplet code is of non overlapping type. Gamow's idea of the direct translation process became unacceptable when Crick proposed adoptor hypothesis. As per this hypothesis during the translation process some mediator molecules intervene between amino acids and nucleic acids. It is now known that the intervening molecules are tRNA.

Experimental evidence for the existence of genetic code may be obtained by two basic approaches - *in vitro* approach and *in vivo* approach. In *in vitro* method nucleotide sequence are compared with the amino acid residues of the proteins in a culture medium. In *in vivo* method proteins and nucleic acids are obtained from intact cells and then compared.

The *in vitro* approach has the following four methods

1. Genetic analysis of deletion mutatants of a cistron
2. Incorporation of amino acids by synthetic polyribonucleotides
3. Binding of known triplets of aminoacyl tRNA.
4. Amino aicd sequence generated by mRNA transcribed from synthetic tRNA of known nucleotides sequences

1. Genetic analysis

Genetic analysis of deletion and insertion mutants (spontaneous as well as induced) in phage T4 was reported by Crick *et al* in 1961. This deletion or insertion of a nucleotide in a DNA chain leads to a frame shift mutation after the point of deletion or insertion. As a result of this, the reading of the codon subsequent to the mutation gets altered. Obviously then the amino acid composition of the resultant protein would be different. As a result, the protein may be functionally inactive. All the mutations tested by Crick *et al* produced a phenotype called mutant II. It was further observed by Crick *et al* that a series of three mutations was necessary to supress the effects of a previous mutation. In other words three supressor mutations would nullify the effect of three frame shift mutations and bring back the original phenotype. This is a clear evidence to show that the altered codon reading would return to the normal state after three point mutations thus indicating that the codon must have three nucleotides. The study also indicates that the code is commaless. If there were to be a punctuation mark between the two codes a deletion or insertion of a base would change the number of bases between two commas adjacent to the point of insertion or deletion.

2. Incorporation of amino acids by synthetic polyribonucleotides

Synthetic polyribo nucleotides were used as mRNA by Nirenberg and Matthai to determine the meaning of codons. They produced a synthetic RNA by using the enzyme polynucleotide phosphorylase discovered in 1955 by Grunberg. Manago and Ochoa. This enzyme does not use DNA as a template and can incorporate ribonucleotides in a random order. The first synthetic RNA used for protein in a cell-free system (the *in vitro* system consists of ribosomes, tRNA's and other necessary factors for protein synthesis) was poly U i.e., the mRNA consists of only one nucieotide namely U in a series. This can be called a homopolymer as it is a polymer of only one nucleotide. The poly U mRNA was able to synthesis a homopolymer of a polypeptide chain containing only phenylalanine residues. This proved that the triplet code UUU specifies phenylaline. Subsequently many such homopolymer species of mRNA were synthesized and were directed to produce polypeptide chains. For example AAA was shown to code for lysine and CCC specified proline.

Nirenberg and Ochoa subsequently used heteropolymeres of mRNA to produce polypeptide chains. A heteropolymer includes a mixture of two, three or all the four nucleotides in mRNA. This yielded further information on the possible codons for certain amino acids. For example poly CG directs the incorporation of arginine and alanine in addition to the proline coded by poly C. Some additional information on the assignment of codons to individual amino acids was obtained by varying the ratios of different nucleotides in the synthetic mRNA. In this experiment the code assignment was at best ambiguous since all the four bases were not required for the incorporation of any of the amino acids.

3. Binding of known triplets to aminoacyl tRNA

A triribonucleotide cannot direct the synthesis of a polypeptide but it leads to binding of specific aminoacyl tRNA to the ribosomes. Nirenberg *et al* used this property for assigning specific codons of specific amino acids. Their technique is known Nirenberg - Leder technique. Conventionally a trinucleotide with 5' terminal phosphate are more active and those lacking this phosphates in inducing the binding of aminoacyl tRNA to ribosomes. On the otherhand those with 3' terminal phosphate are inactive. This shows that the codons must be triplet. In the Nirenberg - Leder technique trinucleotides with 5' terminal phosphate are added to a cell free system containing a mixture of 20 amino acids bound to their respective tRNA. One of the 20 amino acids had radioactive carbon. For different experiments different labelled amino acids were used but with the same trinucleotide after the synthesis of protein the cell free mixture containing ribosomes were analysed to detect radioactivity confirming the binding of aminoacyl tRNA to the ribosomes. These studies showed that UUU, CCC and AAA induced the binding of phenylalnine, proline and lycine respectively to ribosomes. Using this technique Nirenberg *et al* assigned most of the codons to their respective amino acids.

4. Amino acid sequence generated by mRNA transcribed from synthetic tRNA of known nucleotides sequences

H.G. Khorana and his associates used the above technique for codon assignment in *in vitro* conditions. Known as Khorana technique in this method long specific DNA sequences are first synthesized. These are then transcribed to mRNA molecules using the DNA directed RNA polymerase enzyme. These mRNA molecules which have known sequences of nucleotide are directed to synthesize polypeptides. The polypeptides were then subjected to sequential degradation to know the amino acid sequences. The nucleotides sequence is then compared with the amino acid sequences. The nucleotide sequence is to deduce specific codons for different amino acids. Using this technique almost all the codons for amino acids were designated. In this technique Khorana and associates also confirmed many other characters of the code such as degenerecy, commaless nature, non overlappingand triplet nature also.

In Vivo codon assignment

In the cell free systems, genetic codes have been deciphered with reference to their amino acid affinity but a doubt may arise whether these codes are the same in living systems also. In otherwords whether the living systems use the same codes for protein synthesis. In order to prove the similarity of codes in both in *vitro* and *in vivo* three different techniques were used by three different molecular biologists.

(i) *Amino acid replacement studies*

(ii) *Frame shift mutation studies*

(iii) *Comparison of a nucleotide sequence of a cistron and the amino acid sequence of the polypeptide produced by it*

FIRST LETTER	SECOND LETTER U	C	A	G	SECOND LETTER
U	UUU—Phen.	UCU—Ser	UAU—Tyr.	UGU—Cys	U
	UUC—Phen.	UCC—Ser	UAC—Tyr.	UGC—Cys	C
	UUA—Leu	UCA—Ser	UAA—Ochre	UGA—Nonsense	A
	UUG—Leu.	UCG—Ser	UAG—Amber	UGG—Trip.	G
C	CUU—Leu	CCU—Pro	CAU—His	CGU—Arg.	U
	CUC—Leu	CCC—Pro	CAC—His	CGC—Arg.	C
	CUA—Leu	CCA—Pro	CAA—Glu	CGA—Arg.	A
	CUG—Leu	CCG—Pro	CAG—Glu	CGG—Arg	G
A	AUU—Ile.	ACU—Thr	AAU—AspN	AGU—Ser	U
	AUC—Ile	ACC—Thr	AAC—AspN	AGC—Ser	C
	AUA—Ile	ACA—Thr	AAA—Lys.	AGA—Arg	A
	AUG—Met	ACG—Thr	AAG—Lys	AGG—Arg	G
G	GUU—Val.	GCU—Ala	GAU—Asp	GGU—Gly	U
	GUC—Val.	GCC—Ala	GAC—Asp	GGC—Gly	C
	GUA—Val.	GCA—Ala	GAA—Glu	GGA—Gly	A
	GUG—Val.	GCG—Ala	GAG—Glu	GGG—Gly	G

Fig. 21.1 Genetic code

The codon dictionary

Amino acids replacement studies on the synthesis of tryptophan synthetase in *E.coli* (Yanofsky, 1963) have clearly indicated that the codons for the replaced amino acids and for the amino acid replacing it prove the specificity of a codon for a particular amino acid. Frame shift mutations on lysozol enzyme of T4 bacteriophages have also shown the affinity between a specific codon and a particular amino acid.

The direct in vivo evidence for the genetic code comes from a comparison of the base sequence of a cistron and the amino acid sequence in a polypeptide chain produced by it. Such a study has been done with RNA bacteriophage MS2 which has only three cistrons. One of these cistrons produces the coat protein having 129 amino acids. The cistron responsible for it has 387 nucleotides. There is a total agreement between amino acid sequence expected from the base sequence of the cistron and the actual amino acid sequence of the head protein. In othewords each codon of the cistron precisely designates the same amino acids as expected from the coding dictionary.

Wobble hypothesis - for details see protein synthesis

Colinearity of the gene structure and its product

The gene is an unbranched chain of nucleotide pairs in a particular sequence. The polypeptide chain is also unbranched and is made up of amino acid residues attached in a linear series. Thus a linear relationship between the gene and the polypeptide chain produced by it can be indicated. Yanofsky *et al* (1964) established this colinearity in their study of tryptophan synthetase enzyme of *E coli.*

The enzyme tryptophan synthetase is made up of two chains α and β. A genetic map of the α chain gene was established by determining the recombination percentage of mutations for the amino acid tryptophan. The complete amino acid residues of the α chain of the enzyme tryptophan synthetase was thus established. Subsequently the location of amino acid substitutions in the chain for 10 mutations was determined. It was then found out that the order of mutations corresponded with the order of amino acid substitutions in the α chain of the polypeptide.

Sarabhai et al (1964) came to the same conclusion in their study of nonsense mutations in bacteriophage T4.

Pattern of the genetic code

A distinct pattern emerges when one analyses the amino acid structure and the type of codons that specify them. Amino acids with similar structural properties tend to have related codons. For example codons for aspartic acid (GAU and GAC) are similar to the codons for glutanic acid (GAA and GAG). Same relationship is found between the codons of phenylalanine, tyrosine and tryptophan - all these begin with uracil. It is believed that this feature of the code is evolved to minimise the consequences of mistakes that may arise during translation and base substitutions due to mutation. In other words if an amino acidin a protein is replaced by another with almost similar properties the protein may still remain functional.

Another pattern that is seen in the code isthat for synonym codons (multiple codes for the same amino acid) the first two bases are always constant whereas only the third one varies. The flexibility of the third nucleotide of the codon is another method to help minimise the harm posed by mutations. Crick (1966) has called this wobbling and proposed a hypothesis called Wobble hypothesis (see protein synthesis for details).

New genetic codes in mitochondria and ciliate protozoa

It has been mentioned above that the code is universal and ever since it originated it has remained static. However during 1980s it was observed that the genetic codes of mitochondrial DNA in yeasts may be different from the others. In yeast mitochondria UGA codes for tryptophan while in nuclear genes in the same organism UGA is the termination codon. This however is not to say thatin all mitochondria of all organisms the UGA codes for tryptophan. For example in maize cytochrome oxidase tryptophan is coded by CGG. This code in nuclear genes codes for arginine. There are also indications that AUA which codes for isoleucine in nuclear genes codes for methionine in *mit* DNA of mammals, yeast, *Drosophila* etc.

A different genetic code in DNA is perhaps necessiated because of the small *mit* genome size. T.H. Jukes (1966) presented a concept called *archetypal code* which will allow one anticodon (of tRNA) to pair with 4 different codons due to wobbling at the first base of anticodon and the third base of codon.

In 1986 a different genetic code was shown to be working in a ciliate member of Mycoplasma (*Mycoplasma capricolum*). In the genetic code of this organism UAA and UAG are not nonsense codons but code for glutamine.

Evolution of genetic code

Many of the features of the genetic code like degeneracy, non overlapping, commaless nature etc have led to a suggestion that the present version of the genetic code has evolved from a more primitive version which are perhaps a doublet than a triplet. It is believed that this shifting from doublet to triplet occurred very earily in the origin of cellular life. It is not certain however whether all the codes for all the amino acids were evolved simultaneously or the codon dictionary gradually arrived at the present stage of 64 is not certain. One thing however is certain eversince it came into existence in the present form the biological necessity of it imposed such a selection pressure on it that it has remained unchanged upto to the present day.

PROTEIN SYNTHESIS

Protein Structure And Classification

Of the two important macromolecules that mastermind all the activities of the living systems, proteins are one, the nucleic acids being the other. Proteins are present all over the cell and are important both structurally and functionally.

The name protein was first suggested by Berzelius (1838). *Protein* in Greek means 'First rank'. Proteins are highly complex organic molecules with molecular weight ranging upto many millions. Chemically proteins are madeup of C, H, O and N. In addition to these major elements, some proteins may also haave sulphur, phosphorous, iron zinc, iodine etc as components. The ratio of major elements in proteins are C= 50-55%, O=20.23%, N=15-18% and S=0-4%.

Proteins are made up of building blocks called amino acids (i.e. on hydrolysis, proteins yield these units). The idea that proteins are built up of amino acids was first proposed by Emil Fischer.

Amino acids are carboxylic acids which have atleast one COOH group and one NH_2 group. Other functional groups may also be present. There are atleast 22 different types of amino acids encountered in plants.

The amino acids are bound to one another in a protein chain by means of linkages or bonds called peptide linkages. Peptide linkages are formed between the COOH group of one amino acid and the NH_2 group of the other, with the elimination of a molecule of water. Several amino acids combine together with the peptide bond and form a peptide chain. Several of these peptides constitute a polypeptide chain. The polypeptide chain folds and refolds in a characteristic fashion to form a protein molecule.

Peptides

These are made up of amino acids linked with CONH (peptide) bonds. Peptides are intermediate compounds produced during the synthesis or degradation of proteins.

Amino acids ——— peptides ——— proteins

Proteins

These are thefinal productsin the N_2 metabolism, and are made up of folded peptide chains which have many additional linkages (depends upon the point of folding) in addition to the CONH bond. Proteins have a three dimensional configuration and have very high molecular weight reaching upto 10^4 - 10^-.

Although proteins contain not more than 20 amino acids, they are of immense variety and complexity due to variations in combination of amino acids. In general the three following aspects pertaining to the amino acids decide the type of protein. These are - the quality, quantity and sequence of amino acids in the peptide chains.

In other words, the type of amino acid, its quantity and the sequence in which these acids are arranged forms the basis for the categorization of proteins. Besides, proteins may differ in the number of peptide chains per molecule, type of secondary linkages (SH) etc.

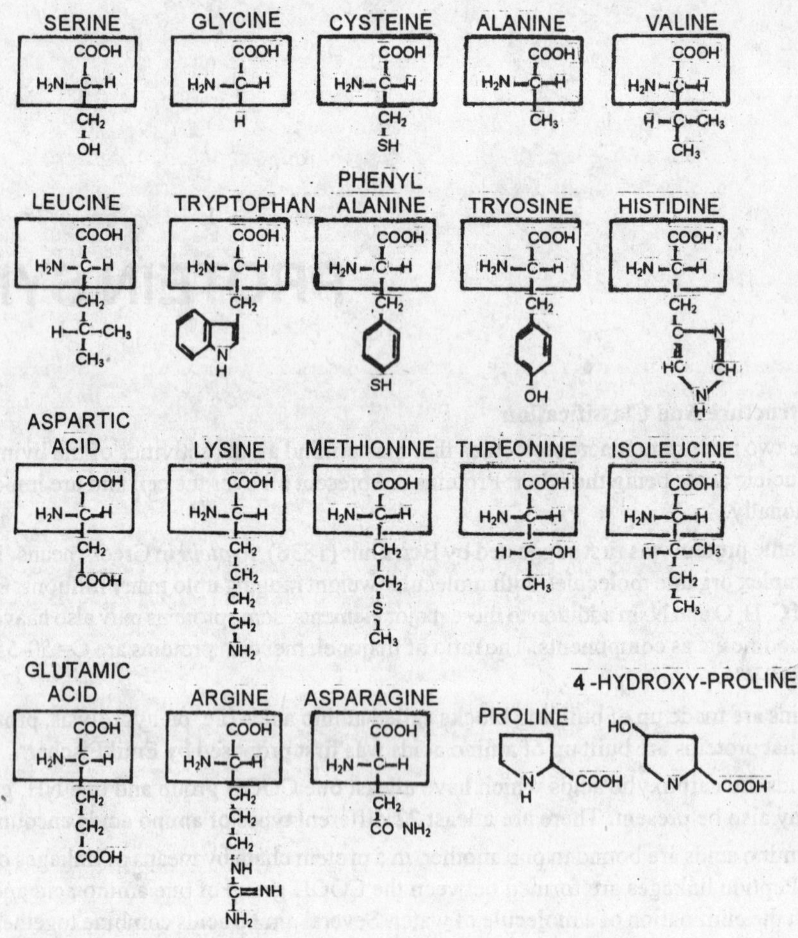

Fig. 22.1 Protein synthesis
Structure of the amino acids

Structure of proteins

Modern biochemists regard four structural levels in the protein molecule which accounts for its complexity. These levels are - primary. Secondary, tertiary and quaternary structure, in the order of increasing complexity.

Primary structure

This is the simplest and the most basic structure and refers to the linear peptide chain. Besides the CONH bond, disulphide bonds (-S-S-) are found or among the peptide chain linkage in the sulphur containing amino acids.

Secondary structure

The primary polypeptide chain spirally coils forming the L helical chain. The characteristic bonds of the secondary structures are hydrogen bonds between the carbonyl and amino groups of peptide links, salt linkages and Vander Wall's forces which help maintain the helical structure.

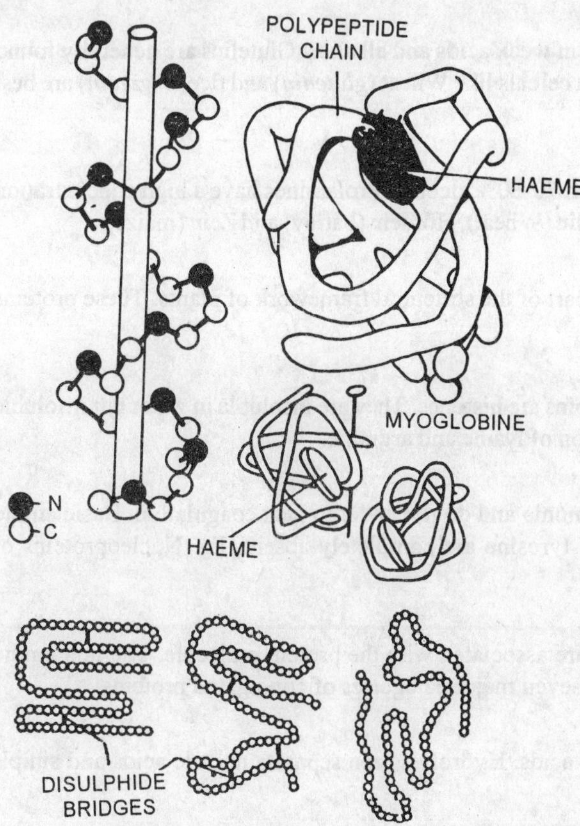

Fig. 22.2 Protein synthesis
Protein structure, **A.** Polypeptide chain with helical configuration, **B.** Tertiary and quaternary structure of myoglobin, **C.** Denaturation of proteins

Tertiary structure

Folding and superfolding of the helical chain in a three dimensional orientation constitutes the tertiary structure. Interaction between the amino acid residues situated far apart in the chain brings about the folding.

Quaternary structure

This consists of more than one identical sub units (poly peptide chain) which associate with one another to form a functional unit. Individually each subunit is non functional. Such complicated structures are usually seen in enzymes. For example the enzyme phosphorylase has two sub units, which are inactive individually, but as dimers, they are active. A protein with identical sub units in the quaternary structure is known to be homogenous or if the subunits are dissimilar it is known as heterogenous. A protein with more than one sub unit (monomer) is referred to as an oligomeric protein.

Classification of Proteins

As the proteins possess, a very complex strucuture, their classification also poses a problem as many categorizations are possible based on structure, function etc. The following is a generalized classification. The major groups are recognized in proteins based on their solubility and chemical properties.

I. Simple proteins
II. Conjugated proteins and
III. Derived proteins.

I. SIMPLE PROTEINS

Chemically these have only amino acids. No additional compounds are present in the molecule. Simple proteins are of the folowing kinds -

1. Albumins

These are soluble in water, dilute salts, acids and alkalies. Heat denatures the protein structure by coagulation. Some of the commonly occuring albumins are Legumelin (from legume seeds), Amylase (from barely), Leucosin (from cereals) etc.

2. Globulins

Less soluble in water than albumins but soluble in less concentrated salt solutions. Heat coagulation is seen in globulins also. Eg. Legumin (pears), Tuberin (potato) etc.

3. Glutelins

Insoluble in water and salt solutions, but soluble in weak acids and alkalies. Glutelins are generally found in plants and are heat immune. Seed proteins found in cereals like Wheat (*glutenin*) and rice, *oryzenin*) are best examples of glutelins.

4. Prolamines

These proteins are water insoluble, but dissolve in 70-80% alcohol. Prolamines have a high concentration of the amino acid proline and other amides. E.g. Gliadin (Wheat), Hordein (barley) and Zein (maize)

5. Scleroproteins

These are fibrous proteins found as an integral part of the structural framework of plants. These proteins are insoluble in most of the solvents.

6. Histones

Most of the nucleoproteins, hemoglobin and globins are histones. They are insoluble in water but insoluble in dilute ammonia. Histones have a high concentration of lysine and arginine.

7. Protamines

These are soluble in water, dilute acids and ammonia and do not undergo heat coagulation. Basic amino acids are rich in protamines, while tryptophan and tyrosine are completely absent. Eg. Nucleoproteins of sperms in fish.

II. CONJUGATED PROTEINS

In these proteins, non amino acid components are associated with the protein molecule. The non amino components are called prosthetic groups. There are seven major categories of conjugated proteins.

1. Nucleoproteins

Proteins are found in combination with nucleic acids. Hydrolysis can separate nucleic acids and simple proteins. Nucleoproteins are water soluble.

2. Glycoproteins

Also called mucoids, these are conjugated with carbohydrates like macromolecular polysaccharides, sugars, sugar phosphate etc. Glycoproteins are found in cell membranes.

3. Chromoproteins

These constitute the pigments and are coloured, the coloured appearance is due to the presence of metals such as Cu, Fe, Mg etc. Chlorophylls, flavoproteins, carotenoids, cytochromes etc are typical examples of chromoproteins.

4. Lipoproteins

Proteins in conjugation with lipids are called lipoproteins. Lecithin, Cephalin etc are the lipids found in combination with proteins. Lipoproteins are an integral part of all cell membranes.

5. Metalloproteins

Certain types of enzymes which require metals as activators belong to this group. Many of the respiratory enzymes belong to this group.

6. Phosphoproteins

These are proteins with phosphoric acid as the prosthetic group. Phosphoproteins are insoluble in water, but soluble in alkalies. E.g. milk and egg protein (casein and viteline).

7. Lecithoprotein

Proteins with lecithin as the prosthetic group (E.g. white of egg) belong to this group.

III. DERIVED PROTEINS

These are degradation products of proteins. Primary proteins on hydrolysis with acids or alkalies yield derived proteins.

(A) Primary derived proteins and

(B) Secondary derived proteins

A. Primary derived Proteins

There are two categories of primary derived proteins. These are metaproteins and Coagulated proteins.

(i) Metaproteins : These are insoluble in water but dissolve in dilute salt solutions, acids and alkalies. Hydrolysis of natural proteins on long treatment with alkalies or acids yield metaproteins.

(ii) Coagulated proteins: Treatment with heat or alcohol of natural proteins yields these proteins.

B. Secondary derived proteins

Three types of secondary derived proteins have been identified. These are protoeses, peptones and peptides.

(i) Proteoses : Prolonged hydrolysis of metaproteins yield proteoses (eg. Albuminoses from albumin). These are water soluble but immune to heat treatment.

(ii) Peptones : Peptones are obtained by continued hydrolysis of proteoses. Like proteoses, these are also water soluble and heat insensitive.

(iii) Peptides : Prolonged hydrolysis of natural proteins yields peptides. These are also water soluble and not coagulated by heat.

Functions of proteins

Most of the vital activities of the cell are controlled by the two master molecules namely proteins and nucleic acids. Proteins are important both from the point of view of structure and function of the cell. Structurally proteins constitute in integral part of membranes, pigments etc. Functionally in the form of enzymes, proteins play such a vital role in cell physiology, that without proteins (enzymes) no metabolic reaction is possible. Even the synthesis of DNA is regulated by proteins. In the form of Chromoproteins (Chlorophyll, Cytochromes etc), proteins mediate all energy transformation reaction. It is because proteins are so important that the nucleic acid directly regulates the protein synthesis.

Protein structure is highly complex (as has been pointed out earlier) as well as specific. The primary structure is determined by the quality, quantity and the sequence of amino acids. Even an alteration of the position of one amino acid would alter the structure of the protein, and if it happens to be an enzyme regulating an important reaction, the entire reaction is disturbed leading to several abnormalities. Hence it is but natural that there is direct control of DNA in the synthesis of proteins. DNA, RNA and ribosomes of the cell are involved in the synthesis of proteins.

The central dogma : The basic mechanism of protein synthesis is that the nuclear DNA regulates the assembly of amino acids in the ribosomes to produce a polypeptide chain. Since DNA cannot move out of the nucleus it sends a messager - mRNA which carries the message in the form of sequence of nucleotides. This mRNA forms a template on the ribosomes. Another type of RNA, tRNA found in cytoplasm (whose nucleotide sequence is complementary to that of mRNA), picks up the amino acids and assembles them on the ribosomes to produce a peptide chain.

This may be briefly expressed as follows :

DNA Replication → DNA Transcription → RNA Translation → Proteins translation. Transcription is the copying of a complementary messenger RNA strand on a DNA strand. In translation, the genetic information present in mRNA (in the form of sequence of nucleotides) directs the sequence of amino acids to be assembled. We shall study these events in detail.

Transcription

Copying of a complementary strand of mRNA requires the following - a template, activated precursors, a divalent metal ion and RNA polymerase.

DNA molecules serves as a template for the production of mRNA. Of the two strands of DNA only one strand called the 'sense' or the 'coding' strand transcribes mRNA. Single stranded DNA also can serve as a template.

The activated precursors required for mRNA synthesis are - Adenine triphosphate, Guanine triphosphate, Uracil triphosphate and Cytosine triphosphate (ATP, GTP, UTP and CTP)

Mg^{++} or Mn^{++} are the divalent metal ions required for mRNA synthesis.

The enzyme RNA polymerase is necessary for transcription. The nucleotides of mRNA are assembled together with the help of RNA polymerase. The enzyme consists of two parts - a core enzyme and a sigma factor. The sigma factor initiates transcription of mRNA on the DNA template and the core enzyme continues transcription. The DNA base sequence are thus transcribed into mRNA base sequence.

Genetic Code

This is the message in mRNA in the form of nucleotide sequence. Three nucleotides code for one amino acid. It is called a triplet code or codon. There are 64 codons for the 20 odd amino acids. i.e., there is more than one code for each amino acid. These alternative codes are helpful in the cae of random mutations which may alter a code (nucleotide).

The nucleotides of mRNA are complementary to the nucleotides of tRNA (present in cytoplasm). The triplet nucleotides of tRNA are called anticodons. Each tRNA has three nucleotides. The pairing between the codon and anticodon (mRNA and tRNA) follows the A-U, G-C combination. For example if mRNA consists of the first code UUU, it pairs with a tRNA having an anticodon. AAA. Thus the series of codons of mRNA determine the series of anticodons of different molecules of tRNA and hence the amino acids (Each tRNA molecule is specific to a particular amino acid). Since the triplets of mRNA are determined by the base sequence of DNA, it follows that the DNA molecule decides the sequence of the amino acids and hence the structure of the protein molecule.

Some triplets do not code for any amino acid (UAA and UAG). These are called nonsense codes. The so called nonsense codes are actually terminator codons and are now called ochre and amber (UAA and AUG). Code AUG is called initiator codon since it initiates the synthesis of a polypeptide chain. (See earlier in the same chapter for details)

The triplet codes for all amino acids collected together is called a codon dictionary. The dictionary was first proposed by Nirenberg (1965) and subsequently worked out in further detail by Khorana H.G.

The mRNA is released out of the nucleus into the cytoplasm where it gets attached to the smaller sub unit of the ribosomes. Many ribosomes are connected by a single mRNA molecule, constituting a polysome or polyribosome. The mRNA attaches itself to the ribosomes in such a way, that its bases remain free (to be able to undergo paring with the bases of tRNA). The whole structure (polyribosomes + mRNA) is called mRNA complex.

tRNA (transfer RNA) : This is present freely in cytoplasm. Also called sRNA (soluble RNA) is the smallest among the various types of RNA molecules found in the cell. Although the tRNA is single stranded like other types of RNA, it appears double stranded because of coiling. Complementary base pairs of the same strand form pseudo bondings of hydrogen due to folding. There are two regions which are free without bonding - one at the acceptor and which always has CCA base pair sequence and the other in the middle region of the molecule called the anticodon region, where there bases turn out of the folded portion and remain free. The types of tRNA is based on the types of these free bases. tRNA has the following functions.

(i) Attach to a particular amino acid at the acceptor and (CCA) with adenine nucleotide. The choice of amino acid however depends on the type of free bases present at the anticodon end.

(ii) to attach itself at a specific site on the mRNA and

(iii) to free the amino acid after it gets attached to another amino acid by means of peptide linkage.

The Wobble hypothesis

The triplet code or codon is degenerate i.e., there are many more codons than there are amino acids. Only 20 amino acids are involved in protein synthesis, while there are 64 possibilities or types of codons (43). Hence more than one codon can code for an amino acid. It is interesting to find out as to why the codes are degenerate? There could have been only so many codons as there are amino acids. But there is problem here. If there were to be only one code for an amino acid, what happens if there is a mutation at that site in DNA? Obviously the code changes and the amino acid cannot be incorporated in the protein leading to abnormalities. On the other hand, if there are alternate codes, they act as a protection against mutation in a specific base. Hence degeneracy of codes is a must.

Regarding tRNA present in cytoplasm, since they have to bind themselves to a specific codon, there should have been as many tRNA's as there are codons. But the number of tRNA is only as many as there are amino acids. This means, the anticodons of the tRNA's must be able to 'read' more than one codon of mRNA. How is this reading possible when bonding between base pairs is highly specific?

Crick (1966) proposed **'the Wobble hypthesis'** in order to solve this apparent dilemma. According to this hypothesis, only the first two bases of the codon have a precise pairing with the bases of the anticodon while the third one may Wobble (non specific). The pairing in the third base is ambigous. Thus a single tRNA can pair (bind) with more than one mRNA codon differing in only the third base. For example the anticodon UCG of tRNA can read codons AGC and AGU of mRNA. The pairing between UGC and AGU is a perfect pairing as per the Crick and Watson pattern. But in the second case i.e., between UGC and AGU, the pairing between the first two bases is normal, while between G and U is against the normal pairing pattern. This unsual bonding G and U is called Wobble pairing.

The degeneracy of the code is not totally randon. Multiple codons for an amino acid always carry the same bases in the first two positions in a triplet with only the third 'Wobbling'. For example all the four codons for Valine have the first two bases - GU and for alanine it is GC.

Valine - GUU, GUC, GUA and GUG.

Alanine - GCU, GCC, GCA and GCG.

Translation

The process of translation consists of activation of amino acids, transfer of activated amino acid to tRNA, initiation of synthesis of polypeptide chain, chain elongation and chain termination.

Activation of amino acids

Before amino acids (20 of them) get themselves attached to the specific type of tRNA they have to be activated (energized). The activation is carried out by the ATP molecule in the presence of the specific enzyme called aminoacyl synthetase with Mg as the cofactor. The reaction produces aminoacyl adenylate (AAA) or aminoacyl AMP. Pyrophosphates arereleased in the process.

Amino acid + ATP + Enzyme ———— Enzyme AA> AMP complex + Ppi

Transfer of activated amino acids to tRNA

The activated amino acid (aminoacyl AMP) is transfered on to a specific tRNA, during which AMP and the enzyme (aminoacyl synthetase) are released. The amino acid gets attached to the CCA and tRNA.

Enzyme AA. AMP + tRNA ———— AA. tRNA + AMP + Enzyme

Initiation of Polypeptide chain synthesis

This involves two steps viz., preparation of the site of protein synthesis and initiation of the polypeptide chain.

(i) Preparation of the site of synthesis

Ribosomes in cytoplasm are the actual sites of protein synthesis. These are usually separated into their subunits (50s and 30s), when they are not participating in protein synthesis. The mRNA usually attaches to the 30s subunit through its 5' end carrying the AUG codon. In the process of attachment certain substances called initiation factors (IF) help. These factors are of three types in prokaryotes (IF1, IF2, IF3) and are of six types in eukaryotes (eIF2, eIF2al, eIF2a2, eIF2a3, eIFa4 and eIF3).

The attachment of the mRNA to the 30s submit requires the help of IF3 and Mg^{++}. The IF3 probably unwinds the rRNA (ribosomal RNA), so that it can bind to mRNA. After the attachment of mRNA, an mRNA - 30s complex is formed.

PEPTIDE BOND

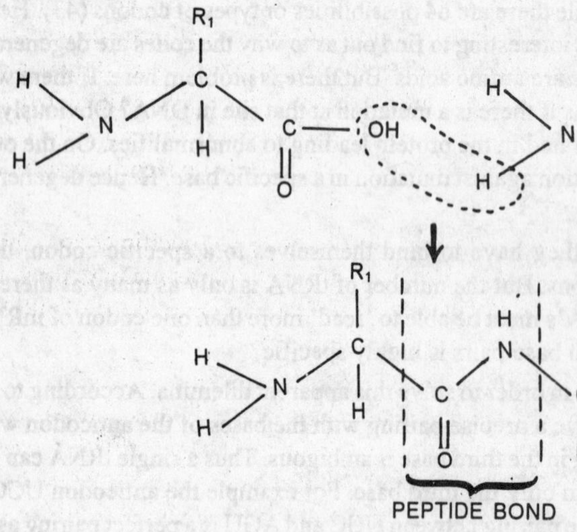

Fig. 22.3 Protein synthesis
Formation of peptide linkages

The starting amino acid in a polypeptide is methionine *(met)* in the case of eukaryotes and N-formly methionine *(f-met)* in the case of prokaryotes. The amino acid (met or f-met) tRNA complex attaches to the initiation codon (AUG) on the mRNA 30-s complex through its anticodon (UAC). This forms the 30s initiation complex and requires the participation of IF2, IF3 and GTP (energy donor).

$$\text{(i)}\quad 30s + mRNA \xrightarrow{\text{IF3}} mRNA - 30s \text{ complex}$$

$$\text{(ii)}\quad mRNA - 30s \text{ complex} + met\ tRNA \longrightarrow 70s - mRNA - met - tRNA$$

During the formation of the initiation complex the 50s subunit combines with the 30s subunit to form the 70s unit. This union is facilitated by the participation of IF1. With the formation of the 70s ribosome) the initiation of the polypeptide chain gegins.

(ii) Initiation of the polypeptide chain

Each ribosome has two cavities - P site (polypeptide site) and Asite (aminoacyl sites). AA tRNA complex occupies the A site, while the met - tRNA (tRNA with the initiator aminoacid - methionine) will be at the P site. This occupation requires a molecule of GTP and the participation of IF_1. The energy of GTP is utilized for the formation of the peptide linkage.

Peptide linkage occurs when the carboxyl (COOH) group of peptidyl tRNA (met. tRNA) present at the P site reacts with the amino group (NH_2) of AA tRNA at the A site. The enzyme peptidyl transfers catalyses the formation of peptide bond. Later the AA+tRNA moves from A site to the P site in the presence of GTP and IF_2

Formation of the polypeptide chain

A specific type of tRNA (which has bases complementary to the codon of mRNA) linked to a particular amino acid gets attached to a specific site on mRNA when its anticodons can form a pairing with the first codon of the mRNA. Soon after a second tRNA with its amino acid joins the second site (triplet) on mRNA. The first and the second amino acids attached to different tRNA form a peptide linkage as both are linked to mRNA *via* the anticodons of tRNA. A molecule of water is released during the formation of the peptide linkae

$$COOH + NH_2 \longrightarrow CONH + H_2O$$

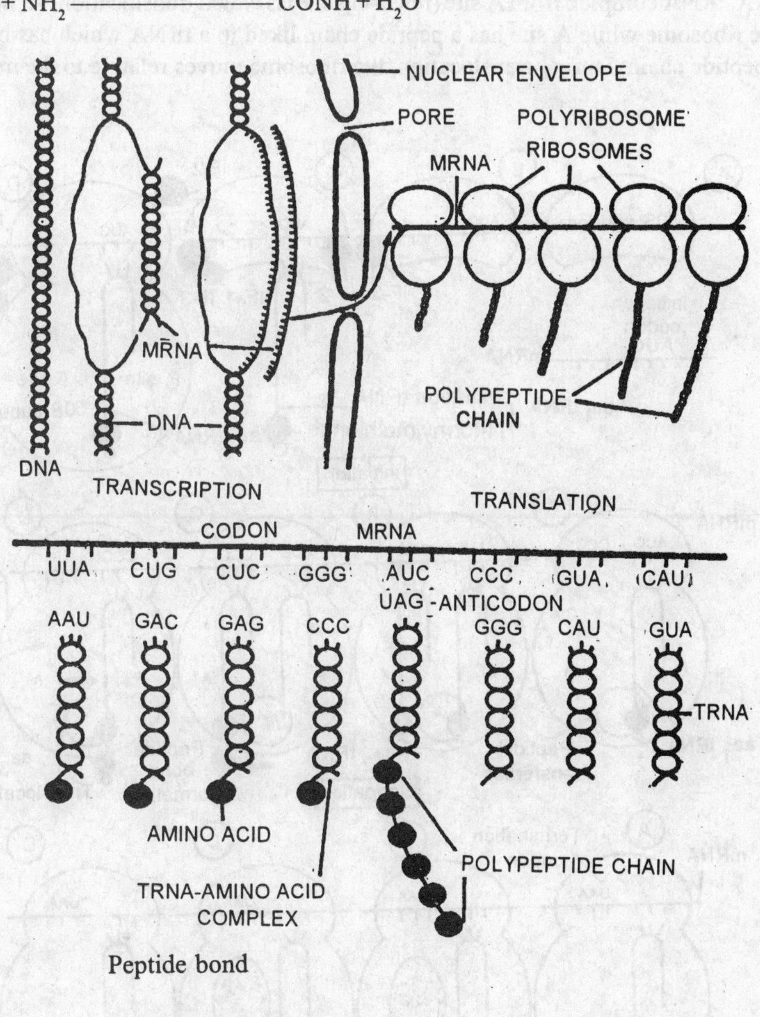

Peptide bond

Fig. 22.4 Protein synthesis
General scheme of protein synthesis

As soon as the peptide linkage is formed the tRNA is made free and comes back to cytoplasm. This whole process in which a specific tRNA carries its amino acid to a specific site is called translation i.e. the code (triplet) is translated into an amino acid. Here the sequence of bases in tRNA will be identical with the base sequence of that part of DNA coding for mRNA.

Chain elongation

Elongation of the chain requires the participation of certain factors called elongation factors (EF). These are EF-Tu, EF-TS and EF-G in prokaryotes and EF-1 and EF-2 in eukaryotes. After the first amino acid tRNA complex has moved from the A site of the ribosome to the P site, a second AA. tRNA complex now occupies the

A site. The codons are recognized enzymatically. This requires the participation of Ef-Tu, GTP and Mg^{++}.

Peptide linkage takes place between the two amino acids and the AA. tRNA complex migrates to the P site rendering the A site vacant. Now a third amino acid tRNA complex occupies the A site and theprocess continues. The shifting of th AA. tRNA complex from A site (releasing site) is called translocation. At the P site the tRNA is released from the ribosome while A site has a peptide chain liked to a tRNA which has brought in the last amino acid of the peptide chain. During translocation, the ribosome moves relative to the mRNA in the 5'-3' direction.

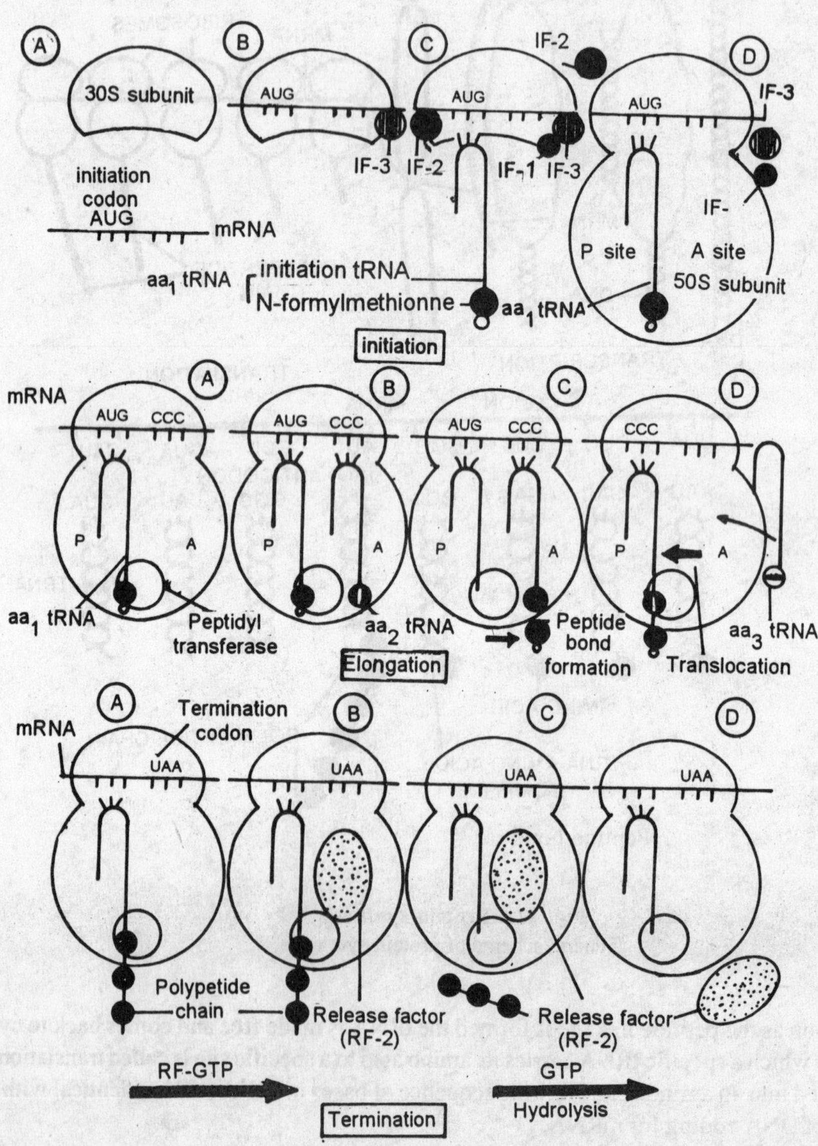

Fig. 22.5 Protein synthesis
Different stages of translation

Chain termination

The elongation of the chain continues by the alternate occupation and vacancy at the A site by the AA. tRNA complexes until a termination codon (UAA, UAG or UGA) reaches the ribosomes. The termination codon sends signals to the ribosomes to attach release factors (RF-1, RF-2 and RF-3 in prokaryotes and RF in eukaryotes).

The release factors interact with the enzyme peptidyl transferase (which catalyses the formation of peptide bond) resulting in the hydrolysis of the bond between the polypeptide chain and the tRNA (linked to the last amino acid of the peptide chain at the site) and the former is released from the ribosome. Other events that occur at the termination of the polypeptide chain are -

(i) Release factors get disassociated from the ribosomes due to the hydrolysis of GTP.

(ii) tRNA is also released.

(iii) Ribosomal sub units (30s and 50s) get disunited.

(iv) The polypeptide chain is processed by the removal of the methionine residue from the chain. Depending upon the circumstances the chain may get cleaved into two or more pieces.

In eukaryotes, the polypeptide chain passes through a tunnel between the two subunits of the ribosome into the cavity of the cisterna of the endoplasmic reticulum.

Rate of Protein synthesis

The rate of transcription in *Escherichia coli* DNA for the enzyme controlling trypotophan synthesis is 28 nucleotides per second at 39°C. The speed of translation is 7 amino acids per second. For the haemoglobin molecule, the rate of translation is 4 aminoacids per second.

DIFFERENCE IN PROTEINS SYNTHESIS BETWEEN PROKARYOTES AND EUKARYOTES.

No.	Character	Prokaryotes	Eukaryotes
1.	Ribosomes involved in protein synthesis	70s, with 30s and 50s sub units.	80s with 40s and 60s sub units.
2.	Occurrence of transcription and translation	Translation begins even while transcription is still going on in mRNA	Translation begins after complete transcription and mRNA comes out of the nucleus.
3.	Migration of mRNA	Ribosomes pull out mRNA from the DNA	mRNA moves out of the nucleus through the pores in the nuclear membrane.
4.	Sites of initiation of protein synthesis in mRNA	many	one
5.	Initiator amino acid for protein synthesis.	N-formyl-methionine	Methionine
6.	Initiator factors	IF1, IF21 and IF3	eIF2, eIFwa1, eIF2a2 eIF2a3 and eIF3
7.	Elongation factors	EF-Tu, Ef-Ts and Ef-G	Ef-1, Ef-2
8.	Release factors	Rf-1, Rf-2 and RF-3	Rf (single factor)

23

GENETIC REGULATION

All the activities of an organism, be it unicellular or multicellular, are ultimately controlled by the genes. In multicellular organisms, where cells differ widely in structure and function all of them carry the same genes, since all of them are traceable to a single cell viz.. zygote. This poses an interesting question as to the function of genes. How do similar genes function differently? The answer to the question lies in *differential gene action*. In other words net all genes are active at all times. The genes are *switched* and *switched off* at different times. This differential gene action is also responsible for the synthesis of different types of proteins (enzymes) at different times. Not all enzymes are required at all times. Enzymes required during early stages of development may not be required at later stages in the life cycle and *vice versa.* The metabolic products produced may induce or repress the enzyme synthesis.

Genetic control on protein synthesis is exercised at three stages - transcription stage, translation stage and at the release of polypeptide chain (post translation stage).

The control mechanism ensures that the cell is not unnecessarily flooded by the enzymes which are required only at particular times and only in particular quantity. Depending on the requirement the enzyme synthesis can either be induced or suppressed. Genetic regulation is seen both in prokaryotes and in eukaryotes. While the purpose of genetic regulation is same in both prokaryotes and eukaryotes, the mechanisms differ. In prokaryotes, the mechanism of genetic regulation is simple when compared with the eukaryotic organisms. The reasons for this are not far to understand. As the eukaryotic organisms are far more complex they do require a very sophisticated mechanism of genetic regulation. In this chapter we will discuss genetic regulation in both prokaryotes and eukaryotes.

GENETIC REGULATION IN PROKARYOTES

Induction and Repression

In *E.coli* the synthesis of the enzyme β *galactosidase* has been studied in considerable detail. This enzyme breaks up lactose into glucose and galactose. If lactose is not supplied to *E.coli* cells, the presence of the enzyme β galactosidase in hardly detectable, but as soon as the substrate (lactose) is added the production of the enzyme β *galactosidase* increases by as much as 10,000 times. The production of enzyme again gets reduced if lactose is removed from the system. Such enzymes whose sythesis can be increased many times by the presence of the substrate are called **inducible enzymes.** The genetic system responsible for synthesis of such enzymes is called the *inducible system.* The inducible enzyme is an example of adaptive enzyme in the sense its (enzyme) synthesis gets increased due to the presence of the substrate.

In some other systems the reverse of the inducible system occurs for instance in a culture of *E.coli* if no amino acids are supplied from outside the cells can produced all the enzymes required for the production of amino acids. On the otherhand if a particular amino acid like histidine is added to the system the production of histidine synthesising enzymes will fall down. In these systems the addition of a product of a synthetic pathway will check the production of enzymes responsible for the synthetic pathway. In otherwords increased presence of a substrate would reduce the synthesis of an enzyme which has the enhancing effect of the production of that substate. Systems such as these are called *repressible systems* and the enzyme whose synthesis gets checked is known the as the **repressible enzyme.**

Inducer and Co-repressor

It has been mentioned above that an enzyme gets induced or repressed by the presence of a particular substrate. The substance which induces the synthesis of an enzyme is called the inducer and the substance which supresses the synthesis of the enzyme is called the repressor. In the synthesis of β *galactosidase,* lactose acts as the inducer and in the supression of histidine synthesising enzyme histidine acts as the repressor.

Enzyme induction or repression is generally brought about by the substrate on which an enzyme can act. But in some instances there will be molecules which resemble the substrate which also can act as the inducers. A typical example of this kind of induction is IPTG (isopropyl thiogalactoside). IPTG resembles lactose and can induce the production of the enzyme β *galactosidase.* Such inducers which are not substrates themselves but can bring about induction because of similarities with the substrate are known as **gratuitious inducers.**

In the above paragraphs we have discussed as to how in the absence of lactose there is no synthesis of β *galacotosidase.* This means then in the absence of the inducer the genes responsible for the production of the enzyme protein do not function. It will be interesting to find out as to why the enzymes do not function in the absence of lactose and how in the presence of lactose the genes regulating the enzyme synthesis get activated. It has been found out that a group of molecules called *repressors* are present in the cell and they check the activity of the enzymes. In the presence of the inducers the repressors get inactivated. In some instances ordinarily the repressors are inactive but they become active in the presence of another group of molecules called co-repressors. The induction and repression is supposed to work in the following manner.

1. Inducible system : active repressor + inducer = inactive repressor
2. Repressible system : inactive repressor + corepressor = active repressor

The operon concept

F.Jacob and J.Monod (1961) the Nobel Laureates have studied in detail the inducible system in the synthesis of the enzyme β *galactosidase* in *E.coli.* In order to explain the induction and repression of enzyme synthesis they have proposed the involvement of a group of genes along with the presence of cytoplasmic inducers and repressors. Their model of gene action is popularly called the operon model. This scheme of gene action proposed by Jacob and Monod as originally conceived by them consisted of an operator gene and a number of structural genes. As per the current understanding however the operon system consists of a structural gene, operator gene, promoter gene, regulator gene and repressor, inducer and co-repressor. The following is an account of the components of the operon model.

Structural genes

There are many structural genes associated in an operon system. These direct the synthesis of the mRNA and govern the sequence of amino acids in a protein molecule. Each structural gene might produce a particular kind of protein or all structural genes might regulate the production of a single protein. The activities of the structural gene(s) are controlled by the promoter and operator site of the operon system. The most well studied structural genes (*z,y* and *ac*) are those of the lactose operon system in *E.coli.*

Operator gene

The operator gene is situated adjacent to the first structural gene. It switches on or switches off the functioning of the structural gene (protein synthesis). In case a structural gene has to be suppressed, a repressor attaches itself to the operator to form an Operator-repressor complex. In the case of protein synthesis, the operator-repressor complex prevents the transcription by blocking the movement of RNA polymerase.

Promoter gene

The promoter gene is continuous with the operator gene and is believed to lie left to it. It is suggested that RNA polymerase binds to the promoter site during transcription. Three regions have been recognised in the promoter site. These are (a) recognition site, initial binding site and the mRNA initiation site (operator site).

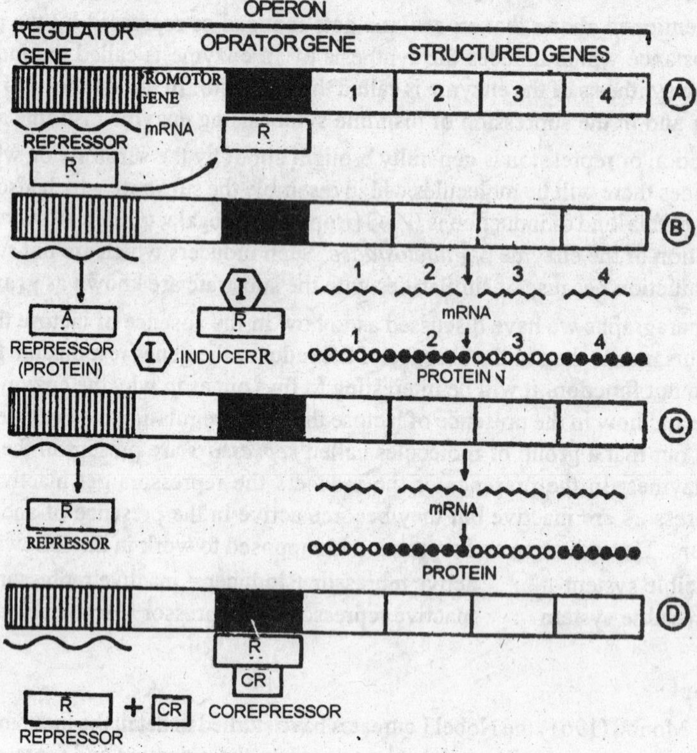

Fig. 23.1 Genetic regulation
Genetic regulation of protein synthesis in prokaryotes
(the Operon system of Jacob and Monod, 1961)

(i) Recognition site

Also called the *cga* site (catabolic gene activator site), it consists of certain palindromic sequence of DNA. These symmetrical sections of DNA are recognised by proteins having symmetrically placed sub units. This site, also called the CRP site (cyclic AMP receptor protein site) binds a Carotenoid protein to the promoter gene and thus facilities the binding of the enzyme RNA polymerase. It has been found that in *E.çoli*, CRP combines with cAMP (cyclic adenosine monophosphate) forming a CRP + cAMP complex which binds to the promoter enhancing the binding of RNA polymerase and activates transcription. This regulation is called positive control.

(ii) Initial binding site

This consists of seven bases (DNA) to which the RNA polymerase binds.

(iii) RNA initiation site

The site where transcription begins is called initiation site. This is the region overlapping with the operator region.

Regulator gene

The regulator gene directs the activity of the operator gene by producing inhibitor proteins called repressors.

This repressor protein binds to the operator gene and blocks the path of RNA polymerase, thus preventing transcription. If an inducer is present in the system, it binds to the repressor which undergoes conformational change and becomes inactive. As the inactive repressor cannot bind to the operator, the structural genes get activated and protein synthesis continues.

In a repressible system, if only a repressor is present, it is inactive by itself and cannot attach to the operator gene and hence transcription continues. However when another protein called corepressor is present it attaches to the repressor forming a *repressor-corepressor* complex, which blocks the operator gene thus preventing transcription.

Types of regulatory mechanism

Earlier in the chapter it has been mentioned as to how the presence or absence of a substrate can induce or supress the synthesis of the enzyme protein. Here we shall study in some more detail the various regulatory mechanisms that are involved in the synthesis of enzymes.

Induction. It has already been mentioned that in inducible systems the inducible enzymes are normally present in small quantities and their synthesis can be increased manifold by the presence of the inducer. Such an example of inducer in *E.coli* is lactose and the inducible enzyme is β *galactosidase.* In the absence of lactose only one ore two molecules of the enzymes β *galactosidase* may be present in the system. When the inducer is added, within three minutes about 3,000 molecules of the enzyme are synthesised by the gene.

When the inducer is absent the repressor protein will block the production of the enzyme by not allowing transcription of mRNA by the structural genes. The repressor protein actually blocks the operator gene. When the inducer is present the active repressor binds to the inducer to form a repressor inducer complex. As a result the repressor gets inactivated. An inactivated repressor cannot bind to the operator and hence the operator will allow the structural genes to transcribe mRNA and the enzyme is synthesised.

Repression. The presence of a metabolite inhibiting the synthesis of an enzyme in the system is called repression. It has been mentioned already that histidine acts as a repressor and blocks the production of the enzyme responsible for histidine synthesis. This is called enzyme repression. Another example of repression is the blocking of the production of the enzyme tryptophan synthetase by the substrate tryptophan. The enzyme tryptophan synthetase is produced only when the bacterium (*E.coli*) is cultured in a tryptophan free medium.

In the repressible system, the protein produced by the regulator gene is an inactive repressor (*aporepressor*). On combining with a corepressor (effector) the aporepressor molecules get activated due to a conformational change. As a result of this the aporepressor becomes active and binds to the operator gene. As a result of the operator genes getting blocked the structural genes cannot transcribe mRNA. However when the corepressor is absent, the aporepressor is inactive and by itself cannot bind to the operator gene. The operator gene being free will allow the structural genes to transcribe mRNA. The following is a summary of the regulation of enzyme synthesis is inducible and repressive systems.

I. Inducible system

1. ***Inducer present :*** Regulator gene → Active repressor + inducer → repressor inactive → operator gene free → structural genes transcribe mRNA → synthesis of enzyme protein.
2. ***Inducer absent :*** Regulator gene → active repressor → binds to the operator gene which becomes inactive → no mRNA transcription → inhibition of enzyme synthesis.

II. Repressible system

1. ***Corepressor present :*** Regulator gene → inactive repressor + corepressor → active repressor → binds and inactivates operator gene → structural genes do not transcribe mRNA → inhibition of enzyme synthesis.
2. ***Corepressor absent :*** Regulator gene → inactive repressor → operator gene not blocked → structural genes transcribe mRNA → enzyme protein synthesised.

Negative control

In the inducible and repressive systems, protein synthesis is allowed to take place when the operator gene is free and it (protein synthesis) stops if the operator gene is blocked. In otherwords, gene expression is possible only when the operator is free. Such a controlled mechanism of protein synthesis is said to be of a negative type. A typical example of negative control is the production of the enzyme protein in the presence of lactose and in the absence of glucose. The presence of glucose actually inhibits the production of the enzyme β *galactosidase* while the presence of lactose promotes the production of the eyzyme.

In negative control the regulator protein acting as the repressor prevents the transcription of mRNA by structural genes. The main controlling factor is the operator gene.

Positive control

In this control, the regulator protein functions as an activator and increases the enzyme synthesis. The activator binds itself to DNA in a location called the initiator site. Here the initiator site need not necessarily be the operator gene. As the regulator protein enhances the transcription of mRNA by structural genes this kind of regulation is called a positive control. A typical example of a positive control is the arabinose operon seen in *E. coli* (see later for details).

Replication of DNA is also an example of positive control. The bacterial genome is made up of replicating units called replicons which are capable of independent duplication. One of the structural genes of a replicon is known to produce an activator protein which in turn is induced by some cytoplasmic factors. The activator binds to a specific site on the replicon called the replicator and initiates replication from this point.

Catabolic repression

The works of Magasaine and others have shown that the presence of glucose inhibits the synthesis of enzymes involved in the breakdown of sugars such as lactose, arabinose or glactose. This phenomenon of repression is known as catabolic repression. This has been seen in many microorganisms including *E.coli,* yeasts etc. For instance when a culture of *E.coli* maintained on lactose is supplemented with glucose it has been seen that the synthesis of β *galactosidase* enzyme becomes suppressed as long as glucose is present in the medium. The inhibitory effect of glucose is due to a marked drop in the intercellular levels of a nucleotide–cyclic AMP. The studies of Zubay and Beckwith (1970), have shown that glucose suppresses the level of cyclic AMP which in turn inhibits the transcription of mRNA.

In vitro studies of Eron, et al (1971) have indicated that lactose operon transcription requires not only cyclic AMP but also a protein called CAP (catabolic activator protein). CAP binds to the cyclic AMP. Thus catabolic repression has two steps at the first stage CAP binds to cyclic AMP and at the second stage this complex (CAP cyclic AMP) exerts its effect on the transcription process.

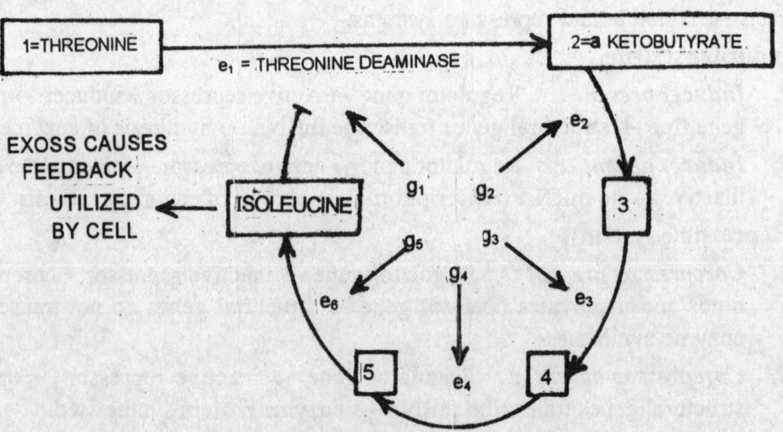

Fig. 23.2 Genetic regulation
Feedback inhibition in *E.coli* (after Stansfield, 1969)

Feed back inhibition

In some instances the final product of a particular biosynthetic pathway gets accumulated if it is not utilised for metabolic activities. Such a kind of accumulation many a time will cause the stoppage of further synthesis of the substance. For instance if the normally growing *E.coli* cells are supplied with isoleucine (an amino acid), it has the effect of blocking the isoleucine synthesising enzyme. This inhibition is due to the binding of isoleucine to enzyme *threonine deaminase.* This kind of inhibition is called feed back inhibition. Feed back inhibition may be defined as the inhibitory effect of the final product on the enzyme involved in the first step of production when a synthetic pathway is made up of several steps and is controlled by the activity of specific enzymes. Feed back inhibition has an advantage in the sense if the enzyme of the first step is blocked there is no necessity to block the intermediary enzymes as in any case they would not get the substrate.

Feed back inhibition is mostly through the allosteric transformation of the enzyme so that its active site gets inactivated. Feed back inhibition is different from repression in that in the later there is no mRNA synthesis and hence no enzyme synthesis. Feed back inhibition is actually a case of control of gene expression after translation.

Role of Leader sequences in genetic regulation

Gene expression in prokaryotes is controlled both at the level of transcription as well as after the transcription. In the mechanisms which operate after transcription a prominent role is played by leader sequences and attenuators. For example in the tryptophan operon of *E.coli* the mRNA of gene *trp E* (gene for anthranilate synthetase) has a sequence of 162 bases between initiator codon and the promoter operator region of this gene. This sequence of 162 bases is called leader sequence and is thought to exercise post transcriptional control. This leader sequence of the gene *trp E* has a mechanism which allows transcription when synthesis is needed and stop the transcription whenever synthesis is not needed.

Regulation by antisense RNA

Some of the recent researches have shown that regulation of protein synthesis can be controlled at the translational level by using RNA which has complementary bases to mRNA. When such a RNA is introduced it binds to mRNA forming RNA mRNA hybrid and prevents the translation, it (RNA) is this RNA is in complementary to mRNA and interferes in the translation which and it (RNA) is called antisense RNA. Antisense RNA can be used in experimental conditions to stop the translation of mRNA. In *in vivo* conditions also antisense RNAs are known to regulate the translation of mRNA. In transposon Tn 10, the production of the enzyme *transposase* is prohibited by antisense RNA. The production of antisense RNA is controlled by a promoter which directs transcription in a direction opposite to the transposase gene thus producing antisense RNA.

Role of antisense RNA in regulating gene expression has also been demonstrated in the case of the two genes involved for the synthesis of outer membrane protein.

Examples of a few operon systems

The lac-operon of E.coli. The lac operator consists of three structural genes.

1. Gene *z*. This is made up of 3,063 base pairs and codes for enzyme β *galactosidase* which is active as a tetramer and breaks lactose into glucose and galactose.
2. Gene *y*. This is made up of 800 base pairs and codes for the enzyme β *galactoside permease* which is a membrane bound protein and helps in the transportation of metabolites.
3. Gene *ac*. This consists of 800 base pairs and codes for β *galactoside transacetylase* an enzyme that transfers an acetyl group from acetyl coA to β *galactosides.*

In addition to the structural genes the lac operon consists of a regulator gene *i,* an operator gene *o* and a promoter gene *p*. These six adjacent cistrons (genes) constitute the lac operon system. An operon can be defined as an integrated genetic unit.

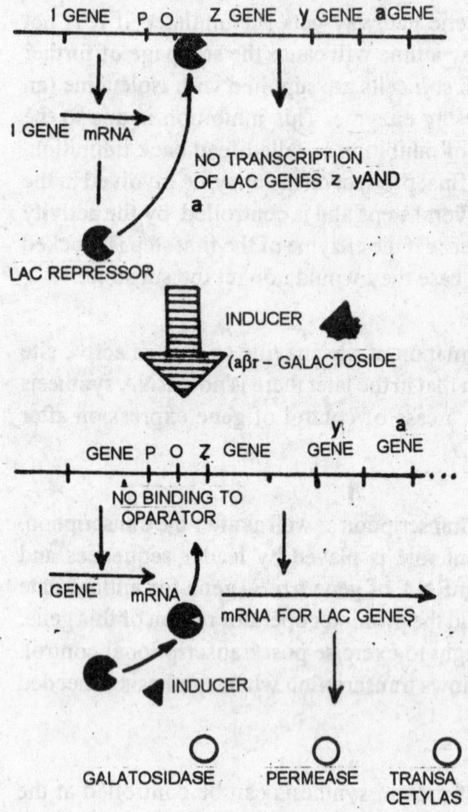

Fig. 23.3 Genetic regulation
Lac-operon model

In lac operon, the gene *i* produces a repressor (protein) which inactivates the operator by binding with it. Subsequently an inducer such a lactose binds with the repressor molecule there by inactiviting it. As a result the operon gene *o* is free from the repressor and becomes active. When the operator gene is active mRNA is produced from the structural genes *z, y* and *ac.* The production of mRNA starts from the promoter gene. It has been found at that all the three structural genes produce a single mRNA molecule which codes for the sequence amino acids for all the three enzymes - β *galactosidase, permease* and *transacetylase.* Thus equal amounts of all the three enzymes are produced and the first message translated is for β *galactosidase;* this is followed by the synthesis of *permease* and then followed by *transacetylase.* Thus the operons show polarity. This has been proved by mutational studies. Mutations in the *z* region will reduce the quantities of all the three enzymes. While mutations in *y* locus will have no effect on β *galactosidase* but will reduce the quantities of other two enzymes. Such mutations are said to be **polar mutations.**

The gal operon system

In the gal operon systen in *E.coli* the sequence of three structural genes are *gal E, gal T* and *gal K.* These genes are directly responsible for the sequence of the enzymes produced namely *galactokinase, galactose-1- phosphate uridyl transferase,* and *uridinediphospho- galactose-4- epimerase.* These enzymes are responsible for the degradation of galactose into glucose-1-phosphate and UDP glucose.

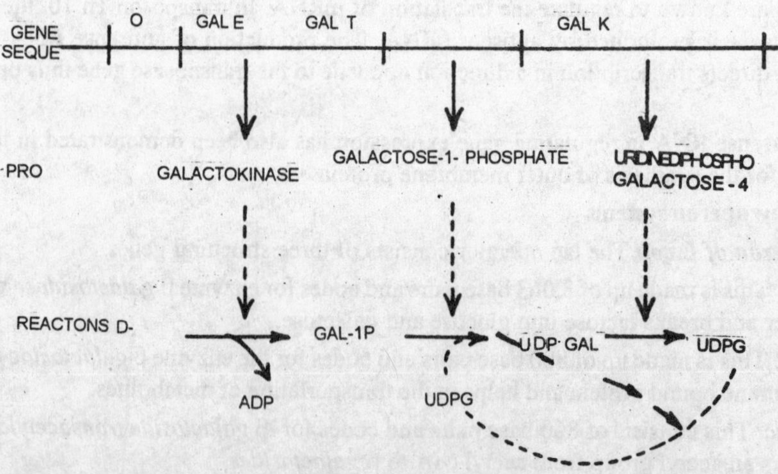

Fig. 23.4 Genetic regulation

The galactose (gal) operon of E.coli. Abbreviations : O-perator, Gal-1-galactose-1-phosphate, UDPG-uridine diphosphoglucose; UDP GAl - uridine diphosphogalactose; Glu-1 P=glucose-1-phosphage (after Goodenough and Levine, 1974)

Tryptophan operon system (trp operon)

The *trp* operon is also seen *E.coli* and consists of five structural genes ABCD and E. These genes code for the following enzymes :

1. Genes A and B code for *tryptophan synthetase.*
2. Gene C codes for *PRA isomerase* and *IGP synthetase.*
3. Gene D codes for *phosphoribosyl pyrophosphate anthranilate transferase*
4. Gene E codes for *anthranilate synthetase.*

The levels of these enzymes is low when the amount of tryptophan is more and the level high when the amount of tryptophan is less. All these enzymes are regulated in a coordinated fashion. The operator and promoter are overlapping and are located next to *trp* E gene. The regulator gene (R) is located away from the operon and the repressor produced by it is inactive until it combines with tryptophan. All the enzymes are produced constitutively if the repressor is either absent or fails to combine with tryptophan. All the structural genes are translated together and form a polycistronic mRNA. This polycistronic mRNA besides the transcription of the bases of structural genes has a 160 nucleotide long leader sequence. This contains the attenuator site to which the substrate tryptophan can get attached and thus reduce the rate of transcription. Thus in *trp* operon there are two sites of regulation - the regulator gene and the attenuator site in the mRNA itself.

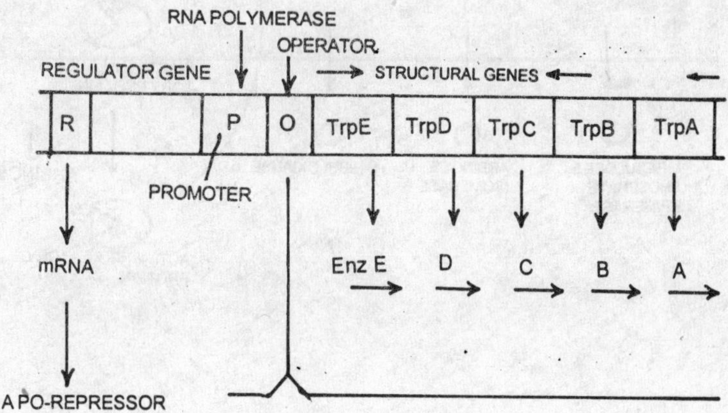

Fig. 23.5 Genetic regulation
Tryptophan operon

The *trp* operon thus seems to have a dual control-a control by regulator and also by attenuator. In addition to this feed back regulation is also seen in *trp* operon.

Arabinose operon

This is also seen in *E.coli* and is an example of positive control. Like the lac operon the arabinose operon also consists of three strcutural genes. These are

(i) *ara* A gene codes for *arabinose isomerase*
(ii) *ara* B gene codes for *L-ribulokinase*
(iii) *ara* B gene codes for *L-ribulose-5-phosphate epimerase*

These structural genes are located together on the *E.coli* chromosome. Their expression is under the regulation of a regulator gene C. This gene C called activator must be present for transcription of the arabinose operon. The product of C gene binds to the operator in the presence of arabinose and forms the activator protein. As a result of this, transcription takes place in the arabinose genes. If the activator is not bound to the arabinose or if the arabinose is absent, the operon is not expressed because the regulator protein binds to the operator gene (*ara* o) and functions as a repressor. In arabinose operon the gene C acts both as an activator and repressor. In this way arabinose operon is different from lac operon.

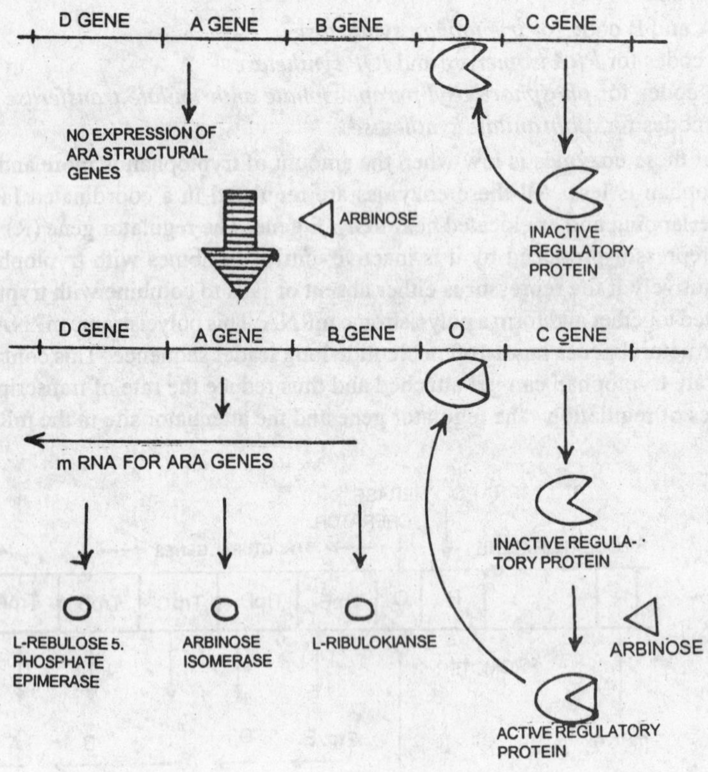

Fig. 23.6 Genetic regulation
The L-arabinose operon of E.coli

Gal genes in yeasts

In yeasts galactose if fermented by the activity of three enzymes-kinase, transferase and epimerase. These are coded by the three structural genes GAL 1, GAL 7 and GAL 10. The regulator gene is located away from the structural genes. If a mutation occurs in a regulator gene all the three enzymes are synthesised without control indicating thereby that the product of the regulator gene is a repressor. The regulator or gene of gal in yeasts is very much similar to the gals in *E.coli*. A detailed genetic analysis has revealed that the product of the regulatory gene controls the functioning of an unlinked gene GAL 4. The product of GAL 4 is a positive regulator for the three linked genes. The GAL 4 gene has an operator gene *(c)* also. The regulator gene is *i,* the product of this gene exerts a negative control over gene c (operator). As a result transcription of GAL 4 is reduced. When GAL 4 is free it produces a product which stimulates the transcription by the structural genes.

Regulation of transcription in λ phage DNA

Regulation of the λ phage DNA has both negative and positive control mechanisms. The phage can exist in a lytic or lysogenic state and this is controlled by some regulatory proteins. Depending on when the genes get activated during infection, multiplication or muturation phae, the λ genes can be classified into *early genes* and *late genes*. The early genes as the name indicates are required at the beginning of the infection cycle while the late genes are required for the assembly and maturation of phages. The following table gives a list of genes identified on the λ genome along with their functions.

List of genes of the λ phage genome

A,W,B,C,D,E and F	-	Head genes
Z,U,V,G,T,H,L,K,Iand J	-	Tail genes
att λ	-	site of attachment and insertion of the phage DNA into the host DNA
int	-	integration of λ DNA e
xis	-	excision of λ DNA
red α	-	exonuclease
red β	-	λ recombination system
c III	-	required for the expression of the c I gene
N	-	required for the expression of early gene
OL	-	operator of a set of early genes
rex	-	restriction of trowth of T4 R II
c I	-	λ repressor
OR	-	operator of another set of early genes
tof	-	negative regulator of c I transcription
c II	-	required for the expressions of the c I gene
ori	-	point of origin of replication
O	-	required for λ replication
P	-	required for λ replication
Q	-	stimulates transcription of late genes
p1	-	promoter of late genes
S	-	required for host lysis
R	-	required for host lysis

The cI gene codes for the repressor which binds to the operator genes thereby blocking the transcription of both the early operons. The repressor can bind to the operons blocking its integration and replication resulting in immunity due to a negative control exerted by a repressor. The product of gene N however acts as a positive regulator and allows transcription. The product of the N gene is known as antiterminator because instead of stopping it allows transcription. Another example of positive regulation is the influenceof Q gene product on late genes. This product combines with the coat transcriptase enzyme of *E.coli* and acts favourably on the promoter of the late genes. Thus in the λ phage there are both negative and positive regulatory mechanisms to control the initiation and completion of lytic or lysogenic cycle. This is very necessary for an organism which has to survive in a continuously changing environment such as the bacterial cell.

GENETIC REGULATION IN EUKARYOTES

The mechanism of gene regulation of protein synthesis explained above illustrates the phenomenon in prokaryotes, whereby a repressor protein transcribed by regulator genes may activate or prevent the protein synthesis depending on the presence of inducer or corepressor.

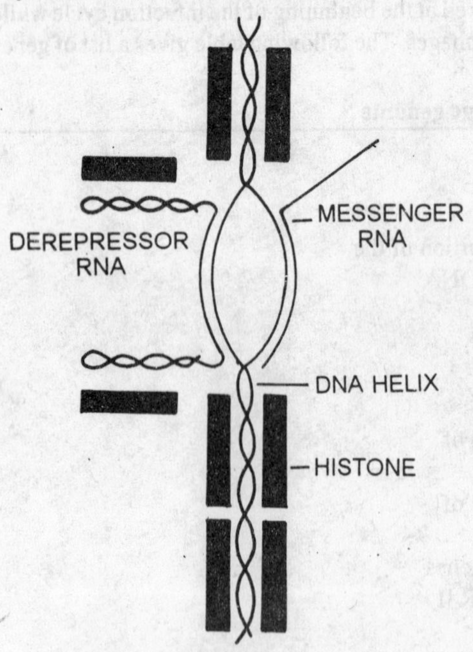

Fig. 23.7 Genetic regulation
Genetic regulation of protein synthesis in Eukaryotes0Frenster's model of gene regulation

In Eukaryotes, genetic regulation takes place by a slightly different mechanism. Two types of proteins associated with the DNA of the chromosomes possibly play a role in gene regulation. These are histones and non histone chromosomal proteins (NHC proteins). Since histones of different organisms have similar amino acid composition it is unlikely that in they can selectively repress gene action. At best, histones may probably bring about a general (non specific) repression. It is believed that histones under certain situations bring about supercoiling of DNA preventing the movement of RNA-polymerase and thus blocking transcription. This is known by the fact RNA precursors are incorporated only in those regions where chromatin is diffuse (no super coiling).

Three models have been proposed by various scientists to account for gene regulation in eukaryotes. These are - 1. Frenster's model, 2. Non histone derepressor model and Britten and Davidson's model.

Frenster's model

This is also called gene specific derepressor RNA model. As has been pointed out earlier, histones bring about non specific repression of transcription. Derepression (initiation of transcription) is brought about when nuclear polyions separate histones from DNA at certain specific parts (of DNA) and form complexes with histones.

Non histones derepressors

The model proposed by Paul et al (1971), is similar to Frenster's model except the derepression is brought about by NHC acid proteins instead gene specific RNA. The fact that NHC proteins are of immense variety in different organisms, and also at different stages of growth even in the same individual goes to show that they can undertake selective derepression at specific gene loci. Derepression is brought about when NHC proteins bind themselves at specific loci to histones forming complexes. This complex separates itself from the DNA and allows it to transcribe, only at those specific loci.

Britten Davidson Model

Also known as operon operator model, this is somewhat different from the other two models. As per this model, four types of genes are involved in the regulation process. These are sensor genes, producer genes, integrator genes and receptor genes.

1. *Sensor genes*

Comparable to the promoter genes these are sensitive to the metabolic status of the cell. The sensor gene is stimulated by various metabolic substances.

2. *Producer genes*

These regulate the output of metabolism and control the formation of enzymes, cell organelle, membranes etc.

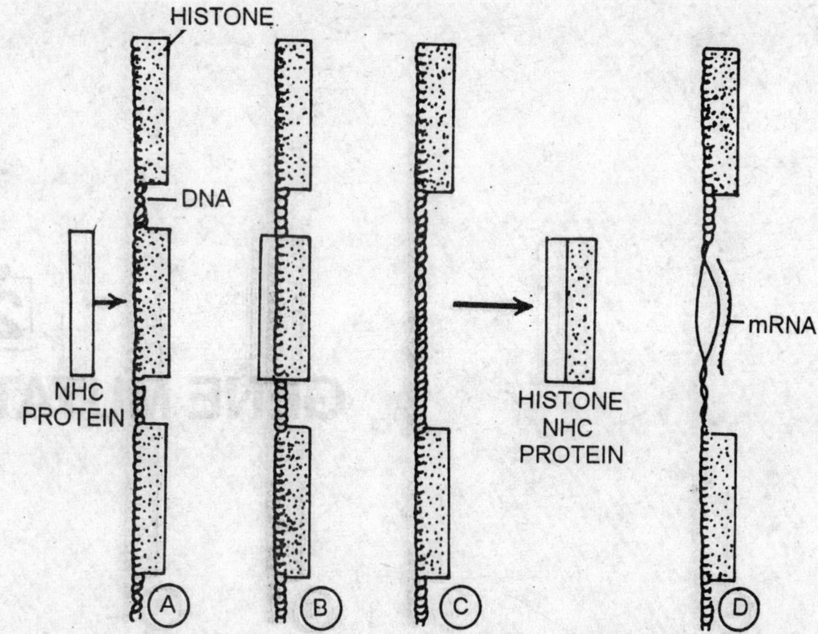

Fig. 23.8 Genetic regulation

Gene regulation by NHC proteins, **A.** Non specific repression by histones, **B.** Specific binding of NHC Proteins to histones, **C.** Removal of histone by NHC proteins, **D.** De repression of DNA, leading to transcription.

3. *Integrator gene*

As the name suggests, the gene integrates the activities of other genes. The integrator gene sends specific molecular signals to other genes as to whether the sensor gene is activated or not. Each sensor gene may be associated with a single or a group of integrator genes. In the latter case, a sensor gene may send a variety of molecular signals to other genes through its various integrator genes.

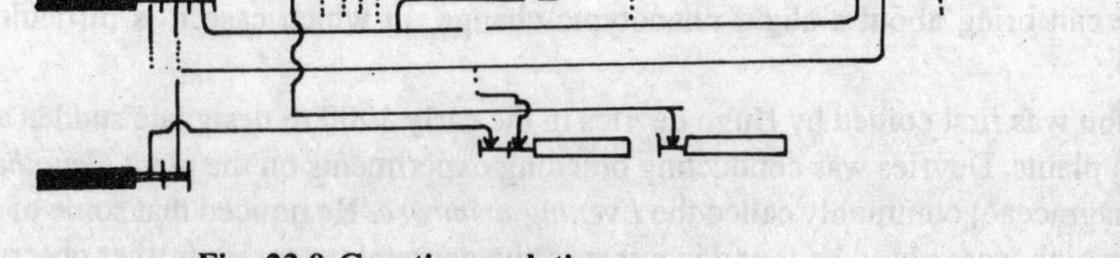

SENSOR GENE
INTEGRATOR GENE
MOLECULA SIGNAL
RECEPTOR GENE
PRODUCER GENE

Fig. 23.9 Genetic regulation

Britten Davidson's Operon-Operator model

4. *Receptor gene*

This acts as an intermediate between the integrators and the producers. When the molecular signal reaches the receptor gene, it inturn activates the producer gene. A producer may have one or a group of receptor genes associated with it in the same way as sensors have seveal integrators. This helps, not only in the variety of origin of molecular signals, (from sensors) but also variety in reception of these signals (by receptors). Perhaps multintegrator gene complex and multireceptor gene complex may have a specific multichannel relationship.

24

GENE MUTATION

Evolution in the biological world is primarily dependent on two events to bring about hereditary variations. These are - **mutation** and **recombination**; while mutation provides the raw material for the changed potentiality of inheritance, recombination permits permutation combinations in the gene pool, thus stabilizing the mutational change in the environment.

What are mutations? To understand this we will briefly analyse how characters are trasmitted from parents to offspring. We know that the characters are trasmitted from parents to offspring through the genes. But the offspring will not have the same genes as those of the parents, rather they have copies of them. In other words during reproduction, parental genes produce copies of themselves to be passed on to the offspring. Gene copying or gene reproduction is a precise process and generally there are no mistakes. Occasionally however, mistakes may occur while copying, so that the newly formed gene gets an altered structure. This modified gene then goes on multiplying and produces its own copies just as the original gene does. This alteration in the changed gene structure is called **gene mutation.**

Broadly speaking, mutation includes all those phenotypic variations, which arise due to altered gene structure and are heritable. Hence the term encompasses, small and minute changes brought about by altered individual genes (gene mutation) and large changes brought about by the altered chromosome structure or number. These latter types of changes can also be called chromosome mutations, and they have been described under the chapter chromosomal aberrations. In this chapter we will study gene mutations. Gene mutations are also called point mutations as they occur at a specific gene locus altering its molecular structure. Sometimes chromosomal aberrations also can bring about a slight phenotypic change, in which case it is difficult to distinguish between these two.

History : The term mutation was first coined by Hugo devries in the early 1900 to designate sudden and drastic changes that occured in plants. Devries was conducting breeding experiments on the plant *Oenothera lamarckiana* (of the family Onagraceae) commonly called the *Evening primrose.* He noticed that some of the progeny possessed characters which were not to be found in either of the parental types. He further observed these changed types bred true. He therefore described these sudden changes as mutations and ascribed them to some changes occuring in the germplasm of parents. Even Charles Darwin was aware of mutations, even though he did not use the term (mhtations). He was aware of sudden changes occuring in organisms both plants and animals and had used the term sports, to describe them.

Fig. 24.1 Gene Mutation
Ancon sheep with ram and ewe

The first recorded instance of mutation dates back to 1791, to the pen of Seth Wright, a New England farmer, where appeared a male lamb with short bowed legs. Wright reared this lamb and it bred true. A lamb with short legs was of great advantage as it could not jump even a low level fence. This short legged breed was called **Ancon breed.** This breed however became extinct after about eight years. But the same Ancon sheep appeared some fifty years later in the flock of a Norwegian farmer, representing probably the repetition of the samemutation. Since then a number of mutant types have been discovered. Hornless individuals in cattle, double toed cats, albino rats, mule footed swine, dwarf cupid sweet pea, double flowered, white flowered varities of many plants are all mutants.

The scientific study of mutations began with the work of Morgan on *Drosophila* in 1910. He observed that among a population of red eyed flies, some individuals with white eye appeared and they were predominant among the male individuals. He further found out that the white eyed gene is recessive to red eye and located on the X chromosome. Since the male individuals possessed only one X chromosome and the females had two X chromosomes, the recessive character appeared more frequently among males than among females.

After the original discovery of the white eyed mutant in *Drosophila* by Morgan and his co-workers, a detailed analysis of millions of these fruit flies for over 17 years was carried out by different geneticists and nearly 500 gene mutations have been discovered. Simultaneously other organisms like maize, rodents, pea, snap dragons, poultry, man etc. also have been investigated and scores of mutations have been discovered. Even in microorganisms also mutations have been detected.

Size or range of mutations : Size of mutation varies in different organisms and in different genes. These may be very minute and may not be detectable at all or as in some cases (Ancon breed), so striking that they can be easily made out. In some microorganisms like Bacteria, Fungi *(Neurospora),* Blue green Algae etc, mutations may not bring out any morphological effect but may bring about some changes in their nutritional requirements. In some cases mutations may be so large as to even result in lethality to the individual.

Frequency of mutations : The frequency is generally less unless there is a drastic change of the surroundings. According to Muller, in *Drosophila,* the mutation rate for any gene is less than one in one million. It has also been calculated that a particular gene in *Drosophila* may mutate only once in every 40,000 years.

The rate of mutation also varies between different individuals, organisms and sometimes even under different ages of the same organism. J.B.S. Haldane, a renowned geneticist has estimated that in a comparison between man and *Drosophila,* the former has less chances of mutation as most of them are harmful so there is a selection pressure in man against mutations. If the mutations in man were to take place at the same rate as in *Drosophila,* the whole genetic system of the human race would be impaired. As against the unaltered life expectancy of 40,000 years for a *Drosophila* gene, the human gene has a life expectacy of 25,000,00. In one of the case studies of Chondrodystrophic dwarfness (in these human beings, arms and legs are short, while other parts are normal), it was estimated that one in every 11,500 births haye this genetic defect (the data obtained from a Denmark hospital). This gives a rate of one mutation for every 23,000 genes at this locus. If we assume, the average age of the parents to be 30, who give birth to dwarf children, then one can estimate that the gene mutates once every 6,900,00 years. The following two tables give the frequency of mutations in man and Corn.

Table : Rates of mutation in some human genes

Sl. No.	Phenotypic effect	Genetic condition	Frequency in %once	Occurrence nce in every gametes
1.	Colour cannot be distinguished	Colour blindness	0.0028	35700
2.	Blood clots very slowly	Hemophilia	0.0032	31250
3.	No pigmentation in hair, skin etc.	Albinism idiocy	0.0028	35700
4.	Mental faculties dull after birth	Amaurotic	0.0011	90909
5.	Limbs and arms dwarf	Chondrodystrophic dwarfism	0.0070	14300
6.	Face with red patches, tumours in Kidney, brains etc., at later stages	Epiloia	0.0012	83333
7.	Iris in the eyes absent	Aniridia	0.0005	2,00,000
8.	Abnormal WBC resulting in reduced immunity	anomaly	0.0080	12,500

Table : Rates of mutation in maize (After Stadler 1943)

Sl. No.	Gene	Phenotype	Frequency (1 in one million gametes)
1.	R	Plant colour	492
2.	Cl	Chlorophyll	106
3.	Dr	Red aleurone	11
4.	WX	Normal endosperm	0
5.	Su	Sugary endosperm	2
6.	Sh	Full endosperm (not shrunken)	1
7.	Ci	Aleurone colour inhibitor	2
8.	Y	Yellow endosperm	2

Mutational Unit : Earlier it was not known as to what constitutes a mutation in a gene. But recent researches have shown that the smallest amount of DNA within which mutation can take place is called a *muton.* At the molecular level mutation involves alteration in the base pair sequence.

Mutator genes and mutable genes : It is difficult to assess (among plants and animals) as to which gene frequently mutates and which gene is comparatively stable. But in some of the orgamisms like maize (Emerson), it has been observed that some genes change quite frequently, w-hile others are not so. Barbara McClintock (Nobel laureate, discovery of jumping genes) has shown in maize that some genes are not only stable, but influence others to mutate. The genes that influence others (to mutate) are called mutator genes and the genes that are so influenced are called mutablc genes. These genes, mutator and mutable are more frequent in plants than in animals.

Classification of mutations : Broadly speaking, mutations are of two types - Chromosomal and gene. Chromosomal mutations actually do not change the gene structure, they bring about change in the phenotype either due to loss, addition, altered seqeuence or trasferred position of the genes. Gene mutation however changes the phenotype by basically altering the gene structure itself at the molecular level.

Gene mutations are basically of two categories - **Spontaneous** and **induced**. Spontaneous mutations apparently arise without the help of any identifiable external agency. They arise due to some internal changes. Induced mutations on the other hand are traceable to some external agency like radiation, temperature, chemicals, gas etc.

Mutations may also be further classified into various categories based on the effect, cell type inwhich it occurs, direction of mutation etc. The following is a brief classification of such mutations.

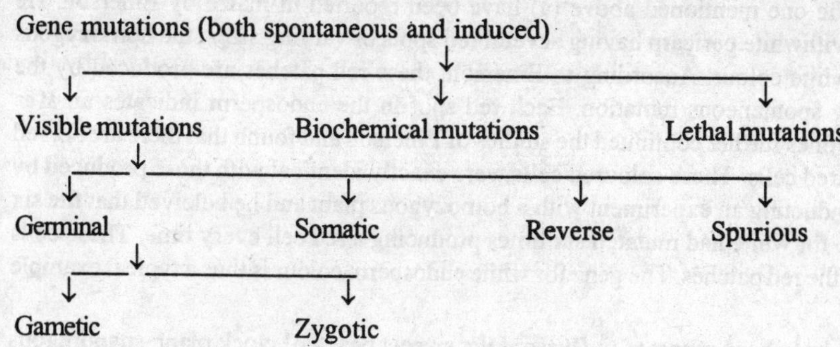

We shall briefly study a few examples of spontaneous and induced mutations and then study a few instances of the specific types of mutations above.

SPONTANEOUS MUTATIONS

As has already been pointed out, the agency which brings about spontaneous mutations is not known. They apparently occur due to natural causes. The first recorded instance of a spontaneous mutation is that of the Ancon breed sheep. A detailed study of the spontaneous mutation in *Drosophila melanogaster,* is that of T.H. Morgan, where he described the sudden appearance of white eyed flies in a population of red eyed individuals. Some of the spontaneous mutations noticed in plants and animals are as follows.

Spontaneous mutations in plants : 1) Double petunia, 2)Double rose, 3) Shirley poppy, 4) Single flower mutant in sunflower, 5) Multileaf formation in tobacco reported by Hayes, 6) Reduction in ray size in the ray florets in *Bidens pilosa* of Compositae(reported by the the author).

Spontaneous mutations in Animals : 1) Polydactyly in man, sheep, cats etc. 2) Multinippled condition in sheep, 3) Silky fowls, 4) Albinism in man, pig, pigeon, rodents etc. 5) Double cared cattle, 6) Mule footed swine, 7) White canaries, 8) Hornless cattle, 9) Hares' lip in man, 10) Widows peak in man, II) Lobster claws in man, 12) Cleft palate in man etc.

Spontaneous mutations are generally rare and may be of two types dominant and recessive. Usually recessive mutations are more frequent than dominant mutations.

Factors affecting spontaneous mutations : The rate of spontaneous mutations are known to change under the influence of various external and internal factors. One of the external factors which might bring about spontaneous mutations is a change in the temperature range (normally a higher temperature makes the genes susceptible for change). Disturbed metabolic rates may also bring about mutational changes. Age and sex may also have a bearing on the mutational rate. Increased age may render the genes prone for mutational change. *In Drosophila,* it has been observed that males are more susceptible to mutational alterations than the female.

One of the important factors bringing about spontaneous mutations may be, specific mutation inducing and mutation succeptible genes. As has already been pointed out, these are called mutator and mutable genes.

The mutator genes have been discovered by Me Clintock in maize. She described some genes which had the capacity to influence other genes to mutate. Rhodes has described the role of a mutator gene which influences the functioning of a colour gene. In maize the colour is governed by three recessive genes *a, c,* and *r,* while the dominant alleles of these govern the colour other than green. In addition to these there is a mutator gene)((dotted colour) which controls the variegated colour in the grain kernel. This gene influences the gene *a* to mutate so that red patches are produced. Usually the plants with the genotype *aa cc RR* will be green (all the three dominant gene pairs are required to produce colour other than green), but the presence of *Dt* makes it variegated.

Mutable genes other than the one mentioned above (a) have been reported in maize by Emerson. He investigated a variety of maize withwhite pericarp having several red spots of varying sizes; the homozygous recessive allele producing the white colour. According to Emerson, these red patches are produced by the unstable white gene undergoing spontaneous mutation. Each red spot in the endosperm indicates an area where a mutation has occurred. Jones further continued the studies of Emerson and found that there are six red spots, each having 10 - 45 coloured cells. These coloured cells were exactly identical with those produced by the pure red gene. Jones was conducting an experiment with a homozygous plant and he beleived that the six red spots indicated that the gene for white had mutated six times producing a red cell every time. These cells further multiplied and produced the red patches. The gene for white endosperm colour is thus a typical example of a mutable gene.

Several such mutable genes have been reported *in Drosophila,* sweeet pea, 4 o' clock plant, snapdragons etc. The genes for miniature wings, gene for purple eye colour in *Drosophila* are known to be mutable genes. In maize plant itself, Stadler has observed that the gene R for aleuron colour mutated some 492 times in about a million gametes that he studied.

INDUCED MUTATIONS

Various external agencies are known to bring about mutational changes in genes. All these are included under the general category of induced mutations. H.J. Muller (1927) may be regarded as pioneer in radiation genetics. In his studies on *Drosophila,* he showed that flies subjected to X radiation undergo mutations. Similarly Stadler (1928) demonstraded the mutational effects of X rays in barley and maize. Various kinds of agents causing mutational changes in organisms are known. These are of the following types, a) Radiation b) Temperature changes and c) Chemicals.

Radiation induced mutations : Various types of radiation (excluding visible light) are known to cause mutations in different intensity. Radiations are of two categories depending upon their capacity of penetration into a substance -*Ionizing* and *non ionizing*. Generally speaking, non ionizing radiations are not of a high potentiality to bring about mutations at a rate which is brought about by ionizing radiation. The Alpha rays, Beta rays and the gamma rays and X rays come under the category of ionizing radiations while low penetration rays lik U.V. rays belong to non ionizing radiation.

Mayor (1920) experimented on the effects of X radiation *in Drosophila* (and similar other radiations) to study the effects on germplasm. He demonstrated that as a result of radiation, there was non disjunction of the X chromosome resulting in a XXY individual, which was female phenotypically.

H.J. Muller also worked on the effects of X rays on *Drosophila* and published a classical paper 'The artificial transmutation of the gene'. He also devised a technique to detect induced mutations (see later in the chapter, for detailed explanation).

Stadler studied the effects of X rays on barley plant, while Gager and Blackslee induced mutants in the jimson weed *Datura stramonium.* It should be noted however that X rays are lethal and the dosage of the rays given to the individual has to be carefully monitored. On an average it has been found out that 80% of the X ray induced mutations are lethal.

Oliver, Timofeeff Ressovsky and many others have worked out the relationship between the frequency of mutation and the amount of X radiation absorbed by the organism. The intensity of the radiation is measured in terms of Roentgen units or *r-* units. The frequency of mutation seems to be proportional to the amount of *r-* units absorbed by the plant. The duration of exposure or the distance between the radiation source and the experimental organism, however do not seem to affect the frequency of mutations. The following table gives the rates of mutation in *Drosophila melanogaster* due to two different doses of X rays. In the table below t4 dose is twice as large as the t2 dose. (The data in the table below are based on Muller's experiment, quoted from Principles of Genetics, Sinnott Dunn and Dobzansky).

Experiment	Number of chromosomes observed	Number of mutations		
		visible	semilethal	lethal
Control	198	0	0	0
Xrays t_2	676	1	44	9
Xrays t_4	772	3	12	89

Some of the mutations on exposure to X radiation in *Drosophila* are as follows.

Gene wt to W (white), W^e (eosin), and W^a (apricot). Genes coral (W^{co}) buff (W^b), cherry (W^c) and apricot W^a) can mutate into white (W).

Chlorophyll deficiency has been induced in barley due to X radiation (Stadler). X rays can also induce mutations in the wasp *Habrobracon.* Among the other types of ionizing radiation Alpha, Beta, and Gamma rays are important. The First two however do not seem to have a profound mutational role as they cannot penetrate the body cells as deeply as gamma rays.

Mechanism of action of ionizing radiation : There are two theories which account for the mechanism of mutagenic effect of ionizing radiation. These are the **direct hit theory** and the **indirect hit theory**. According to the direct hit theory, the radiation directly strikes the genes bringing about a change in the chemical composition. The electrons that. strike the atoms of genes bring about a change in the chemical composotion, thus altering the gene. But in several experiments, it has been seen that , the same dose of radiation does not result in the same frequency of mutation indicating that direct hit theory cannot account for all radiation induced mutations.

According to the indirect hit theory, gene alteration is brought about by changing the molecular environment of the surroundings. Muller is a supporter of this theory. According to him when the high speed protons of X ray collide with the molecules of the cell, electrons are ejected out of the atoms rendering them positively charged. The electrons that come out move at high speeds and knock out other electrons that come in their way resulting in an environment which is ionized (positively charged). The electrons moving freely may attach themselves to other atoms rendering them negatively charged. Various chemical reactions and shifts take place to bring both the positive and negatively charged atoms to their normal state. During this process, alteration in the chemical composition takes place, thus mutating the gene. At the molecular level, base pair sequences are altered in the DNA molecule (for more details see mutation at the molecular level, in the same chapter).

Effect of non ionizing radiation : Ultra violet rays are non ionizing because they have a long wave length and therefore have less energy. They are generally non penetrative. They may bring about many changes in the traits, but are not as effective as X rays. Altenberg was able to accelerate the rates of mutations by using U.V. rays. Stadler also achieved pollen mutations induced by U.V. rays in maize; the same is the case in *Antirrhinum* as reported by Noethling and Stubbs. The U.V. rays are known to cause alterations in the bond characters of purines and pyrimidines. Between the two, the latter are more prone to mutational changes.

Chemical mutagens : A number of chemicals are known to have mutagenic properties. It was C.Auerbach who discovered during second world war that chemicals could be used to induce mutations. Thorn and Steinberg (1939) had also reported the chemical induction of mutations. They treated the Fungus *Aspergillus* with nitrous acid. Since then a long list of mutangenic chemicals have been identified. Some of the common chemicals used as mutagens are given in the following table.

Table : Some mutagenic chemicals

Class of chemicals	Name of individual compound	Common Abreviation
1 Aziridine	1 Ethyleneimine	EI
2 Mustards	2 Nitrogen mustard	
	3 Sulphur mustard	
3 Nitrosamines	4 Dimethylnitrosamine	DMN
	5 Nitrosoguanidine	NG
	6 Nitrosomethyl urea	NMU
4 Epoxides	7 Ethylene oxide	Eo
	8 Diepoxybutane	DEB
5 Alkyl sulphonates	9 Diethyl sulphate	dES
	10 Methyl methane sulphonate	MMS
	11 Ethyl methanal sulphonate	EMS
6 Others	12 Nitrous acid	
	13 Colchicine	
	14 Maleic hydrazide	
	15 Hydrazine	
	16 Hydroxylamine	
	17 Caffeine	
	18 Hydrogen peroxide	
	19 Formaldehyde	
	20 Phenol	
	21 Urithane	

Hydrogen peroxide is known to bring about mutations in *Neurospora.* Urithane and phenol are capable of inducing mutations in higher plants and *Drosophila.* Formaldehyde may induce mutations during spermatogenesis in *Drosophila.*

Colchicine is an alkaloid extracted from the bulbs of *Colchicum autmnale* of the family Liliaceae. This is known to induce polyploidy by suppressing spindle formation. Colchicine is not known to bring about any point mutation. Mustard gas is another chemical widely employed as a mutagen. Its mutagenic effect has been tested with reference to *Drosophila, Neurospora,* bacteria, barley, mice etc.

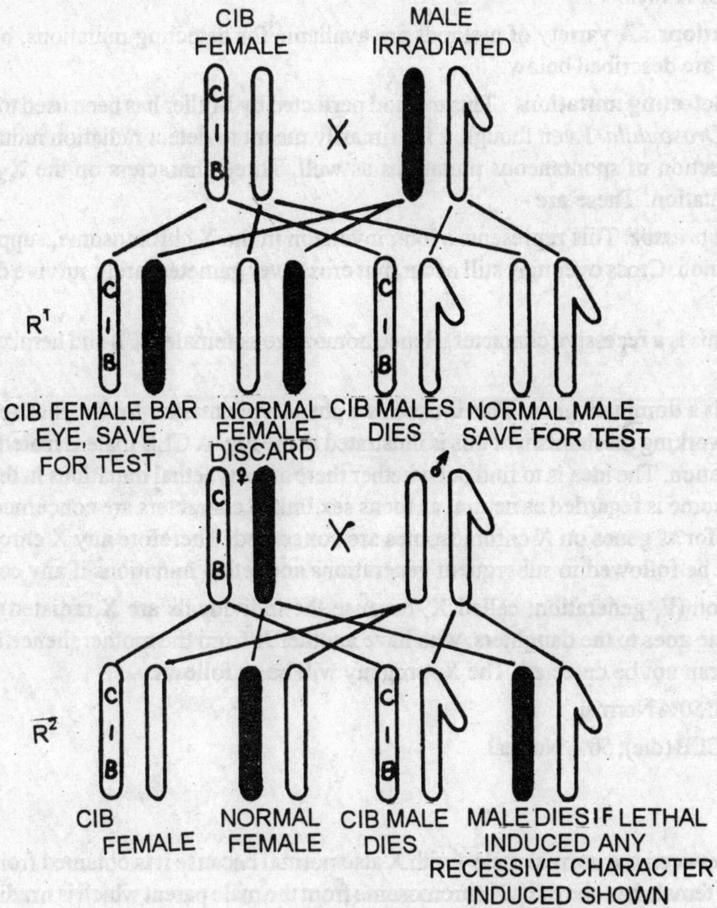

Fig. 24.2 Gene Mutation
Muller's CIB method to detect induced mutations drosophila melanogaster

Temperature induced, mutations : Child, Plough and Ires have studied the effect of temperature as a mutagenic agent in *Drosophila.* A range of temperature from 4°C to lethal range was used to find out the possible effect of temperature on the rate of mutation. A higher range of temperature is known to accelerate the rate of mutation. But it has to be borne in mind that a gradual increase in temperature might not be an effective mutagen. Any sudden increase or decrease in temperature might alter the gene copying mechanism thus altering the gene. The temperature effects largely remain as a somatic influence.

Vegetable oils as mutagenes : In one of their experiments concerning oil yielding plants, Swaminathan and Natarajan (1959) noticed that seeds with high oil content are generally resistant to mutagens. They soaked the cereal grains in oils of ground nut, mustard, and castor and wanted to know whether these oils provide any mutation resistance to the grains. Much to their surprise, the results were startling. Instead of the oils providing resistance to the cereals, they (oils) themselves had acted as mutagens as noticed by marked reduction in rates of germination, chromosome irregularities and many ear head variation mutations in wheat, one of the cereals used in the experiment. It was assessed that in wheat *(Triticum astivum),* the rate of mutation, induced was much higher than that of X rays.

Detection of mutations : A variety of methods are available for detecting mutations, both in plants and animals. Some of these are described below.

CLB method for detecting mutations : This method perfected by Muller has been used to detect mutations in X chromosomes of *Drosophila.* Even though it is primarily meant to detect radiation induced mutations, it can be used in the detection of spontaneous mutations as well. Three characters on the X chromosome are chosen to study the mutation. These are -

C = Cross over suppressor. This represents a long inversion in the X chromosome, suppressing the cross over in the inverted portion. Cross over may still occur, but cross over gametes rarely survive due to deficiences and duplications.

L = Lethal gene. This is a recessive character. Hence homozygous females (l/l) and hemizygous (l/y) males do not survive.

B = Bar eye. This is a dominant gene. This is used as a phenotypic marker for the flies required to mate in the F_1 generation. The working mechanism of this is illustrated in the Fig. A CLB male is mated to a normal male which is subject to radiation. The idea is to find out whether there are any lethal mutations in the X chromosome of the male. Y chromosome is regarded as neutral as for as sex linked characters are concerned, hence the male will be hemizygous as for as genes on X chromosomes are concerned. Therefore any X chromosome of male which is irradiated can be followed in subsequent generations and lethal mutations if any can be detected.

In the X_1 generation (F_1 generation; called X_1 because the individuals are X radiated) the irradiated X chromosome of the male goes to the daughters, who have another X from the mother; hence in this generation any mutation induced can not be detected. The X_1 progeny will be as follows.

Females : :50% CLB; 50% Normal

Males : 50% CLB (die); 50% Normal.

A cross is now made between the normal male (with X also normal because it is obtained from female) and the CLB female. This CLB female has one of its X chromosome from the male parent which is irradiated and the other is CLB. In this (X_2 or F_2) generation all the males get one of the X chromosome, they die (50%) because there is already a lethal gene. If they get the irradiated X chromosome, and if a new lethal mutation is induced then the other 50% also will die. Hence if there is a lethal mutation on the X chromosome in the F_2, no male progeny will survive and all will be females. This is a clear cut evidence for the presence of a lethal mutation on the irradiated X chromosome. In the X_2 generation the progeny will be as follows.

Females : 50% CLB; 50% Normal (with an irradiated X chromosome)

Males : 50% CLB (die); 50% carry irradiated X chromosome; die if there is a lethal mutation.

The Muller 5 method : .Like the CLB method this is also devised by Muller. The experiment involves crossing a homozygous Muller 5 female *Drosophila* with an irradiated male. The Muller 5 female *Drosophila* has two marker genes - apricot eye (w^a/w^a) and Bar eye (B/B) in a long inversion so that there is no cross over. A homozygous apricot (eyed) Bar eyed female is crossed with an irradiated male. As shown in the figure, the

F_1 generation female offspring which gets the irradiated X from the male parent is crossed with a wild male. If a lethal mutation has occurred in the X chromosome, no wild types of male would survive, because they would get the irradiated X.

The absence of wild males in the F_2 is an indication of lethal mutation. Muller was awared Noble prize in Medicine in 1946 for this and many other discoveries that he had made in Genetics.

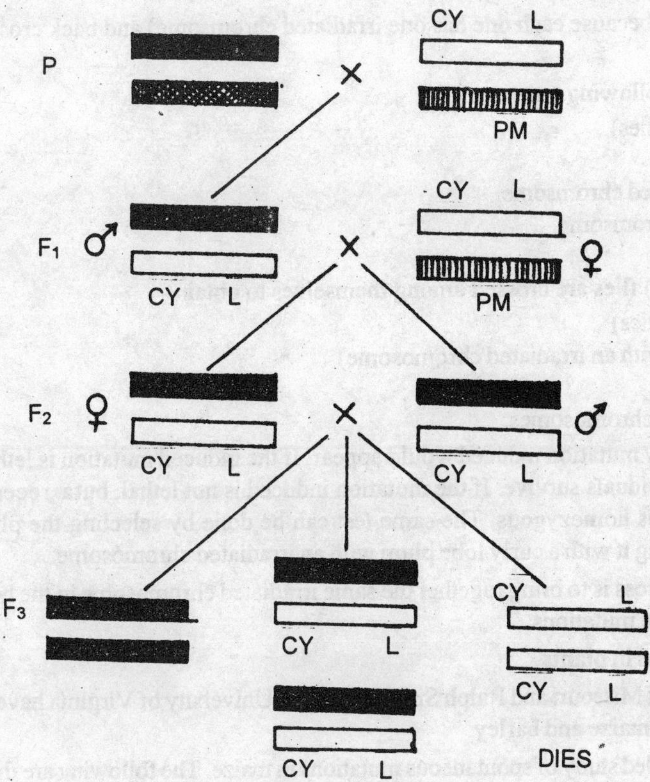

Fig. 24.3 Gene Mutation
Deletion of lethal mutations in a balanced lethal system

Detection ofAutosomal mutations (Balanced lethals)

Autosomal mutations can be detected by making use of balanced lethal genes in organisms. A balanced lethal stock is one which carries a lethal gene in one of the homologues and its wild allele on the other homologue. Homozygous individuals will die, so what survives is always heterozygous. Balanced lethals can be studied by using a specific strain of *Drosophila melanogaster* called the **curly lobe plum.**

In this, a stock of flies carry on one of the autosomes two dominant genes *C* (curly wing) and *L* (lobed eye) and a recessive lethal gene. An individual with two *CYL* chromosomes will not survive because the recessive lethals will be in double dose and its homologue carried the *Pm* (plum coloured eye) gene which is also dominant.

When two individuals with the genotype *{(CYL/Pm)* X *(CyllpM)}* are crossed, the heterozygous individuals only would survive, as homozygous for *(CYL/CYL)* would carry lethal genes in double dose, because *CYL* chromosome carries a lethal gene. There is a cross over suppressor to prevent the transfer of the gene *CY to* the chromsome carrying *Pm* gene.

A female *CYL/Pm* is crossed to another normal male which is irradiated.

Four types of progeny are produced in the F_1. These are -

(a) Curly lobe (with an irradiated autosome A)

(b) Curly lobe (with another irradiated autosome B)

(c) Plum (with an irradiated autosome A)

(d) Plum (with another irradiated autosome B)

One of these flies is selected (because each one has one irradiated chromsome) and back crossed to curly lobe plum.

The progenies will be of the following types.

(a) homozygous curly lobe (dies)

(b) Curly lobe plum.

(c) Curly lobe with an irradiated chromsome.

(d) Plum with an irradiated chromsome.

Among the above progeny, (c) flies are crossed among themselves to obtain -

(a) homozygous curly lobe (dies)

(b) Heterzygous curly lobe (with an irradiated chromosome)

(c) same as b.

(d) A fly with both irradiated chromosomes.

It is in the (d) category that any mutation induced would appear. If the induced mutation is lethal, it also dies and only the curly lobed individuals survive. If the mutation induced is not lethal, buta recessive trait, that would be expressed, since it is homozygous. The same test can be done by selecting the plum with an irradiated chromosome and crossing it with a curly lobe plum with an irradiated chromosome.

The significant aspect of this cross is to bring together the same irradiated chromosome in the homozygous condition to detect all the recessive mutations.

Detection of induced mutations in plants

L.J. Stadler of the University of Missouri and Ralph Singleton of the University of Virginia have worked on the methods to detect mutations in maize and barley.

Stadler (1942) has made a detailed study of spontaneous mutations in maize. The following are the important steps.

(a) A genetic stock dominant for a number of genes is grown as the female parent. The tassels are removed before pollen discharge (to avoid self pollination).

(b) A multiple recessive plant in every fifth row is used as the pollen providing parent for (a) plants.

(c) Seeds were observed in large numbers for endosperm characters. Most of them show dominant characters. If any recessive character (of the male parent) appeared it is due to mutation in the female gametes.

Singleton (1962) used a techinque similar to that of Stadler do detect both spontaneous and induced mutations in the pollen of maize. His method involves collecting pollen from plants dominant for many genes and cultivated in a field irradiated with cobalt 60 and transferring them (pollen) on to the silks of plants with all recessive traits and cultivated in a normal field. The seeds of the progeny would normally show dominant characters. If any mutation has been induced in a dominant gene, the recessive trait would appear in the progeny. In this way Singleton studied four genes *Su -su,* on chromosome 4, *Pr - pr* on chromosome 5, *Y-y* on chromsome 6 and *Sh -sh* on chromosome 9. The phenotypes of these genes are *Su* - normal endosperm, *su* - sugary endosperm; .Pr- Purple aleurone,pr-red aleurone; *Y*- yellow endosperm, *y* - white endosperm and *Sh* - full endosperm, *sh* - shrunken endosperm.

Apart from spontaneous and induced, mutations may also be classifed (see the beginning of this chapter) into various categores based on different criteria. Some of the examples are discussed briefly below.

Somatic mutations : These are sudden alterations appearing in the somatic characters of an individual. Such mutations normally do not affect the entire plant and express themselves only in those regions where a cell has undergone change. Certain varieties of apples and navel orange are examples of this.

Somatic mutations also occur in Snapdragons, when a white flowered variety produced red flowers in a single blossom.

Another kind of mutation, which has much significance from the poin of view of Horticulture is the *bud mutation.* In this, few cells of the bud or the entire bud (if mutation has occured quite early) shows mutated characters, when grown into a flower or a vegetative branch. If the mutation has occurred a little late during the development of the bud, the flower or the vegetative branch will be a mosaic of normal and mutant characters. Such a combination of two genetically non identical tissues is called a **Chimaera.**

Chimaeras have been seen in *Geranium (Pelargonium zonale)* in which a branch bears both green and white leaves. Another example of bud mutation is that of nectarine mutation in which the skin of the fruits is smooth instead of being rough. Smooth skin is recessive to rough skin.

There are numerous examples of somatic mutations in animals also; for e.g. white eyes, vestigial wings etc. of *Drosophila.*

Germinal mutations : Mutations occuring in the reproductive cells are called germinal mutations. They are heritable and consequently they affect the whole of the organism in the next generation. Germinal mutations are called gametic when they occur in gametes and zygotic when they occur in the zygote. The gametic mutation will express itself readily when either the mutation is dominant or if recessive two mutant gametes should unite. Zygote mutations on the other hand express themselves readily. The ancon breed sheep is a typical example of germinal mutation.

Biochemical mutations : Mutations bringing about alterations in metabolic pathways, nutritional requirements etc. are called biochemical mutations. These are mostly noticed in microorganisms. Biochemical mutations in *Escherichia coli* can be induced by repeated sub culturing. In *E.coli,* the normal strains are called *Paratrophs* while the mutants are called *Auxotrophs*. These mutations are for some of the nutritional requirements like Biotene, Methionine, Leucine etc. The paratroph is called genetically B+, M+ and L^+ because it has the capacity to synthesize all these. On mutation the auxotroph loses the capacity to synthesize these and consequently it is known as $B^- M^- L^-$

Several metabolic disorders due to biochemical mutations have been reported in man also. They are commonly called **Inborn errors in metabolism'**. The first such biochemical disorder to be reported was Alkaptoneuria (Garrod 1902). This relates to a metabolic disorder concerning the amino acid phenyl alanine. **Alkaptaneuric** individuals have a mutated gene which does not produce the .enzyme homogentisate oxidase, resulting in the accumulation of homogentisic acid in the cell, which is excreted along with the urine. Homogenticic acid is a strong reducing agent and gets oxidised in the presence of air into a black pigment, resulting in the black colouration of the urine.

Another biochemical mutation in man is phenyle ketoneuria. It is caused by a recessive mutant gene in which due to failure of conversion of phenylalanine to tyrosine, the former accumulates in the blood and excreted through the urine. This leads to mental retardation. In addition to the above, Tyrosinosis, Albinism and Cretinism are the other metabolic disorders due to biochemical mutations.

Reverse mutations : The back mutation of a gene to come back to its wild state is known as reverse mutation. In *Drosophila,* the wings are normally long and due to a mutation, the wing becomes miniature. In some instances, all of a sudden, a few long winged flies appear in a population of miniature winged flies. This is due to the reverse mutation of the miniature winged gene back to the wild state. Another reverse mutation in

Drosophila is the mutant, fork bristles back to normal bristles. There are many such reverse mutations in bacteria and fungi.

Lethal mutations : Mutations of gene, from normal to lethal leads to the death of the individual. Normally lethal genes are recessive and nothing happens to the organism in the heterozygous condition. But occassionally there may be dominant lethals in which case individuals do not survive even in the heterozygous condition. The latinum colour in Fox, yellow coat colour (see chapter on Interaction of factors for a detailed discussion) in mice are such examples.

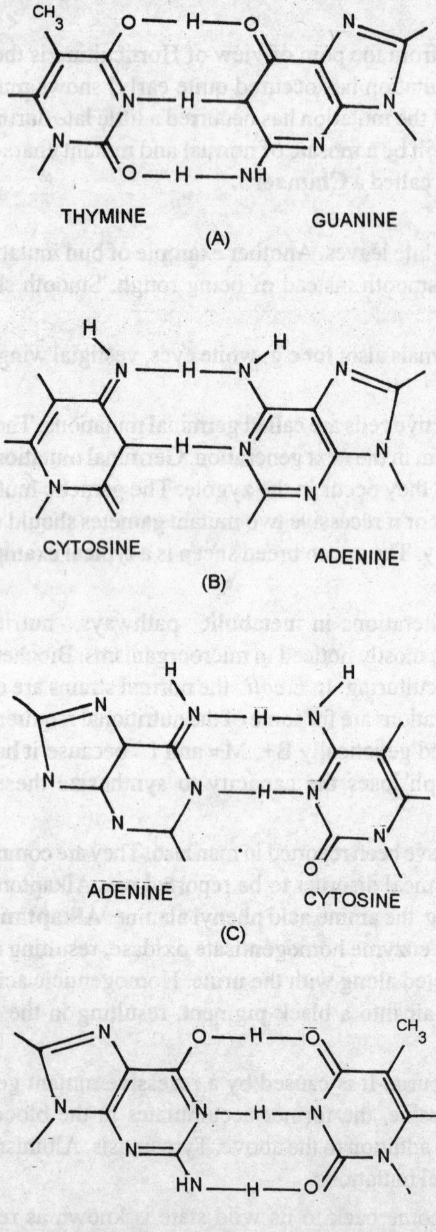

Many lethal mutations are also known in plants like*Antirrhinum majus* (Snapdragon). The golden plant stock has variegated leaves. On self pollination this variety produces Golden and green in the ratio 2:1 (not 3:1), obviously the golden is heterozygous and green breeds true. The other one (golden) is missing because pure golden can not exist due to homozygosity of lethal genes. Lethal genes are also known in paddy and maize plants.

MUTATIONS AT THE MOLECULAR LEVEL

The basic hereditary material namely DNA is a highly stable biological molecule. The hereditary information is present in the form of sequence of the four nucleotides

Adenine, Thymine, Guanine and Cytosine. During the inheritance of characters from one individual to the other, the sequence is duplicated and transmitted in a precise and accurate manner.

A slightest mistake at this stage will alter the base pair sequence and results in an entirely different type of genetic code resulting in the formation of an altered protein which may have far reaching effects on the organism. Such mistakes or faults occuring in the DNA molecule during duplication are ultimately the basis of mutation at the molecular level. The following two basic types of sub nucleotide changes are noticed in the DNA molecule - Frame shift mutation and substitution mutation.

Frame shift mutations : Addition or deletions of nitrogen bases either from DNA or mRNA are called frame shifts, because, from the site of the alteration, the reading frame of the message alters. In some cases frame shifts are neutralized ie. deletion may be neutralized by the addition at the same site. Frame shift mutations are of two categories. These are deletion mutations and insertion mutations. In deletion a single nucleotide in a triplet in deleted. These are found frequently in T_4 bacterophages. In insertion mutations an extra

Fig.24 .4 Gene Mutation
Uncommon forms of DNA bases

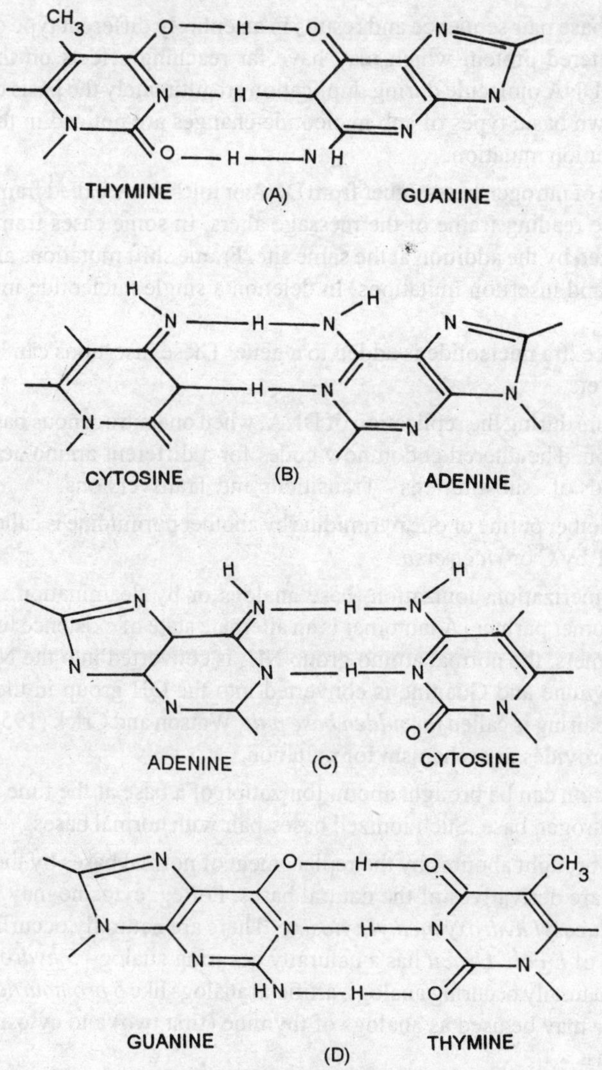

CH3 O—H—O N N H N N N O—H—N H
THYMINE (A) GUANINE

H H N—H—N N H N N N N
CYTOSINE (B) ADENINE

H H N N—H—N N N—H—N N N
ADENINE (C) CYTOSINE

N O—H—O CH3 N N—H—N N N HN—H—O
GUANINE (D) THYMINE

Fig. 24.5 Gene Mutation
Unusual pairing of
tautomeic bases in DNA

nucleotide is added to a gene. These insertions can be induced by chemicals such as acirdine, Proflavin etc.

Substitution mutations : This generally occurs during the replication of DNA, when one nitrogenous base is replaced by another changing the triplet codon. The altered codon now codes for a different amino acid resulting in an altered protein. There are two kinds of substitutions -

Transitions and Transversions.**Transitions :** The replacement of one purine by another purine or one pyrimidine by another pyrimidine is called trasition. In other words A is replaced by G and T by C or *vice versa.*

Transitions may be brought about by tautomerization, ionization, base analogs or by deamination. In tautomerization, a normal base, pairs with its tautomer partner [A tautomer is an alternate state of existence in a molecule. For e.g. In cytosine and adenine tautomers, the normal amino group NH_2 is converted into the NH (imino) group and the keto group (C = 0) of thymine and Guanine is converted into the OH group in their tautomers. Normally this does not happen]. Such pairing is called *forbidden base pair.* Watson and Crick (1953) opined that the existence of a tautomer to DNA provides a mechanism for mutation.

Ionization is another mechanism by which transition can be brought about. Ionization of a base at the time of DNA replication involves the loss of H from a nitrogen base. Such ionized bases pair with normal bases.

Another method by which transition may be brought about is by the replacement of normal bases by their analogs. These compounds called *base analogs* are derivatives of the natural bases. For eg. cytosine may be replaced by *5 hydroxyl methyl cytosine* or by 5 *glucosyl hydroxy methyl cytosine.* There are naturally occuring base analogs of cytosine found in T even phages of *E.coli. Uracil* has a naturally occuring analog - *5 hydroxy methly uracil* found in some Viruses. Besides the naturally occuring analogs, artificial analogs like *5 bromouracil, 5 iodouracil, 5 bromocytosine, 5 methyl cytosine* may be used as analogs of thymine (first two) and cytosine (last two) which may be used to induce mutations.

Deamination is another method of induction of transition, by replacing the NH_2 group of bases with OH group. This may be induced by chemials such as nitrous acid, hydroxylamine, diethyl sulphate etc. Deamination of cytosine leads to the formation of uracil, and that of adenine forms hypoxanthine. During replication U pairs

A slightest mistake at this stage will alter the base pair sequence and results in an entirely different type of genetic code resulting in the formation of an altered protein which may have far reaching effects on the organism. Such mistakes or faults occuring in the DNA molecule during duplication are ultimately the basis of mutation at the molecular level. The following two basic types of sub nucleotide changes are noticed in the DNA molecule - Frame shift mutation and substitution mutation.

Frame shift mutations : Addition or deletions of nitrogen bases either from DNA or mRNA are called frame shifts, because, from the site of the alteration, the reading frame of the message alters. In some cases frame shifts are neutralized ie. deletion may be neutralized by the addition at the same site. Frame shift mutations are of two categories. These are deletion mutations and insertion mutations. In deletion a single nucleotide in a triplet in deleted. These are found frequently in T_4

bacterophages. In insertion mutations an extra nucleotide is added to a gene. These insertions can be induced by chemicals such as acirdine, Proflavin etc.

Substitution mutations : This generally occurs during the replication of DNA, when one nitrogenous base is replaced by another changing the triplet codon. The altered codon now codes for a different amino acid resulting in an altered protein. There are two kinds of substitutions - Transitions and Transversions.

Transitions : The replacement of one purine by another purine or one pyrimidine by another pyrimidine is called trasition. In other words A is replaced by G and T by C or *vice versa.*

Transitions may be brought about by tautomerization, ionization, base analogs or by deamination. In tautomerization, a normal base, pairs with its tautomer partner [A tautomer is an alternate state of existence in a molecule. For e.g. In cytosine and adenine tautomers, the normal amino group NH_2 is converted into the NH (imino) group and the keto group (C = 0) of thymine and Guanine is converted into the OH group in their tautomers. Normally this does not happen]. Such pairing is called *forbidden base pair.* Watson and Crick (1953) opined that the existence of a tautomer to DNA provides a mechanism for mutation.

Ionization is another mechanism by which transition can be brought about. Ionization of a base at the time of DNA replication involves the loss of H from a nitrogen base. Such ionized bases pair with normal bases.

Another method by which transition may be brought about is by the replacement of normal bases by their analogs. These compounds called *base analogs* are derivatives of the natural bases. For eg. cytosine may be replaced by *5 hydroxyl methyl cytosine* or by 5 *glucosyl hydroxy methyl cytosine.* There are naturally occuring base analogs of cytosine found in T even phages of *E.coli. Uracil* has a naturally occuring analog - 5 *hydroxy methly uracil* found in some Viruses. Besides the naturally occuring analogs, artificial analogs like 5 *bromouracil, 5 iodouracil, 5 bromocytosine, 5 methyl cytosine* may be used as analogs of thymine (first two) and cytosine (last two) which may be used to induce mutations.

Deamination is another method of induction of transition, by replacing the NH_2 group of bases with OH group. This may be induced by chemials such as nitrous acid, hydroxylamine, diethyl sulphate etc. Deamination of cytosine leads to the formation of uracil, and that of adenine forms hypoxanthine. During replication U pairs with A and hypoxanthine pairs with C. As a result A = T is substituted by G = C and G=C byA=T.

Trasversions : Unlike m transition, a purine is replaced by a pyrimidine and *vice versa* ie. CG replaced by GC and AT by TA. This change can be induced by alkylating agents like ethylmethane sulphate (EMS) and Methylmethane sulphate (MMS). The alkylating agents alkylate the nitrogen in the 7th position in a purine and lead to its separation from the DNA strand. This gap can be filled by any of the four bases at the time of replication. Trasversions were first postulated by E.Freese in 1959.

PRACTICAL APPLICATION OF MUTATIONS

Generally mutations are lethal and recessive and possibly of no help except as a scientific curiosity in understanding the basic funcitoning of the hereditary material. But there are specific instances where both spontaneous and induced mutations are of selective value from the point of view of Evolution and also from the point of view of human beings. Ancon breed sheep is one such example of a spontaneous mutation. Mutations are also of value in plant breeding, when mutant varieties can be stabilized in the gene pool by cross breeding. Gustaffson has estimated that about one in thousand mutations may be used in breeding experiments. Several mutations in wheat, like branched ears, resistance to lodging, high lysine content, amber seed colour, awned spikelets have been useful inbreeding. The amber seed colour which is originally seen in the mexican variety has been induced in wheat of Indian varieties.

Many useful mutations in rice are also known. Some of these are *Reirnei* variety mutant, for reduced duration of the crop. In barley also mutants like *erectoids* and *eceriferum* have been induced for increased yield. The *Todds Mitchan* variety of mint plant *and-Aruna* variety of castor serve as good examples of the role of mutations in crop improvement. The above mentioned variety of mint has disease resistance as well as high oil content. In the castor variety duration of the crop was reduced by 130 days (originally 270 days). In a survey conducted by FAO in 1974 it has been estimated that 98 varieties of crops have been improved by induced mutations.

25

GENETIC ENGINEERING

Recombinant DNA technology or Genetic engineering as it is popularly known, is the newest tool that is available in the human arsenal of knowledge in man's quest for a better human life. What is genetic engineering or recombinant DNA technology?

Imagine a situation of chronic diabetic patients being able to get back the capacity of synthesis of insulin; patients suffering from incurable diseases getting an injection of gene therapy-lo and behold they are cured of their diseases; consider a situation which is at present in the realms of fantasy of making man producing his own food just like the green plants do from natural sources; or visualize a situation wherein revolution in agricultural technology would make the use of chemical fertilizers completely redundant or going to the extreme, culturing a human being of the desired trait on a medium of nutrients i.e., man made to order. The scope is fantastic; possibilities are immense; it is as if the role of creator has been for once given to human beings. Well this is genetic engineering or recombinant DNA technology, the most powerful biological technique which the human beings can use either to make or mar the society. It can be used either for progress or for destruction. Using the techniques of genetic engineering you can create deadly pathogenic bacteria and unleash them on human beings; you can produce varieties of agricultural crops which are disease free or you can introduce pathogens into the crops and totally destroy them. Genetic engineering used judiciously can be a harbinger of new revolution but used injudiciously may produce many prototype of criminals, antisocial elements, boot leggers etc. In this chapter we shall briefly discuss some of the fundamental aspects of genetic engineering and how to use them for betterment of human life.

Genetic engineering may be defined as the method of artificial synthesis of new genes and their subsequent transplantation in the new genome of an organism, by molecular biological techniques. Recombinant DNA is achieved by constructing molecules outside the living cells by joining natural or synthetic DNA molecules that can replicate inside a living cell. Recombinant DNA technology involves several stages or steps.

The main aspects of genetic engineering (gene cloning) are the following -

1. Identification and isolation of the gene to be cloned.
2. Insertion on the gene into another piece of DNA called vector which will allow it to be taken up by microorganisms like bacteria and replicate within them as the cells divide.
3. Transfer of the recombinant vectors into bacterial cells.
4. Selection of those cells which have the desired recombinant vectors.
5. Growth of the microorganisms to get increased quantity of cloned DNA.
6. Expression of the desired gene to obtain the desired product.

Outline procedure of Genetic engineering

Genetic engineering of a desired gene and gene product involves the following procedure. First of all the desired gene has to be obtained from a source which then has to be identified and isolated.

(i) The living cells from which the genetic material is to be obtained has to be broken open. One popular method is to shear the cells in a blender and treat them with a suitable detergent.

(ii) The genetic material from the cell is to be removed. Several techniques are available to remove the DNA. The DNA molecules can even be spooled on to a glass rod. The glass rod bearing the DNA molecules is then lifted up from the debris of the cells.

(iii) Thè next step is to cut away the specific genes of interest from the DNA molecule. Restriction enzymes popularly called molecular scissors are used to cut and separate the desired gene from the DNA helix.

(iv) In the next step the specific sections of DNA are incorporated into cloning vectors or vehicles such as plasmids, phages etc. These cloning vehicles are then introduced into a rapidly multiplying living system such as bacteria where multiple copies of the cloned DNA can be obtained. The cloned DNA also called chimeric DNA contains the desired foreign gene. The coloned DNA is also called recombinant DNA.

(v) In the final step which is a very delicate process the cloned gene is allowed to express itself in the bacterial cell. Gene expression here means production of a desired gene product. For instance if a human insulin gene is coloned in a plasmid and is inserted into a bacterial cell, the bacterial cell along with its metabolic products also will produce insulin. This insulin may than be separated by the normal methods of chemical purification and isolation.

RESTRICTION ENZYMES (restriction endonucleases)

The first account of restriction enzymes was given by Werner Arber while studying the bacteriophages. He found that in some instances as soon as the viral DNA entered into a bacterial cell it (viral DNA) was cut into bits and destroyed. Based on this observation Arber opined that the bacterial cell must possess certain specific DNA destroying enzymes that could infect only the viral DNA but not the host DNA. In otherwords the enzyme should be such that it should be able to distinguish between foreign DNA and its own DNA. He gave the name restriction enzymes and hypothesized that these restriction enzymes can recognise and act at specific sites on the viral DNA.

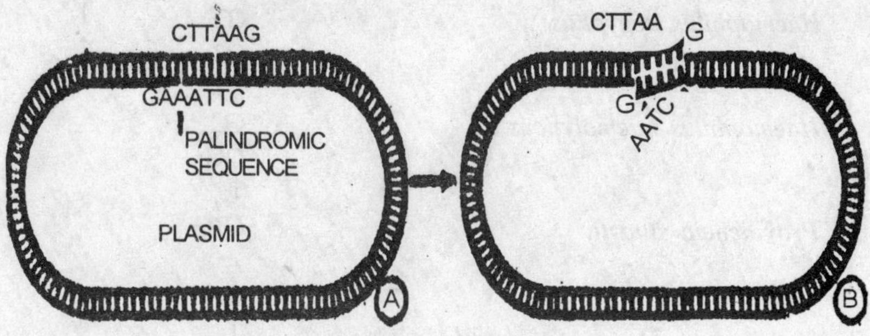

Fig. 25.1 Genetic Engineering
Action of restriction enzymes

The next step in the discovery of restriction enzymes was taken by Smith *et al* (1970) when they isolated a restriction enzyme from the bacterium *Haemophilus influezae.* Smith showed that the restriction enzyme can cut the viral DNA at specific sites. All the DNA termini were pGpTpPy-3' and 5' pPupApCp, showing that these were cut from the double stranded DNA having the following sequence

5'pGpTpy | pPupApCp-3'
3'pGpApPu | pPypTpGp-5'

The above example shows that the restriction enzyme from *Haemophilus influenzae* can cut the DNA molecule which exhibit a two fold symmetry i.e., with a palindromic sequence (same sequence running in opposite directions). Some restriction endonucleases, there source, recognition sequence and sites of cleavage (indicated by arrow) are given below.

Enzyme	Source	Recognition sequence and clearing site		
		↓		
Eco R1	*Escherichia coli* Ry13	G.AA		TTC
		GTT		AAG
Hind II	*Haemophilus influenzae*	GTPy	↓	PuAC
		CAPu	↓	PyTG
		↓		
Hind III	*Haemophilus influenzae* Rd	A AG		CTT
		TTC	↓	GA A
				↑
Hpa I	*Haemophilus parainfluenzae*	GTT	↑	AAC
		CAA	↓	TTG
Hpa II	*Haemophilus parainfluenzae*	CC		GG
		GG		C C
		↓		↑
Bam HI	*Bacillus amyloliquefaciens*	GGA		TCC
		CCT		AGG
		↓		↑
Bgl III	*Bacillus globigi*	A GA		TCT
		TCT		AG A
			↓	
Hae II	*Haemophilus aegyptius*	Pu GG		GCPy
		Py CG	↑	CGPu
Hae III	*Haemophilus aegyptius*	GG	↓	CC
		CC	↑	GG
		↓		
Hha I	*Haemophilus haemolyticus*	GC		GC
		CG		CG
		↑		
Pst I	*Providencia stuartii*	CTG		CA G
				↓
		GAC		GTC
		↑	↓	
Sma I	*Serratia marcescens*	CCC		GGG
		GGG		CCC
		↓	↑	
Taq I	*Thermus aquaticus*	T C		GA
		AG		CT
				↑
Bal I	*Brevibacterium albidum*	TCG	↓	CCA
		ACC	↑	GGT
		↓		
Sal I	*Streptomyces albus* G	G TC		GAC
		CAG		CT G

Enzyme	Source	Recognition sequence		
Xor II	*Xanthomonas oryzae*	CGA GCT ↑		↓↑ TCG AGC
Alu I	*Arthrobacter luteus*	AG TC	 ↓	CT GA
Msp I	*Moraxella* sp.	CC GG ↓	 ↑	GG CC
Mbo I	*Moraxella bovis*	GA		TC
Sau 3 Al	*Staphylococcus aureus* 3A	GA CT	↓ ↑	TC AG

Subsequent work by Daniel Nathans (1971) showed that the restriction enzyme discovered by Smith can cut the monkey virus DNA into eleven well defined pieces. By using specific restriction enzymes it is also possible to map the DNA which they cut as their specificity is known. Arber, Nathans and Smith were awarded the 1978 Noble prize in physiology and medicine for their work on restriction enzymes.

In many microorganisms restriction enzymes are a means of protection against intrudent viral DNA. In the working of the restriction enzyme the nucleotide sequence of the host DNA is identified and methylated to avoid degradation. As a result of this the DNA gets the protection while the foreign DNA with the same sequence will not be methylated and consequently subjects itself for degradation. The whole mechanism is called the RM system (restriction modification) and is wide spread in bacteria.

Classification of restriction enzymes

Restriction enzymes are able to recognise specific nucleotide sequence in the double stranded DNA and cleave both the strands. To date more than 150 restriction enzymes have been discovered which are known to recognise 40 different nucleotide sequence. Some restriction enzymes however may not cut DNA at specific sites. Based on this restriction enzymes are classified into two categories type 1 and type 2. Type 1 restriction enzymes recognise a specific sequence of DNA molecule but the cut is made elsewhere whereas type 2 enzymes make a cut only within the restriction site and produce two single strand breaks, one break in each strand. Type 2 enzymes are most important because they produce broken chain of nucleotides with a palindromic sequence. The following table gives a summary of the difference between type 1 and type 2 enzymes.

	Type 1	Type 2
1.	Non-specific in cleavage	Cleavage is site-specific
2.	M.W 300,000	Lower MW (20,000 - 1,00,000)
3.	Has non-identical subunits	Has two identical subunits
4.	Require ATP, Mg ++ and generally S-adenosyl	Require only Mg ++ as co-factor
5.	Examples : EcoB, EcoK, EcoPi	Example : EcoRI

Nomenclature of restriction enzymes

The names are designated according to the following principles.

(1) Name of the organism is identified by the first letter of the gene's name and the first two letters of the species name to form a three letter abbreviation in italics. For example *E.coli* = *Eco* and *H.influenzae* = *Hin* etc.

(2) A strain or type identified is written as a subscript. For example *E.coli* strain k is written as Eco_k.

(3) In cases where the restriction modification systems are genetically specified by a virus or plasmids the

extra chromosomal element is identified by a subscript as *Eco RI, Eco PI* etc.

When a particular strain has several restrictions and modification systems these are identified by Roman numerals. For example Hin dI, Hin dII, Hin dIII etc.

Cleavage brought about by restriction enzymes

Restriction enzymes have an ability to recognise a particular sequence and cleave DNA molecule at that site. Most of the sequences recognised by these enzymes are tetranucleotide or hexanucleotide pairs which show a characteristic two fold symmetry i.e., palindromic sequence. The Eco RII site is very unique as it contains 5 base pairs with the axis of symmetry passing through the middle AT pair as shown below.

CCTGG

CCACC

Some times restriction enzymes from unrelated species can recognise the same nucleotide sequence. For instance GGCC sequence specific enzymes have been discovered in *Bacillus subtilis, Haemophilus aegypticus* etc. Such enzymes are called isoschizomers.

Cleavage sites may be even or staggered. In even cleavage the breaks on the two strands are exactly opposite to each other. Whereas in staggered breaks the breaks on the two strands are a little away from each other. Even breaks donot produce unpaired terminal nucleotides whereas staggered breaks may produce sticky ends.

Use of restriction enzymes

The use of restriction enzymes has proved to be a highly useful tool for the sequence analysis of DNA. It also plays a significant role for gene cloning and amplification in genetic engineering.

The DNA molecule from bacteriophages and many other viruses consists of 5,000 to 50,000 base pairs. It is of fundamental importance to know the sequence of these bases to understand the gene structure and regulation and also for decoding the information stored in these genes. By the use of several restriction enzymes it is possible to cut fragments of DNA as small as 300 bp. These fragments can be used for sequence analysis and mapping.

The DNA segments produced by restriction enzymes can be covalently linked to the plasmid DNA and such recombined DNA can be inserted into bacteria such as *E.coli* for transformation. The transformed cells can now be cultured as separate colones.

The ability of restriction enzymes to locate and cut at a particular sequence is of great help in identifying and isolating a particular gene.

Identification of a required DNA piece (gene)

Theoritically, identification of a gene involves, the reading of the message back in protein synthesis. During protein synthesis DNA produces mRNA and this in turn identifies and allows the alignment of amino acids to synthesize a protein. In the identification of a gene responsible for the production of a particular protein, the process has to be followed backwards. Let us take an instance of the human insulin which is a protein.

The steps involved in the identification of the gene or segment of DNA responsible for insulin are as follows.

(a) Isolation and purification of the protein

(b) Sequential degradation of the protein : In sequential degradation, not only the qualitative composition of the amino acids are known, but even their sequence of arrangement in the polypeptide chain will be known. After ascertaining the sequence of amino acids the next step would be :

(c) Identification of codes for each amino acid : This can be done easily as we know the triplet codes for each amino acid. This would give us a length of the mRNA segment equivalent to the length of the polypeptide chain. (For example if the protein consists of 25 amino acids, the length of the mRNAshould be 75 nucleotides).

The nucleotide sequence in mRNA can be easily substituted for the complementary DNA segment. We now have a gene or a segment of DNA which governs the production of insulin.

ISOLATION OF SPECIFIC DNA SEGMENT (GENE)

The DNA of an organism whose specific gene is required for the purpose of expression in another organism contains thousands of genes. The first problem therefore is to separate the required gene from the DNA of the organism. There are several techniques to isolate the required gene from the DNA molecule.

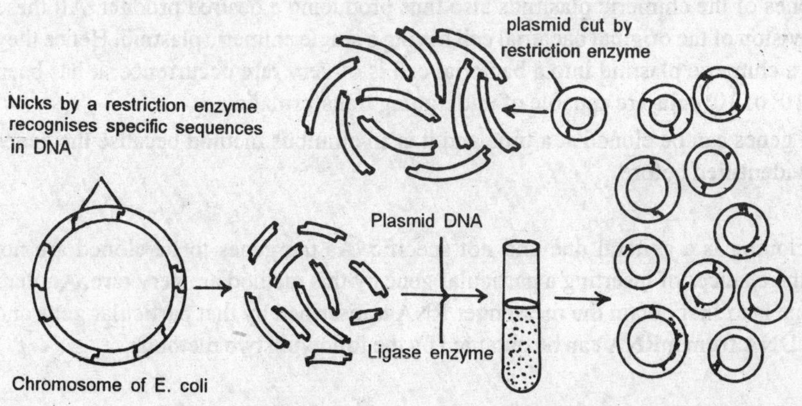

Fig. 25.2 Genetic Engineering
Isolation and recombination of genes using shot gun method

Shotgun method

In this, the entire DNA of the organism is subjected to the action of restriction enzymes and is broken down into a number of fragments. Each of these fragments is then combined with the DNA from a suitable vector to produce recombinant DNA molecule. Subsequently these recombined DNA molecules are inserted randomly into the host cells such as *E.coli*. These host cells are cultured separately and grown into colonies. The following are the steps of the shotgun method.

1. Using specific restriction enzymes the isolated DNA from a particular organism is cleaved into many bits. As the restriction enzymes may bring about even or staggered cuts many sticky ends are left. The resultant DNA fragments are of different length.

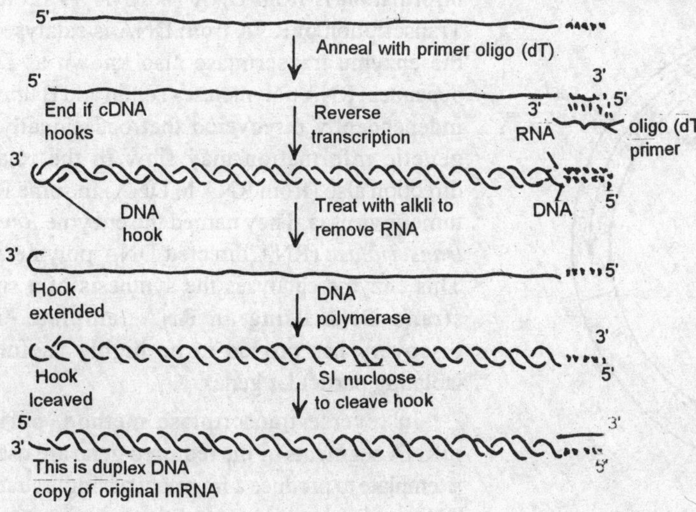

Fig. 25.3 Genetic Engineering
Obtaining cDNA from mRNA

2. Host cells of *E.coli* are selected and the outer membrane is first dissolved by a detergent and its DNA is released. The plasmids of *E.coli* are then separated from the main genome by ultracentrifugation. The plasmids are then kept in a solution containing the same restriction enzyme which will cut the plasmid circle into a linear form. Linear plasmid DNA also has overlapping sticky ends.

3. In this step isolated DNA and linear plasmid DNA are mixed. As a result random association occurs between the foreign DNA and the plasmid

DNA. Subsequently the foreign DNA gets integrated into plasmid DNA to form a long linear DNA molecule which then becomes a circle due to covalent joining. The resultant DNA is a recombinant or chimeric DNA.

4. The chimeric DNA molecules are now placed in a solution of calcium chloride which also has *E.coli* cells. On heating suddenly upto 42°C the membranes of bacteria become permeable to chimeric DNA molecules. Thus the *E.coli* cells incorporate the chimeric DNA plasmids.

5. The bacterial cells containing the chimeric plasmids are then cultured on a medium. The cells express not only their genes but the genes of the chimeric plasmids also thus producing a desired product. All these cells are produced by the division of the original bacterial cell having a single chimeric plasmid. Hence they are called clones. Insertng a chimeric plasmid into a bacterial cell is of very rare occurrence; it has been estimated that only one in 10^5 of 10^6 cells are capable of undergoing transformation.

In shotgun method several genes can be cloned at a time and it is an omnibus method because the genes to be cloned are not picked and identified before.

Obtaining DNA from mRNA

In shotgun method gene cloning is a general one and not specific. As the genes to be cloned are not identified and isolated before, the chances of inserting a particular gene by this method are very rare. Another specific method of isolating gene is to first obtain the messenger RNA transcribed by that particular gene and then obtain DNA from mRNA. DNA from mRNA can be obtained by the following two methods.

Hybridization method

When DNA transcribes and forms mRNA a complex is formed. The hybridization method depends on this mechanism. At the first step the total DNA from an organism is isolated and treated with heat or alkali to convert it into a single stranded form. These single strands are then mixed with the specific mRNA obtained from the gene. Obviously the mRNA chains pair with that section of DNA which has complimentary nucleotides to it thus forming aDNA-RNA hybrid. This complex can be isolated and the DNA can be separated from the RNA. The single stran DNA can be converted into a double strand DNA by enzymes such as polymerase I. This method is highly useful for isolating ribosomal genes.

Reverse transcriptase method

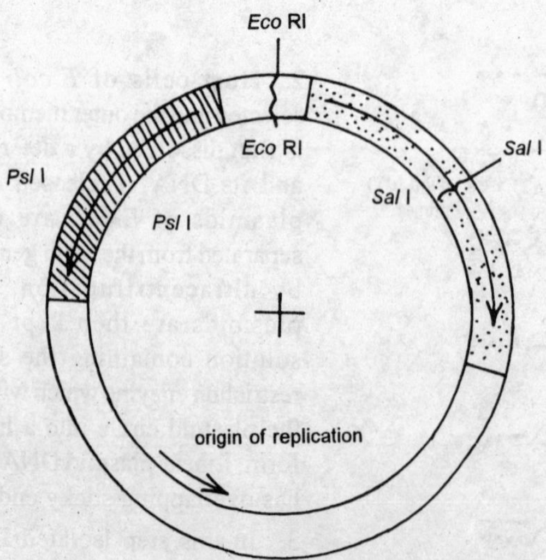

Fig. 25.4 Genetic Engineering
Genetic map of plasmid pBR 322

In the normal transcription the flow of information is from DNA → RNA → proteins. Transcription of RNA from DNA is catalysed by the enzyme transcriptase also known as DNA dependent RNA polymerase. Temin and Baltimore independently discovered that occasionally the genetic information may flow in the reverse direction also (from RNA to DNA) in some RNA tumour viruses. They named the enzyme *Reverse transcriptase* (RNA directed DNA polymerase). This enzyme catalyses the synthesis of a single strand DNA using an RNA template. This discovery has proved to be highly useful for isolating particular genes.

In reverse transcriptase method, purified mRNA segments of the required gene are used as a template to produce a fragment of single stranded DNA under the influence of the enzyme reverse transcriptase. As this single strand DNA is complementary to mRNA it is called complimentary DNA or cDNA. Hydrolysis of RNA template will separate a single strand DNA. This single strand DNA is converted into double strand using the enzyme DNA polymerase. Another enzyme S1 nuclease breaks the covalent linkage between two DNA strands.

VECTORS (Carriers or Vehicles)

Vectors are defined as those DNA molecules that can carry a foreign DNA fragment inserted into them without any disturbance in their expression. Vectors are also called vehicle DNAs. Vectors are obtained from a number of sources. These are grouped into categories such as plasmids, bacteriophages, cosmids and phasmids.

Plasmids

These are extra chromosomal, self replicating and double stranded covalentaly closed circular DNA molecules present in the bacterial cell. A number of plasmids are present in bacterial cells that have sufficient genetic information for their replication. Eventhough they are not necessary for the survival of the bacteria they carry such necessary information as drug resistance, metal resistance, toxin production, nitrogen fixation etc. Naturally occuring plasmids can also be modified by *in vitro* techniques such as code transformation. Use of plasmids as a vector for DNA cloning was reported for first time by Cohen et al (1973).

Plasmids are used as cloning vehicles when they posses the following features.

1. It can be readily isolated from the cells
2. It possesses a single restriction site for one or more restriction enzyme(s)
3. Insertion of a linear molecule at one of these sites does not alter its replication properties.
4. It can be reintroduced into a bacterial cell and cells carrying the plasmid with or without the insert can be selected or identified (Bernard and Helinski, 1980).
5. They do not occur free in nature but are found in bacterial cells.

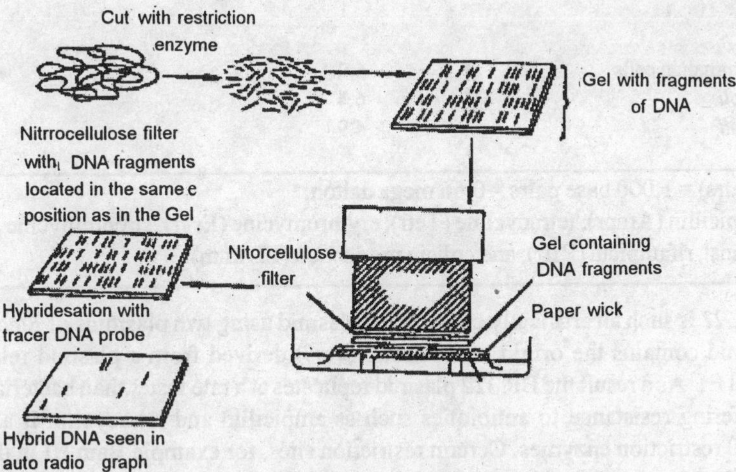

Fig. 25.5 Genetic Engineering
The Southern blotting technique

In many instances the main objective of coloning is the insertion of a particular fragment of DNA into a suitable plasmid vector and its amplification. Amplification is a process of increasing the number of plasmids in a bacterial cell. In this process, a cell containing a relaxed plasmid is treated with a drug to inhibit protein synthesis. Consequently replication stops and the relaxed plasmid (eg pBR 322) continues to replicate despite the drug treatment. Addition of chloramphenicol causes the plasmid BR 322 to increase its number to 3,000 per cell. In order to replicate and clone a piece of foreign DNA, it is necessary to possess a sequence of nucleotides which is recognised by the host bacterium as origin of replication. The origin of replication occurs naturally in plasmids. In naturally occurring plasmids due to exchange of genetic information between plasmids and bacterial chromosome new genes originate in the plasmids. Therefore plasmids have been artifically developed with specific sequence of genes so that they would integrate foreign DNA without any problem.

Some cloning vectors

Cloning vectors	Natural occurrence	Size (Kb)*	Selective marker**
Plasmids			
pACYC 177	*E.coli*	3.7	Amp^r, Kan^r
pBR 322	*E.coli*	4.0	Amp^r, Tet^r
pBR 324	*E.coli*	8.3	Amp^r,, tet^r. El imm
pMB9	*R.coli*	5.8	Tet^r
pRK 646	*E.coli*	3.4	Amp^r
pC194	*Staphylococcus aureus*	3.6	Ery_r
pSA 0501	*S.aureus*	4.2	Str^r
pBS 161-1	*Bacillus subtilis*	3.65	Tet^r
pWWO	*Pseudomonas putida*	117	Kan^r
Cosmids			
pJC 74	Derived plasmid from ColEL	16	Amp^r El imm
pJC 720	do	24	Amp^r, Tet^r
pHC 79	Derivative of PBR 322	6	Amp^r, Tet^r
Virusus			
SV40	Mammalian cells	5.2	-
Phage M13$^+$	*E.coli*	6.4	-
Phage	*E.coli*	4.9	-

* 1Kb (Kilobase pairs) = 1,000 base pairs = 0.66 mega dalton;

**. Resistance of ampicillin (Ampr), tetracycline (Tetr), erythromycine (Eryr), streptomycine (Sttr), Kanamycin (Kanr), rifampicin (Rifr), and colicin production (EL imm).

The plasmid BR 322 is such an artifically constructed plasmid using two plasmids of *E.coli* - PBR 318 and PBR 322. This plasmid contains the origin of replication *(ori)* derived from a plasmid related to naturally occurring plasmid Col E1. As a result the BR 322 plasmid replicates at a rate faster than bacterial genome. It also possesses genes confering resistance to antibiotics such as ampicillin and tetracyclin. It also has a unique recognition site for 20 restriction enzymes. Certain restriction sites, for example Bam HI in the *tet* gene of the plasmid is present within the gene in such a way that the insertion of foreign DNA will inactivate the *tet* gene. As a result, the recombinant plasmid will allow the cells to grow only in the presence of ampicillin but will not protect them against tetracyclin. This antibiotic resistance will allow the selective isolation of cells which can be cultured on a medium supplemented with ampicillin.

In addition to the *E.coli* plasmids a plasmid of gram negative bacteria called RK2 plasmid has been developed which has a very broad host range and has a single restriction site which can be acted upon by several endonucleases.

Isolation of bacterial plasmids

The following steps are ncessary for isolating the bacterial plasmids

1. Treat the prokaryotic cells with detergents which will release plasmid DNA and chromosome from the bacterial cell along with other molecules. This mixture is known as lysate.
2. Treat the lysate with potassium acetate or acetic acid solution. This precipitates chromosomal DNA and some protein molecules.
3. Remove the precipitate from the lysate by high speed centrifugation. The supernatant contains plasmid DNA along with RNA, protein and some DNA debris of the chromosome.
4. Treat the lysate with RNAase, this will digest the RNA.
5. The lysate is treated with phenol and as a result two layers are formed - phenol layer and water layer. The water layer contains plasmid DNA and the other contaminants are found in the phenol layer.

Bacteriophages

A virus which attacks the bacteria is called a bacteriophage. A number of bacteriophages with different genetic materials have been reported so far from a variety of bacteria. Some of these phages are ϕ x 174, phages λ, M13, Fd11, R209 etc. Phages also can be used as cloning vectors like plasmids. Of all the phages the λ phage perhaps is the one that has been used very frequently as a cloning vector. The λ phage contains a proteinaceous head and a long tail. The head portion has 50 genes in its 49 kb genome of which about half of the genes are essential. On attachment to the host (*E.coli*) the linear DNA molecule is injected into the cell. The linear double stranded molecule of DNA becomes a circle by joining through the single strand of 12 nucleotides commonly known as *cos* site. This *cos* site is a very important feature of λ DNA.

The λ phage has two kinds of life cycles - the lytic phase and the lysogenic phase in the bacterial cell. In the lytic pathway early in the infection the circular DNA replicates by a rolling circle mechanism and produces long molecules of DNA. At the same time the phage DNA directs the synthesis of proteins to produce the head. In order to pack the DNA into the head it undergoes cleavage to yield the fragments of such sizes as to fit in the head. Eventually a tail is attached to the head. In the lysogenic pathway the phage genome integrates with the bacterial genome at an attachment site (*att*) where it replicates along with the bacterial genome.

The following are the advantage of using phage as cloning vectors - 1. DNA can be packed *in vitro* into phage particles and transduced into *E.coli* with high efficiency, 2. Foreign DNA upto 25 kb in length can be inserted into phage vector, and 3. Screening and storage of recombinant DNA is easier (Dahl et al., 1981). Before using the λ phage vector it will be essential to remove the genome from the restriction sites for the enzymes commonly used for cloning. Two types of phage cloning vectors have been constructed - these are the *insertion vector* and the *replacement vector*.

Insertion vector. These have unique cleavage sites into which a relatively small piece of foreign DNA can be inserted. These fragments do not affect the functioning of the phage. The upper and lower limits of DNA that may be packed into the phage particles is between 35-53 kb. The maximum size of a foreign DNA that can be inserted is about 18kb.

Replacement vector. These vectors have cleavage sites present on either side of the length of non essential DNA. As a result of cleavage left and right arms are formed and each arm has a terminal *cos* site and a longer stiffer region (the non essential region) which can be substituted by foreign DNA. The maximum size of foreign DNA that can be inserted into this vector depends on how much the phage DNA is non essential. It has been estimated that about 25-35kb of genome can be replaced with foreign DNA segments. Example of substituted vectors are-gt, WES and λ 1059.

Cosmids

These may be defined as the hybrid vectors derived from the plasmids which contain *cos* site of λ phage.

Reported for the first time by Colins and Hohn (1975) cosmids lack genes for encoding viral proteins. Therefore neither viral particles are formed in the host cell nor cell lysis occurs. Special features of cosmids similar to plasmids are origin of replication, a marker gene coding for antibiotic resistance, a special cleavage site for the insertion of foreign DNA and the small cells. Characters of cosmids dissimilar to plasmids are presence of extraphage DNA, and the *cos* site.

Phasmids

When plasmids are inserted into the genome of λ phage they are said to be phasmids. The process of insertion of the plasmid DNA into the phage genome occurs in the same way as a phage genome is inserted into the bacterial chromosome. Insertion of plasmid into λ phage is done with a view to have a specific site responsible for recombinational insertion of the phage into bacterial chromosome.

Phasmids may be used in many ways. They can be used as a phage cloning vector from which a recombinant vector may be released.

CLONING ORGANISMS

The vectors that are chosen as vehicles to carry the foreign DNA must be cloned in an organism that multiplies rapidly thereby producing the gene product in a large quantity. Such organisms into which the vectors are inserted for the purpose of gene expression are known as cloning organisms. A number of microorganisms have been used for the purpose of cloning. The most important among them are the various strains of the colon bacterium *E.coli.* The strains of *E.coli* used as cloning organisms are HB101, h303 and RR1. The cloning organisms should normally be deficient in DNA recombination as it will reduce undesirable hybridization of DNA. Another essential feature of the cloning organism is it should lack restriction enzymes that might degrade foreign DNA. Strains of bacteria that can bud off mini cells are highly useful if the gene expression is the purpose of experiment (mini cells are those which contain all the elements of gene expression except chromosome).

For a long time *E.coli* has been the most favourite cloning organism in recombinant DNA studies. Of late yeasts are also being used as cloning vehicles and they have some advantages over bacteria. They do not live in human beings and are non pathogenic. Fink et al (1974) for the first time introduced a bacterial gene for the production of leucine into baker's yeast. A snail enzyme was used to induce the yeast to take up bacterial DNA.

Techniques used for identification of clones

Several techniques are employed to identify the isolated clones using a specific molecular probe. Some of these techniques are described below :

1. Polyacrylamide gel electrophoresis

When the genomic DNA extracted from a particular source is digested by a specific restriction enzyme it is broken into several fragments. These fragments of different sizes can be separated through a technique called gel electrophoresis. This separation is necessary to identify a specific gene that is required for use. After these are separated a specific molecular probe can be used to detect the segments which have sequences similar to those in the probe. For instance if a gene with a specific nucleotide sequence is required to be separated from a mixture a probe is prepared with nucleotide sequence complementary to the gene. These (molecular probes) are then allowed to hybridize and the particular gene to be identified.

In gel electrophoresis the DNA molecules migrate from one end of the gel to the other. The gel is usually a cylindrical structure or occasionally even may be a slab. The rate of movement of fragments of DNA on the gel is inversely proportional to the size of the molecules. Heavier fragments will remain at the beginning while lighter fragments will move away. Fragments of different sizes appear as discrete bands on the gel and can be isolated for further study.

Separation of DNA molecules using PFGE (pulsed field gel electrophoresis)

Gel electrophoresis is suitable for only small fragments of DNA which can easily get separated. Large size molecules of DNA however cannot be separated using this technique. A new technique called PFGE is used for the separation of large sized DNA fragments, some times representing even whole chromosomes. Using the PFGE technique separation of DNA molecules belonging to each of the sixteen chromosomes of yeast has been made possible. In this technique short pulses of electric current are used in two different directions and the DNA is embedded in Agarose plugs to avoid fragmentation. Using this technique genomes of several fungi have been resolved into chromosomal bands.

Blotting techniques

Gel electrophoresis helps in the separation of fragments of DNA or RNA. The separated bands can be stained and absorbed on the gel. This technique however will only help in the separation of fragments of DNA but will not tell as to which fragment contains the desired gene. For this purpose a molecular probe (which is a DNA fragment) consisting of nucleotides complimentary to the gene are used for hybridization and the required gene is identified. For easy identification and detection, the probe is radioactively labelled. To facilitate hybridization the bands of the genome are usually transferred to a nitrocellulose membrane in a technique called blotting. Blotting technique was first invented by E.M. Southern and is known as the Southern blotting technique. Two other blotting techniques discovered subsequently are named Northern blotting technique and Western blotting technique. These are described briefly below.

Southern blotting technique. The DNA sample is first cleaved with a restriction enzyme and the sample is loaded on to gel electrophoresis for separation. The DNA bands that appear in the gel are denatured into single strands using an alkali solution. Later the gel is laid on top of a buffer saturated filter paper placed on a glass plate with its two edges immerged in the buffer. A sheet of nitrocellulose membrane is placed on top of the gel and a stack of many paper towels along with a weight are placed on this membrane (see figure). The buffer solution is drawn up by the filter paper folds and passes through the gel and to the nitrocellulose membrane. Finally buffer moves up to the paper towels. While passing through the gel the buffer carries with it the fragments of single stranded DNA and binds them on to the nitrocellulose membrane as it passes through it (membrane) to the paper towels above. For the complete migration of the buffer it may take a few hours. Subsequently the paper towels are discarded, the nitrocellulose membrane with the single stranded fragments of DNA attached to it in different positions is baked at 80°C to fix the strands permanently on it. This membrane now has the fragments of single stranded DNA fixed on to it in the same position in which they were on the gel. In otherwords the nitrocellulose membrane is the replica of the gel with its attached DNA fragments. The nitrocellulose membrane is now dipped in a solution containing labelled gene that is present on the membrane. The radio labelled probe hybridizes with the complimentary fragment of DNA on the membrane. The filter subsequently is removed from the solution and washed thoroughly to remove any unbound fragments of DNA. Later the filter is placed against a photographic film. Wherever the probe has hybridized with the DNA fragments there will be markings like black dots. This can now be compared with the original gel and the corresponding DNA band can be identified as the required gene.

Northern blotting technique. This is used to identify the required RNA bands. The technique of southern blotting used for DNA transfer from the gel to the nitrocellulose membrane could not be used for transfer of RNA because RNA would not bind with the nitrocellulose membrane. Alwin *et al* (1979) devised a technique in which the mRNA bands from the gel were blot transfered to a chemically reactive paper prepared by diazotization of aminobenzyloxymethyl paper. RNA fragments can bind themselves on to this paper and the paper can be used for hybridization with radiolabelled DNA probes. Subsequently by autoradiography the hybridized bands can be identified. Some of the researches of Thomas (1980) however have shown that mRNA bands can also be blotted directly on to nitrocellulose membrane using special techniques.

Western blotting technique. This technique is used to detect proteins of a particular speciality. Discovered by Towbin *et al* (1979) in this technique newly coded proteins by a transformed cell can be identified. In this method however nucleic acid probes are not used. This technique involves the following steps:

(a) Electrophoretic separation of the required protein on a polyacrylamide gel

(b) Transfering of the separated peptide chains on to a nitrocellulose membrane

(c) Using radiolabelled antibodies as probes the required peptide chain may be hybridized. (The probe consists of antibodies which bind to the antigenic protein)

(d) Autoradiographic identification of the required hybridized protein

CONSTRUCTION OF CHEMERIC DNA (insertion of DNA into vector)

Chimeric DNA molecules are produced by the insertion of a foreign DNA segment into a DNA molecule of the vector. These chimeric DNA molecules are the means of the cloning genes or other small DNA sequences. Basically production of chimeric DNA molecules involves breaking open the vector DNA and insertion of foreign DNA followed by joining of the ends. There are three techniques available for the construction of chimeric DNA. These are - pallindromic sequences (cohesive ends) and staggered cleavage. dAdT tailing and blunt end ligation.

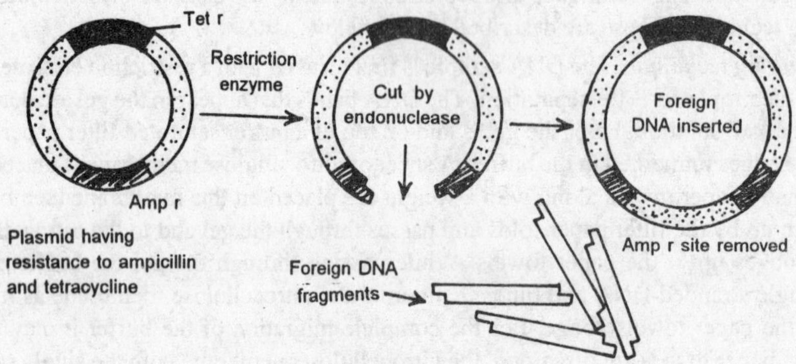

Fig. 25.6 Genetic Engineering
The concept of cloning of foreign DNA in a vector

1. Pallindromes and staggered cleavage (cohesive ends). It has already been explained that a pallindromic sequence has complimentary sequences at the two ends of a single strand. Restriction enzymes cutting the pallindromic sequences may have an even cut or a staggered cut. Result of the staggered cut is DNA fragments with sticky ends are formed. When another DNA molecule (a foreign DNA molecule) is cut similarly and when the two DNAs are mixed both DNAs will combine forming a chimeric DNA

Discovered for the first time by Boyer and Cohen, this mechanism has an advantage of easily combining two different DNA molecules. Further it also produces two restriction sites in the chimeric DNA. There are some disadvantages in this method. For instance, the two sticky ends of a vector DNA or of a foreign DNA may themselves join without allowing for the insertion of one into the other.

dA - dT tailing (the terminal transferase method)

In this method DNA can be cut at the desired position both in the vector and in the clone without staggered cleavage. Using the precursor dATP polydA is added at both the 3' cut ends of the vector with the help of the enzyme terminal transferase. Sin.ilarly using the precursor dTTP poly dT is added at both the 3' cut ends of the DNA sequence to be cloned. The vector and clone can then be joined by annicaling the polydA with polydT tails and then lygating them using the DNA lygase enzyme. In this technique there is no possibility of the two cut ends of the same DNA molecule joining together because they have same nucleotides. There is one disadvantage in this technique because it will not be easy to retrive the cloned DNA as the recognition site of the enzyme is lost due to the addition of poly dA and poly dT. It has been found out that use of poly dG and poly dC will help in the restoration of the recognition site.

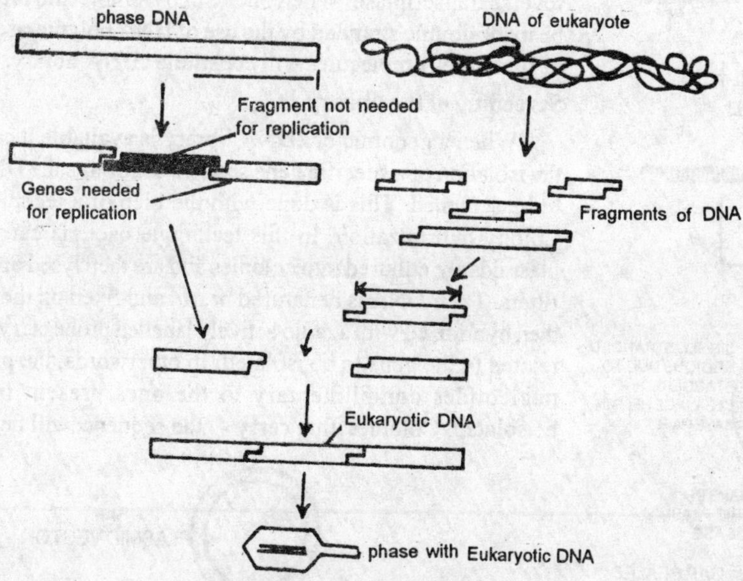

Fig. 25.7 Genetic Engineering
Use of λ phage for cloning

Blunt end ligation

In this technique a restriction enzymes is used to cut a duplex DNA at the same site in both the strands. The broken ends are then used for joining with the two ends of foreign DNA molecule irrespective of the sequences present at the broken ends of the two molecules. The blunt end ligaton is made possible by the use of the T4 DNA lygase. This technique also has a disadvantage in that the broken ends of the same DNA molecule may join together.

Construction of genomic and cDNA libraries

In order to isolate required gene or genes from a genome a mixture of clones each carrying DNA fragments derived either from the natural genome or from cDNA is prepared. This mixture contains thousands of clones and is collectively called genomic library. Similarly when cDNA clones are collected it is known as cDNA library.

Construction of genomic library

Cloning an entire genome in the form of fragments mixed randomly is known as genomic library. This may be done by a method known as shotgun method (see earlier in the same chapter for details). With a genome library clones can be perpetuated indefinetely in a plasmid vector and retrived when ever necessary.

Genomic libraries can also be prepared by using a number of restriction endonucleases one at a time so that one will have fragments of different sizes with cuts at different places. This has a disadvantage in that many a time the inciscion site many be within a gene so that complete genes may be difficult to obtain. This may be overcome by using restriction enzymes which have very short recognition sequences so that the fragments that are obtained will be long and the chances of splicing within a gene are avoided.

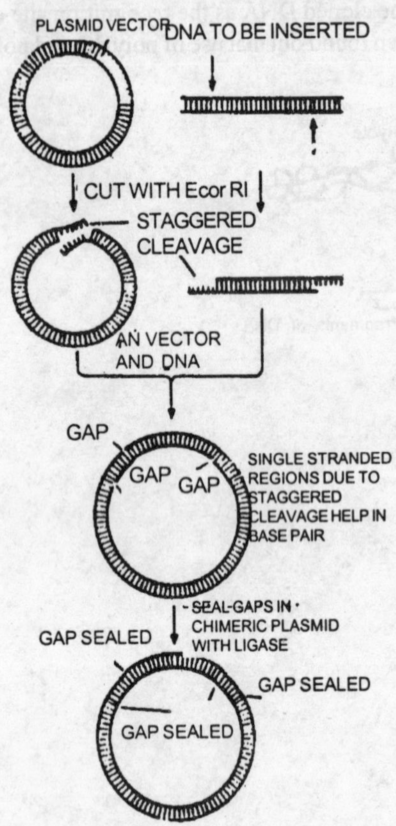

Fig. 25.8 Genetic Engineering Cloning of a segment of DNA by staggered cleavage

cDNA library

It has already been explained above as to how to obtain a cDNA using its complimentary mRNA under the influence of the enzyme reverse transcriptase. When once a cDNA molecule is obtained it can be made double stranded by the use of DNA polymerase. A collection of such cDNA molecules will constitute cDNA library.

Screening of the library

When a genomic or cDNA library is available it can be used for the isolation of a specific gene sequence. For this the DNA library has to be screened. This is done with the hlep of a technique known as Colony hybridization. In this technique bacteria carrying chimeric plasmids are cultured into colonies and are then lysed on nitrocellulose filters. Their DNA is denatured *in situ* and fixed on the filter which is then hybridized with a radio actively labelled probe carrying a sequence related to the gene to be isolated. In otherwords, the probe will have nucleotides complimentary to the ones present in the gene to beisolated. Colonies that carry the sequencewill be identified by dark spots after radioautography so that the colonies carrying the vector with the desired gene sequence can be recovered from the master plate and used for further experiment.

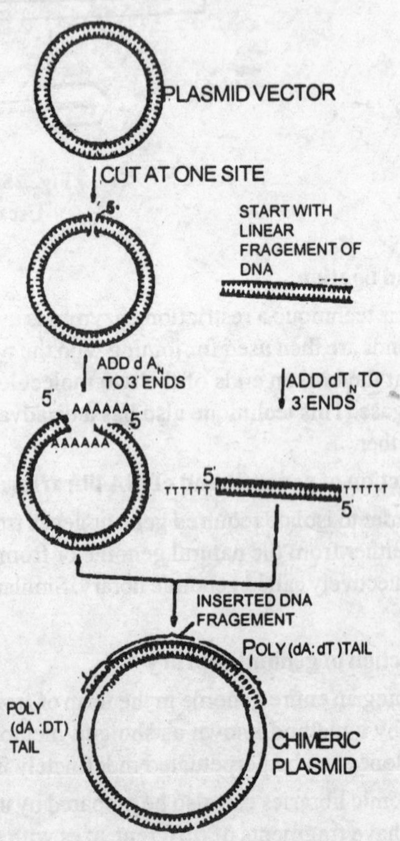

Fig. 25.9 Genetic Engineering Cloning of DNA segment by dA-dT tailing

Transfer of the recombinant DNA into a cloning organism

When once a mixture of recombinant DNA is obtained it has to be transfered to a suitable cloning organism for the purpose of expression. Several techniques are used to transfer the recombinant DNA into suitable cloning cells. Transforming the cloning cells with recombinant DNA is a difficult process as the cells of bacteria *(E.coli)* have restriction enzymes which will immediately degrade the foreign DNA. To escape from degradation exponentially growing culture of *E.coli* is pretreated with $CaCl_2$ at low temperature and thereafter the DNA is mixed up. This treatment is known to sufficiently prevent the action of restriction enzymes. When the recombinant plasmids get established in the cells of *E.coli* the cells are said to be transformed.

Instead of plasmids even λ phages could be used as transforming elements. The entry of the phage into *E.coli* is usually termed transfection (a combination of transfermation and infermation).

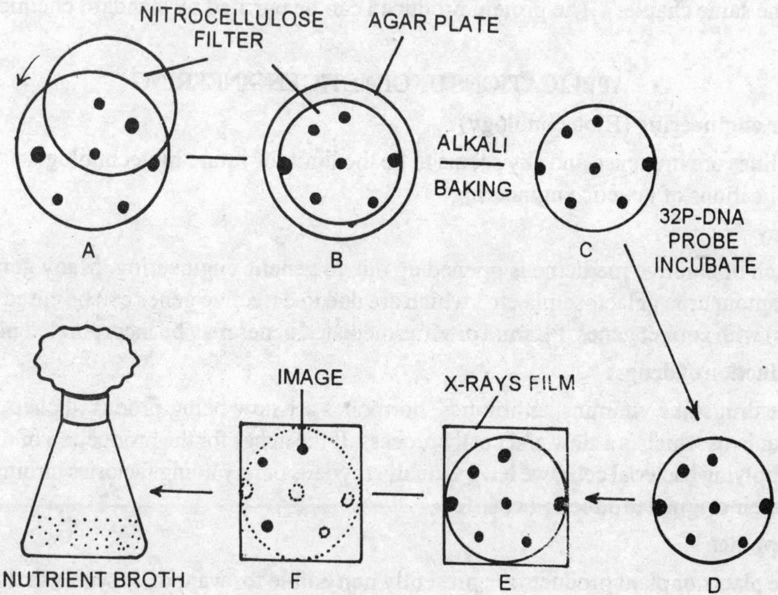

Fig. 25.10 Genetic Engineering

Colony hybridization technique

Not all the cells will accept recombinant DNA and get transformed. It has been estimated that about 10^5 transformats per microgram of cloned circular plasmid can be generated. The cells which have been transformed will grow into colonies along with other non transformed ones. Colonies with transformed cells can be identified with suitable techniques.

Expression of cloned DNA

The purpose of genetic engineering is not just to separate and clone the genes in a suitable organism. Isolation and multiplication of genes is only a means ultimately to obtain the gene product namely the required protein. Hence the recombinant DNA molecules should be allowed to express in the new environment (*E.coli* cells) and produce what ever protein they were producing. For instance if an insulin gene from human beings is transfered into *E.coli* the gene should be able to produce insulin molecules using the protein synthesizing machinery of the bacterium. Expression of a cloned gene is a verydifficult process. This is more so when a eukaryotic gene is inserted into a prokaryotic organism. As the protein synthesizing genetic machinery of prokaryote is different from eukaryotes, the eukaryotic gene should be provided with the necessary ingredients of a prokaryote to produce the protein. The factors that are necessary for eukaryotic gene to express in the prokaryotic environment are as follows :

(i) Supply of prokaryotic promoter necessary for expression of eukaryotic gene

(ii) Supply of ribosomal binding sites for the cloning vector

(iii) Removal of introns from mRNA obtained from the eukaryotic genes

(iv) Inhibition of bacterial gene responsible for degradation of the protein produced by the eukaryotic gene.

Recovery of the gene product

This is the final step in genetic engineering techniques. After all the hurdles are passed and the eukaryotic gene has successfully expressed itself in the prokaryotic cell, the required product namely the eukaryotic protein is produced. This can be identified by antigen antibody reaction (for details see genetically engineered insulin later in the same chapter). The protein produced can be purified by standard chemical processes.

APPLICATIONS OF GENETIC ENGINEERING

Scope of genetic engineering (Biotechnology)

The possibilities are immense and sky seems to be the limit for future biotechnologists. The following are some of the applications of genetic engineering.

I. Gene therapy

A new branch of curative medicine is opened up due to genetic engineering. Many genetic disorders like hemophilia, alkaptoneuria, galactosemia etc., which are due to defective genes can be cured by replacing them (defective genes) with correct genes. Plasmid or virus mediated genes may be incorporated into human systems.

II. Cheap production of drugss

Many of the drugs like vitamins, antibiotics, hormones are now being produced chemically or produced from higher organisms which is a slow and costly process. If the genes for the production of these are introduced into rapidly multiplying bacterial cells, we have virtually myriads of tiny living factories turning out the medicines at a fraction of their original production cost.

III. Gene for new diet

Some of the plants or plant products are presently non edible for want of enzymes in the human system to digest them. For ex wood, grass, etc. If the genes that code for the digestive enzymes (for materials like wood etc) could be introduced into the human system, wherein there would virtually be no famine on earth. Another possibility, (which sounds unbelivable at present) could be to introduce genes that can produce the chlorophyll pigment in the animal or human system, wherein we would be solving the food problem once for all !.

IV. Role of genetic engineering in Agriculture

The following are some of the advantages of gene recombination in crop improvement.

(a) Production of new varieties

New varieties of plants that are not present in nature can be produced by the introduction of new genes.

(b) Somatic hybridization

Two crop plants with important characters, which could not be hybridized sexually due to genome incompatibility can be hybridized through protoplast fusion. Production of *Pomato* (Tomato x Potato) is one such example.

(c) Increasing photosynthetic efficiency

Ribulose diphosphate carboxylase (Rubisco) is a key enzyme in the dark fixation of CO_2. Modification of *Rubisco* to reduce photorespiration (Osmond 1981) by elimination of oxygenase activity would increase-the CO_2 fixation rate. The enzyme has two sub units and is known to be coded by seven genes. If these genes could be altered, resulting in the changed *Rubisco,* it will increase the photosynthetic efficiency.

(d) Nitrogen fixation

The roots of leguminous plants normally possess nitrogen fixing nodules, induced by various strains of *Rhizobium*.m As a result, leguminous plants require little or no nitrogen fertilizer. Arable crops require about 100 kg of nitrogen per hectare. If this can be replaced by N_2 fixing genes, it would lead to a lot of saving in terms of fertilizer cost. Nitrogen fixation is catalyzed by the ATP dependent electron reduction of dinitrogen to ammonia by nitrogenase. If it is possible to introduce the gene to higher plants, it would bring about a revolution in agriculture. This may be achieeved by one of the following ways.

1. Transfer of nitrogen fixing gene (*nif*) from the living bacteria like *Klebsiella, Azatobacter* etc., to higher plants. This would enable the plants to fix the N_2 directly. This is still in the experimental stage, as the *nitrogenase,* produced by the *nif* gene is highly oxygen liabile and gets degraded in higher organisms.
2. Expansion of host range of symbiosis to crops other than legumes. At present non leguminous plants do not have symbiosis with *Rhizobium.* If it is possible to expand the host range for the bacteria, a lot more plants will have the capacity of N_2 fixation.
3. Increasing the efficiency of symbiotic bacteria to N_2 fixation. By gene manipulation, it may be possible to increase the N_2 fixing capacity of the symbiotic bacteria, which will benefit the plants.

(e) Herbicide resistance, disease resistance etc

By gene manipulation, it will be possible to introduce this character into plants.

In conclusion it may be said, that genetic engineering is a highly potential tool by which man can alter the macromolecule of inheritance and thereby change the biological environment to his choice. But the new technique has to be used judiciously, otherwise it can also create untold harm to human society, if unscrupulous scientists use it to create destructive genes.

A few of the specific applications of genetic engineering are discussed below in some detail.

Nitrogen fixing genes (*nif* genes)

Nodule formation and nitrogen fixation are two of the important biological processes which are genetically controlled in diazotrophic bacteria. From the genome of *Rhizobium,* several genes coding for nodulation (*nod* genes) and nitrogen fixation (*nif* genes) have been identified and isolated. In free living and symbiotic nitrogen fixers, nodule forming and non-nodule forming *nif* genes are present either on the genome or on the plasmids in their cells.

Nod genes

Nodule formign species of *Rhizobium* contain in their cells an extremely large plasmid called *mega plasmid* which consists of a number of *nod* genes and *nif* genes (Rosenberg et al 1981). Bot the genes are closely located.

Kondorosi *et al* (1982) were able to transfer the megaplasmids from one species of *Rhizobium* to other simultaneously transfering the ability of nodule formation.

The nod gene region is of 8.5 kb in nucleotide sequence length and proteins encoded by the 8.5 kb fragment when transfered to *E.coli* has been determined. The *nod* gene is a cluster made up of 4 genes designated as *nod a, b, c* and *d.* A comparison of *nod* genes between different species of *Rhibozium* shows that about 70% of the nucleotides are homologous.

***Nif* genes**

The *nif* genes are present in several species of *Rhibozium* and are located on a megaplasmid adjacent to the *nod* genes. *Nif* genes have also been identified in many cyanobacteria such as *Anabena,* where the *nif* genes are located on the main genome.

Early studies on *nif* genes were carried out in *Klebsiella pneumoniae* and the function of these genes was confirmed by transfering into *E.coli* cells. Researches have also been conducted to identify the product of *nif* gene cluster. The *nif* gene cluster consists of 7 operons.

In cyanobacteria *nif* genes have been located both in vegetative cells as well as in heterocysts. But there is some difference in the nature of the *nif* gene between the vegetative cells and the heterocysts.

Cloning of *nif* genes

Researches are being conducted in many countries to transfer the *nif* genes to higher plants especially to monocots. If this is successful it will be indeed a great boon to agriculture as there will be plenty of saving in nitrogen fertilizers for the plants themselves will be able to fix atmospheric nitrogen. Gene expression of *nif* genes in eukaryotic cells is very difficult because the nitrogenase enzyme required for fixing the nitrogen and controlled by the *nif* gene would get denatured in the preence of oxygen. If this can be overcome *nif* genes can be easily transfered to eukaryotes. *Nif* genes however have been cloned in *E.coli*. One of the methods of transfering *nif* genes is to transfer them into the chloroplast of eukaryotic cells as the translational and transcriptional processes of chloroplasts resemble prokaryotic features. These attempts might succeed because the chloroplasts are able to produce the ATP and reducing power both of which are required for nitrogen fixation (Merrick and Dixon, 1984). But there are major problems for such a transfer of *nif* gene to chloroplasts. These are

1. Lack of chloroplast transfering technique
2. Protection of nitrogenase from oxygen evolved during photosynthesis.

Genetically engineered Insulin

Insulin is an important hormone that plays a vital role in the metabolism of carbohydrates. In patients suffering from diabetes insulin has to be injected subcutaneously to regulate the blood sugar level. Insulin producing genes can be transfered to bacteria by techniques of genetic engineering and large quantities of insulin may be extracted. This process is cheaper and convenient than the conventional one. We will briefly discuss below the method of procuring insulin by genetic engineering techniques.

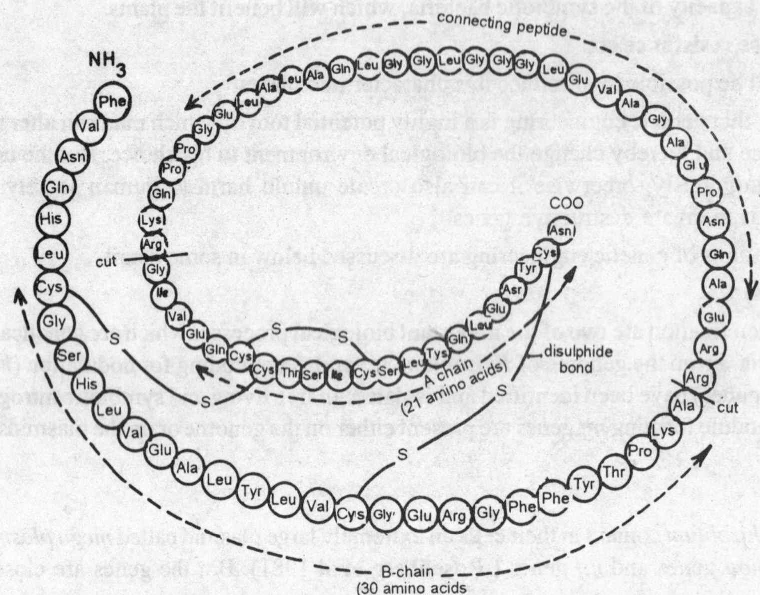

Fig. 25.11 Genetic Engineering
Structure of Insulin molecule

Structure of human insulin

Insulin molecule is composed of two chains - α and β chains of aminoacids. Initially it is produced as a single long chain called *Pre-proinsulin* which is later cut into a shorter chain called *proinsulin* (composed of 84 aminoacid units). Proinsulin then breaks into α and β chains when a piece from the middle is cut off. Chain α is composed of 21 aminoacids and chain β has 30 amino acids - The remaining 33 aminoacids act as a connecting peptide chain.

F

Production of Insulin in *E.coli*

The genes for insulin chain α and β are identified and located. This may be done first by identifying the codons and then the corresponding mRNAs can be reconstructed. Using these mRNA molecules and with the help of the enzyme *reverse transcriptase* the corresponding segmentas of double stranded DNA molecules are produced. Then with the help of the enzyme *terminal transferase,* the ends of the insulin genes (DNA strands) are extended with short sequence of identical bases.

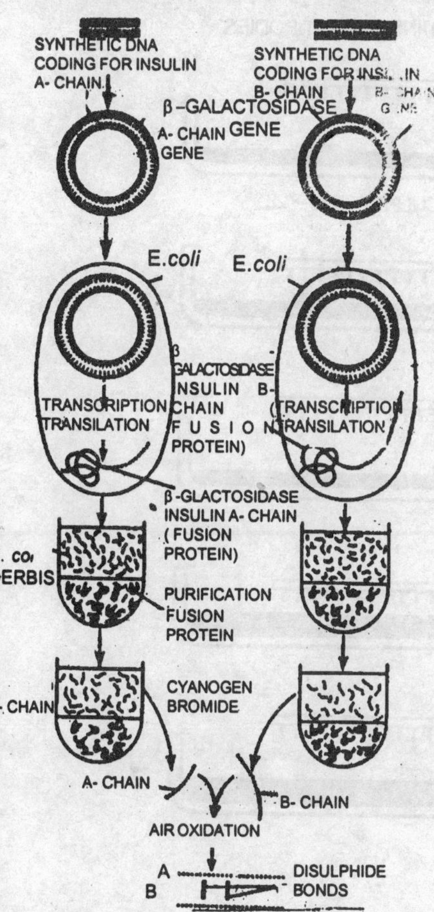

Fig. 25.12 Genetic Engineering
Technique of producing genetically engineered insulin in E.coli

Plasmids obtained from *E.coli* are opened up at specific portions by the use of restriction enzyme. The insulin genes (DNA strands) of α and β chains are added to the plasmids separately forming recombinant DNA plasmids. Gaps if any in the circular plasmids of α and β chains are filled by lac operon gene for switching on the synthetic process. The plasmids with recombinant DNA for α and β chains are inserted into *E.coli* finally.

The recombinant DNA plasmids multiply in *E.coli* and trigger the synthesis of α and β chains of insulin as a genetic expression. Following large scale production cultures, the insulin chains are obtained following specific purification processes. Finally the α and β chains are chemically bonded by disulphide bonds to produce the complete molecule of insulin.

Gilbert and Villakomaroff (1980) isolated mRNA for insulin from β cells of the pancreas of rat and inserted it into PBR 322 plasmid in the middle of a gene normally coding for the enzyme penicillinase and incorporated it into the cloning organism (*E.coli*) *E.coli* cells produced the hybrid protein consisting of penicillinase + proinsulin. Proinsulin was then separated from penicillinase.

Immunological tests to identify insulin producing colonies

On a culture medium there will be colonies which are transformed as well as non transformed. In otherwords colonies with a recombinant DNA and those without recombinant DNA will be there. Colonies with recombinant DNA and successfully expressing the gene will produce the insulin protein. These colonies have to be identified so that they can be picked up for further multiplication. This is done by two immunological techniques - radioactive antibody test and immunoprecipitation test.

Radio active antibody test

In this test a plastic disk to which antibodies of the gene product (insulin) are bound is pressed on to the colonies. If the protein (insulin) is present in any colony it will bind to the antibody and to the disk. The disk is then placed in a solution containing radio active antibody (to insulin). If any required proteins (insulin) are present on the disk the radioactive antibody sticks to the proteins on to the disk. The disk is then washed and the autoradiograph will identify colonies if any producing insulin.

Immunoprecipitation test

This test is also used to identify a protein (insulin) producing colony. In this technique insulin antibodies are added to the agar medium. When the bacterial colonies grow if any of the colonies contain insulin it gets percipitated surrounding the colonies due to the action of the antibodies.

Human insulin humulin produced by the genetic engineering technique was made available to Diabetic patients in September 1982. It was the first therapeutic product to be produced by genetic engineering.

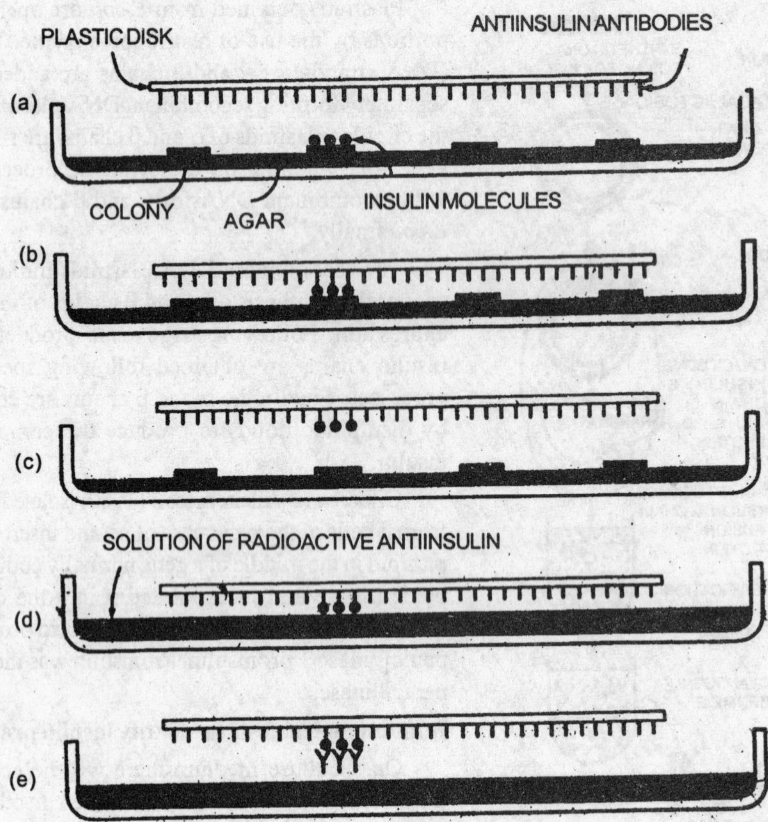

Fig. 25.13 Genetic Engineering
Radioactive antibody test

Human growth hormone (HGH)

Also known as Somatotrophin, this hormone is secreted by the anterior part of the pituitary and is made up of 190 aminoacid residues. The deficiency of this hormones results in dwarffism. For the conventional treatment of dwarfism HGH is injected into the patients. But the extraction of hormone from the pituitary glands cannot meet the ever increasing demand for this hormone. Genetic engineering procedures have successfully produced somatotrophin.

Double stranded cDNAs were obtained from the mRNA precursors of HGH and then incorporated into bacterial cells where they expressed in non precursor form. The bacteria were unable to convert the peptides of HGH into biological active forms. A recombinant plasmid containing a full length of HGH cDNA is cleaved with restriction enzymes that release fragments containing the complete HGH coding sequence after codon 24. Several overlapping complimentary oligo nucleotides were joined to form a synthetic strand of a small DNA fragment which contains the coding sequence for the first 24 aminoacids of mature HGH. The synthetic DNA and cDNA molecules are joined together to produce a new fragment which has the complete coding sequence of HGH. This is then ligated to a restriction site a little below the *lac* promoter operator gene cloned on a plasmid. These plasmids are cloned in *E.coli*. Synthesis of HGH is induced by an inducer (IPTG) of *lac* operon. The HGH thus obtained is later purified by chemical methods.

About 1 lakh molecules of the hormone per cell of *E.coli* have been produced by genetic engineering techniques. One of the problems in this is the attachment of an extra methionine to the molecule. This has to be removed to make the molecule biologicaly active.

Vaccine for Rabies virus (RV)

Hydrophobia caused by made dog bite used to take a heavy toll of the human life. The credit of discovering a method to cure hydrophobia or rabies goes to Louis Pasteur. Even now however ignorance and not taking the vaccine in time, is causing a number of death of humans and animals in many Asian and African countries. The conventional antivirus vaccine is difficult to obtain and also a tedious process. Researches are being conducted to produce genetically engineered rabies vaccine using *E.coli.* Attempts are made to isolate the mRNA coding for the viral protein from the RV infected cells. The genes coding for the production of a virus glycoprotein coat has been successfully transfered to the cell of *E.coli.* This seems to the first step in the production of antirabies vaccine because this glycoprotein stimulates the production of antibodies in affected individuals. The Eister Institute in Philadelphia USA has developed a new genetically engineered vaccinia virus by inserting a small fragment of foreign DNA. This genetically engineered virus could synthesise antirabies vaccine. The virus however does not cause rabies in those animals receiving the rabies genome but instead activates the immune system against infection. The vaccine has been certified that it can be administered orally to animals.

BIOHAZARDS OF GENETIC ENGINEERING

Recombinant DNA technology popularly called genetic engineering is one of the most potent tools in the arsenal of human knowledge that can be used either for good or for bad. Ever since human civilisation evolved, two developments in the field of science have had the effect of changing the human destiny. One development that took place in the field of physical sciences is that of atomic fission which gave the man a vast amount of energy which when produced can be used for improvement or for destruction. A development similar to this or much more potent than this has taken place in the biological sciences in the form of recombinant DNA technology. With the power and ability to synthesise the gene, to order the alignment of gene and to regulate its expression man indeed has been handed over the very basis of origin and differentiation of life. With this enormously potential tool he can create different types of organisms, vastly modify the existing organisms for his betterment or he can create organisms which are more harmful than the existing ones and can even spell the deathknell of the very nature which has created him.

The brief discussion above will clearly point out to the fact that research in genetic engineering has to be carefully controlled, manipulated and should be encouraged only when it is absolutely necessary and that too with plenty of safeguards. In the paragraphs below we will briefly outline the biohazards of uncontrolled genetic engineering research and what safeguards have to be followed while following the techniques of recombinant DNA.

Harmful aspects

Genetic engineering involves introducing a foreign DNA into another organism so that it gets transformed. Many a time when a gene segment is transfered one may be aware of its expression in the original organism. But one is not too sure as to what kind of expression it may have in the new environment i.e., in the new organism. This kind of recombination many a time creates a dangerous new form either deliberately or accidentally. A harmless microorganism might turn out to be a very harmful one after it acquires new genes from another organism. Eventhough it is true that the genes introduced by themselves may not be harmful in their original combination, but when they are introduced into a new set up they might bring about the potentially dangerous expression. If by chance a transformed organism comes to carry a deadly disease due to genetic manipulation and if that organism escapes out of laboratory one can imagine the uncontrolled horror that the organism might bring about. The genetically manipulated organism might be so potent in its pathogenicity that any of the existing drugs may not be able to combat it. For instance *Streptococcus* a pathogenic organism was never resistant to penicilin originally in nature. For the purpose of genetic engineering research a plasmid with a desired gene is introduced into *Streptococcus* and at the same time to identify the transformed organisms a marker gene such as penicillin resistant gene is itnroduced, when the organism becomes penicillin resistant, the antibiotic now would be useless. If such an organism even though created in the laboratory were to escape one can imagine the dangers caused by it.

Escape of transformed microbes into environment

Inspite of precautions being taken their always exists a possibility that a recombnant orgnanism might escape from laboratory and get permanently entrenched in the ecological system. If the organism in question were to be a parasite it would cause immense havoc to allthose organisms it parasitizes. Microorganisms that live in special ecological niche like hot spring or salt water might not pose such a great problem as they cannot spread widely.

The escape of microorganisms from the laboratory may be by design or by accident. If it is by design then one has to understand that microbes are allowed to escape deliberately to cause harm to the environment outside. Accidental escape of chimeric microbes from the experimental laboratories may take any one of the following routes.

(1) Laboratory personnel such as scientists, attendants, visitors, sweepers etc might inadvertentaly carry the microbes and spread them outside.

(2) Microbes might also find their way out through laboratory facilities such as glassware, drainage of chemicals, spillage of media and such other inanimate objects.

In addition ot the above, non scientific personnel in the laboratory many a time may not be aware of the stringent precautions that have to be mandatorily taken while dealing with pathogenic microbes. Many a time even if they are aware they might just not follow the precautions as they are quite tedious and cumbersome. Laboratory personnel while constantly dealing with pathogenic microbes many a time might acquire immunity and may not suffer from the consequences of contact with pathogens but they act as carriers and others who come in contact with them will have to suffer the consequences.

Unintended escape of microbes and possible creation of new microbes posses perhaps the greatest danger to the environment. In this context it is worthwhile mentioning the opinion of three international experts on AIDS - Dr.John Seale from London. Dr.Robert Strecker from California and Dr.Jacob Segal from Berlin. These scientists claimed in 1987 that the AIDS virus was artificially created by American scientists during laboratory experiments when they were trying to discover a cure for cancer by combining parts of Maedy - visna virus found in the sheep and human T-cell Leukaemia virus type 1. They argued that the scientists or other personnel in the laboratory could have been infected with the new virus through a minute cut or even by inhaling (Ignacimuthu, 1995).

Plus and Minus points of genetic engineering research

A great controversy is going on currently in the International Community involving biologists, social scientists, politicians and religious people regarding the desirability or otherwise of conducting genetic engineering research. While many people argue that genetic engineering research would be a harbinger of prosperity others are equally vehement in opposing that research in genetic engineering would cause untold harm to human life as it is trying to tinker with the very basis of life and evolutionary process. The arguments for and against recombinant DNA technology are summarised below.

Arguments favouring genetic engineering research

1. The transfer of genetic material from eukaryotes to prokaryotes and *vice versa* have been going on even in

nature perhaps on a very small scale. If nature were to intend to prevent the transfer of a genetic material, evolution would have surely brought in a perfect genetic barrier preventing such a transfer. In fact ever since the origin of life evolution could proceed only because nature took to genetic engineering.

2. The dangers of genetic engineering research perhaps will receed with time as people and scientists become experienced and accumulate knowledge.
3. Enough safe guards are already available in research laboratories dealing with pathogenic organisms. They can be made much more stringent. In any case the beneficial aspects of combating the disease perhaps far more outweigh the possible dangers.
4. Genetically transformed microbes have been designed with a special purpose. Hence the argument that they would compete with the other normal microbes and perhaps might be picked by natural selection appears to be imaginary. Further with attenuated characters the transformed organisms will not be able to survive unless special conditions such as the laboratory environment are provided with.
5. The idea that by accuring a new gene the microorganisms become parhogenic is not true because in any genetic transformation the host organism will always reject a virulent gene combination and the unwanted genes are generally lysed by restriction enzymes and even if the genes express themselves the gene product namely the foreign protein gets degraded by proteolytic enzymes.
6. All types of genetic engineering research need not be dubbed as harmful and dangerous. For instance research in the field of agriculture, dairy and others will not pose any problem. Researches in the medical field however can be continued with stringent regulations and control. At the same time however one cannot simply push aside the vast benefits that have accrued to man kind because of improved health care products from genetic engineering. Hence total stoppage of genetic engineering research would weaken the ability of man to understand and control nature and ultimately it is not in the interest of man himself.

Arguments against genetic engineering research

1. Unintended or accidental escape of microbes from the laboratory is possible through many routes. This escape of the transformed organism into the environment will pose great danger.
2. Genetic engineering research actually tinkers with the very basis of the evolutionary process. Nature indeed has been a very efficient genetic engineer and it has been doing this work very slowly and steadily over a period of millions of years. Transfer of genetic material and production of new traits in nature takes millions of years to be achieved. If this is sought to be achieved in a very short period it amounts to short circuiting the normal process and the consequences could be really hazardous. When an individual is fashioned out of natural process it will be a package of both weakness and strength. If we were to tinker with this process it would naturally disturb the fine tuned balance and relationship that is existing between the organisms since millions of years.
3. The intended benefits of genetic engineering research can be actually achieved through other research methodologies. To achieve these goals one need not resort to such dangerous techniques lke genetic engineering. The benefit of genetic engineering research does not justify the cost (not just in terms of money) involved.
4. With the kind of diverse political set up prevailing in different countries of the world where dictators, fundamentalists etc still have a very powerful say, this technique can definitely be misused and there is every possibility that these people might unleash a biological warfare on countries whom they consider inimical to them. For instance water bodies of a country could be infected with new versions of pathogens or their environment might be sprayed with pathogenic microbes. All these possibilities should make one think whether it is worthwhile indulging in this kind of research.

Safe guards in genetic engineering research

The scientific committee at large and biologists in general have been thinking for a very long time as to how to introduce some sense into genetic engineering research. While it may not be prudent to indulge in genetic engineering for problems which can be solved otherwise it is equally true that genetic engineering cannot be totally given up as it indeed has the potential of solving some of the serious problems confronting the human race which other research methodologies cannot hope to solve. The only answer to this being, research in genetic engineering should be properly regulated, monitored and if necessary with stringent international control.

A committee headed by Paulberg (1973-74) an eminent scientist held discussions with other interested persons and published a small bullettin outlining the risks involved in genetic engineering research. In 1975 an International group of scientists who were very much concerned with genetic engineering research convened a meeting in California USA and considered the ethical and safety implications of genetic engineering research. An outcome of this meeting was the formulation of a series of guidelines that sought to regulate and in some cases even prohibit research in recombinant DNA. In the USA, the National Institute of Health (NIH) issued guidelines in the year 1976 for US scientists involved in genetic engineering research. The guidelines required among other things that any research proposal involving recombinant DNA should be submitted to a special committee of NIH for approval. NIH also graded the laboratories into four levels (P1 to P4) where genetic engineering research of different levels were recommended. For instance experiments involving DNA from primates and animal viruses involving harmful genes could be carried out only in extra secure P4 laboratories. In 1982 an advisory panel of NIH - the recombinant DNA advisory committee (RAC) published revised guidelines. According to these new regulations recombinant DNA experiments were graded into 5 classes and of these it is essential that three levels of research must obtain NIH approval even before initiation. These include researches involving drug resistance of microorganisms, toxin synthesizing genes and operations involving release of genetically engineered organisms into environment (as those used for pest control).

On the 26th January 1982 a meeting was held at the parlimentary assembly of the council of Europe in Strasbury and recommended certain guidelines aimed at protecting the individual from the inheritant dangers of genetic manipulation. Three areas were mentioned in particular - genome map of human cells, gene therapy of somatic cells and gene therapy of germ cells. The report of the council concluded that inspite of the possible advantages of genetic engineering research, man made decisions should not substitute the free play of nature and everyone's right not to be genetically manipulated should be affirmed (Ignacimuthu, 1995).

Outside USA, the regulations on genetic engineering research are not actually as stringent as they ought to be. France and Germany are considering whether to restrict research in genetic engineering because of the ethical questions involved. Recombinant DNA guidelines of France are ambiguous about the deliberate release of genetically altered organisms. In Germany however release into environment of genetically altered organisms are strictly prohibited. Many social activist groups are opposing gene therapy and such other genetic engineering parctices in terms of ethical and moral parameters.

In India, genetic engineering research is still in infacy, yet considering the importance of it, Government of India has constituted four committees for this purpose.

1. Recombinant DNA advisory committee (RDAC). To evolve guidelines.
2. Institutional Biosafety Committee (IBSC). For implementation of guidelines.
3. Review Committee of Genetic manipulation (RCGM). To recommend special suitable conditions for experiments and trial.
4. Genetic Engineering approval committee (GEAC). For large scale manufacturing, environmental release of genetically altered organisms/products and import/export of genetically altered organisms.

The followng may be considered as some of the safeguards that should involve any research in genetic engineering.

1. Harmful genes such as those coding for toxins and antibiotics should not be cloned without extreme precautions.

2. Animal genes, genes of tumour viruses should not ordinarily be cloned in bacteria or others microbes.
3. Laboratory facilities should be such that there should be virtually no possibility of the escape of genetically altered microbe. In addition to scientists even the other ordinarily personnel of the laboratory should be thoroughly briefed about the dangers of carelessness before they are employed. Workshops on precautionary measures should be conducted periodically so that no one violates them. There should be mandatory check of laboratory equipments of cabinets, hoods and other glassware to see that they are microfree when they are taken out of the laboratory. Laboratory should also possess special equipments like negative pressure rooms, safety cabinets, special traps and drinage etc., to prevent the possible escape of the microbe.
4. Use of microbes which can live only in special environmental conditions should be picked up for genetic engineering research because even if these organisms escape into the environmental after genetic alteration they will not survive in the normal environmental conditions so that the chances of their rapid spread are very less.
5. As far as possible cloning vectors should be non conjugative plasmids because even if they are uncontrolled they will not be able to promote their own transfer by conjugation.
6. The cloned vector should be sufficiently modified or altered in such a way that it will not be able to express itself in any environment but can do so only under special circumstances like a particular host strain that are normally not available The genetic alteration must be such that in case it is transfered to another microbe it should prove lethal to it.
7. The cloning organisms should be modified in such a way that after cloning they depend upon very special conditions of existence which can be provided only in a laboratory and are not available in the environment.
8. Attempts are being made to produce "crippled microbes" for use in genetic engineering research. The genome of these microbes is altered and disabled in such a way that if they escape from the special laboratory conditions they will not survive. The following are some of the instances of crippled microbes for use in genetic engineering research.

Diaminopimelic acid (DAP) is an important component of the cell membrane in *E.coli*. Genetically altered *E.coli* were produced which were DAP deficient. These cannot survive unless supplemented by DAP. Hence these organisms even if they escape outside cannot survive. Such organisms are ideal tools for genetic manipulation. Precautions are however necessary because many a time microbes are known to undergo mutation very fast. In fact in the above example the DAP deficient *E.coli* very soon mutated back and was able to produce DAP. Scientists however discovered that there was more then one gene involved in the production of DAP and deleted that also. But the bacteria would not yield so easily to the manipulation of the microbiologists. The DAP deficient bacteria again survived by producing a sticky substance called Calanic acid around their protoplast. But then the calanic producing gene was again removed. This will clearly reveal that our knowledge on the genome of microbes would always remain incomplete because we will be able to identify a gene only when it express itself after natural selection; for instance calanic acid producing gene was not selected to express when DAP producing gene is there. In the absence of the later the former was selected. Who knows in the absence of calanic acid producing gene which other bacterial gene which is hybernating may get selected and express itself. Afterall life is a struggle and no individual will give up its existence so easily.

Genetic engineering research and Ethics

Man is a product of Evolution and however much hey may think that he is different from other organisms because of his intelligence, the stark truth is that he is definitely a part of biological spectrum whether he likes it or not. Man is neither the first step nor the last step in the evolution of life. The process of Evolution which has produced so many diverse types of organisms has also produced man and all these organisms that from the biological spectrum were not produced within a day. Even the Evolution of human beings took several millions of years. It is quite true that the millions of organisms in the present day biological spectrum are produced out

of nature's genetic engineering. But one must realise that these have been crafted painstakingly and patiently over the course of millions of years. Man's attempt of fast genetic engineering seeks to destabilise the delicate balance that has been inbuilt into the relationship of all organisms. Further, the diversity of life and its glorious uncertainty is the spice that makes life meaningful. Present day world consists of beneficial organisms, harmful organisms, bad individuals, good individuals etc. Variety should be there, only then one can understand the difference. If every thing is sought to be made uniform then the very essence of life is lost. Man should seriously think whether it is in the interest of nature and his own interest to pursue with unbridled genetic engineering research. Because as per the process of natural selection the genes that are not expressing in a given set of environmental conditions might express in a new environment and who knows what do they contain.

IMMUNOLOGY

All living organisms get infected and suffer from malfunctioning of the body resulting in the appearance of diseases. In some cases however, organisms even when they are exposed to infectious agents remain healthy and do not contract the disease. Some organisms on the other hand get diseases even on a brief exposure to infectious agents. In other words some organisms are resistant to diseases, while some are susceptible or sensitive to the diseases. The question that has to be answered here is what makes some organisms to be resistant or *immune* to the diseases, while others are not so ?

Immunology is a branch of microbiology that tries to answer the questions posed above. It deals with the defence mechanism of organisms against the infectious agents that keeps the organisms healthy. Immunology analyses, the biochemical or molecular basis of immunity or the absence of it (in special cases as in AIDS). Immunology also stuides the genetic basis of the origin of immune system. Lastly Immunology also deals with offering immunity to succeptible individuals so that they can remain healthy due to 'acquired immunity'.

Immunology thus plays a significant role in human health being very important in the treatment of various systematic and infectious diseases. Of late techniques of Immunology are applied to industry, agriculture, horticulture etc. where they have has proved their role in the diagnosis of many problems.

Types of Immunity

The protection or immunity to the organism against the infective agents is of two types - *non specific resistance* and *specific resistance*. Non specific resistance provides protection to the individual against a wide range of pathogens. In otherwords, the immunity is broad based and not specific to any particular agent. Specific resistance on the other hand provides immunity against particular pathogens and has a well defined defence mechanism. It is this kind of resistance that forms the base of Immunological studies.

Non specific resistance

This kind of resistance is very broad based and involves a wide range of factors such as heredity, environment, nutrition, age, sex etc. Some individuals are nonsusceptible ie., they are totally immune to a number of diseases. This resistance depends on the general well being of the individual and proper functioning of the body organs. Non specific resistance or immunity can be called Natural immunity or innate immunity. This is of two types - *species immunity* and *racial immunity*.

Species immunity refers to a situation where a host species will be completely resistant to a pathogen. A disease affecting one species will not affect the other. For instance mumps which affects human beings does not cause any disease in cats and dogs. Similarly human beings do not contract hog cholera, while hogs do not get many human diseases. This kind of species specific resistance is due to their physiology, anatomy and hereditary factors.

Another kind of non specific immunity is racial immunity which exists among various races of the world. Such immunities have developed over the course of several hundreds of years that go to characterize a particular race.

In addition to the above types of non specific immunities that are based on common ground involving a large number of individuals, the body also possesses a number of general defence mechanisms to ward off the diseases. Some of these are -

1. Mechanical and chemical barriers

The best mechanical barrier against infection is the skin outside and mucous membrane inside. Unless there is a cut or abrasion in the skin, microbes cannot get into the body. Cells of the mucuous membrane along the lining of the respiratory tract secret the fluid - mucus, which traps heavy particles and microbes in the air. Cilia present in some cells

push them up to the throat where it is swallowed and is pushed into the stomach. Stomach acids destroy the microbes. Stomach acid with a pH of about 2.0 is a natural microbicide in the gastrointestinal tract. Bile from the gall bladder also inhibits microbes.

Another non specific chemical inhibitor in human beings is the enzyme lysozyme present in tear and saliva. It can digest the cell wall of bacteria.

Nonspecific mechanical and chemical barriers to disease

Resistance mechanism	Activity
Skin layers	providea protective covering to all body tissues.
Mucous membrane of body cavities	trap airborne particles in mucus
Acidity in the vagina and stomach	Acidic pH toxic to microbes
Bile	Inhibits most microorganisms
Duodenaal enzymes	Digest structural and chemical components of microbes
Lysozyme in tears, saliva, secretions	Digest cell walls of Gram-positive bacteria
Interferons	Inhibit viral replication

2. Phagocytosis. Elie Metchinkoff, a Russian scientist made a chance discovery to explain, how cells protect themselves against the invading microbes. He observed that motile cells in the larvae of starfish gathered around a wooden splinter placed in a mass of cells. He opined that the cells engulfed foreign particles and called the phenomenon *phagocytosis*. Metchinkoff received Nobel prize in 1908 for his theory of phagocytosis. Presently phagocytosis is regarded as a major type of non specific defence mechanism of the body. The cells are called phagocytes and are of two types - *monocytes* and *reticuloendothelial system cells*. (RES cells).

3. Inflammation. After a physical injury or exposure to agents that cause damage to the tissue, a swelling develops as a result of dilation of the blood vessels. Redness appears due to increased flow of blood to the region. Neutrophils of the blood vessel move close to the injury and begin phagocytosis.

4. Fever. Unusally high body temperature called fever which is often mistaken for a disease is actually an immune response to provide resistance to the disease. Increase of body temperature, increase in the rate of metabolism and blood vessels constrict restricting blood circulation to the skin. Increased blood supply within the body enhances the rate of phagocytosis.

Specific resistance :

Specific resistance to a disease is also called acquired immunity. This may be defined as an immunitv that an individual develops during his life time to a specific pathogen. Acquired immunity is usually based on the presence of certain chemical agents called antibodies which react against the chemical (antigen) induced into the body by the pathogen. Acquired immunity is the most specific, most efficient and target oriented disease resistance. Acquired immunity is classified into two types namely *actively acquired* and *passively acquired.*

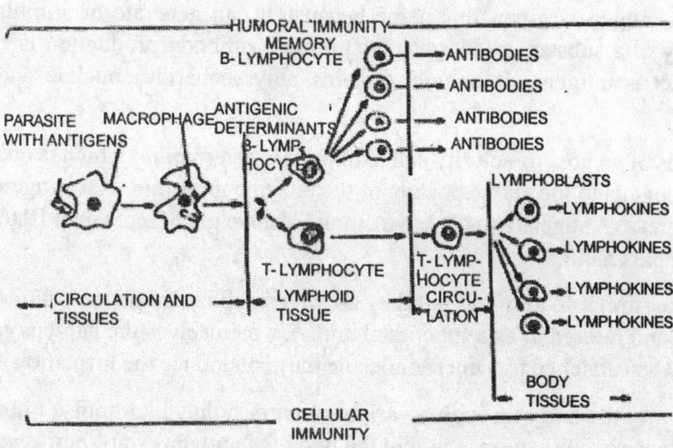

Fig. 26.1 Immune response to an antigen - through humoral immunity (B cells) and through cellular immunity (T-cells)

Actively acquired immunity. This may be either artificial or natural. Naturally acquired immunity results from any infection from which a person recovers. During the infection antibody production in the body is increased against the specific pathogen. As a result of this if the individual contracts the same pathogen the body will not get the disease as the antibodies already present destroy the pathogen.

Actively acquired artificial immunity is introduced into the body through a process called immunisation or vaccinisation. In this process, the pathogens (heat killed or weakened) are introduced in a quantity so as not to bring about a fulfledged disease. But the presence pathogen is sufficient to induce the production of specific antibodies. As a result of this the body will have sufficient antibodies ready to combat the disease if and when the pathogens enter the body.

Passively acquired immunity. This may be induced either by natural or artificial means. The best example of passive immunity is the transfer of antibodies from mother to the child through the placenta before birth. Artificialy acquired passive immunity involves the transfer of antibodies produced in one individual to the other through injections.

There are some basic concepts in the immunological response of the individual, these are -

1. Specificity. This refers to a particular response of the body to a particular pathogen.

2. Memory. This refers to the response of the body to a particular infective agent during the course of a second infection. This can be explained that when once an individual suffers from a disease and recovers from it subsequently the antibodies already present will prevent second infection.

3. Recognition of foreign particles. By means of chemical reactions the body has the ability to recognise agents that come from outside. Soon after recognition this foreign agent will be destroyed.

In specific immunity certain chemicals are involved (which are mainly proteins) which provide the immunological defence to the body. These are the antigen and the antibodies.

Antigens

These are chemical substances that are introduced into the body by the pathogen. They may be also called toxic substances. An antigen is given this name because it can generate or stimulate the production of antibodies. The ability of a substance (antigen) to stimulate antibody production is called *antigenicity*. A variety of substances act as antigens. It includes proteins, polysaccharides, nucleic acids, complex lipids and even viral particles.

The antigen consists of an area of activity called *antigenic determinant* which is necessary for its activity. This is some what analogous to the active portion of the enzyme molecule. An antigen may have one or two antigenic determinant areas. Antigens usually have a molecular weight of more than 10,000 daltons. Their large size indicates their complex nature.

of the **Haptenes** are similar to antigens but they are of a small molecular configuration. Usually they get attached to an antigen and refered to as a functional unit. By themselves the haptens cannot induce antibody formation. However when attached to a micromolecule they can induce the formation of a specific antibody.

Antigens usually enter the body through a variety of entry points like injury, bites of insects, abrasions, through the nasal passage etc. When once within the body the antigens start their reactions. There are three types of antigens these are -

1. **Autoantigens**. These are very rarely found. In this self is recognised as nonself and the body's own chemicals are destroyed by the antibody. this is called autoimmune disease.

2. **Alloantigens**. These are antigens introduced from oustide eg. Rh antigen.

3. **Heterophile antigens**. These are antigens which are similar but found in unrelated species of organisms.

Antibody

These are proteins synthesised by the body of organism as a result of the presence of an antigen. Antibodies are highly specific to antigens. All antibodies are proteins and have a special affinity to the foreign substances that have caused their production. They go in search of the antigens and destroy them. Antibodies are actually modified globulin proteins of the blood. Globulins present in the plasma of the blood called imunoglobulins are mainly responsible for the antibody activity.

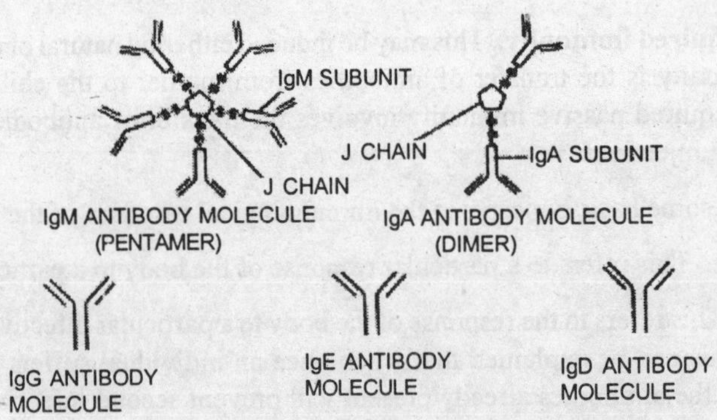

Fig. 26.2 The five types of antibodies

A basic antibody molecule consists of 4 polypeptide chains. Of these two are heavy (H) chains and two are light (L) chains. These chains are joined together by disulphide linkages to form a Y shaped structure. Each H chain consists of 400 amino acid residues while a L chain consists of 200 aminoacid residues.

Five types of antibodies have been identified so far based on the differences in the H chains. These five classes of antibodies (immunoglobulins) are **IgM, IgG, IgA, IgE** and **IgD**. IgM is made up of five subunits and its segments are connected by a glycoprotein called the J chain. IgA has two subunits and the two units are connected by a J chain. The remaining three globulins have only one subunit each.

IgM is the first antibody to appear in circulation as an immune response. It is the largest antibody molecule. IgG is also called the gamma globulin and is the main antibody in circulation. It comprises of upto 80% of the total antibody in the serum. IgG appears about 24 - 48 hours after antigenic stimulation. It is also called the secondary antibody response. IgG is also the main antibody that is given to the child from the mother when the foetus is in the womb. IgA comprises 10% of the total antibody component in the serum. IgE plays an important role in allergic reactions by sensitising the cells to antigens. The exact role of IgD is not properly understood at present. It may perhaps stimulate the B - lymphocytes.

Antigen - Antibody reactions

In order to provide immunity the anitbody should react with the antigen and alter its make up in such a way that it is incapactated i.e., rendered inactive. An antibody inactivates the antigen in a number of ways. The following are some of the mechanisms involved in the incapacitation of the antigens.

1. Neutralization. Certain antibodies called neutralizing antibodies react with antigens such as viruses and prevent them from attaching to the host cells. Some neutralizing antibodies clump viruses together and allow them to be engulfed by the phagocytes.

2. Other reactions. Some antibodies called agglutinins react with the surface of the antigen resulting in their clumping or agglutination. This will facilitate the phagocytes to absorb them. A typical example of agglutination, reaction can be seen when blood groups of different types are mixed.

There are some antigens called precipitins which react with the liquid antigens and convert them to solid precipitates. *Opsonins* are the antibodies that stimulate phagocytosis by direct interference.

3. Complement system. In this kind of reaction a group of proteins called complement system bring about antigenic inactivation. Described in 1895 by Jules Bordet, this system involves 11 proteins that function in a series of reactions. This exists in all normal sera and is activated by IgM or IgG.

TYPES OF IMMUNE SYSTEMS

The immune mechanism involves special types of cells of the blood called Lymphocytes. There are two kinds of immune systems both connected to lymphocytes. These are *humoral immune system* or *humoral immunity* and *cell mediated immune system* or *cellular immunity*. In humoral immunity the lymphocytes produce antibodies which bring about immunity while in cell mediated immunity the lymphocytes directly react with the antigenic material.

Structure of Lymphocytes

There are two kinds of lymphocytes namely the B-lymphocytes and the T-lymphocytes. In the adult organism both types of lymphocytes are derived from the common primordial cells of the bone arrow called the *stem cells*. In the foetus they are formed in the yolk sac, the liver and possibly in the thymus also.

The T-lymphocytes also called T-cells (thymus dependent lymphocytes) migrate to the thymus where they come under the influence of the hormone *thymosin*. Interaction with thymosin makes these lymphocytesimmunologically competent. T-cells are mainly involved in cell mediated immunity like rejection of foreign tissue.

The B-lymphocytes are so called because in birds they develop in a mass of lymphoid tissue (bursa of Fabricius) near the cloaca. In mammals however there is no such tissue but the corresponding tissue may be the lymphoid tissue of the intestine, appendix and tonsils. The B-cells are mainly involved in humoral immunity.

Humoral immune system

It has already been pointed out that in this system the lymphocytes destroy the antigens through agents called antibodies. When B-lymphocytes come into to contact with antigens they synthesise RNA and differentiate themselves into *immunocytes* or *plasma cells*. These cells produce immunoglobulins a particular immunocyte can secrete only one type of antibody.

The conversion of B-lymphocytes into plasma cells is through several successive cell generations. The immature plasma cells called *plasmablasts* undergo 8 cell generations and ultimately they get differentiated with a specific type of cisternae of endoplasmic reticulum filled with antibody molecules. The plasma cell contains a unique protein synthesing machinery that can turn out only one kind of a protein.

The immune response in a humoral immunity is explained on the basis of **Clonal selectional mode**. The essential features of the model are as follows -

1. The aminoacids sequence in the immunoglobulin chain is determined by a unique base sequence of the DNA in the globulin producing cell.

2. The kind of antibody produced is determined before the cell comes in contact with an antigen.

3. As the maturation to plasma cell proceeds antibodies produced will attach to the cell membrane acting as receptors.

4. A mature cell is stimulated on coming into contact with an antigen. The antigen molecules are attracted towards the immunoglobulin receptors on the cell surface and get clumped. The cell continues to produce more quantities of antibody. The daughter cells produced from this form a clone and secrete the same antibody until the antigen is destroyed.

5. The cloanl cells remain ever after the antigen disappears. In future if they come into contact with the same antigen antibodies are produced readily.

Cell mediated immune system

The presence of an antigen might induce a lymphocyte to proce antibodies or it may sensitize the lymphocytes to act directly. In cell mediated immunity sometimes called the tissue immunity the lymphocytes react with the antigenic material directly. Certain T-lymphocytes become senstitizied to particular antigens and when they come into contact with them produce substances called lymphokines, which are low molecular weight proteins. These substances draw phagocytes to the area where antigens are present. Lymphokines have several chemical factors. One factor called the Chemotactic factor (CF) attracts the phagocytes while another factor called migration inhibition factor (MIF) prevents macrophages from moving away. A third factor called macrophage aggregation factor causes thephagocytes to clump together at the site. A fourth kind of factor called macrophage activating factor (MAF) increase the mobility of the phagocytes and induces secretion of lysozyme enzymes in them. The overall influence of lymphokines is to increase the efficiency of phagocytosis. Interleukin is a lymphokine produced by the white blood cells. Some of the interlukins have the effect on other white blood cells like stimulating their maturation. Interlukin2 is supposed to activate the rapid growth and division of T-lymphocytes. This interlukin has practical application in the treatment of tumours.

Lymphokines usually disappear soon after the elimination of antigens but a few cells remain to combat any future infection. These are called memory T-lymphocytes.

Cellular immunity is a chief method of resistance to bacterial diseases such as leprosy and tuberculosis, fungal disease like candidiasis and to many protozoan and helminthic parasitic diseases.

Applications of immuno techniques

Immunological techniques have a wide range of application in medicine, industry and also in agriculture. Some of these techniques are monoclonal antibody or hybridoma technology and ELISA technique which has wide applications in medical diagnostics as well as agriculture.

Monocloanl antibodies (hybridoma technology). Whenever any micro or foreign material enters our body, as a defence mechanism our body responds by the production of antibodies. Antibodies are manufactured by specialised cells called B-lymphocytes. These antibodies circulate in the body and destroy the antigen. Antibodies are extremely specific in their interaction and react with the specific antigen only.

For the purpose of human health care and research study of antibodies has become extremely important. Production of specific antibodies have wide applications in the production of vaccines and also in diagnosis.

One method of studying antibodies is collecting the serum and obtaining antibodies from them. The serum however contains and number of antibodies. For the purpose of research the antibodies are usually obtained from hyper immunising the animal with antigens and obtaining the antibody from the serum of such animals. As has been pointed already antibodies obtained by this method are heterogenus i.e., they are a mixture of several antibodies. Such sera cannot be used for the study of a specific problem because of the interference of several antibodies. Specific problems however can be studied by the use of a pure preparation only one kind of antibody. These are called monoclonal antibodies so that they have a specific kind of a reaction to an antigen. A monoclonal antibody is produced by a clone of cells and is a collection of pure antibodies specific against a particular antigen.

The ideal of monoclonal antibodies however is difficult to achieve because the clonal cells producing a particular kind of antibody may not divide at a desirable rate so that the quantity of antibody produced would be insufficient for the purpose of study. There is a practical problem of making the specific B-lymphocytes to divide at a rapid rate on an artificial medium. Soon after they are removed from the body the lymphocytes die hence it is difficult to make them survive and divide on a culture medium.

Hybridoma technology. Kohler and Milstein have found out an ingenious method of making the lymphocytes survive and divide on the artificial medium. Their technique was to fuse the lymphocyte cells with cancer cells to obtain hybrid cells which can remain alive for long and keep on giving pure antibodies. This technology called hybridoma technology involves fusion of a lymphocyte cell with a single myeloma cell (bone marrow tumour cell) which is capable of multiplying indefinitely. These fused cells are hybridoma have the ability of the production of antibody obtained from lymphocytes and the ability to grow indefinetely inherited from the cancer cells.

Following is a summarised method for the production of monoclonal antibodies through hybridoma.

1. Isolate the antigen against which monoclonal antibodies are to be raised.
2. Immunise the mouse with the specific antigen of interest for raising immune response.
3. Harvest antibody producing lymphocyte cells from the spleen of an immunised mouse.
4. Mix these immune lymphocyte cells with suitable myeloma cells and add cell fusion agent like polyethylene glycol.
5. Distribute the fused cells in separate selective tissue culture conditions to raise hybridised cells.
6. Screeing and identification of cultures containing (monoclonal) antibodies of a specific type.

7. Repeated cloning of cells from these cultures and production of particular monoclonal antibodies.

8. Freezing stocks of these cultues in liquid nitrogen for permanent preservation and future use.

9. Growing cloned cell lines as hybridomas and deriving the monoclonal antibodies as and when needed.

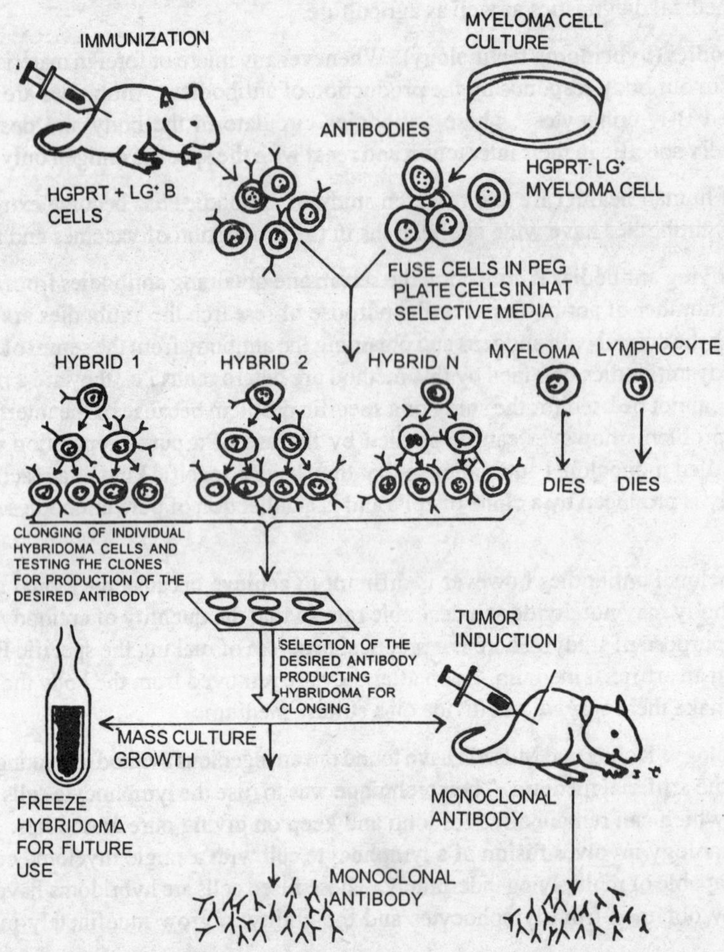

Fig. 26.3 Various steps involved in the production of monoclonal antibodies using hybridoma technology

Monoclonal antibodies can also be produced without involving hybridoma technology. These techniques involve the isolation and identification of antibody producing genes from lymphocytes of immunised animals and cloning them in a vector. These antibodies produced as a result of genetic engineering can be screened for binding to specific antigens.

Application of monoclonal antibodies. Monoclonal antibodies find applications in several fields such as -

1. For diagnostic purposes.

2. For boosting natural defences of patients.
3. In cancer therapy.
4. For immunopurification of drugs (before giving to patients).
5. For reducing transplant rejections.

1. Diagnosis. Monoclonal antibodies can be used for the identification of blood groups, for the detection of viruses for immuno purification purposes and also for therapy. Pregnancy can be detected by assaying of hormones with monoclonals. Similarly monoclonals can be used for detecting pathogens.

2. Cancer therapy. Monoclonal antibodies are used for diagnosisng and monitoring of the effectiveness of different types of cancer therapy. The success of therapy is detected by measuring the decreasing of cancer markers in the blood sample using monoclonal antibodies which are specific to different cancer marking antigens.

3. Organs transplants. According to Bill Stimson monoclonal antibodies will be able to improve the chances of survival of transplanted organs. Transplant patients produce lymphocytes against the transplanted organ. Monoclonal antibodies can be used specifically to block the killer lymphocytes which bring about organ rejection.

4. Purification of drugs. Monoclonal antibodies can help in high level purification of medically important sustances. Immuno afinity columns containing monoclonal antibodies coupled to cyanogen bromide activated sepharose can help in higher purification of interferons.

ELISA technique (enzyme linked immunosorbent assay)

This is a technique based on the antigen antibody reactions using 'monoclonal antiboes'. Before the emergence of ELISA, there were several other techniques-Radio immuno assay, radioallergosorbent test etc. which used monoclonal antibodies for various tests. These tests however require use of radio active labelled compounds and though sensitivie, were costly and sometimes hazardous also.

The main principle of ELISA is based on the ability of low molecular weight antibodies to couple with enzymes to produce enzymatically active immulogical compounds. This allows the detection of immune reaction with histochemical staining techniques. In this method antigens or antibodies are attached to a solid surface and the coated surface are combined or immunosorbed with the test material. An enzyme system with the antibody is linked to the complex. After the experiment the enzyme system is washed away and the extent of the enzyme activity is measured. This indicates the presence or absence of antigens or antibodies in the material.

In the detection of antibodies using ELISA test a monoclonal antibody is coated on to the surface of the solid substance. This is then allowed to react with a test compound supposed to contain the specific antigen. This complex is then allowed to interact with a specific antibody coupled to an enzyme. The antibodies combine with the antigen if present and accumulates on the surface along with the enzyme. This complex is now allowed to interact with the substrate of the enzyme upon which a specific color reaction is produced as the substrate gets converted. Based on this reaction the presence of antibody for a specific antigen can be confermed for instance the antigen present is not specific to the antibody the enzyme coupled antibody will not react with the antigen, as a result the enzyme will not stick to the surface and does not produce any color reaction to the substrate because the enzyme will not be there. This is an indirect method of proving the presence of an antibody by observing the reaction of the enzyme with the substrate.

ELISA technique is very simple, efficient and inexpensive. A number of variations of the basic technique mentioned above have been in practise and are used in various applications. Specific commercial ELISA kits are available for various kinds of tests.

Applications of ELISA

1. Detection of diseases via antibodies. ELISA technique is employed in the detection of several diseases including AIDS. In the diagnosis of AIDS the blood of the suspected patient is tested for the presence of HIV antibodies. The presence of HIV antibodies is an indirect proof of the presence of HIV virus. ELISA can also be used for the detection of hepatitis B surface antigen. The following is one of the ELISA methods to detect the presence of gonorrhea.Plastic beads coated with monoclonal antibodies of gonococci are combined with a sample from the suspected patient. If the sample consists of gonorrhea antigen it combines with the antibodies of gonorrhea on the surface of the bead. The beads are then incubated with the antibodies of gonorrhea linked to the enzyme horse raddish peroxidase. The antibodies linked to the enzyme combine with the antigen on the bead surface along with the enzyme. Next the substrate of the enzyme is added to the incubated system. As the substrate gets converted a yellow orange color develops indicating that the enzyme has converted the substrate. If the enzyme has to remain on the surface of the bead it should have specific antibodies for the gonococci antigen. Hence the enzyme reaction indirectly proves the presence of antibody. If no color develops that means the enzyme does not remain on the surface of the bead because there was no specific antigen to the antibody linked to it (enzyme).

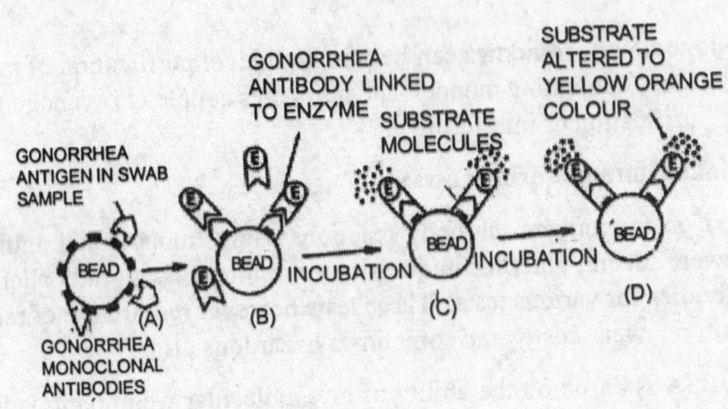

Fig. 26.4 ELISA technique to detect gonorrhea

2. *Application in the field of plant pathology.* ELISA has been used widely for detecting plant viruses. This is of great use while identifying symptomless viruses which will otherwise cause diseases at a later stage. In a large scale multiplication such as micropropagation if the material has latent infection all the seedlings obtained from it will develop disease at a later stage. This will cause a great loss to the horticulturist as he is unware that he is propagating a disease carrying material.

A selection of enzymes & substrates suitable for ELISA

Enzymes	**Detection systems and selected substrates**		
	Visual	**Fluorescent**	**Radioactive**
Alkaline Phosphate	P-nitrophenyl phosphate	Methylumbelliferly phosphate	Tritiated adenosine monophosphate
Peroxidase	3, 3, 5, 5 tetra methyl benizidine	nictoinamide adenine dinucleotide	
B-galactosidase	nitrophenyl galactose	methylumbelliferyl galactose	tritiated galactose-6 phosphate
Glucose Oxidase	Glucose + 5 amino salicylic acid		
Urease	Urea		

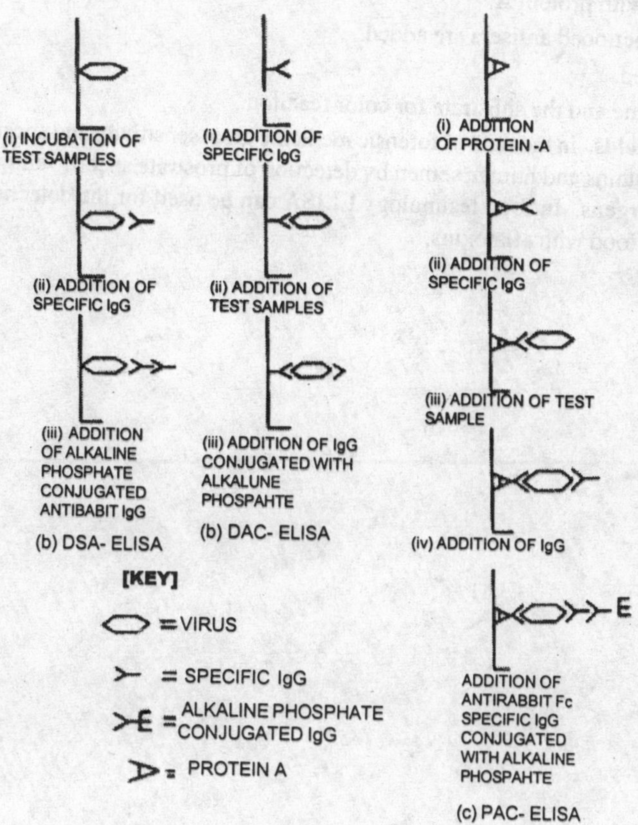

Fig. 26.5 Different types of ELISA for the detection of virus infections

At least three different **ELISA** techniques are available for detection of plant viruses. These are - **DAS-ELISA, DAC-ELISA** and **PAC-ELISA.** In all these techniques a polystyrene microtitre plate is used.

DAS-ELISA. In this technique called double antibody sandwitch ELISA the wells of the plate are coated with antibody prepared from antisera. The test sample is added to the well to allow trapping of the virus antigen. Into the well next is added Ig combined enzyme which will attach to the virus antigen. A substrate of the enzyme is now added to the well to produce color reaction. The intensity of the color if any produced can be measured spectrophotometrically which is an indication of the presence of the virus.

DAC ELISA. This is called the direct antigen coating ELISA. In this, plant extracts are obtained in a carbonate buffer and are introduced into the well directly. In the next step diluted, unfractionated antiserum is added. Ig antibody attached to the virus antigen is detected by the addition of enzyme Ig conjugates.

PAC ELISA. This is called protein A coating techniques and involves four steps.

1. The wells are coated with protein A.
2. High dilution of unfractioned antisera are added.
3. Test samples are added.
4. Addition of the enzyme and the substrate for color reaciton.

Applications in other fields. In the field of forensic medicine great sensitivity and specificity is achieved for the identification of blood stains and human semen by detection of prostrate specific antigen. ELISA can also be used to test specific allergens. In food technology ELISA can be used for the detection of adulteration of meat and contamination of food with aflatoxins.

27

ANALYTICAL INSTRUMENTS

In the previous chapter we have studied various types of microscopes that form a key components of microbiological instruments. In the microbiology laboratory, there are various other analytical instruments which are equally indispensible like the microscope. These instrument are necessary to understand the various parameters of microbial life. Some of these instrument are

1. Centifuge
2. pH meter
3. Colourimeter
4. Nephelometer etc.
5. Spectrophotometer

CENTRIFUGE

A centrifuge is an instrument which can produce centrifugal force because of accelerated rotation around an axis. This instrument is used in the microbiology laboratory mainly to separate out particles or heavier or lighter liquids that are mixed with the original liquid. Essentially, the centrifuge uses the principal of centrifugal force to separate components of varying densities. Apart from the biology laboratories, centrifuge finds applications in many industries like food processing industry to separate the unwanted particles from food. The washing machine used for washing clothes also makes use of the centrifugal force in the spin dry cycle.

Principle of Working

It has been mentioned above that centrifuge makes use of centrifugal force and brings about sedimentation (separation) of particles based on their density. In order to understand the principles of working of a centrifuge we must first understand what is meant by sedimentation and centrifugal force.

Sedimentation : When you put a measure of sand particles into a beaker containing water, the particles sediment at different rates depending on their sizes and density. Instead of water, if oil used, the rate of sedimentation is lowered. By this we understand that sedimentation depends on

(i) the density of the particle

(ii) the size of the particle and

(iii) the viscosity of the medium

In addition to the above, one more factor viz., gravitational pull is also involved in sedimentation. Under normal circumstances, gravitational pull is in the order of 980 cm/sec^2 or 1g unit. This gravitational pull can be increased by applying a force called centrifugal force.

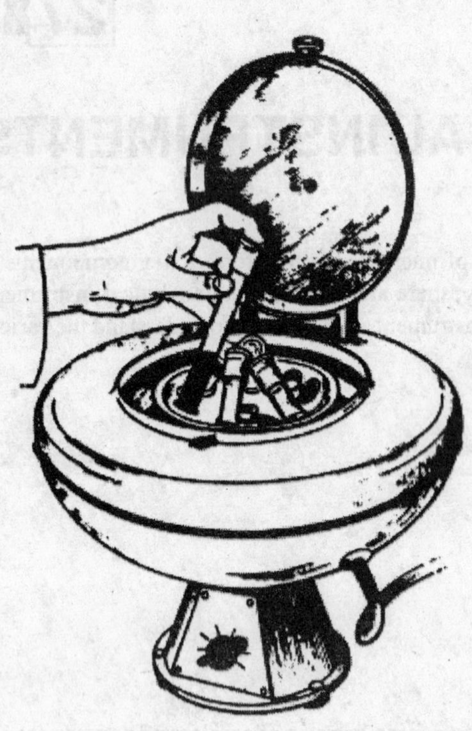

Fig. 27.1 Placing tubes in a clinical centrifuge

Centrifugal force :

Centrifugal force may be difined as the force with which a body going round in a circle or curve wants to fly away from the centre of the circle. There are a number of familiar examples of centrifugal force. In an automobile speeding around a curve, one feels pulled over to oneside. The classical sling shot subjects the stone to a centrifugal force.

Issac Newton while formulating the laws of centrifugal force, stated that a moving body tends to travel in a straight line and if diverted to a curved path will exert a force against the restricting vector, in an attempt to revert to a straight tangential course. Centrifugal force measured in dynes, acts along the radius of the path in which the mass is being carried at a given speed. This speed or force can be calculated as follows.

$CF = w^2 \times r$

where CF = Centrifugal Force

w = angular velocity and

r = radius

Angular velocity is related to rotations per minute (rpm) as per the following formula

$$w = \frac{2\pi \times rpm}{60} \text{ radians per second}$$

The centrifugal field is usually expressed as relative centrifugal field in g units. This is calculated as follows.

$$RCF = \frac{4\pi^2 (rpm)^2 r}{3600 \times 980} \text{ g units}$$

(RPM refers to the revolutions per minute of any object, while g units refer to the extra gravitational force (over and above the normal g force of 980 cm/sec^2) that the object is subjected to due to centrifugal force)

To the biologist, the important aspect of centrifugal force is the haviour of particulate matter suspended in a liquid. Particles denser than the suspending liquid will migrate towards the periphery of the centrifugal field. The speed of migration is related to the strenth of centrifugal force and also the size and density of the particles besides the density of the liquid.

Parts of a centrifuge

Laboratory centrifuges vary considerably in size and shape. The simplest ones are hand driven two tube models while the complicated moderncentrifuges are motor driven, automated and also refrigerated.

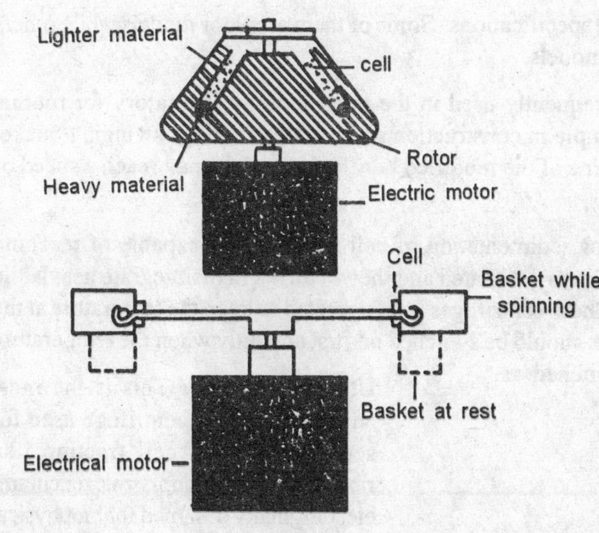

Fig. 27.2 Centrifuges with rotor and with baskets

The parts of a common centrifuge include the chamber which encloses the internal parts, a cover with a latch, the centrifuge head with rotors, the shaft which turns the rotor and a motor drive assembly. Most of the centrifuges have a power switch, a brake assembly, speed control, automatic timer with switch of facility and sometimes a tachometer also.

Centrifuge motors are general DC motors which turn faster with the increase in voltage. Occassionally AC motors are also used and speed control is achieved through step wise reduction of number of poles in the magnetic field. The motors employ a direct drive system to the head through the shaft.

The speed of the centrifuge is controlled through a potentiometer which increases or decreases the voltage supplied to the motor. The calibrations provided on the speed control of a centrifuge are often only voltage increments and are not accurate indicators of speed.

Some centrifuges have Techometer to indicate the speed of the rotor in RPM. A fexible cable or shaft attached to the motor spindle rotates inside a flexible housing to which the Tachometer is attached. As the cable turns the needle moves up in the scale. In some of the centrifuges, electric tachometers measure the speed. Many modern centrifuges have a digital display of the speed.

A timer is uaually provided in all the modern centrifuges. Most of them have a spring driven clock work mechanism that shuts off power supply at the end of a fixed period. Some centrifuges may have an electronic timer also.

High speed centrifuges have braking devices to provide rapid deacceleration of the rotor. Braking devices may be mechanical or electronic Mechanical braking device physically applies pressure on the motor while the electronic one, reverses the polarity of the current to the motor and slows it down.

Rotors : These are attached to the shaft of the motor and are meant to keep the samples for spinning. Rotors are basically of two types - Anylytical and Preparative. Rotors used for analytical purposes are constructed in such a way so that light, which is used for determining the position of the solute boundary can pass through the sample and the container when the rator is spinning.

Preparative rotors are of three types - (a) Angle, (b) swinging bucket and (c) Zonal. The first two are more frequently used than the third. In the angle rotor, sample tubes are inserted into cavities in the rotor, that are constructed at an angle to the axis of rotation. The rotor head is a solid block of metal, usually made of aluminium. In the swinging bucket rotor, the sample tubes are placed in holders which is attached to the rotor body by a pair of yokes. The metal holders are allowed to swing. While at rest, the container tubes are paralled to the rotor axis, but at operational speeds, they will be perpendicular to the rotor axis.

The rotors are designed in such a way that they can withstand the centrifugal force. They also should be light in weight. Another important criterion is that the opposite sides of rotor must be properly balanced in weight.

Types of centrifuges

Centrifuges are available in various models and specifications. Some of them are floor models while others can be kept on tables. These are called bench top models.

The bench top model centrifuge is the one frequently used in the microbiology laboratory for routine spinning work. A standard bench top model is simple in construction and usually employs swinging bucket rotors, capable of acheiving 3000 -5000 RPM. Some of the modified bench top models may reach a speed of 16,000 RPM.

Advanced centrifuges which are necessary for sedimentation of cell fractions are capable of reaching 20,000 RPM. But at this speed, due to friction the rotors get heated and they in turn will disintegrate heat labile samples like enzymes. Hence the rotor chamber in these centrifuges is refrigerated to keep the temprature at the required level. In these contrifuges, the refrigerators should be switched on first and only when the temperature reaches the required range, the rotor should be switched on.

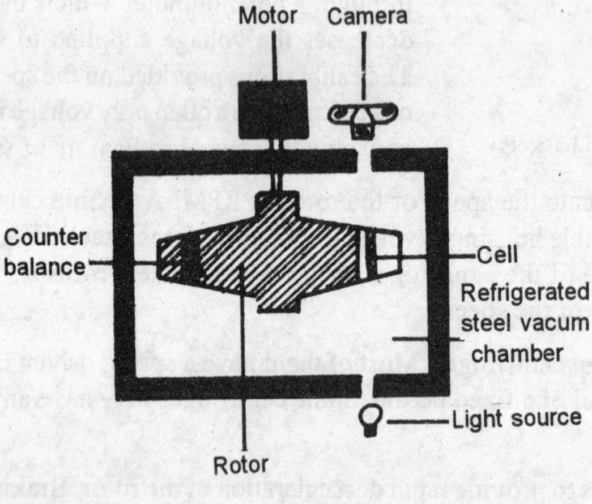

Fig. 27.3 Analytical ultracentrifuge

Ultracentrifuge : This is the most advanced type of centrifuge used for sedimentation of cell fraction like ribosomes, ER (endoplasmic reticulum) etc. Originally designed for biophysical work to study the sedimentation rates of protein molecules, the ultracentrifuge is now a common instrument even in microbiology laboratory.

The ultracentrifuge was first designed and constructed by Theodor Svedberg in 1926 and since then has undergone many modifications and refinement. The modern ultracentrifuge is capable of attaining 50 milion g's with rotor speed greater then one million RPM. There are two basic types of ultracentrifuges *Preparative and Analytical*

The preparative ultracentrifuge is basically a conventional centrifuge modified to attain speeds upto 75,000 RPM and forces up to 15,000 g's. These centrifuges are used for the separation of minute components from a mixture. Density gradients are also run in preparative ultracentrifuges. Ultracentrifuge has to be mandatorily refrigerated and the rotor chamber should be ecacuated to avoid air friction. The rotor heads are sealed and hold tubes upto 250ml of the sample.

The analytical ultracentrifuge posseses a tiny rotor that can accomodate only mili iter quantity of the sample. It is used for quantitative measurement of sedimentation rates. Because of the frictionless bearing, the unit is capable of reaching very high speeds not possible even in preparative ultracentrifuge. The centrifuge is equipped with optical system with optical windows and photographic equipment which permit observation and recording of the sedimentation rate in a time sequence. Stroboscopic synchronization of the rotor head gives the visual effect that the rotor is stationary.

Care of centrifuge

1. Balance the container with sample before placing them in the opposite cavities of the rotor.
2. The sample tubes should not be overfilled
3. The rotor shuld be checked for any damage before spinning
4. The speed given for a rotor should not be exceeded 6
5. The unit should be switched off (power) should any irregularity noticed during operation.
6. Rotors and chambers must be absolutely clean.
7. Rotors should be checked for any metal fatigue. After a certain amount of use, the rotors should be used at less than the allowed speed.

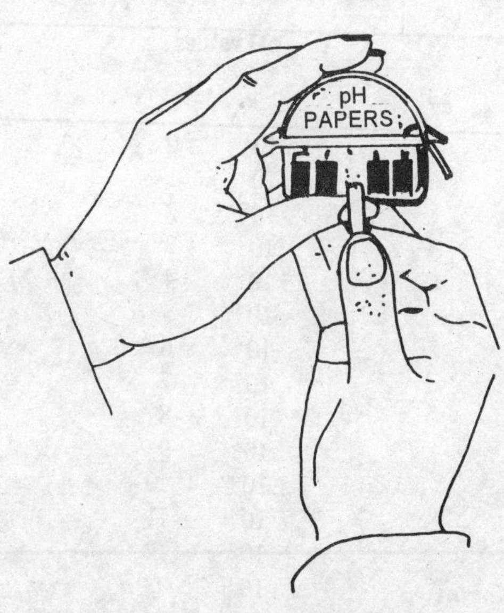

Fig. 27.4 Taking pH of a liquid by using pH paper.

pH METER

In many microbiological experiments, the culture medium must be of a specific type with reference to alkalinity or acidity. This is necessary to culture specific microbes. Similarly during enzymatic studies, the enzymes are to be harvested and studied in a particular medium with a specific pH, otherwise it (enzyme) will be inactive. When buffer solutions are prepared care is taken to see that they have a specific pH and the pH is checked with the help of an instrument called pH meter. We will first understand what is pH and its importance before studying the working of a pH meter.

What is pH ?

The hydrogen ion concentration of any solution is called the pH. It was Arrhenius who advanced the idea of electrolytic distociation of acids and bases in water to form ions. But the whole of acid or base will not be present in ionic form. Some quantity is still present in molecular form. for instance, a normal solution of HCl has only 78% molecules in ionized condition. The term normal is applied to the whole of hydrogen whether present in the form of ions or in the combined form. Hence in terms of hydrogen ions a solution of HCl is only 0.78 N.

All aqueous solutions and water normally contain some concentration of H^+ions. Pure water is (1/ 10,000,000 N) with respect to H^+ions at - 22^0 c. In other words, the hydrogen ion concentration of pure water is 10^{-7}. If we add some acid to water, the hydrogen ion concentration increases. For example the concentration may increase to 10^{-2}. In such a situation the OH^- concentration gets reduced to

10^{-12}

The hydrogen ion concentration is generally expressed as 10^- Sorensen suggested that the term pH can replace 10. The term pH stands for potential of hydrogen. pH is defined as the negative logarithm of the hydrogen ion concentration. The pH scale covers a range of values from 0-14. A litre of pure water has a hydrogen ion concentration of 0.0000001 or 10^{-7}. Since pH is equal to the negative logarithm of the hydrogen ion concentration, then

$$pH = -\log 10^{-7}$$

$$pH = -\log \frac{1}{10^{-7}} = 7$$

Pure water then has a pH of 7 and is considered neutral values of pH below 7 are acidic and above 7 are basic

A solution with a pH of 8 has hydrogen ion concentration ten times less than a solution with a pH of 7 i.e. its hydrogen ion concentration is 0.00000001 or 10^{-8}. The values always differ by a factor of 10. A chart of pH values is given in the following table.

Table pH scale

Hydrogen ion Concentration in terms of normality	pH values	
1	10^{0}	0
0.1	10^{-1}	1
0.001	10^{-2}	2
0.0001	10^{-3}	3
0.00001	10^{-4}	4
0.000001	10^{-5}	5
0.0000001	10^{-6}	6
0.00000001	10^{-7}	7
0.000000001	10^{-8}	8
0.0000000001	10^{-9}	9
0.00000000001	10^{-10}	10
0.000000000001	10^{-11}	11
0.0000000000001	10^{-12}	12
0.00000000000001	10^{-13}	13
0.000000000000001	10^{-14}	14

Measurement of pH ?

The measuren ent of pH for the determination of acidity or alkalinity of a solution were formerly accomplished with colour changing indicators (pH papers).

Many a time however the shades of colour or a subtle change is difficult to determine. Electronic instrumentation however has eliminated this drawback by sensing the ion concentration and indicating pH value on a meter. Some of the modern instruments have a digital display of pH. The instrument for the measurement of pH is called the pH meter.

A pH meter essentially consists of a pair of electrodes (fig 4.5) connected to a meter scale indicating the pH value. The pH value is determined by measuring the direct current voltage produced between these electrodes when immersed in a solution. The electrode pair is made up of a reference electrode (calomel electrode), that produces a constant potential, independent of the pH of the solution and a measuring electrode (glass electrode) in which the potential svaries with the pH of the solution.

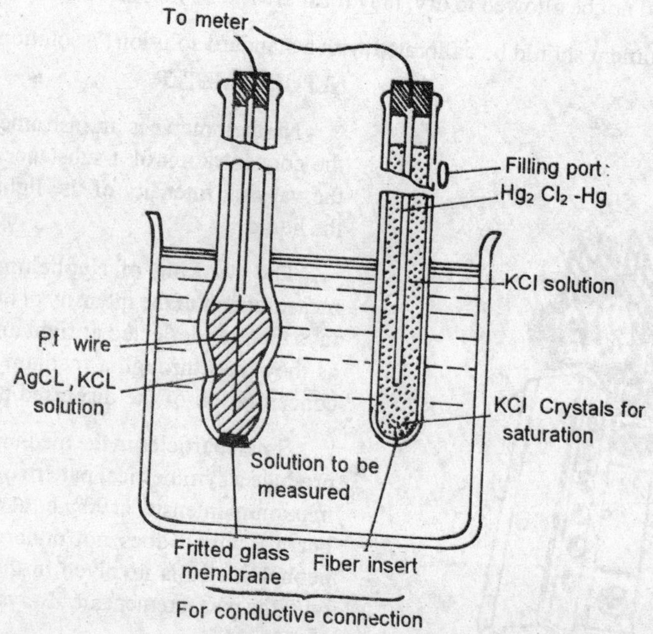

Fig. 27.5 pH meter electrode system. The electrode measure the acidity or alkalinity of a solution through a range of 1 to 14.

The calomel electrode consists of a glass tube containing liquid mercury and mercurous chloride (calomel) in a reservoir filled with a saturated solution of potassium chlorode. The glass electrode contains a silver chloride coated platinum wire immersed in a potassium chloride and silver chloride solution. Electrical contact is made by each of the electrode combinations with the solution at the tube end.

As the electrode pairs are in contact through the solution (whose pH is to be measured) the electrical output of the complete unit is the algebraic sum of the two electrode potential of the calomel electrode is constant that of the glass electrode varies depending on the solution. This potential changes 59 millivolts per pH unit of the unknown solution. Thus when the voltage flow is directed to a suitable multivoltameter, the pH value is directly indicated.

Electrodes are of different sizes depending on the quantity of the sample to be checked. While standard pH meter might have electrodes for use in 10-15 ml of a solution, some micro electrodes are very tiny and they can check samples as little as 1 to 0.5 ml.

Care of the pH meter

1. Glass electrodes when used for the first time should be dipped in distilled water or 0.1 mol / Hcl solution for several hours before use.

2. The solution should be throughly stirred before measuring the pH.
3. The temperature of the solution should be kept constant as pH is heat sensitive.
4. After every measurement, the glass electrode must be throughly washed with distilled water.
5. The electrodes should not be allowed to dry; they must always be immersedin distilled water.
6. Before use, the instrument should be calibrated with a standard solution (a solution of known pH).

Fig. 27.6 A pH meter

NEPHELOMETER

Nephelometer is an instrument used for analysing the concentration of a substance in a liquid, based on the varying intensity of the light as it passes through the liquid.

The working of Nephelometer is based on the measurement of the intensity of the scattered light (light rays that get deflected at right angles to the direct rays as they pass through a medium) as a function of the concentration of the dispersed phase.

Small particles in the medium undergo scattering to produce a symmetrical pattern of secondary rays with a maximum intensity at 90^0 i.e., at right angles to the direct beam (which does not undergo deflection). Since nephelometry is involved in the analysis of deflected rays, the measurement are always done at 90^0 (secondary rays)

In Nephelometry, measurements are conducted based on the comparison between the intensity of light scattered by the sample and the light scattered by a standard suspension under identical conditions. Higher intensity of the scattered light as it passes through the medium indicates higher turbidity i.e., increased concentration of the sample. The standard (reference) turbidity indicator used in most nephelometric measurements is a Formazin polymer compound. The choice of this compound is due to its easily dispersible nature in water and uniform light scattering properties when compound with clay and other natural turbid materials.

The instrument essentially consists of a light source, a sample tube to hold the turbid suspension and a photocell.

The light source is a ungsten lamp, the light rays from which are focussed on to the sample tube holding the suspension to be analysed. As the light passes through the suspension, two kinds of light rays are produced- direct rays and deflected or scattered rays. Direct rays are absorbed by a light shield as they are not measuresed by nephelometer. The scattered rays which undergo scattering (by the suspension) to the extent of 90^0 are sensed by a phototube which is kept at right angles to the sample tube. The amount of scattered light that comes out of the sample suspension is derectly proportional to the concentration of the sample compound.

and separating the hyphae or filaments in a preparation. It can also be used for inoculating a stab culture.

COLONY COUNTER

In many microbiological experiments it will be necessary to count the number of viable spores, cells or colonies in a culture. This is required to find out whether a microbial colony can grow on multiply on a particular type of medium. In order to count the colonies on a petri plate an instrument called colony counter is used.

By counting the number of viable colonies one can find out the number of viable spores or cells in a particular culture since each cell or spore that is viable forms a colony.

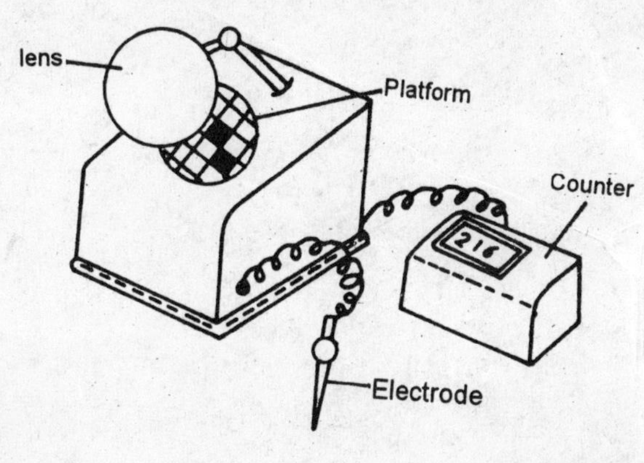

Fig. 27.14 An Electric colony counter

Construction of colony counter

Colony counters are of two types manual colony countes and electrical colony counters. The manual colony counter is made up of small cabinet having a circular grid in the center. There is a provision to illuminate the grid from below by a built in light source. Above the grid there is a magnifying glass which can magnify upto 1.5 times. This facilitates better counting.

Electrically operated colony counters are easy to use and avoid the possible mistakes in a manual counter. In most of the electrical operated colony counters there is a pencil shaped electrode and an LCD disply which immediately registers the number of colonies counted. There is a reset button which brings the display back to zero.

Procedure for using colony counter

1. Petriplate containing the colonies is placed on the circular grid of the colony counter.

2. Power is switched on to illuminate the grid.

3. Colonies are then counted manually in a manual counter.

4. In the electrically operated counters the pencil shaped electrode is removed from its position and each colony on the petriplate is touched.

5. As each colony is touched by the pencil electrode it is automatically registered and the number is displayed digitally.

In a special type of electrical colony counter called Qubec colony counter there is a counter switch which when depressed immediately displays the number of colonies on the petriplate on a digital read out.

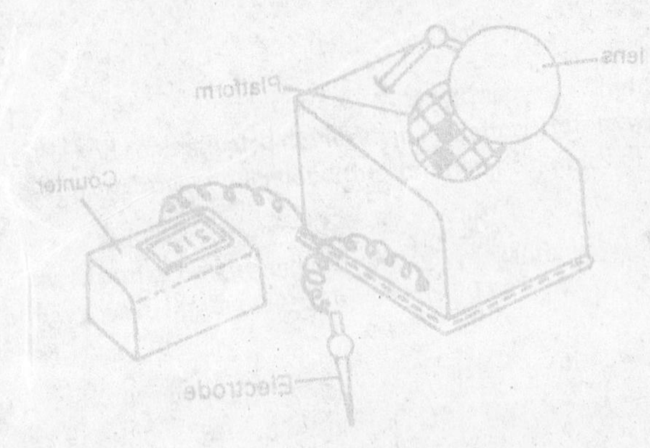

Measurement of turbidity.

The following reagents are required for measurement of turbidity.

1. Turbidity free water : Highly purified distilled water can be obtained by passing distilled water through a membrane filter with pore sizes not exceeding 0.2 micrometer. The water must have turbidity not more than 0.02 NTU (Nephelo turbidity unit)

2. Standard turbidity suspension (stock solution)

The stock solution is prepared with fornazin particles with a strength of 400 NTU. The stock solution is prepared as follows.

(I) Solution A : Dissolve one gram of hydrazine sulphate [$(NH_2)_2 H_2So_4$] in distilled water and make the volume to 100 ml in a volumetric flask.

(II) Solution B : Dissolve 10 grams of hexamethylene tetramine [$(CH_2)_6 N_4$] in distilled water and make the volume to 100 ml in a volumetric flask Standard turbidity stock solution is prepared as follows.

Mix 5 ml of solution A and 5ml of solution B in a 100 ml volumetric flask and allow it to stand for 48 hours at room temperature Dilute the solution with distilled water up to 100ml (up to the mark in the flask). This suspension will have a turbidity of 400 NTU and it can be used for 4-6 months.

3. Standard turbidity suspension can be prepared by taking 25 ml of stock and diluting it with distilled water to make the volume to 100 ml.

This is suspension will have a strength of 100 NTU. Turbidity readings are taken as follows.

1. Switch on the instrument and allow a warm up period of 10-15 minutes.
2. Select the turbidity range (i.e 0-100 NTU) using the selection knob.
3. Insert the sample tube with turbidity free distilled water into the holder and cover with light shield.
4. With the set zero control, adjust the meter so that it reads '0'
5. Remove the tube and replace it with standard turbidity suspension. Adjust the control so that the matter indicates 100 NTU, the actual strength of the suspension.
6. Replace the standard suspension with the unknown sample and read the turbidity directly from the meter reading in NTU scale.

In case the turbity of the sample is more than 100, use suitable dilutions with distilled water and calculate NTU as follows

$$NTU = \frac{A \times (B+C)}{C}$$

Where A = NTU found in diluted sample

B = Volume of distilled water (in ml) added to the sample

C = Volume of sample used for dilution

For examples -

A = 30

B = 80

C= 20

$$NTU = \frac{30 \ X \ (80+20)}{20}$$

= 150 NTU

COLOURIMETER/SPECTROPHOTOMETER

Colourimeter and Spectrophotometer are used to measure the quantity or concentration of a substance in a liquid sample by anylysing the intensity of light rays that come out through the sample. The amount of light absorbed is a measureof the concentration of the substance (Note in Nephelometry only the scattered light is analysed). When light rays strike a substance some amount is absorbed and some amount is transmitted. In both colourimeter and spectrophotometer, the amount of light absorbed is calculated as optical density (OD) and this is an indicator of the concentration of the absorbing substance.

Spectrophotometer and colourimeter both work on the same principles of light absorbance / transmission, but the former is a more versatile and sophisticated device than the latter. Before we understand the working of these instruments and application, we must understand the common principles of behavious of light rays and photometry.

Light and energy : Light is a form of energy and is propagated through waves. It is also made up of particles called Photons. Visible light has a wave lingth ranging berween 4000 A^0 8000 A^0. AS the wavelength increases the energy potential decreases.

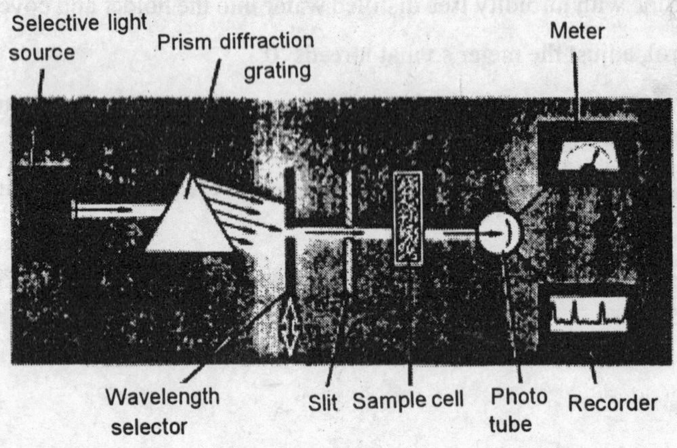

Fig 27.7 Fundamental system of a spectrophotometer

Absorption of light : Visible light is also a form of energy and it interacts with matter and gets absorbed. Due to this absorption, the electrons in the matter get pushed to a higher orbit or sometimes they might even escape. The absorption of light and its effect on the matter is not uniform. It depends on two factors.

i) Chemical nature of the matter and

ii) Wavelength of light

For example some substances may not absorb any light, they simply transmit the whole of incident light, example - water. Water is colourless because the whole of light passes throug without being absorbed. On the other hand light passing through a chlorophyll suspension will have all the rays transmitted except the green and that is why chlorophyll looks green. In other words a substance has a colour because it allows all other rays except rays of that particular colour. The intensity of the colour depends on the amount of absorbing substance. For example a dilute chlorophyll suspension looks pale green, because it cannot absorb all the green rays and some may be transmitted, While a concentrated suspension absorbs all green rays, hence it is deep green. From this we can come to the following conclusions

1. A substance can absorb light energy if its chemical structure favours it.
2. Absorption of light is only at a specific wavelength.
3. The extent of absorption of light rays at a specific wave length is dependent upon the concentration of the absorbing substance.

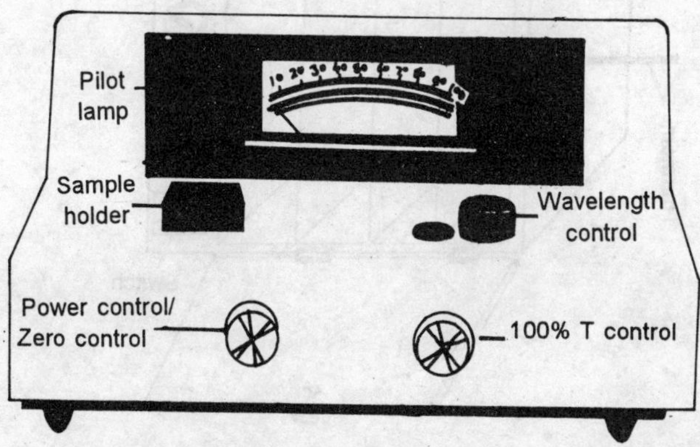

Fig. 27.8 A Spectronic 20 Spectrophotometer

Photometry : This refers to the measurement of light transmitting power of a solution in order to determine the concentration of (light absorbing) substances present in it. In photometric measurements a monochromatic light is always employed. The fact that a beam of white light can be broken into colours as it passes through a prism was first found out by Leonaroto da Vinci in 1500. Subsequently Issac Newton (1665) repeated the experiment and explained that as light passes from one medium (air) into another (prism), the rays bend in varying degrees propprtional to their wavelength and form a spectrum. Another device that can break up the white light like prism is diffraction grating developed in the early 20th century. The difraction grating consists of thousands of engraved parallel slits which can disperse the beam of light into its individual rays.

The absorption of monochromatic light by substances and its bearing on the concentration has been explained on the basis of the following laws.

Beer's law : The law states that when a beam of monochromatic light enters an absorbing medium, the intensity of light that comes out decreases exponentially with the increase of concentration of the light absorbing substance in the medium.

Lambert's law : According to this law, under identical conditions the intensity of the light decreases exponentially with increase in the length of the medium through which light passes. Colourimeter and spectrophotometer are the two instruments constructed which can measure the quantity of a substance in a medium.

Colourimeter : It is simpler in construction when compared to a spectrophotometer and has a limited application. It consists of a light source, a filter to produce monochromatic light (blue, green, red etc) and a photocell with detector to measure the light intensities. The tube (cuvette) containing the solvent is place on the light path. The solvent chosen will have 100% transmittance of light. It is then replaced by the sample tube and the light that is transmitted is analysed in a multivoltmeter which shows how much amount of light is transmitted. The meter consists of a scale which shows percentage light absorbed and percentage light transmitted. The amount of ligh absorbed is measured in units of optical density (OD) and is a direct indication of concentration (Note : A colourimeter has only filters and everytime they have to be changed if wave length of light is to be changed)

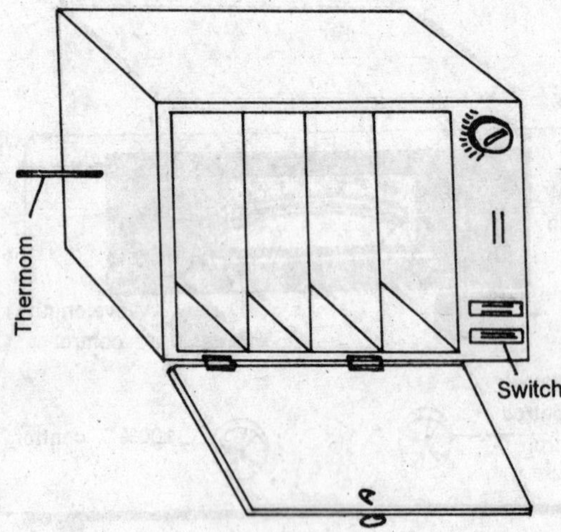

Fig. 27.9 Incubator

Spectrophotometer : This is constructed on the same principle as the colourimeter, but may be said to be a modified version. The major difference between the two is the ability to select a wide range of wavelength within the visible spectrum. A spectrophotometer uses a grating instead of filters. A knob is provided to manipulate the grating to provide the monochromatic light of the desired wavelength; the range that can be selected is as low as 1- 2 nm.

In a spectrophotometer the source of light is a tungsten filament incandescent bulb for visible range, and a hydrogen gas discharge tube for light in the UV region. Glowing rods of silicon carbide or zirconium oxide produce wave lengths in the infrared region. Since glass is not suitable for UV region, optical system in spectrophotometer is usually made of quartz. The sample tubes (Cuvettes) are also made up of quartz.

The monochromatic beam is selected using the selection knob. The rays then pass through the sample tube and then are focussed on to the photocell which converts the light intensity into electrical impulses and shows it on a multivoltmeter which functions in the same way as in a colourimeter. In some modern spectrophotometer the optical density is displayed digitally.

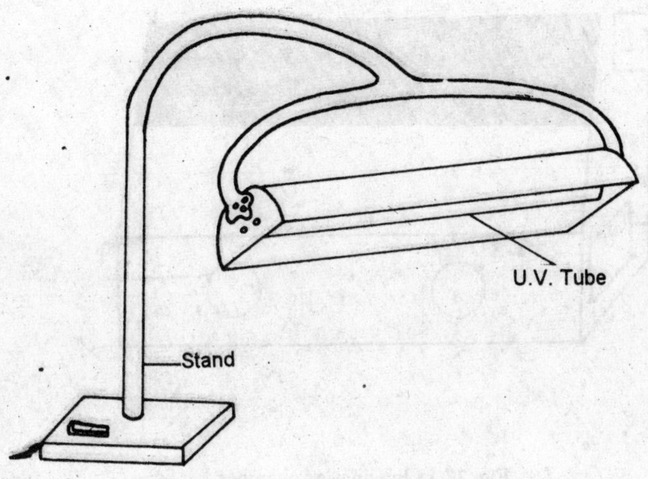

Fig. 27.10 Ultra violet lamp

In some of the sophicticated spectrophotometers, a continuous absorption spectrum of the compound may be obtained. This is made possible by the use of a motor driven wavelength scanning mechanism. Called ratio recording spectrophotometer, in this instrument a graphic presentation of the absorption spectrum showing the peak may be obtained. Absorption peak is a characterstric of a compound and this may be used for the identification of an unknown compound.

Quantification of the compound.

After obtaining the OD of a compound, quantification can be done by comparing it with a standard curve. A standard curve is prepared by taking series of concentrations of a standard compound and noting down the ODs. The ODs and concentration are then plotted on a graph sheet to give a straight line (standard curve). OD of a test compound can then be compared with the standard curve and the concentration calculated.

INCUBATOR

The culture of microorganisms in the laboratory is an extremely dilicate one as the microbes require various physical, chemical and nutritional parameters. Among the physical parameter a suitable temperature range is perhaps the most important factor governing the development of colonies on a culture medium. Many a time the room temperature may not be suitable for the culture of certain microbes. For this purpose an instrument which maintains a constant temperature throughout is used to culture the microbes. This instrument is called incubator.

Fig. 27.11 Inoculation chamber

Construction :

An incubator is basically similar to hot air oven and is made up of an insulated metal cabinet. The opening is controlled by two doors - an inner glass door and an outer asbestos insulated metal door. There is an indicator which turns on as soon as the incubator is switched on. There is another indicator which turns on in the beginning but goes off as soon as the required temperature is reached. This is regulated by a thermostat switch.

The main difference between hot air oven and incubator is the limit of the temperature in the later. An incubator operates between room temperature and 50^0 C.

Procedure for use :

Switch on the incubator and fix the thermostat switch for the required temperature.n Open the doors of the incubator and place the culture glassware (tubes, flasks, petriplates etc) on the racks and close both the doors. At the end of the required period remove the culture material and observe the growth.

Some precautions

1. Unnecessarily donot open the doors of the incubator during incubation period. If necessary only the outer door should be opened and the culture material can be observed through the inner glass door.

2. Donot overcrowd the racks with the culture material. Allow sufficient space between them for hot air to pass through.

3. Never increase the temperature beyond the required limit.

ULTRA VIOLET LAMP (UV LAMP)

One of the methods of sterilization of equipments and other microbiological materials is sterilization by radiation. Both ionizing as well as non ionzing radition can be used for sterilization. Normally however ionozing radiation is not used in the laboratory as it is not only expensive but also very hazardous and requirs very careful handling. Non-ionizing radiation however is less hazardous, specific and easy to handle. UV radiation is the most popular non-ionizing radiation used in the microbiological laboratory. For this purpose table top UV lamps are available. Using this, room, chamber, air, tools etc an be sterilized.

A UV lamp is almost like a table lamp and has a special filter over the source of elimination to emit UV light. Articles that need to be sterilized are kept in the path of light and lamp is switched on for the required period. At the end of this period the UV light is switched off and the materials taken out.Care should be taken to see that bare hands ans skin are not exposed to direct UV light as it may be injurious in the long run.

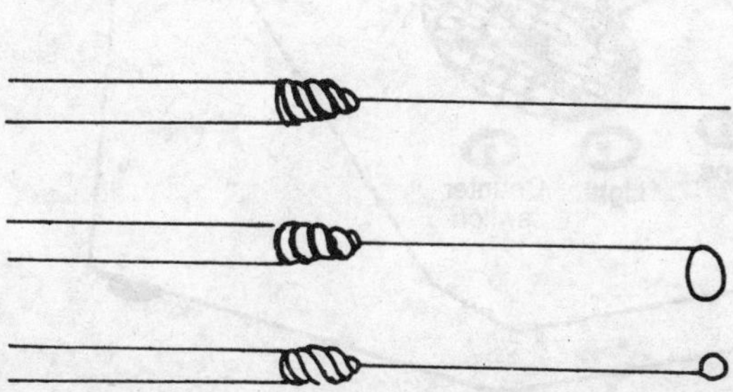

Fig. 27.12 Inoculation loops; small above, large below and needle

INOCULATION CHAMBER

Many a time in the microbiological laboratory several experiments like transfering cultures from one medium to another for subculting and isolation is necessary. For all these purpose maintenance of aspectic condition is necessary. An equipment called inoculation chamber permits these operations under aseptic conditions.

Construction of inoculation chamber

Inoculation chamber is a cabinet and has four doors. Of these two doors are on the opposite sides and they can be used to place the cultures etc into the chamber. The front panel has two circular doors which allows the insertion of the hands of the operator into the chamber to conduct inoculation, transfer of culture etc. The cabinet on top is provided with an UV lamp as well as an ordinary tube light. The UV lamp is used for sterilizing the internal environment while the tube light provides illumination if and when necessary.

Procedure for use

1. Clean the interiors of inoculation chamber using a suitable disinfectant such as 70% alcohol.

2. Through the doors present on the opposite sides of the chamber keep the materials required for experimentation inside the chamber.

3. Switch on the UV light for about 15 minutes to sterilize the interior of the chamber.

4. Transfer of cultures inoculation etc can be carried out after this period.

5. It is preferable to keep a spirit lamp which can be used for flaming the loops before and after inoculation.

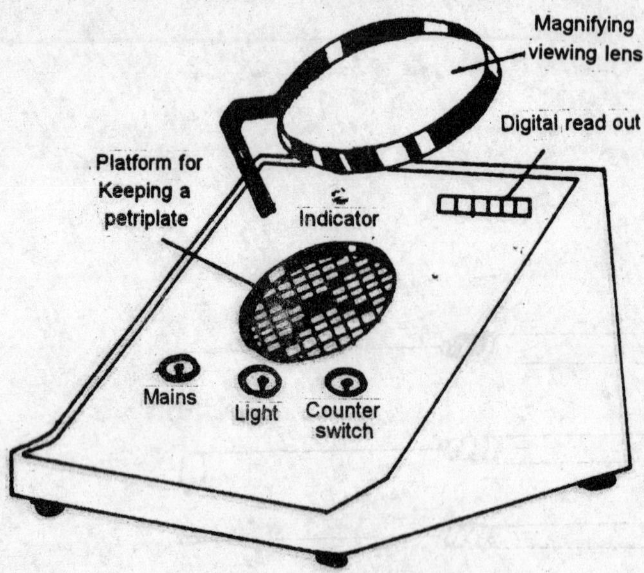

Fig. 27.13 A Quebec colony counter

INOCULATION LOOPS AND NEEDLES

In the laboratory one of the most frequently conducted operations are inoculation of the medium and the transfer of the culture from the source to the other. Many a time it will also be necessary to separate the colonies of bacteria, actinomycetes etc to stain and study them properly. For this purpose inoculation loops and needles are very essential. Inoculating loops consist of a handle made of glass on to which is fitted a specific length of platinum wire, nichrom or eureka wire. The free end of the platinum loop is bent in the form of a small loop. Usually the platinum wire along with the loop measures about 8 cms in length. Usually two types of loops are used in the laboratory. These are - the smaller loop and the larger loop. The smaller loop has a diameter of 3mm and the loop is on the same axis as that of the platinum wire. The larger loop however has a 5mm diameter and is bent at right angles to the main axis of the wire. These loops can hold a thin film of culture.

Use of the loops

Small loops are used to transfer extremely small quantities of the culture either from liquid or solid medium. The large loops are used to transfer larger quantities like entire colonies from solid media.

Needles

These are usually mounted straight pointed needles. The free end is very sharp. It is used either for teasing and separating the hyphae or filaments in a preparation. It can also be used for inoculating a stab culture.

COLONY COUNTER

In many microbiological experiments it will be necessary to count the number of viable spores, cells or colonies in a culture. This is required to find out whether a microbial colony can grow on multiply on a particular type of medium. In order to count the colonies on a petri plate an instrument called colony counter is used.

By counting the number of viable colonies one can find out the number of viable spores or cells in a particular culture since each cell or spore that is viable forms a colony.

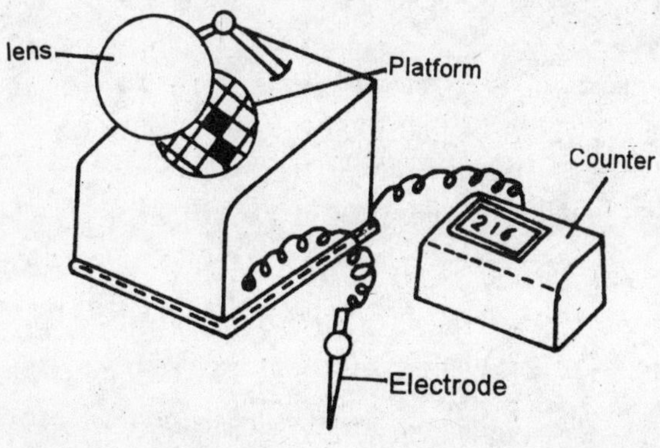

Fig. 27.14 An Electric colony counter

Construction of colony counter

Colony counters are of two types manual colony countes and electrical colony counters. The manual colony counter is made up of small cabinet having a circular grid in the center. There is a provision to illuminate the grid from below by a built in light source. Above the grid there is a magnifying glass which can magnify upto 1.5 times. This facilitates better counting.

Electrically operated colony counters are easy to use and avoid the possible mistakes in a manual counter. In most of the electrical operated colony counters there is a pencil shaped electrode and an LCD disply which immediately registers the number of colonies counted. There is a reset button which brings the display back to zero.

Procedure for using colony counter

1. Petriplate containing the colonies is placed on the circular grid of the colony counter.

2. Power is switched on to illuminate the grid.

3. Colonies are then counted manually in a manual counter.

4. In the electrically operated counters the pencil shaped electrode is removed from its position and each colony on the petriplate is touched.

5. As each colony is touched by the pencil electrode it is automatically registered and the number is displayed digitally.

In a special type of electrical colony counter called Qubec colony counter there is a counter switch which when depressed immediately displays the number of colonies on the petriplate on a digital read out.

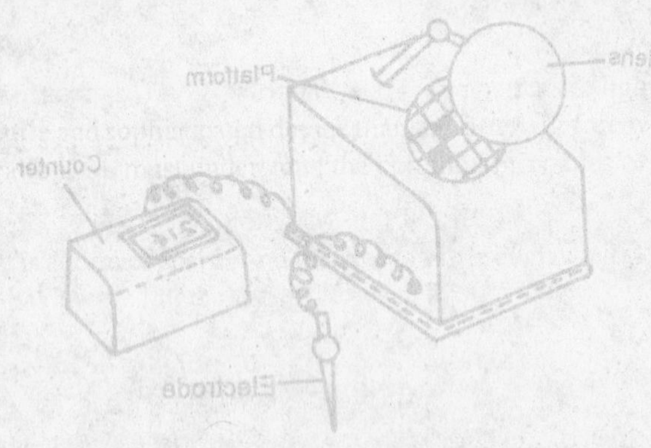